Image Processing and Analysis with Graphs

THEORY AND PRACTICE

Digital Imaging and Computer Vision Series

Series Editor

Rastislav Lukac

Foveon, Inc./Sigma Corporation
San Jose, California, U.S.A.

Computational Photography: Methods and Applications, *by Rastislav Lukac*

Super-Resolution Imaging, *by Peyman Milanfar*

Digital Imaging for Cultural Heritage Preservation: Analysis, Restoration, and Reconstruction of Ancient Artworks, *by Filippo Stanco, Sebastiano Battiato, and Giovanni Gallo*

Visual Cryptography and Secret Image Sharing *by Stelvio Cimato and Ching-Nung Yang*

Image Processing and Analysis with Graphs: Theory and Practice *by Olivier Lézoray and Leo Grady*

Image Restoration: Fundamentals and Advances *by Bahadir Kursat Gunturk and Xin Li*

Perceptual Digital Imaging: Methods and Applications *by Rastislav Lukac*

Image Processing and Analysis with Graphs

THEORY AND PRACTICE

EDITED BY

OLIVIER LÉZORAY • LEO GRADY

CRC Press
Taylor & Francis Group
Boca Raton London New York

CRC Press is an imprint of the
Taylor & Francis Group, an **informa** business

CRC Press
Taylor & Francis Group
6000 Broken Sound Parkway NW, Suite 300
Boca Raton, FL 33487-2742

First issued in paperback 2017

Version Date: 20120525

ISBN 13: 978-1-4398-5507-2 (hbk)
ISBN 13: 978-1-138-07176-6 (pbk)

Library of Congress Cataloging-in-Publication Data

Image processing and analysis with graphs : theory and practice / editors, Oliver Lézoray, Leo Grady.
    p. cm. -- (Digital imaging and computer vision series)
    Summary: "The first book to serve as a comprehensive review of digital imaging and computer vision, this book begins with an introduction chapter to ease readers unfamiliar with concepts into following topics. The book is divided into two parts that focus on the processing of functions on graphs, graph-based image processing, and the representation and analysis of objects on graphs, graph-based image analysis. Each chapter provides a comprehensive review on a specific topic, which ranges from research challenges to industry trends, and provides numerous examples to illustrate how the proposed methods can be used in practice. A companion website is available"-- Provided by publisher.
    Includes bibliographical references and index.
    ISBN 978-1-4398-5507-2 (hardback)
    1. Computer vision--Mathematics. 2. Image processing--Digital techniques. 3. Graph theory. I. Lézoray, Olivier. II. Grady, Leo.

TA1634.I49 2012
006.6--dc23
                                                                            2012016078

Dedication

To my beloved wife Sophie, daughter Manon, and son Quentin.

Their love and patience are an important support.

Olivier Lézoray

To my very patient, understanding and loving wife Amy,

my daughter Alexandra and son Leo.

Leo Grady

vi

Preface

The last two decades have witnessed the explosive growth of image production from digital photographs to the medical scans, satellite images, and video films. Consequently, the number of applications based on digital images has drastically increased, including multimedia integration, computer animation, video games, communication and digital arts, medicine, biometry, etc. Although being very different from one another, all these application areas rely on similar image processing and analysis techniques. The field of image or video processing analysis is very broad, encompassing a wide variety of research issues from low-level processing (such as image enhancement, restoration, and segmentation) to high-level analysis (semantic object extraction, indexing databases of images, and computer-human interaction).

Recently, graphs have emerged as a unified representation for the processing and the analysis of images. The number of concepts that can be defined with graphs is very large. In particular, many real-world problems have been successfully modeled using graphs. Consequently, graph theory has found many developments and applications for image processing and analysis, particularly due to the suitability of graphs to represent any discrete data by modeling neighborhood relationships. Different graph models have been proposed for image analysis, depending on the structures to analyze. However, graphs are not only of interest for *representing* the data to process, but also for defining graph-theoretical algorithms that enable the processing of functions associated with graphs. Additionally, representing problems with graphs makes it possible to draw on the rich literature of combinatorial optimization to produce highly efficient solutions. This research topic is timely, very influential in computer science and has led to many applications in denoising, enhancement, restoration, and object extraction. Consequently, graphs have become indispensable for the development of cutting-edge research and applications in image processing and analysis.

With the rapid development of graphs in image processing and analysis, the book aims at providing a comprehensive overview of the current state-of-the-art. The book not only covers the theoretical aspects of image processing with graphs but also demonstrates how these concepts can be used to design cutting-edge solutions to real world applications. Due to the wide variety of problems being solved with graphs in image processing and computer vision, the book has the form a contributed volume in which each chapter addresses a specific technique or application and is written by renowned experts in the field.

The intended audience for the book is graduate and postgraduate students, researchers, and practitioners. The aim is first to provide students and researchers with a state-of-the-art view of the important ideas involved in the use of graphs in image processing and analysis. Secondly, the book provides application examples showing how the theoretical algorithms can be applied in practice. Therefore, the book can serve as a support for graduate courses in image processing and computer vision as well as a reference for practicing engineers for the development and implementation of image processing and analysis algorithms.

The book begins with an introduction chapter to graph theory basics to make the book self-contained, establish notation and address some particulars of how graphs are used to model images. The rest of the book contains chapters that focus on either techniques for representing images with graphs or on the use of graph-based algorithms to solve problems in image processing and computer vision.

The first part of the book therefore starts with chapters on low level processing with graph-based image processing.

Many of the problems that arise in machine vision can be naturally expressed in terms of energy minimization. For instance, many computer vision problems require the assignment of labels to pixels (e.g., image restoration, disparity estimation and object recognition) and finding the best labelling can be seen as an optimization problem. Minimum cut/maximum network flow algorithms have emerged as an efficient and useful tool for exact or approximate energy minimization. Chapters 2, 3, and 4 are dedicated to this topic. Chapter 2 describes a class of combinatorial optimization methods called graph cuts. After introducing the energy minimization formalisms used in the image-related areas, the graph-cut algorithm is described for the binary case, which is the most fundamental one where the energy-minimization problem directly translates to the minimum-cut problem. Next, the case of graph cuts where there are more than two labels to be assigned to each pixel is considered. Chapter 3 considers labeling problems modeled via Markov Random Fields (MRF). The standard MRF formulations for such problems employ pair-wise potentials defined over pairs of random variables and are often insufficient to completely model the complexities of a problem. Novel families of higher order potentials, i.e., potentials defined over multiple variables, have therefore higher modeling power and lead to more accurate models. This chapter presents provides an overview of various types of higher-order random field models that have been considered in the past. Patch or region-based potentials, and global potentials enforcing topology constraints and label statistics are described, as well as methods to perform inference in these models. Chapter 4 presents an efficient algorithm for minimizing the total variation plus a convex and separable data fidelity term (not necessarily differentiable). The approach relies on mapping the original problem into a parametric network flow problem that can be solved using efficient maximum-flow algorithms. Extension to non separable convex data fidelity terms, such as the convolution with a linear operator (for deconvolution problems), is also considered.

Chapters 5, 6, 7, and 8 focus on defining graph-theoretical algorithms that enable the processing of functions associated with graph vertices and/or edges for image processing tasks. Chapter 5 deals with targeted image segmentation to localize a single, particular object. Methods for specifying the target object fit naturally with graph techniques, which optimize a combination of a unary term and a binary term. The chapter reviews alternative methods for specifying a target object from known information in the context of a graph algorithm and the use of graph-based algorithms for producing the segmentation from the target information. Chapter 6 describes how mathematical morphology can be defined on edge-weighted graphs. First, lattices making it possible to compare weighted and non weighted edges and vertices are introduced, and morphological operators and filters on graphs are detailed. Second, it is shown that the watershed has very strong links with the minimum spanning tree and can be used for building hierarchies of segmentations as well as an optimization tool. Chapter 7 presents a framework called partial difference equations on graphs to solve partial differential equations (PDEs) on graphs. The proposed framework mimics on graphs well-known PDEs variational formulations under a functional analysis point of view, and unifies local and nonlocal processing. In particular, a nonlocal discrete regularization on graphs of arbitrary topologies is introduced as a framework for data simplification and interpolation (label diffusion, inpainting, colorisation). Chapter 8 focuses on constructing wavelet transforms of functions defined on graph and introduce the spectral graph wavelet transform (SGWT). The SGWT construction is based on defining scaling using the the graph analogue of the Fourier domain, namely, the spectral decomposition of the discrete graph Laplacian. Using the SGWT construction with a nonlocal image graph gives a nonlocal graph wavelet transform well-suited to solve inverse problems such as denoising.

Chapters 9 and 10 present how the graph-theoretical algorithms can be adapted and used in practice for specific imaging applications in computational photography and medical imaging. Chapter 9 considers image and video matting, which is the problem of accurate foreground estimation. Matting is one of the key techniques in various image editing and film production applications. A comprehensive review of recent state-of-the-art approaches are described and how they use various graph regularization methods to achieve accurate and stable results. Chapter 10 describes a general methodology for optimal segmentation of single or multiple interacting surfaces that is applicable to n-D image data. The utility of the reported segmentation methods is demonstrated in practical examples from medical image analysis in which accurate and reliable image segmentation is of paramount importance.

The second part of the book continues with chapters on high-level processing using graph-based image analysis.

Chapter 11 presents a rigorous review of regular and irregular pyramids to encode image content with hierarchical graph encoding. The search for efficient data structures encoding pyramids is vital, as the complexity of the data structure is a limiting factor to the utility of pyramids in applications. After describing the main properties and drawbacks of regular pyramids, irregular pyramids are introduced as well as the associated construction schemes (e.g., maximal independent set, maximal independent directed edge set, and data-driven decimation). The construction scheme of a pyramid determines its decimation ratio and can be understood as the pyramids vertical dynamic. Then, three types of irregular pyramids are described using a generic notion of contraction kernel: simple graphs, dual graphs, and combinatorial pyramids. The choice of a given graph model determines which set of topological and geometrical image properties can be encoded at each level.

Graphs are a general and powerful data structure for the representation of objects and concepts. In a graph representation, the nodes typically represent objects or parts of objects, while the edges describe relations between objects or object parts (as detailed in Chapter 11). In applications such as pattern recognition and computer vision, object similarity is an important issue. If graphs are used for object representation this problem turns into determining the similarity of graphs, which is generally referred to as *graph matching*. For large graphs, graph matching often relates to graph embedding, which is a mapping from a graph or a set of graphs to a vectorial space. The purpose of graph embedding and more generally of graph-based methods for dimensionality reduction is to represent each vertex of a graph as a low-dimensional vector that preserves similarities between the vertex pairs, where similarity is measured by a graph similarity matrix that characterizes certain statistical or geometric properties of the data set. Chapters 12, 13, 14, and 15 deal with both graph-based dimensionality reduction and graph matching. Chapter 12 reviews some of the most prominent dimensionality reduction methods that rely on graphs. These include several techniques based on geodesic distances, such as Isomap and its variants. Spectral methods involving a graph Laplacian are described as well. More biologically inspired techniques, such as the self-organizing map, identify a topographic mapping between a predefined graph and the data manifold. Chapter 13 reviews the graph edit distance (GED) from three different points of view, namely, theory, algorithms, and applications. Indeed, two approaches to graph matching can be found: exact and inexact (or error-tolerant) graph matching. The graph edit distance (GED) is the basis of inexact graph matching, and has emerged as an important way to measure the similarity between pairs of graphs in an error-tolerant fashion. The chapter introduces fundamental concepts related to GED with an emphasis on the theory underlying the edit distance cost function, gives an introduction to GED computation (exact and approximate), and concludes with some GED applications. Chapter 14 reviews the role of graphs in matching shapes and in categorization. This chapter explores the value of graphs in computer vision from the perspective of representing both structural quantum variations as well as metric changes, as applied to the matching and categorization problems. It provides an overview on the use of shock graphs in matching and recognition, and proximity graphs in categorization. Chapter 15 addresses the problem of 3D shape registration and presents a technique based on spectral graph theory and probabilistic matching. This chapter extends the spectral graph matching methods to very large graphs by combining spectral graph matching with Laplacian embedding.

Chapters 16 and 17 conclude the book and concentrate on Graphical Models and Kernels methods in computer vision. Chapter 16 discusses mathematical models of images that can be represented with graphs, known as graphical models. The chapter reviews the fundamental theory, problems, and algorithms involved in using graphical models to solve computer vision problems. Graphical models are a powerful tool for constructing mathematical models that are rich enough to express to the complexity of the world pictured in each image, but are also computationally tractable and easy to design. Moreover, they make it possible to construct global models out of a collection of local relationships. Kernel-based methods have also proven to be highly effective in many applications because of their wide generality. As soon as a similarity measure leading to a symmetric positive-definite kernel can be designed, a wide range of learning algorithms working on Euclidean dot-products of the examples can be used, such as support vector machines. A recent line of research consists in designing kernels for structured data such as graphs. Chapter 17 presents a family of positive-definite kernels between images, making it possible to compute image similarity measures respectively in terms of color and of shape. The kernels consist of matching subtree-patterns called "tree-walks" of graphs extracted from the images. Computationally efficient kernels that can be computed in polynomial-time in the size of the graphs are also detailed.

Complementary material is available on the book companion website at

`http://greyc.stlo.unicaen.fr/lezoray/IPAG`

We would like to thank all the contributors for their effort in preparing chapters as well as incorporating suggestions from us and colleagues in the field who served as reviewers. We also thank CRC Press for giving us the opportunity to edit a book on image processing and analysis with graphs. In particular, we would like to thank Dr. Rastislav Lukac, editor of the Digital Imaging and Computer Vision book series, for initiating this project, and Nora Konopka for her support and assistance.

Olivier Lézoray and Leo Grady

Université de Caen Basse-Normandie
Siemens Corporate Research

The Editors

Olivier Lézoray (http://www.info.unicaen.fr/~lezoray) received his B.Sc. in mathematics and computer science, M.Sc. and Ph.D. degrees from the Department of Computer Science, University of Caen, France, in 1992, 1996, and 2000, respectively. From September 1999 to August 2000, he was an assistant professor with the Department of Computer Science at the University of Caen. From September 2000 to August 2009, he was an associate professor at the Cherbourg Institute of Technology of the University of Caen, in the Communication Networks and Services Department. In July 2008, he was a visiting research fellow at the University of Sydney. Since September 2009, he has been a full professor at the Cherbourg Institute of Technology of the University of Caen, in the Communication Networks and Services Department. He also serves as Chair of the Institute Research Committee. In 2011 he cofounded Datexim and is a member of the scientific board of the company, which brought state-of-art image and data processing to market with applications in digital pathology. His research focuses on discrete models on graphs for image processing and analysis, image data classification by machine learning, and computer-aided diagnosis. He is a contributor to four books, and he has published over 100 papers in refereed journals and conference proceedings in the areas of image processing and machine learning. In 2008, he received the IBM Best Paper Student Award at ICIP 2008 with V-T. Ta and A. Elmoataz. Dr. Lézoray is a member and Local Liaison Officer of the European Association for Signal, Speech and Image Processing (EURASIP), the IEEE Signal Processing Society's, the IEEE Communication Society Multimedia Communications Technical Committee, and the International Association for Pattern Recognition (IAPR). He was the program cochair of the International Conference on Image and Signal Processing (ICISP) in 2008, 2010, and 2012. He was a guest coeditor of the *EURASIP Journal on Advances in Signal Processing* for a special issue on Machine Learning in Image Processing, of *Signal Processing* for a special issue on the Processing and Analysis of High-Dimensional Masses of Image and Signal Data, and of *Computerized Medical Imaging and Graphics* for a special issue on Whole Slide Microscopic Image Processing. He serves as a reviewer for various scientific international journals.

Leo Grady (http://cns.bu.edu/~lgrady) received his B.Sc. degree in electrical engineering from the University of Vermont in 1999 and a Ph.D. degree from the Cognitive and Neural Systems Department at Boston University in 2003. Since autumn 2003 he has been with Siemens Corporate Research in Princeton, where he works as a Principal Research Scientist in the Image Analytics and Informatics division. The focus of his research has been on the modeling of images and other data with graphs. These graph models have generated the development and application of tools from discrete calculus, combinatorial/continuous optimization, and network analytics to perform analysis and synthesis of the images/data. The primary applications of his work have been in computer vision and biomedical appli-  cations. This work has led to the publication of the book *Discrete Calculus* with Jonathan Polimeni (Springer 2010), which provides an introduction to the topic of discrete calculus and details how it may be applied to solve a wide range of applications drawn from different scientific domains. In addition to this book, he has published 10 journal papers, more than 20 full-length papers in top peer-reviewed conferences, as well as invited chapters to three books and papers in smaller conferences. He currently holds 25 granted patents with more than 40 additional patents currently under review. Dr. Grady has served on grant panels for the U.S. government and as a program committee member for several international conferences, including Computer Vision and Pattern Recognition (CVPR), European Conference on Computer Vision (ECCV), and Medical Image Computing and Computer Assisted Intervention (MICCAI), as well as several workshops, including Interactive Computer Vision, Perceptual Organization for Computer Vision, Structured Models in Computer Vision, Information Theory in Computer Vision, and Pattern Recognition. He has given invited lectures on his research at over 30 international academic and industrial laboratories, as well as plenary talks for the Workshop on Graph-Based Representations in Pattern Recognition, International Symposium on Mathematical Morphology (ISMM), and the International Conference on Image and Signal Processing (ICISP). Dr. Grady has contributed to over 20 Siemens products that target biomedical applications and that are used in medical centers worldwide.

Contributors

Francis Bach
INRIA – Ecole Normale Supérieure
Paris, France

Sébastien Bougleux
Université de Caen Basse-Normandie
Caen, France

Horst Bunke
The University of Melbourne
Victoria, Australia

Luc Brun
EnsiCaen
Caen, France

Antonin Chambolle
École Polytechnique
Palaiseau, France

Jérome Darbon
École Normale Supérieure de Cachan
Cachan, France

Abderrahim Elmoataz
Université de Caen Basse-Normandie
Caen, France

Miquel Ferrer
Universitat Politècnica de Catalunya
Barcelona, Spain

Mona Garvin
University of Iowa
Iowa City, Iowa

Leo Grady
Siemens Corporate Research
Princeton, New Jersey

David K. Hammond
University of Oregon
Eugene, Oregon

Zaid Harchaoui
INRIA Grenoble – Rhône-Alpes
Montbonnot Saint-Martin, France

Radu Horaud
INRIA Grenoble – Rhône-Alpes
Montbonnot Saint-Martin, France

Hiroshi Ishikawa
Waseda University
Tokyo, Japan

Laurent Jacques
Université Catholique de Louvain
Louvain-la-Neuve, Belgium

Benjamin Kimia
Brown University
Providence, Rhode Island

Pushmeet Kohli
Microsoft Research
Cambridge, United Kingdom

Walter Kropatsch
Vienna University of Technology
Vienna, Austria

John Aldo Lee
Université catholique de Louvain
Louvain-la-Neuve,Belgium

Olivier Lézoray
Université de Caen Basse-Normandie
Caen, France

Diana Mateus
Technische Universität München
München Germany

Fernand Meyer
École des mines de Paris
Fontainebleau, France

Laurent Najman
Université Paris-Est - ESIEE
Paris, France

Vinh-Thong Ta
Université de Caen Basse-Normandie
Caen, France

Marshall Tappen
University of Central Florida
Orlando, Florida

Carsten Rother
Microsoft Research
Cambridge, United Kingdom

Avinash Sharma
INRIA Grenoble – Rhône-Alpes
Montbonnot Saint-Martin, France

Milan Sonka
University of Iowa
Iowa City, Iowa

Pierre Vandergheynst
École Polytechnique Fédérale de
    Lausanne
Lausanne, Switzerland

Michel Verleysen
Université Catholique de Louvain
Louvain-la-Neuve, Belgium

Jue Wang
Adobe Systems
Seattle, Washington

Xiaodong Wu
University of Iowa
Iowa City, Iowa

Contents

1 Graph theory concepts and definitions used in image processing and analysis 1

Olivier Lézoray and Leo Grady

1.1 Introduction 2

1.2 Basic Graph Theory 2

1.3 Graph Representation 5

1.4 Paths, Trees, and Connectivity 7

1.5 Graph Models in Image Processing and Analysis 15

1.6 Conclusion 21

Bibliography 21

2 Graph Cuts—Combinatorial Optimization in Vision 25

Hiroshi Ishikawa

2.1 Introduction 26

2.2 Markov Random Field 27

2.3 Basic Graph Cuts: Binary Labels 35

2.4 Multi-Label Minimization 45

2.5 Examples 55

2.6 Conclusion 56

Bibliography 57

3 Higher-Order Models in Computer Vision 65

Pushmeet Kohli and Carsten Rother

3.1 Introduction 65

3.2 Higher-Order Random Fields 67

3.3 Patch and Region-Based Potentials   69

3.4 Relating Appearance Models and Region-Based Potentials   77

3.5 Global Potentials   79

3.6 Maximum a Posteriori Inference   84

3.7 Conclusions and Discussion   89

Bibliography   90

4 A Parametric Maximum Flow Approach for Discrete Total Variation Regularization   **93**

Antonin Chambolle and Jérôme Darbon

4.1 Introduction   93

4.2 Idea of the approach   95

4.3 Numerical Computations   97

4.4 Applications 104

Bibliography 106

5 Targeted Image Segmentation Using Graph Methods   **111**

Leo Grady

5.1 The Regularization of Targeted Image Segmentation 113

5.2 Target Specification 118

5.3 Conclusion 134

Bibliography 135

6 A Short Tour of Mathematical Morphology on Edge and Vertex Weighted Graphs   **141**

Laurent Najman and Fernand Meyer

6.1 Introduction 142

6.2 Graphs and lattices 143

6.3 Neighborhood Operations on Graphs 145

6.4 Filters 149

6.5 Connected Operators and Filtering with the Component Tree . . . 152

6.6 Watershed Cuts 154

6.7 MSF Cut Hierarchy and Saliency Maps 161

6.8   Optimization and the Power Watershed 164

6.9   Conclusion 169

Bibliography 169

7   Partial difference Equations on Graphs for Local and Nonlocal Image Processing   175

Abderrahim Elmoataz, Olivier Lézoray, Vinh-Thong Ta, and Sébastien Bougleux

7.1   Introduction 176

7.2   Difference Operators on Weighted Graphs 177

7.3   Construction of Weighted Graphs 182

7.4 p-Laplacian Regularization on Graphs 185

7.5   Examples 192

7.6   Concluding Remarks 203

Bibliography 203

8   Image Denoising with Nonlocal Spectral Graph Wavelets   207

David K. Hammond, Laurent Jacques, and Pierre Vandergheynst

8.1   Introduction 208

8.2   Spectral Graph Wavelet Transform 210

8.3   Nonlocal Image Graph 216

8.4   Hybrid Local/Nonlocal Image Graph 220

8.5   Scaled Laplacian Model 225

8.6   Applications to Image Denoising 227

8.7   Conclusions 233

8.8   Acknowledgments 234

Bibliography 234

9   Image and Video Matting   237

Jue Wang

9.1   Introduction 237

9.2   Graph Construction for Image Matting 241

9.3   Solving Image Matting Graphs 250

9.4   Data Set 253

9.5   Video Matting 254

9.6   Conclusion 260

Bibliography 261

10 Optimal Simultaneous Multisurface and Multiobject Image Segmentation   265

Xiaodong Wu, Mona K. Garvin, and Milan Sonka

10.1 Introduction 266

10.2 Motivation and Problem Description 267

10.3 Methods for Graph-Based Image Segmentation 268

10.4 Case Studies 290

10.5 Conclusion 299

10.6 Acknowledgments 299

Bibliography 299

11 Hierarchical Graph Encodings   305

Luc Brun and Walter Kropatsch

11.1 Introduction 306

11.2 Regular Pyramids 307

11.3 Irregular Pyramids Parallel construction schemes 310

11.4 Irregular Pyramids and Image properties 324

11.5 Conclusion 344

Bibliography 347

12 Graph-Based Dimensionality Reduction   351

John A. Lee and Michel Verleysen

12.1 Summary 352

12.2 Introduction 352

12.3 Classical methods 353

12.4 Nonlinearity through Graphs 357

12.5 Graph-Based Distances 358

12.6 Graph-Based Similarities 361

12.7 Graph embedding 365

12.8 Examples and comparisons 369

12.9 Conclusions 373

Bibliography 374

13 Graph Edit Distance—Theory, Algorithms, and Applications   **383**

Miquel Ferrer and Horst Bunke

13.1 Introduction 384

13.2 Definitions and Graph Matching 386

13.3 Theoretical Aspects of GED 396

13.4 GED Computation 401

13.5 Applications of GED 407

13.6 Conclusions 417

Bibliography 417

14 The Role of Graphs in Matching Shapes and in Categorization   **423**

Benjamin Kimia

14.1 Introduction 423

14.2 Using Shock Graphs for Shape Matching 426

14.3 Using Proximity Graphs for Categorization 429

14.4 Conclusion 437

14.5 Acknowledgment 437

Bibliography 437

15 3D Shape Registration Using Spectral Graph Embedding and Probabilistic Matching   **441**

Avinash Sharma, Radu Horaud, and Diana Mateus

15.1 Introduction 442

15.2 Graph Matrices 444

15.3 Spectral Graph Isomorphism 446

15.4 Graph Embedding and Dimensionality Reduction 452

15.5 Spectral Shape Matching 458

15.6 Experiments and Results 464

15.7 Discussion 468

15.8 Appendix: Permutation and Doubly- stochastic Matrices 469

15.9 Appendix: The Frobenius Norm 470

15.10 Appendix: Spectral Properties of the Normalized Laplacian 471

Bibliography 472

16 Modeling Images with Undirected Graphical Models   475

Marshall F. Tappen

16.1 Introduction 476

16.2 Background 476

16.3 Graphical Models for Modeling Image Patches 482

16.4 Pixel-Based Graphical Models 483

16.5 Inference in Graphical Models 490

16.6 Learning in Undirected Graphical Models 492

16.7 Conclusion 496

Bibliography 496

17 Tree-Walk Kernels for Computer Vision   499

Zaid Harchaoui and Francis Bach

17.1 Introduction 500

17.2 Tree-Walk Kernels as Graph Kernels 502

17.3 The Region Adjacency Graph Kernel as a Tree-Walk Kernel 506

17.4 The Point Cloud Kernel as a Tree-Walk Kernel 510

17.5 Experimental Results 518

17.6 Conlusion 525

17.7 Acknowledgments 525

Bibliography 525

Index   529

Symbol Description

$\mathcal{S}$	Set	$w(v_i)$ or w_i	Node weights
$\bar{\mathcal{S}}$	Set complement	$w(v_i, v_j)$ or w_{ij}	Edge weights
$|\mathcal{S}|$	Set cardinality	x_i for node v_i	Node variable
$\mathbf{A}$	Matrix	y_i for edge e_i	Edge variable (flow)
$\mathbf{v}$	Vector	$\deg(v_i)$ or d_i	Node degree
$\tilde{p}$	Tuple	$\deg^+(v_i)$	Node in-degree
$\mathcal{V}$	Set of nodes	$\deg^-(v_i)$	Node out-degree
$\mathcal{E}$	Set of edges	K_n	Complete graph of n vertices
$\mathcal{F}$	Set of cycles (faces)		
$\mathcal{G} = (\mathcal{V}, \mathcal{E})$	Graph	$K_{m,n}$	Complete bipartite graph
$v_i \in \mathcal{V}$	Nodes		
$v_i \sim v_j$	Adjacency of two nodes	$\mathbf{A}$	Incidence matrix
		$\mathbf{W}$	Adjacency matrix
e_i	An edge in the context of an element of $\mathcal{E}$ (i.e., $e_i \in \mathcal{E}$)	$\mathbf{D}$	Degree matrix
		$\mathbf{L} = \mathbf{D} - \mathbf{W}$	Laplacian matrix
e_{ij}	An undirected edge between nodes v_i to v_j	$\tilde{\mathbf{L}} = \mathbf{D}^{-\frac{1}{2}}\mathbf{L}\mathbf{D}^{-\frac{1}{2}}$	Normalized Laplacian matrix
$e_{i \rightarrow j}$	Directed edges	$\pi = (v_1, \cdots, v_n)$	Path
		ψ	Coupling

1

Graph theory concepts and definitions used in image processing and analysis

Olivier Lézoray

Université de Caen Basse-Normandie
GREYC UMR CNRS 6072
6 Bvd. Maréchal Juin
F-14050 CAEN, FRANCE
Email: olivier.lezoray@unicaen.fr

Leo Grady

Department of Image Analytics and Informatics
Siemens Corporate Research
755 College Rd.
Princeton, NJ 08540, USA
Email: Leo.Grady@siemens.com

CONTENTS

1.1	Introduction	2
1.2	Basic Graph Theory	2
1.3	Graph Representation	4
	1.3.1 Matrix Representations	5
	1.3.2 Adjacency Lists	7
1.4	Paths, Trees, and Connectivity	7
	1.4.1 Walks and Paths	7
	1.4.2 Connected Graphs	8
	1.4.3 Shortest Paths	9
	1.4.4 Trees and Minimum Spanning Trees	10
	1.4.5 Maximum Flows and Minimum Cuts	11
1.5	Graph Models in Image Processing and Analysis	15
	1.5.1 The Regular Lattice	15
	1.5.2 Irregular tessellations	17
	1.5.3 Proximity Graphs for Unorganized Embedded Data	19
1.6	Conclusion	21
	Bibliography	21

1.1   Introduction

Graphs are structures that have a long history in mathematics and have been applied in almost every scientific and engineering field (see [1] for mathematical history, and [2, 3, 4] for popular accounts). Many excellent primers on the mathematics of graph theory may be found in the literature [5, 6, 7, 8, 9, 10], and we encourage the reader to learn the basic mathematics of graph theory from these sources. We have two goals for this chapter: First, we make this work self-contained by reviewing the basic concepts and notations for graph theory that are used throughout this book. Second, we connect these concepts to image processing and analysis from a conceptual level and discuss implementation details.

1.2   Basic Graph Theory

Intuitively, a graph represents a set of elements and a set of pairwise relationships between those elements. The elements are called **nodes** or **vertices**, and the relationships are called **edges**. Formally, a **graph** $\mathcal{G}$ is defined by the sets $\mathcal{G} = (\mathcal{V}, \mathcal{E})$ in which $\mathcal{E} \subseteq \mathcal{V} \times \mathcal{V}$. We may denote the i-th vertex as $v_i \in \mathcal{V}$, and the i-th edge as $e_i \in \mathcal{E}$. Since each edge is a subset of two vertices, we may also write $e_{ij} = \{v_i, v_j\}$. In this book we do not consider graphs with self-loops, meaning that $e_{ij} \in \mathcal{E}$ implies that $i \neq j$. We also do not consider graphs for which there are multiple edges of the same orientation connecting the same node pair. However, for specific graph encodings, self-loops and multiple edges can be useful (see Chapter 11).

Every graph may be viewed as *weighted*. Given a graph $\mathcal{G} = (\mathcal{V}, \mathcal{E})$, vertex weighting is a function $\hat{w} : \mathcal{V} \to \mathbb{R}$, and edge weighting is a function $w : \mathcal{E} \to \mathbb{R}$. To simplify the notation, we will use w to refer to both vertex and edge weighting. The weight of a vertex is denoted by $w(v_i)$ or w_i, and the weight of an edge incident to two vertices is denoted by $w(v_i, v_j)$ or w_{ij}. If $w_i = 1$, $\forall v_i \in \mathcal{V}$ and $w_{ij} = 1$, $\forall e_{ij} \in \mathcal{E}$, then we may consider the graph to be *unweighted*. If not otherwise specified, all node and edge weights are considered to be equal to unity. We treat $w_{ij} = 0$ as equivalent to $e_{ij} \notin \mathcal{E}$. Intuitively, an edge weight of zero is equivalent to meaning that the edge is not a member of the edge set.

Each edge is considered to be **oriented** and some edges additionally **directed**. An *orientation* of an edge means that each edge $e_{ij} \in \mathcal{E}$ contains an ordering of the vertices, v_i and v_j. An edge e_{ij} is *directed* if $w_{ij} \neq w_{ji}$. A graph for which none of the edges are directed is called an **undirected graph**, and a graph in which at least one edge is directed is called a **directed graph** or **digraph**. A directed edge is represented by the notation $e_{i \to j}$. A directed graph is more general than an undirected graph because it does not require that $w_{ij} = w_{ji}$ for each edge. Consequently, all algorithms for directed graphs may also be applied to undirected graphs, but the converse may or may not be true. Therefore, in this chapter we use digraphs to illustrate the most general concepts.

The edge orientation of an undirected graph provides a reference to determine the *sign* of a flow through that edge. This concept was developed early in the circuit theory literature to describe the direction of current flow through a resistive branch (an edge). For example, a current flow through e_{ij} from node v_i to v_j is considered positive, while a current flow from v_j to v_i is considered negative. In this sense, flow through a directed edge is usually constrained to be strictly positive.

Drawing a graph is typically done by depicting each node with a circle and each edge with a line connecting circles representing the two nodes. The edge orientation or direction is usually represented by an arrow. Edge e_{ij} is drawn with an arrow pointing from node v_i to node v_j.

We consider two functions $s, t : \mathcal{E} \to \mathcal{V}$. Function s is called the **source function**, and function t is called the **target function**. Given an edge $e_{ij} = (v_i, v_j) \in \mathcal{E}$, we say that $s(e_{ij}) = v_i$ is the **origin** or **source** of e_{ij}, and $t(e_{ij}) = v_j$ is the **endpoint** or **target** of e_{ij}. Given any edge, $e_k \in \mathcal{E}$, the vertices $s(e_k)$ and $t(e_k)$ are called the **boundaries** of e_k, and the expression $t(e_k) - s(e_k)$ is called the boundary of e_k.

Given these preliminaries, we may now list a series of basic definitions:

1. *Adjacent*: Two nodes v_i and v_j are called **adjacent** if $\exists e_{ij} \in \mathcal{E}$ or $\exists e_{ji} \in \mathcal{E}$, which is denoted by $v_i \sim v_j$. Two edges e_{ij} and e_{sk} are adjacent if they share a common vertex, that is, if $i = s$, $i = k$, $j = s$, or $j = k$.

2. *Incident*: An edge e_{ij} is **incident** to nodes v_i and v_j (and each node is incident to the edge).

3. *Isolated*: A node v_i is called **isolated** if $w_{ij} = w_{ji} = 0$, $\forall v_j \in \mathcal{V}$. Intuitively, a node is isolated if it is not connected to the graph by an edge (of nonzero weight).

4. *Degree*: The **outer degree** of node v_i, $\deg^-(v_i)$, is equal to $\deg^-(v_i) = \sum_{e_{ij} \in \mathcal{E}} w_{ij}$. The **inner degree** of node v_i, $\deg^+(v_i)$, is equal to $\deg^+(v_i) = \sum_{e_{ji} \in \mathcal{E}} w_{ji}$. Note that, in an undirected graph, $\deg^-(v_i) = \deg^+(v_i)$, $\forall v_i \in \mathcal{V}$. Consequently, we may simply use $d(v_i)$ or d_i to denote the degree of a vertex in an undirected graph. An isolated node v_i has zero degree.

5. *Complement*: A graph $\overline{\mathcal{G}} = (\mathcal{V}, \overline{\mathcal{E}})$ is called the **complement** to graph $\mathcal{G} = (\mathcal{V}, \mathcal{E})$ if $\overline{\mathcal{E}} = \mathcal{V} \times \mathcal{V} - \mathcal{E}$. Therefore, the complement graph has all the same nodes as the original, but each node pair in the complement is connected iff the node pair was not connected in the original graph.

6. *Regular*: An undirected graph is called **regular** if $d_i = k$, $\forall v_i \in \mathcal{V}$ for some constant k.

7. *Complete (fully connected)*: An undirected graph is called **complete** or **fully connected** if each node is connected to every other node by an edge, that is when $\mathcal{E} = \mathcal{V} \times \mathcal{V}$. A complete graph of n vertices is denoted K_n. Figure 1.1 gives an example of K_5.

8. *Partial graph*: A graph $\mathcal{G}' = (\mathcal{V}, \mathcal{E}')$ is called a **partial graph** [11] of $\mathcal{G} = (\mathcal{V}, \mathcal{E})$ if $\mathcal{E}' \subseteq \mathcal{E}$.

9. *Subgraph*: A graph $\mathcal{G}' = (\mathcal{V}', \mathcal{E}')$ is called a **subgraph** of $\mathcal{G} = (\mathcal{V}, \mathcal{E})$ if $\mathcal{V}' \subseteq \mathcal{V}$, and $\mathcal{E}' = \{e_{ij} \in \mathcal{E} | v_i \in \mathcal{V}' \text{ and } v_j \in \mathcal{V}'\}$.

10. *clique*: A **clique** is defined as a fully connected subset of the vertex set (see Figure 1.2).

11. *Bipartite*: A graph is called **bipartite** if $\mathcal{V}$ can be partitioned into two subsets $\mathcal{V}_1 \subset \mathcal{V}$ and $\mathcal{V}_2 \subset \mathcal{V}$, where $\mathcal{V}_1 \cap \mathcal{V}_2 = \emptyset$ and $\mathcal{V}_1 \cup \mathcal{V}_2 = \mathcal{V}$, such that $\mathcal{E} \subseteq \mathcal{V}_1 \times \mathcal{V}_2$. If $|\mathcal{V}| = m$ and $|\mathcal{V}| = n$, and $\mathcal{E} = \mathcal{V}_1 \times \mathcal{V}_2$, then $\mathcal{G}$ is called a **complete bipartite** graph and is denoted by $K_{m,n}$. Figure 1.2 gives an example of $K_{3,5}$.

12. *Graph isomorphism*: Let $\mathcal{G}_1 = (\mathcal{V}_1, \mathcal{E}_1)$ and $\mathcal{G}_2 = (\mathcal{V}_2, \mathcal{E}_2)$ be two undirected graphs. A bijection $f : \mathcal{V}_1 \to \mathcal{V}_2$ from $\mathcal{G}_1$ to $\mathcal{G}_2$ is called a **graph isomorphism** if $(v_i, v_j) \in \mathcal{E}_1$ implies that $(f(v_i), f(v_j)) \in \mathcal{E}_2$. Figure 1.3 shows an example of an isomorphism between two graphs.

13. *Higher-order graphs (hypergraph)*: A graph $G = (\mathcal{V}, \mathcal{E}, \mathcal{F})$ is considered to be a **higher-order graph** or **hypergraph** if each element of $\mathcal{F}$, $f_i \in \mathcal{F}$ is defined as a set of nodes for which each $|f_i| > 2$. Each element of a higher-order set is called a **hyperedge**, and each hyperedge may also be weighted. A k-**uniform hypergraph** is one for which all hyperedges have size k. Therefore, a 3-uniform hypergraph would be a collection of node triplets.

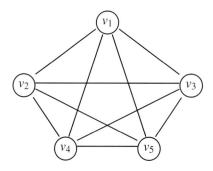

FIGURE 1.1
An example of complete graph K_5.

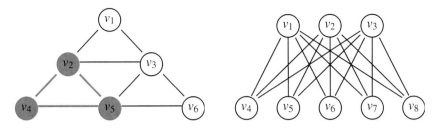

FIGURE 1.2
Left: A clique in a graph (grey). Right: The complete bipartite graph $K_{3,5}$ with $\mathcal{V}_1 = \{v_1, v_2, v_3\}$ and $\mathcal{V}_2 = \{v_4, v_5, v_6, v_7, v_8\}$.

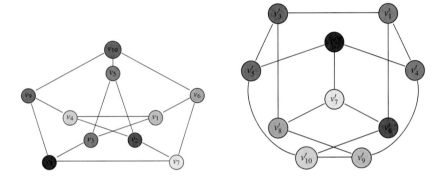

FIGURE 1.3
Example of two isomorphic graphs under the mapping $v_{10} \rightarrow v'_3$, $v_9 \rightarrow v'_5$, $v_8 \rightarrow v'_2$, $v_7 \rightarrow v'_7$, $v_6 \rightarrow v'_8$, $v_5 \rightarrow v'_1$, $v_4 \rightarrow v'_{10}$, $v_3 \rightarrow v'_4$, $v_2 \rightarrow v'_6$, $v_1 \rightarrow v'_9$.

1.3 Graph Representation

Several computer representations of graphs can be considered. The data structures used to represent graphs can have a significant influence on the size of problems that can be performed on a computer and the speed with which they can be solved. It is therefore important to know the different representations of graphs. To illustrate them, we will use the graph depicted in Figure 1.4.

1.3.1 Matrix Representations

A graph may be represented by one of several different common matrices. A matrix representation may provide efficient storage (since most graphs used in image processing are sparse), but each matrix representation may also be viewed as an *operator* that acts on functions associated with the nodes or edges of a graph.

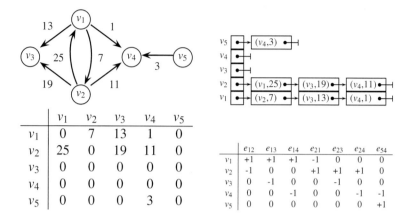

FIGURE 1.4
From top-left to bottom-right: a weighted directed graph, its adjacency list, its adjacency matrix, and its (transposed) incidence matrix representations.

The first matrix representation of a graph is given by the **incidence matrix**. The incidence matrix for $\mathcal{G} = (\mathcal{V}, \mathcal{E})$ is an $|\mathcal{E}| \times |\mathcal{V}|$ matrix $\mathbf{A}$ where

$$\mathbf{A}_{ij} = \begin{cases} -1 \text{ if } s(e_i) = v_j, \\ +1 \text{ if } t(e_i) = v_j, \\ 0 \text{ otherwise.} \end{cases} \tag{1.1}$$

The incidence matrix has the unique property of the matrices in this section that it preserves the orientation information of each edge but not the edge weight. The adjoint (transpose) of the incidence matrix also represents the *boundary operator* for a graph in the sense that multiplying this matrix with a signed indicator vector of edges in a path will return the endpoints of the path. Furthermore, the incidence matrix defines the exterior derivative for functions associated with the nodes of the graph. As such, this matrix plays a central role in *discrete calculus* (see [12] for more details).

The **constitutive matrix** of graph $\mathcal{G} = (\mathcal{V}, \mathcal{E})$ is an $|\mathcal{E}| \times |\mathcal{E}|$ matrix $\mathbf{C}$ where

$$\mathbf{C}_{ij} = \begin{cases} w(e_i) \text{ if } i = j, \\ 0 \text{ otherwise.} \end{cases} \tag{1.2}$$

The **adjacency matrix** representation of graph $\mathcal{G} = (\mathcal{V}, \mathcal{E})$ is a $|\mathcal{V}| \times |\mathcal{V}|$ matrix $\mathbf{W}$ where

$$\mathbf{W}_{ij} = \begin{cases} w_{ij} \text{ if } e_{ij} \in \mathcal{E}, \\ 0 \text{ otherwise.} \end{cases} \tag{1.3}$$

For undirected graphs the matrix $\mathbf{W}$ is symmetric.

The **Laplacian matrix** of an undirected graph $\mathcal{G} = (\mathcal{V}, \mathcal{E})$ is a $|\mathcal{V}| \times |\mathcal{V}|$ matrix **GL** where

$$\mathbf{L}_{ij} = \begin{cases} d_i \text{ if } i = j, \\ -w_{ij} \text{ if } e_{ij} \in \mathcal{E}, \\ 0 \text{ otherwise}, \end{cases} \qquad (1.4)$$

and **G** is a diagonal matrix with $\mathbf{G}_{ii} = w(v_i)$. In the context of the Laplacian matrix, $w(v_i) = 1$ or $w(v_i) = \frac{1}{d_i}$ are most often adopted (see [12] for a longer discussion on this point).

If $\mathbf{G} = \mathbf{I}$, then for an undirected graph these matrices are related to each other by the formula

$$\mathbf{A}^T \mathbf{C} \mathbf{A} = \mathbf{W} - \mathbf{D} = \mathbf{L}, \qquad (1.5)$$

where **D** is a diagonal matrix of node degrees with $\mathbf{D}_{ii} = d_i$.

1.3.2   Adjacency Lists

The advantage of adjacency lists over matrix representations is less memory usage. Indeed, a full incidence matrix requires $O(|\mathcal{V}| \times |\mathcal{E}|)$ memory, and a full adjacency matrix requires $O(|\mathcal{V}|^2)$ memory. However, sparse graphs can take advantage of sparse matrix representations for a much more efficient storage.

An adjacency list representation of a graph is an array L of $|\mathcal{V}|$ linked lists (one for each vertex of $\mathcal{V}$). For each vertex v_i, there is a pointer L_i to a linked list containing all vertices v_j adjacent to v_i. For weighted graphs, the linked list contains both the target vertex and the edge weight. With this representation, iterating through the set of edges is $O(|\mathcal{V}| + |\mathcal{E}|)$ whereas it is $O(|\mathcal{V}|^2)$ for adjacency matrices. However, checking if an edge $e_k \in \mathcal{E}$ is an $O(|\mathcal{V}|)$ operation with adjacency lists whereas it is an $O(1)$ operation with adjacency matrices.

1.4   Paths, Trees, and Connectivity

Graphs model relationships between nodes. These relationships make it possible to define whether two nodes are *connected* by a series of pairwise relationships. This concept of connectivity, and the related ideas of paths and trees, appear in some form throughout this work. Therefore, we now review the concepts related to connectivity and the basic algorithms used to probe connectivity relationships.

1.4.1 Walks and Paths

Given a directed graph $\mathcal{G} = (\mathcal{V}, \mathcal{E})$ and two vertices $v_i, v_j \in \mathcal{V}$, a **walk** (also called a chain) from v_i to v_j is a sequence $\pi(v_1, v_n) = (v_1, e_1, v_2, \ldots, v_{n-1}, e_{n-1}, v_n)$, where $n \geq 1$, $v_i \in \mathcal{V}$, and $e_j \in \mathcal{E}$:

$$v_1 = v_i, v_j = v_n, \text{ and } \{s(e_i), t(e_i)\} = \{v_i, v_{i+1}\}, 1 \leq i \leq n. \qquad (1.6)$$

The length of the walk is denoted as $|\pi| = n$. A walk may contain a vertex or edge more than once. If $v_i = v_j$, the walk is called a **closed walk**; otherwise it is an **open walk**. A walk is called a **trail** if every edge is traversed only once. A closed trail is called a **circuit**. A circuit is called a **cycle** if all nodes are distinct. An open walk is a **path** if every vertex is distinct. A **subpath** is any sequential subset of a path. Figure

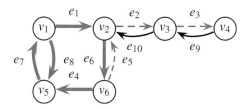

FIGURE 1.5
Illustration of a walk $\pi(v_1, v_1)$ (bolded grey), trail, and path $\pi(v_6, v_4)$ (dashed grey).

1.5 is used to illustrate the concepts of walk, trail, and path. In Figure 1.5, bolded grey arrows provide a walk $\pi(v_1, v_1) = (v_1, e_8, v_5, e_7, v_1, e_1, v_2, e_6, v_6, e_4, v_5, e_7, v_1)$ that is a closed walk, but is neither a trail (e_7 is traversed twice) nor a cycle. Dashed grey arrows provide a walk $\pi(v_6, v_4) = (v_6, e_5, v_2, e_2, v_3, e_3, v_4)$ that is an open walk, a trail, and a path. Black arrows are arrows not involved in the walks $\pi(v_1, v_1)$ and $\pi(v_6, v_4)$.

1.4.2 Connected Graphs

Two nodes are called **connected** if $\exists \pi(v_i, v_j)$ or $\exists \pi(v_j, v_i)$. A graph $\mathcal{G} = (\mathcal{V}, \mathcal{E})$ is called a **connected graph** iff $\forall v_i, v_j \in \mathcal{V}, \ \exists \pi(v_i, v_j), \text{ or } \exists \pi(v_j, v_i)$. Therefore, the relation

$$v_i \ R_w \ v_j = \begin{cases} & v_i = v_j, \\ & \exists \pi(v_i, v_j) \text{ or } \exists \pi(v_j, v_i), \end{cases} \qquad (1.7)$$

is an equivalence relation. The induced equivalence classes by this relation form a partition of $\mathcal{V}$ into $\mathcal{V}_1, \mathcal{V}_2, \ldots, \mathcal{V}_p$ subsets. Subgraphs $\mathcal{G}_1, \mathcal{G}_2, \ldots, \mathcal{G}_p$ induced by $\mathcal{V}_1, \mathcal{V}_2, \ldots, \mathcal{V}_p$ are called the *connected components* of the graph $\mathcal{G}$. Each connected component is a connected graph. The connected components of a graph $\mathcal{G}$ are the set of largest subgraphs of $\mathcal{G}$ that are each connected.

Two nodes are **strongly connected** if $\exists \pi(v_i, v_j)$ and $\exists \pi(v_j, v_i)$. If two nodes are

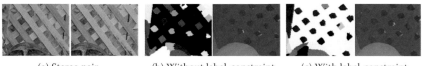

(a) Stereo pair   (b) Without label-constraint   (c) With label-constraint

FIGURE 3.9
Illustrating the label-cost prior. (a) Crop of a stereo image ("cones" image from Middlebury database). (b) Result without a label-cost prior. In the left image each color represents a different surface, where gray-scale colors mark planes and nongrayscale colors, B-splines. The right image shows the resulting depth map. (c) Corresponding result with label-cost prior. The main improvement over (b) is that large parts of the green background (visible through the fence) are assigned to the same planar surface.

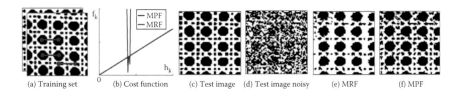

(a) Training set   (b) Cost function   (c) Test image   (d) Test image noisy   (e) MRF   (f) MPF

FIGURE 3.10
Illustrating the advantage of using an marginal probability field (MPF) over an MRF. (a) Set of training images for binary texture denoising. Superimposed is a pairwise term (translationally invariant with shift (15; 0); 3 exemplars in red). Consider the labeling $k = (1, 1)$ of this pairwise term ϕ. Each training image has a certain number $h_{(1, 1)}$ of $(1, 1)$ labels, i.e. $h_{(1, 1)} = \Sigma_i[\phi_i = (1, 1)]$. The negative logarithm of the statistics $\{h_{(1, 1)}\}$ over all training images is illustrated in blue (b). It shows that all training images have roughly the same number of pairwise terms with label $(1, 1)$. The MPF uses the convex function f_k (blue) as cost function. It is apparent that the linear cost function of an MRF is a bad fit. (c) A test image and (d) a noisy input image. The result with an MRF (e) is inferior to the MPF model (f). Here the MPF uses a global potential on unary and pairwise terms.

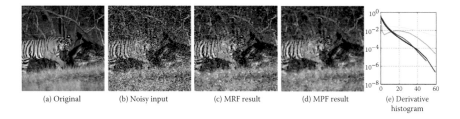

(a) Original   (b) Noisy input   (c) MRF result   (d) MPF result   (e) Derivative histogram

FIGURE 3.11
Results for image denoising using a pairwise MRF (c) and an MPF model (d). The MPF forces the derivatives of the solution to follow a mean distribution, which was derived from a large dataset. (f) Shows derivative histograms (discretized into the 11 bins). Here black is the target mean statistic, blue is for the original image (a), yellow for noisy input (b), green for the MRF result (c), and red for the MPF result (d). Note, the MPF result is visually superior and does also match the target distribution better. The runtime for the MRF (c) is 1096s and for MPF (d) 2446s.

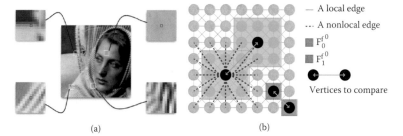

	— A local edge
	⋯ A nonlocal edge
	■ $F_0^{f^0}$
	■ $F_1^{f^0}$
	●—● Vertices to compare

(a)   (b)

FIGURE 7.2
(a) Four examples of image patches. Central pixels (in red) are associated with vectors of gray-scale values within a patch of size 13×3 ($F_6^{f^0}$). (b) Graph topology and feature vector illustration: a 8-adjacency grid graphs is constructed. The vertex surrounded by a red box shows the use of the simplest feature vector, i.e., the initial function. The vertex surrounded by a green box shows the use of a feature vector that exploits patches.

Gray level image   Color scribbles

$p = 1, G_1, F_0^{f0} = f^0$ $p = 1, G_5, F_2^{f0}$

FIGURE 7.9
Isotropic colorization of a grayscale image with $w = g_2$ and $\lambda = 0{:}01$ in local and nonlocal configurations.

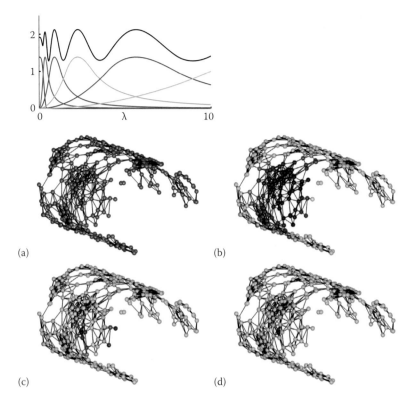

FIGURE 8.1
Left: Scaling function $h(\lambda)$ (blue curve), wavelet generating kernels $g(t_j\lambda)$, and sum of squares G (black curve), for $J = 5$ scales, $\lambda_{max} = 10$, $K_{1p} = 20$. Details in Section 8.2.3. Right: Example graph showing specified center vertex (a), scaling function (b) and wavelets at two scales (c,d). Figures adapted from [1].

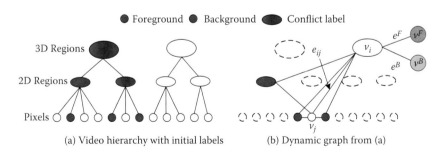

FIGURE 9.8
Video hierarchy and dynamic graph construction in the 3D video cutout system [38]. (a). Video hierarchy created by two-step mean shift segmentation. User-specified labels are then propagated upward from the bottom level. (b). A graph is dynamically constructed based on the current configuration of the labels. Only highest level nodes that have no conflict are selected. Each node in the hierarchy also connects to two virtual nodes v^F and v^B.

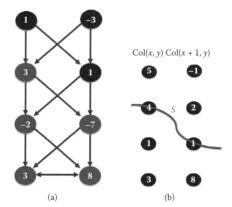

Col(x, y) Col(x + 1, y)

(a)   (b)

FIGURE 10.5
Illustrating the recovery of a feasible surface from a closed set. (a) shows a closed set in the constructed graph in Figure 10.4(d), which consists of all green nodes. The surface specified by the closed set in (a) is shown in (b).

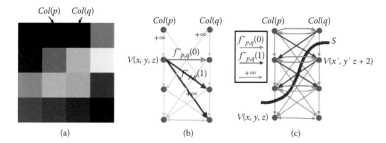

Col(p)   Col(q)

(a)   (b)   (c)

FIGURE 10.9
Graph construction for the convex smoothness penalty function. (a) Example of two adjacent columns. The smoothness parameter $\Delta_x = 2$. (b) Weighted arcs introduced for each node $V(x, y, z)$. Gray arcs of weight $+\infty$ reflect the hard smoothness constraints. (c) Weighted arcs are built between nodes corresponding to the two adjacent columns. The feasible surface S cuts the arcs with a total weight of $f_{p,q}(2) = f''_{p,q}(0) + [f''_{p,q}(0) + f''_{p,q}(1)]$, which determines smoothness shape compliance term of $\alpha(S)$ for the two adjacent columns.

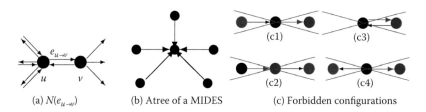

(a) $N(e_{u \to v})$ (b) A tree of a MIDES   (c) Forbidden configurations

FIGURE 11.11
Directed neighborhood $N(e_{u \to v})$ of edge $e_{u \to v}$(a). A tree of a MIDES defining unambiguously its surviving vertex (b). Two edges within a same neighborhood cannot be both selected. Hence, red edges in (c) cannot be selected once black edges have been added to the MIDES.

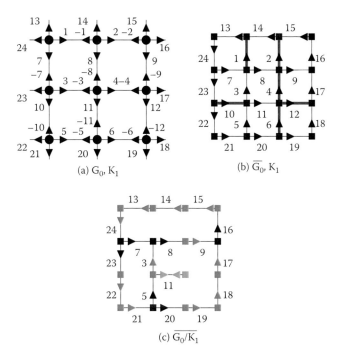

(a) G_0, K_1

(b) $\overline{G}_0$, K_1

(c) $\overline{G_0/K_1}$

FIGURE 11.21
A combinatorial map (a) and its dual (b) encoding a 3 × 3 grid with the contraction kernel $K_1 =$ $\alpha^*(1, 2, 4, 12, 6, 10)$ (■) superimposed to both figures. Only positive darts are represented in (b) and (c) in order to not overload them. The resulting reduced dual combinatorial map (c) should be further simplified by a removal kernel of empty self-loops (RKESL) $K_2 = \alpha^*(11)$ to remove dual vertices of degree 1 (■). Dual vertices of degree 2 (■) are removed by the RKEDE $K_3 =$ $\varphi^*(24, 13, 14, 15, 9, 22, 17, 21, 19, 18) \cup \{3, -5\}$. Note that the dual vertex $\varphi(3) = (3, -5, 11)$ becomes a degree 2 vertex only after the removal of $\alpha^*(11)$.

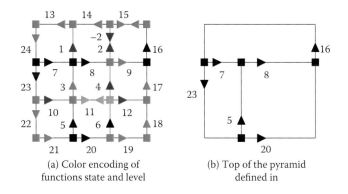

(a) Color encoding of
functions state and level

(b) Top of the pyramid
defined in

FIGURE 11.22
Implicit encoding (a) and top level combinatorial map (b) of the pyramid defined in Figure 11.21. Red, green and blue darts have level equal to 1, 2, and 3 respectively. States associated to these levels are, respectively, equal to *CK, RKESL, RKEDE*.

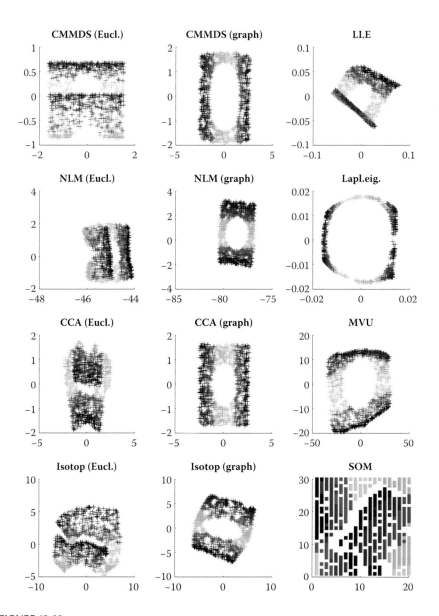

FIGURE 12.03
Two-dimensional embeddings of the Swiss roll with various NLDR methods. Distance-based methods use either Euclidean norm ('Eucl.') or shortest paths in a neighborhood graph ('Graph').

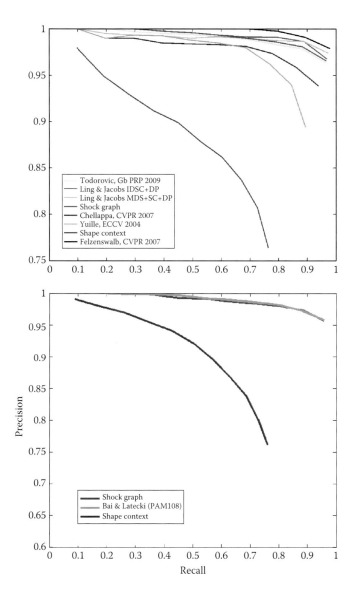

FIGURE 14.6
The recognition performance of the shock-graph matching based on the approach described above for the Kimia-99 and Kimia-216 shape databases [11].

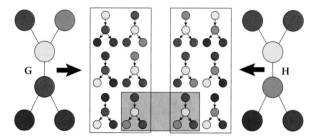

FIGURE 17.3
Graph kernels between two graphs (each color represents a different label). We display all binary 1-tree walks with a specific tree structure, extracted from two simple graphs. The graph kernels are computing and summing the local kernels between all those extracted tree-walks. In the case of the Dirac kernel (hard matching), only one pair of tree-walks is matched (for both labels and structures).

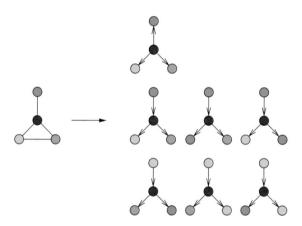

FIGURE 17.5
Examples of tree-walks from a graph. Each color represents a different label.

connected but not strongly connected, then the nodes are called **weakly connected**. Note that, in an undirected graph, all connected nodes are strongly connected. The graph is strongly connected iff $\forall v_i, v_j \in \mathcal{V}$, $\exists \pi(v_i, v_j)$, and $\exists \pi(v_j, v_i)$. Therefore, the relation

$$v_i \; R_s \; v_j = \begin{cases} v_i = v_j \\ \exists \pi(v_i, v_j) \text{ and } \exists \pi(v_j, v_i), \end{cases} \tag{1.8}$$

is an equivalence relation. The subgraphs induced by the obtained partition of $\mathcal{G}$ are the *strongly connected components* of $\mathcal{G}$. Figure 1.6 illustrates the concepts of strongly and weakly connected components.

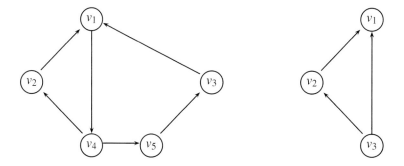

FIGURE 1.6
Left: A strongly connected component. Right: A weakly connected component.

1.4.3 Shortest Paths

The problems of routing in graphs (in particular the search for a shortest path) are among the oldest and most common problems in graph theory. Let $\mathcal{G} = (\mathcal{V}, \mathcal{E})$ be a weighted graph where a weight function $w : \mathcal{E} \to \mathbb{R}$ associates a real value (i.e., a weight also called a length in this context) to each edge.[1] In this section we consider only graphs for which all edge weights are nonnegative. Let $l(\pi(v_i, v_j))$ denote the total weight (or length) of a walk:

$$l(\pi(v_i, v_j)) = \sum_{e_k \in \pi(v_i, v_j)} w(e_k) \, . \tag{1.9}$$

---

[1]Edge weights may be used to represent either *affinities* or *distances* between nodes. An affinity weight of zero is viewed as equivalent to a disconnection (removal of the edge from $\mathcal{E}$), and a distance weight of ∞ is viewed as equivalent to a disconnection. An affinity weight a may be converted to a distance weight b via $b = \frac{1}{a}$. A much longer discussion on this relationship may be found in [12]. In this work, affinity weights and distance weights use the same notation, with the distinction being made by context. All weights in this section are considered to be distance weights.

We will also use the convention that the length of the walk equals ∞ if v_i and v_j are not connected and $l(\pi(v_i, v_i)) = 0$. The minimum length between two vertices is

$$l^*(v_i, v_j) = \arg\min_{\pi(v_i, v_j)} l(\pi(v_i, v_j)). \tag{1.10}$$

A walk satisfying the minimum in (1.10) is a path called a **shortest path**. The shortest path between two nodes may not be unique. Different algorithms can be used to compute l^* if one wants to find a shortest path from one vertex to all the others or between all pairs of vertices. We will restrict ourselves to the first kind of problem.

To compute the shortest path from one vertex to all the others, the most common algorithm is from Dijkstra [13]. Dijkstra's algorithm solves the problem for every pair v_i, v_j, where v_i is a fixed starting point and $v_j \in \mathcal{V}$. Dijkstra's algorithm exploits the property that the path between any two nodes contained in a shortest path is also a shortest path. Specifically, let $\pi(v_i, v_j)$ be a shortest path from v_i to v_j in a weighted connected graph $\mathcal{G} = (\mathcal{V}, \mathcal{E})$ with positive weights ($w : \mathcal{E} \rightarrow \mathbb{R}^+$) and let v_k be a vertex in $\pi(v_i, v_j)$. Then, the subpath $\pi(v_i, v_k) \subset \pi(v_i, v_j)$ is a shortest path from v_i to v_k. Dijkstra's algorithm for computing shortest paths with v_i as a source is provided in Figure 1.7 with an example on a given graph. In computer vision, shortest paths have been used for example, for interactive image segmentation [14, 15].

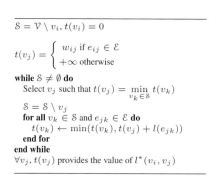

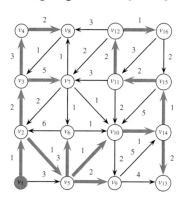

FIGURE 1.7
Left: Dijkstra's algorithm for computing shortest path from a source vertex v_i to all vertices in a graph. Right: Illustration of Dijkstra's algorithm. The source is vertex v_1. The shortest path from v_1 to other vertices is shown with bolded grey arrows.

1.4.4 Trees and Minimum Spanning Trees

A **tree** is an undirected connected graph without cycles (acyclic). An unconnected tree without cycles is called a forest (each connected component of a forest is a tree). More precisely, if $\mathcal{G}$ is a graph with $|\mathcal{V}| = n$ vertices and if $\mathcal{G}$ is a tree, the following properties are equivalent:

- $\mathcal{G}$ is connected without cycles,

- $\mathcal{G}$ is connected and has $n - 1$ edges,

- $\mathcal{G}$ has no cycles and has $n - 1$ edges,

- $\mathcal{G}$ has no cycles and is maximal for this property (i.e., adding any edge creates a cycle),

- $\mathcal{G}$ is connected and is minimal for this property (i.e., removing any edge makes the graph disconnected),

- There exists a unique path between any two vertices of $\mathcal{G}$.

A partial graph $\mathcal{G}' = (\mathcal{V}, \mathcal{E}')$ of a connected graph $\mathcal{G} = (\mathcal{V}, \mathcal{E})$ is called a **spanning tree** of $\mathcal{G}$ if $\mathcal{G}'$ is a tree. There is at least one spanning tree for any connected graph.

We may define the weight (or cost) of any tree $\mathcal{T} = (\mathcal{V}, \mathcal{E}_{\mathcal{T}})$ by

$$c(\mathcal{T}) = \sum_{e_k \in \mathcal{E}_{\mathcal{T}}} w(e_k) . \tag{1.11}$$

A spanning tree of the graph that minimizes the cost (compared to all spanning trees of the graph) is called a **minimal spanning tree**. The minimal spanning tree of a graph may not be unique. The problem of finding a minimal spanning tree of a graph appears in many applications (e.g., in phone network design). There are several algorithms for finding minimum spanning trees, and we present the one proposed by Prim. The principle of Prim's algorithm is to progressively build a tree. The algorithm starts from the edge of minimum weight and adds iteratively to the tree the edge of minimum weight among all the possible edges that maintain a partial graph that is acyclic. More precisely, for a graph $\mathcal{G} = (\mathcal{V}, \mathcal{E})$, one progressively constructs from a vertex v_1 a subset $\mathcal{V}'$ with $\{v_1\} \subseteq \mathcal{V}' \subseteq \mathcal{V}$ and a subset $\mathcal{E}' \subseteq \mathcal{E}$ such that the partial graph $\mathcal{G}' = (\mathcal{V}, \mathcal{E}')$ is a minimum spanning tree of $\mathcal{G}$. To do this, at each step, one selects the edge e_{ij} of minimum weight in the set $\{e_{kl} \mid e_{kl} \in \mathcal{E}, v_k \in \mathcal{V}', l \in \mathcal{V} \setminus \mathcal{V}'\}$. Subset $\mathcal{V}'$ and $\mathcal{E}'$ are then augmented with vertex v_j and edge e_{ij}: $\mathcal{V}' \leftarrow \mathcal{V}' \cup \{v_j\}$ and $\mathcal{E}' \leftarrow \mathcal{E}' \cup \{e_{ij}\}$. The algorithm terminates when $\mathcal{V}' = \mathcal{V}$. Prim's algorithm for computing minimum spanning trees is provided in Figure 1.8 along with an example on a given graph. In computer vision, minimum spanning trees have been used, for example, for image segmentation [16] and image hierarchical representation [17].

1.4.5   Maximum Flows and Minimum Cuts

A transport network is a weighted directed graph $\mathcal{G} = (\mathcal{V}, \mathcal{E})$ such that

- There exists exactly one vertex v_1 with no predecessor called the **source** (and denoted by v_s), that is, $\deg^+(v_1) = 0$,

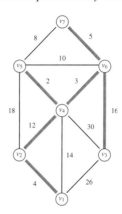

```
𝒱' = {v₁}, ℰ' = ∅
for all vⱼ ∈ 𝒱\𝒱' do
    L(vⱼ) = l(e₁ⱼ)
end for
while 𝒱' ≠ 𝒱 do
    Choose vᵢ ∈ 𝒱\𝒱' such that L(vᵢ) is minimum
    Let eₖ be the associated edge
    𝒱' ← 𝒱' ∪ {vᵢ}
    ℰ' ← ℰ' ∪ {eₖ}
    for all vⱼ ∈ 𝒱\𝒱' do
        if l(eᵢⱼ) < L(vⱼ) then
            L(vⱼ) = l(eᵢⱼ)
        end if
    end for
end while
```

FIGURE 1.8

Left: Prim's algorithm for computing minimum spanning trees in a graph starting from node v_1. Note that, any node may be selected as v_1, and the algorithm will generate a minimal spanning tree. However, when a graph contains multiple minimal spanning trees the selection of v_1 may determine which minimal spanning tree is determined by the algorithm. Right: Illustration of Prim's algorithm. The minimum spanning tree is shown by the grey bolded edges.

- There exists exactly one vertex v_n with no successor called the **sink** (and denoted by v_t), that is, $\deg^-(v_n) = 0$,

- There exists at least one path from v_s to v_t,

- The weight $w(e_{ij})$ of the directed edge e_{ij} is called the **capacity** and it is a nonnegative real number, that is, we have a mapping $w : \mathcal{E} \to \mathbb{R}^+$.

We denote by $\mathcal{A}^+(v_i)$ the set of inward directed edges from vertex v_i:

$$\mathcal{A}^+(v_i) = \{e_{ji} \in \mathcal{E}\} . \tag{1.12}$$

Similarly, we denote by $\mathcal{A}^-(v_i)$ the set of outward directed edges from vertex v_i:

$$\mathcal{A}^-(v_i) = \{e_{ij} \in \mathcal{E}\} . \tag{1.13}$$

A function $\varphi : \mathcal{E} \to \mathbb{R}^+$ is called a **flow** iff

- Each vertex $v_i \notin \{v_s, v_t\}$ satisfies the conservation condition (also called Kirchhoff's Current Law):

$$\sum_{e_{ji}} \varphi(e_{ji}) - \sum_{e_{ij}} \varphi(e_{ij}) = 0, \tag{1.14}$$

which may also be written in matrix form as

$$\mathbf{A}^T \varphi = \mathbf{p}, \tag{1.15}$$

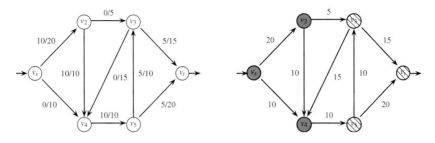

FIGURE 1.9
Left: an (s, t)-flow with value 10. Each edge is labeled with its flow/capacity. Right:
An (s, t)-cut with capacity 30, $\mathcal{V}_1 = \{v_s, v_2, v_4\}$ (grey) and $\mathcal{V}_2 = \{v_3, v_5, v_t\}$
(striped).

where $p_i = 0$ for any $v_i \notin \{v_s, v_t\}$. Intuitively, this law states that the flow
entering each node must also leave that node (i.e., conservation of flow).

- For each edge e_{ij}, the capacity constraint $\varphi(e_{ij}) \leq w(e_{ij})$ is satisfied.

The value of a flow is $|\varphi| = \sum\limits_{e_{ij} \in \mathcal{A}^-(v_s)} \varphi(e_{ij}) = \sum\limits_{e_{ij} \in \mathcal{A}^+(v_t)} \varphi(e_{ij}) = p_s$. A flow φ^*
is a **maximum flow** if its value is the largest possible, that is, $|\varphi^*| \geq |\varphi|$ for every
other flow φ. Figure 1.9 shows an (s, t) flow with value 10.

An (s, t)-cut is a partition $P =< \mathcal{V}_1, \mathcal{V}_2 >$ of the vertices into subsets $\mathcal{V}_1$ and
$\mathcal{V}_2$, such that, $\mathcal{V}_1 \cap \mathcal{V}_2 = \emptyset$, $\mathcal{V}_1 \cup \mathcal{V}_2 = \mathcal{V}$, $v_s \in \mathcal{V}_1$ and $v_t \in \mathcal{V}_2$. The capacity of a
cut P is the sum of the capacities of the edges that start in $\mathcal{V}_1$ and end in $\mathcal{V}_2$:

$$C(P) = \sum_{v_i \in \mathcal{V}_1} \sum_{v_j \in \mathcal{V}_2} w(e_{ij}) , \qquad (1.16)$$

with $w(e_{ij}) = 0$ if $e_{ij} \notin \mathcal{E}$. Figure 1.9 shows an (s, t)-cut with capacity 30. The
minimum cut problem is to compute an (s, t)-cut for which the capacity is as low as
possible. Intuitively, the minimum cut is the least expensive way to disrupt all flow
from v_s to v_t. In fact, one can prove that, for any weighted directed graph, there is
always a flow φ and a cut $(\mathcal{V}_1, \mathcal{V}_2)$ such that the value of the maximum flow is equal
to the capacity of the minimum cut [18].

We call an edge e_{ij} **saturated** by a flow if $\varphi(e_{ij}) = w_{ij}$. A path from v_s to v_t
is saturated if any one of its edges is saturated. The **residual capacity** of an edge
$c_r : \mathcal{V} \times \mathcal{V} \to \mathbb{R}^+$ is

$$c_r(e_{ij}) = \begin{cases} w_{ij} - \varphi(e_{ij}) \text{ if } e_{ij} \in \mathcal{E}, \\ \varphi(e_{ij}) \text{ if } e_{ji} \in \mathcal{E}, \\ 0 \text{ otherwise.} \end{cases} \qquad (1.17)$$

The residual capacity represents the flow quantity that can still go through the edge.
We may define the **residual graph** as a partial graph $\mathcal{G}_r = (\mathcal{V}, \mathcal{E}_r)$ of the initial

$\varphi(e_{ij}) \leftarrow 0, \forall e_{ij} \in \mathcal{E}$
Construct the residual graph $\mathcal{G}_r = (\mathcal{V}, \mathcal{E}_r)$
while $\exists\, \pi(v_s, v_t)$ in $\mathcal{G}_r$ such that $c_r(\pi) > 0$ **do**
    for all $e_{ij} \in \pi$ **do**
        $\varphi(e_{ij}) \leftarrow \varphi(e_{ij}) + c_r(\pi)$
        $\varphi(e_{ji}) \leftarrow \varphi(e_{ji}) - c_r(\pi)$
    end for
end while

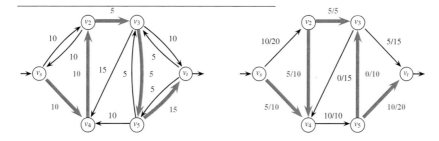

FIGURE 1.10
Top: Ford–Fulkerson algorithm for maximum flow. Bottom: An augmenting path π
in $\mathcal{G}_r$ with $c_r(\pi) = 5$, and the associated augmented flow φ'.

graph where all the edges of zero residual capacity have been removed, that is, $\mathcal{E}_r = \{e_{ij} \in \mathcal{E} \mid c_r(e_{ij}) > 0\}$.

Given a path π in $\mathcal{G}_r$ from v_s to v_t, the residual capacity of an augmenting path is the minimum of all the residual capacities of its edges:

$$c_r(\pi) = \arg\min_{e_{ij} \in \pi} c_r(e_j). \tag{1.18}$$

Such a path is called an **augmenting path** if $c_r(\pi) > 0$. From this augmenting path we can define a new augmented flow function φ':

$$\varphi'(e_{ij}) = \begin{cases} \varphi(e_{ij}) + c_r(\pi) \text{ if } e_{ij} \in \pi, \\ \varphi(e_{ij}) - c_r(\pi) \text{ if } e_{ji} \in \pi, \\ \varphi(e_{ij}) \text{ otherwise.} \end{cases} \tag{1.19}$$

A flow is called a **maximum flow** if and only if the corresponding residual graph is disconnected (i.e., there is no augmenting path). There are several algorithms for finding maximum flows, but we present the classic algorithm proposed by Ford and Fulkerson [18]. The principle of the Ford–Fulkerson algorithm is to add flow along one path as long as there exists an augmenting path in the graph. Figure 1.10 provides the algorithm along with one step of generating an augmenting path. In computer vision, the use of maximum-flow/min-cut solutions to solve problems is typically referred to as *graph cuts*, following the publication by Boykov et al. [19]. Furthermore, the Ford–Fulkerson algorithm was found to be inefficient for many computer vision problems, with the algorithm by Boykov and Kolmogorov generally preferred [20].

1.5   Graph Models in Image Processing and Analysis

In the previous sections we have presented some of the basic definitions and notations used for graph theory in this book. These definitions and notations may be found in other sources in the literature but were presented to make this work self-contained. However, there are many ways in which graph theory is applied in image processing and computer vision that are specialized to these domains. We now build on the previous definitions to provide some of the basics for how graph theory has been specialized for image processing and computer vision in order to provide a foundation for the subsequent chapters.

1.5.1   The Regular Lattice

Digital image processing operates on sampled data from an underlying continuous light field. Each sample is called a **pixel** in 2D or a **voxel** in 3D. The sampling is based on the notion of *lattice*, which can be viewed as a regular tiling of a space by a primitive cell. A d-dimensional lattice $\mathbb{L}^d$ is a subset of the d-dimensional Euclidean space $\mathbb{R}^d$ with

$$\mathbb{L}^d = \{\mathbf{x} \in \mathbb{R}^d \mid \mathbf{x} = \sum_{i=1}^d k_i \mathbf{u_i}, \ k_i \in \mathbb{Z}\} = \mathbf{U}\mathbb{Z}^d , \tag{1.20}$$

where $\mathbf{u_1}, \ldots, \mathbf{u_d} \in \mathbb{R}^d$ form a basis of $\mathbb{R}^d$ and $\mathbf{U}$ is the matrix of column vectors. A lattice in a d-dimensional space is therefore a set of all possible vectors represented as integer weighted combinations of d linearly independent basis vectors. The lattice points are given by $\mathbf{Uk}$ with $\mathbf{k}^T = (k_1, \ldots, k_d) \in \mathbb{Z}^d$.

The unit cell of $\mathbb{L}^d$ with respect to the basis $\mathbf{u_1}, \ldots, \mathbf{u_d}$ is

$$C = \{\mathbf{x} \in \mathbb{R}^d \mid \mathbf{x} = \sum_{i=1}^d k_i \mathbf{u_i}, \ k_i \in [0, 1]\} = \mathbf{U} \cdot [0, \mathbf{u_1}]^d , \tag{1.21}$$

where $[0, \mathbf{u_1}]^d$ is the unit cube in $\mathbb{R}^d$. The volume of C is given by $|\det \mathbf{U}|$. The density of the lattice is given by $1/|\det\mathbf{U}|$, that is, the lattice sites per unit surface. The set $\{C + \mathbf{x} \mid \mathbf{x} \in \mathbb{R}^d\}$ of all lattice cells covers $\mathbb{R}^d$. The Voronoi cell of a lattice is called the unit cell (a cell of a lattice whose translations cover the whole space). The Voronoi cell encloses all points that are closer to the origin than to other lattice points.

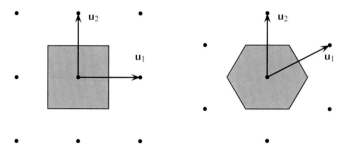

FIGURE 1.11

The rectangular (left) and hexagonal (right) lattices and their associated Voronoi cells.

The Voronoi cell boundaries are equidistant hyperplanes between surrounding lattice points. Two well-known 2D lattices are the rectangular $\mathbf{U}_1 = \begin{bmatrix} 1 & 0 \\ 0 & 1 \end{bmatrix}$ and hexagonal

$\mathbf{U}_2 = \begin{bmatrix} \dfrac{\sqrt{3}}{2} & 0 \\ \dfrac{1}{2} & 1 \end{bmatrix}$ lattices. Figure 1.11 presents these two lattices. By far the most

common lattice in image processing and computer vision is the rectangular sampling lattice derived from $\mathbf{U}_1$. In 3D image processing, the lattice sampling is typically given by

$$\mathbf{U}_3 = \begin{bmatrix} 1 & 0 & 0 \\ 0 & 1 & 0 \\ 0 & 0 & k \end{bmatrix}, \tag{1.22}$$

for some constant k in which the typical situation is that $k \geq 1$.

The most common way of using a graph to model image data is to identify every node with a pixel (or voxel) and to impose an edge structure to define local neighborhoods for each node. Since a lattice defines a regular arrangement of pixels it is possible to define a regular graph by using edges to connect pairs of nodes that fall within a fixed Euclidean distance of each other.

For example, in a rectangular lattice, two pixels $\mathbf{p} = (p_1, p_2)^T$ and $\mathbf{q} = (q_1, q_2)^T$ of $\mathbb{Z}^2$, are called 4-adjacent (or 4-connected) if they differ by at most one coordinate:

$$|p_1 - q_1| + |p_2 - q_2| = 1 , \tag{1.23}$$

and 8-adjacent (or 8-connected) if they differ at most of 2 coordinates:

$$\max(|p_1 - q_1|, |p_2 - q_2|) = 1 , \tag{1.24}$$

In the hexagonal lattice, only one adjacency relationship exists: 6-adjacency. Figure 1.11 presents adjacency relationships on the 2D rectangular and hexagonal lattices. Similarly, in $\mathbb{Z}^3$, we can say in a rectangular lattice that voxels are 6-adjacent,

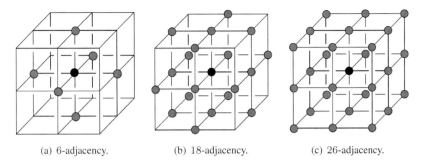

(a) 6-adjacency.   (b) 18-adjacency.   (c) 26-adjacency.

FIGURE 1.12
Different adjacency structures in a 3D lattice.

18-adjacent, or 26-adjacent whether they differ at most of 1, 2, or 3 coordinates. Figure 1.12 presents adjacency relationships between 3D cells (voxels).

If one considers edges that connect pairs of nodes that are not directly (i.e., locally) spatially adjacent (e.g., 4- or 8-connected in $\mathbb{Z}^2$) [21], the corresponding edges are called nonlocal edges [22]. The level of nonlocality of an edge depends on how far the two points are in the underlying domain. These nonlocal edges are usually obtained from proximity graphs (see Section 1.5.3) on the pixel coordinates (e.g., ϵ-ball graphs) or on pixel features (e.g., k-nearest-neighbor graphs on patches).

1.5.2 Irregular tessellations

Image data is almost always sampled in a regular, rectangular lattice that is modeled with the graphs in the previous section. However, images are often simplified prior to processing by merging together regions of adjacent pixels using an untargeted image segmentation algorithm. The purpose of this simplification is typically to reduce processing time for further operations, but it can also be used to compress the image representation or to improve the effectiveness of subsequent operations. When this image simplification has been performed, a graph may be associated with the simplified image by identifying each merged region with one node and defining an edge relationship for two adjacency regions. Unfortunately, this graph is no longer guaranteed to be regular due to the image-dependent and space-varying merging of the underlying pixels. In this section we review common simplification methods that produce these irregular graphs.

One very well-known irregular image tessellation is the region quadtree tessellation [25]. A **region quadtree** is a hierarchical image representation that is derived from recursively subdividing the 2D space into four quadrants of equal size until every square contains one homogeneous region (based on appearance properties of every region of the underlying rectangular grid) or contains a maximum of one pixel.

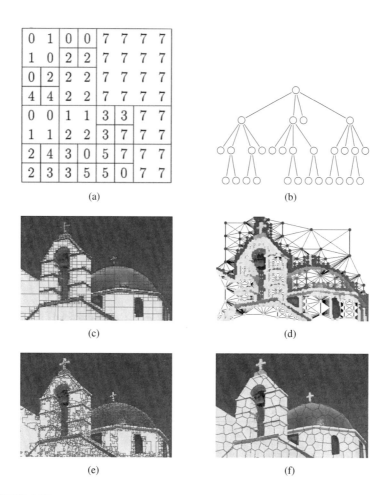

FIGURE 1.13

(a) An image with the quadtree tessellation, (b) the associated partition tree, (c) a real image with the quadtree tessellation, (d) the region adjacency graph associated to the quadtree partition, (e) and (f) two different irregular tessellations of an image using image-dependent superpixel segmentation methods: Watershed [23] and SLIC superpixels [24].

A region quadtree tesselation can be easily represented by a tree where each node of the tree corresponds to a square. The set of all nodes at a given depth provides a given level of decomposition of the 2D space. Figure 1.13 presents a simple example on an image (Figure 1.13(a)) with the associated tree (Figure 1.13(b)), as well as an example on a real image (Figure 1.13(c)). Quadtrees easily generalize to higher dimensions. For instance, in three dimensions, the 3D space is recursively subdivided into eight octants, and thus the tree is called an **octree**, where each node has eight children.

Finally, any partition of the classical rectangular grid can be viewed as producing an irregular tessellation of an image. Therefore any segmentation of an image can be associated with a **region adjacency graph**. Figure 1.13(d) presents the region adjacency graph associated to the partition depicted in 1.13(c). Specifically, given a graph $\mathcal{G} = \{\mathcal{V}, \mathcal{E}\}$ where each node is identified with a pixel, a partition into R connected regions $\mathcal{V}_1 \cup \mathcal{V}_2 \cup \ldots \cup \mathcal{V}_R = \mathcal{V}$, $\mathcal{V}_1 \cap \mathcal{V}_2 \cap \ldots \cap \mathcal{V}_R = \emptyset$ may be identified with a new graph $\tilde{\mathcal{G}} = \{\tilde{\mathcal{V}}, \tilde{\mathcal{E}}\}$ where each partition of $\mathcal{V}$ is identified with a node of $\tilde{\mathcal{V}}$, that is, $\mathcal{V}_i \in \tilde{\mathcal{V}}$. A common method for defining $\tilde{\mathcal{E}}$ is to let the edge weight between each new node equal the sum of the edge weights connecting each original node in the sets, that is, for edge $e_{ij} \in \tilde{\mathcal{E}}$

$$w_{ij} = \sum_{e_{ks}, v_k \in \mathcal{V}_i, v_s \in \mathcal{V}_j} w_{ks}. \qquad (1.25)$$

Each contiguous $\mathcal{V}_i$ is called a **superpixel**. Superpixels are an increasingly popular trend in computer vision and image processing. This popularity is partly due to the success of graph-based algorithms, which can be used to operate efficiently on these irregular structures (as opposed to traditional image-processing algorithms which required that each element is organized on a grid). Indeed, graph-based models (e.g., Markov Random Fields or Conditional Random Fields) can provide dramatic speed increases when moving from pixel-based graphs to superpixel-based graphs due to the drastic reduction in the number of nodes. For many vision tasks, compact and highly uniform superpixels that respect image boundaries are desirable. Typical algorithms for generating superpixels are Normalized Cuts [21], Watersheds [23], Turbo Pixels [26], and simple linear iterative clustering superpixels [24], etc. Figure 1.13(e) and (f) presents some irregular tessellations based on superpixel algorithms.

1.5.3   Proximity Graphs for Unorganized Embedded Data

Sometimes the relevant features of an image are neither individual pixels nor a partition of the image. For example, the relevant features of an image might be the blood cells in a biomedical image, and it is these blood cells that we wish to identify with the nodes of our graph. We may assume that each node (feature) in a dD image is associated with a coordinate in d dimensions. However, although each node has a geometric representation, it is much less clear than before how to construct a meaningful edge set for a graph based on the proximity of features.

There are many ways to construct a proximity graph representation from a set of data points that are embedded in $\mathbb{R}^d$. Let us consider a set of data points $\{\mathbf{x_1}, \ldots, \mathbf{x_n}\} \in \mathbb{R}^d$. To each data point we associate a vertex of a proximity graph $\mathcal{G}$ to define a set of vertices $\mathcal{V} = \{v_1, v_2, \ldots, v_n\}$. Determining the edge set $\mathcal{E}$ of the proximity graph $\mathcal{G}$ requires defining the neighbors of each vertex v_i according to its embedding $\mathbf{x}_i$. A proximity graph is therefore a graph in which two vertices are connected by an edge iff the data points associated to the vertices satisfy particular geometric requirements. Such particular geometric requirements are usually based on a metric measuring the distance between two data points. A usual choice of metric is the Euclidean metric. We will denote as $\mathcal{D}(v_i, v_j) = \|\mathbf{x}_i - \mathbf{x}_j\|_2$ the Euclidean distance between vertices, and as $\mathcal{B}(v_i \ ; \ r) = \{\mathbf{x}_j \in \mathbb{R}^n \mid \mathcal{D}(v_i, v_j) \le r\}$ the closed ball of radius r centered on $\mathbf{x}_i$. Classical proximity graphs are:

- The ϵ-ball graph: $v_i \sim v_j$ if $\mathbf{x}_j \in \mathcal{B}(v_i \ ; \ \epsilon)$.

- The k-nearest-neighbor graph (k-NNG): $v_i \sim v_j$ if the distance between $\mathbf{x}_i$ and $\mathbf{x}_i$ is among the k-th smallest distances from $\mathbf{x}_i$ to other data points. The k-NNG is a directed graph since one can have $\mathbf{x}_i$ among the k-nearest neighbors of $\mathbf{x}_j$ but *not* vice versa.

- The Euclidean Minimum Spanning Tree (EMST): This is a connected tree subgraph that contains all the vertices and has a minimum sum of edge weights (see Section 1.4.4). The weight of the edge between two vertices is the Euclidean distance between the corresponding data points.

- The symmetric k-nearest-neighbor graph (Sk-NNG): $v_i \sim v_j$ if $\mathbf{x}_i$ is among the k-nearest neighbors of y *or* vice versa.

- The mutual k-nearest-neighbor graph (Mk-NNG): $v_i \sim v_j$ if $\mathbf{x}_i$ is among the k-nearest neighbors of y *and* vice versa. All vertices in a mutual k-NN graph have a degree upper-bounded by k, which is not usually the case with standard k-NN graphs.

- The Relative Neighborhood Graph (RNG): $v_i \sim v_j$ iff there is no vertex in

$$\mathcal{B}\left(v_i \ ; \ \mathcal{D}(v_i, v_j)\right) \cap \mathcal{B}(v_j \ ; \ \mathcal{D}(v_i, v_j)) \ . \tag{1.26}$$

- The Gabriel Graph (GG): $v_i \sim v_j$ iff there is no vertex in

$$\mathcal{B}\left(\frac{v_i + v_j}{2} \ ; \ \frac{\mathcal{D}(v_i, v_j)}{2}\right) \ . \tag{1.27}$$

- The β-Skeleton Graph (β-SG): $v_i \sim v_j$ iff there is no vertex in

$$\mathcal{B}\left(\left(1 - \frac{\beta}{2}\right)v_i + \frac{\beta}{2}v_j \ ; \ \frac{\beta}{2}\mathcal{D}(v_i, v_j)\right) \cap \mathcal{B}\left(\left(1 - \frac{\beta}{2}\right)v_j + \frac{\beta}{2}v_i \ ; \ \frac{\beta}{2}\mathcal{D}(v_i, v_j)\right) \ . \tag{1.28}$$

The Gabriel Graph is obtained with $\beta = 1$, and the Relative Neighborhood Graph with $\beta = 2$.

- The Delaunay Triangulation (DT): $v_i \sim v_j$ iff there is a closed ball $\mathcal{B}(\cdot\,;\,r)$ with v_i and v_j on its boundary and no other vertex v_k contained in it. The Delaunay Triangulation is the dual of the Voronoi irregular tessellation where each Voronoi cell is defined by the set $\{\mathbf{x} \in \mathbb{R}^n \mid \mathcal{D}(\mathbf{x}, v_k) \leq \mathcal{D}(\mathbf{x}, v_j) \text{ for all } v_j \neq v_k\}$. In such a graph, $\forall\, v_i,\ \deg(v_i) = 3$.

- The Complete Graph (CG): $v_i \sim v_j \,\forall (v_i, v_j)$, that is, it is a fully connected graph that contains an edge for each pair of vertices and $\mathcal{E} = \mathcal{V} \times \mathcal{V}$.

A surprising relationship between the edge sets of these graphs is that [27]

$$1 - \text{NNG} \subseteq \text{EMST} \subseteq \text{RNG} \subseteq \text{GG} \subseteq \text{DT} \subseteq \text{CG}. \qquad (1.29)$$

Figure 1.14 presents examples of some of these proximity graphs. All these graphs are extensively used in different image analysis tasks. These methods have been used in Scientific Computing to obtain triangular irregular grids (triangular meshes) adapted for the finite element method [28] for applications in physical modeling used in industries such as automotive, aeronautical, etc. Proximity graphs have also been used in Computational and Discrete Geometry [29] for the analysis of point sets in $\mathbb{R}^2$ (e.g., for 2D shapes) or $\mathbb{R}^3$ (e.g., for 3D meshes). They are also the basis of many classification and model reduction methods [30] that operate on feature vectors in $\mathbb{R}^d$, for example, Spectral Clustering [31], Manifold Learning [32], object matching, etc. Figure 1.15 presents some proximity graph examples on real-world data.

1.6   Conclusion

Graphs are tools that make it possible to model pairwise relationships between data elements and have found wide application in science and engineering. Computer vision and image processing have a long history of using graph models, but these models have been increasingly dominant representations in the recent literature. Graphs may be used to model spatial relationships between nearby or distant pixels, between image regions, between features, or as models of objects and parts. The subsequent chapters of this book are written by leading researchers who will provide the reader with a detailed understanding of how graph theory is being successfully applied to solve a wide range of problems in computer vision and image processing.

Bibliography

[1] N. Biggs, E. Lloyd, and R. Wilson, *Graph Theory, 1736–1936*. Oxford University Press, New-York, 1986.

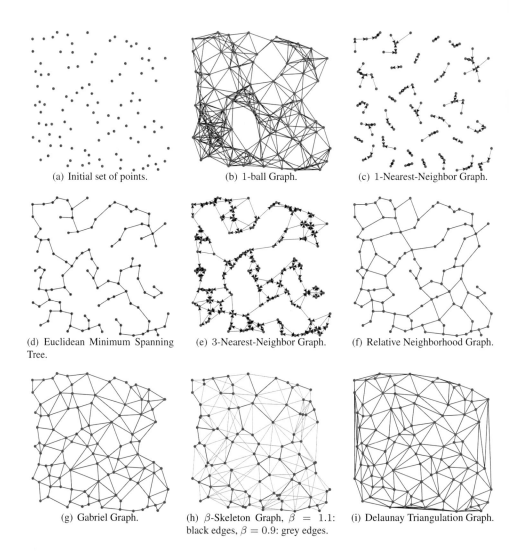

(a) Initial set of points.   (b) 1-ball Graph.   (c) 1-Nearest-Neighbor Graph.

(d) Euclidean Minimum Spanning Tree.   (e) 3-Nearest-Neighbor Graph.   (f) Relative Neighborhood Graph.

(g) Gabriel Graph.   (h) β-Skeleton Graph, $\beta = 1.1$: black edges, $\beta = 0.9$: grey edges.   (i) Delaunay Triangulation Graph.

FIGURE 1.14
Examples of proximity graphs from a set of 100 points in $\mathbb{Z}^2$.

(a) 3D Mesh of a satellite obtained by Delaunay Triangulation.

(b) Nearest Neighbor graph of points in $\mathbb{R}^3$.

(c) Nearest Neighbor graph of an image database (with each image treated as a vector of $\mathbb{R}^{16}$).

FIGURE 1.15
Some proximity graph examples on real-world data.

[2] A.-L. Barabasi, *Linked: How Everything Is Connected to Everything Else and What It Means.* Plume, New-York, 2003.

[3] D. J. Watts, *Six Degrees: The Science of a Connected Age.* W. W. Norton, 2004.

[4] N. A. Christakis and J. H. Fowler, *Connected: The Surprising Power of Our Social Networks and How They Shape Our Lives.* Little, Brown and Company, 2009.

[5] F. Harary, *Graph Theory.* Addison-Wesley, 1994.

[6] A. Gibbons, *Algorithmic Graph Theory.* Cambridge University Press, 1989.

[7] N. Biggs, *Algebraic Graph Theory*, 2nd ed. Cambridge University Press, 1994.

[8] R. Diestel, *Graph Theory*, 4th ed., ser. Graduate Texts in Mathematics. Springer-Verlag, 2010, vol. 173.

[9] J. L. Gross and J. Yellen, *Handbook of Graph Theory*, ser. Discrete Mathematics and Its Applications. CRC Press, 2003, vol. 25.

[10] J. Bondy and U. Murty, *Graph Theory*, 3rd ed., ser. Graduate Texts in Mathematics. Springer-Verlag, 2008, vol. 244.

[11] J. Gallier, *Discrete Mathematics*, 1st ed., ser. Universitext. Springer-Verlag, 2011.

[12] L. Grady and J. R. Polimeni, *Discrete Calculus: Applied Analysis on Graphs for Computational Science.* Springer, 2010.

[13] E. W. Dijkstra, "A note on two problems in connexion with graphs," *Numerische Mathematik*, vol. 1, no. 1, pp. 269–271, 1959.

[14] E. N. Mortensen and W. A. Barrett, "Intelligent scissors for image composition," in *SIGGRAPH*, 1995, pp. 191–198.

[15] P. Felzenszwalb and R. Zabih, "Dynamic programming and graph algorithms in computer vision," *IEEE Trans. Pattern Anal. Mach. Intell.*, vol. 33, no. 4, pp. 721–740, 2011.

[16] P. F. Felzenszwalb and D. P. Huttenlocher, "Efficient graph-based image segmentation," *Int. J. Computer Vision*, vol. 59, no. 2, pp. 167–181, 2004.

[17] L. Najman, "On the equivalence between hierarchical segmentations and ultrametric watersheds," *J. of Math. Imaging Vision*, vol. 40, no. 3, pp. 231–247, 2011.

[18] L. Ford Jr and D. Fulkerson, "A suggested computation for maximal multi-commodity network flows," *Manage. Sci.*, pp. 97–101, 1958.

[19] Y. Boykov, O. Veksler, and R. Zabih, "Fast approximate energy minimization via graph cuts," *IEEE Trans. Pattern Anal. Mach. Intell.*, vol. 23, no. 11, pp. 1222–1239, 2001.

[20] Y. Boykov and V. Kolmogorov, "An experimental comparison of min-cut/max-flow algorithms for energy minimization in vision," *IEEE Trans. Pattern Anal. Mach. Intell.*, vol. 26, no. 9, pp. 1124–1137, 2004.

[21] J. Shi and J. Malik, "Normalized cuts and image segmentation," *IEEE Trans. Pattern Anal. Mach. Intell.*, vol. 22, no. 8, pp. 888–905, 2000.

[22] A. Elmoataz, O. Lezoray, and S. Bougleux, "Nonlocal discrete regularization on weighted graphs: A framework for image and manifold processing," *IEEE Trans. Image Process.*, vol. 17, no. 7, pp. 1047–1060, 2008.

[23] L. Vincent and P. Soille, "Watersheds in digital spaces: An efficient algorithm based on immersion simulations," *IEEE Trans. Pattern Anal. Mach. Intell.*, vol. 13, no. 6, pp. 583–598, 1991.

[24] A. Radhakrishna, "Finding Objects of Interest in Images using Saliency and Superpixels," Ph.D. dissertation, 2011.

[25] H. Samet, "The quadtree and related hierarchical data structures," *ACM Comput. Surv.*, vol. 16, no. 2, pp. 187–260, 1984.

[26] A. Levinshtein, A. Stere, K. N. Kutulakos, D. J. Fleet, S. J. Dickinson, and K. Siddiqi, "Turbopixels: Fast superpixels using geometric flows," *IEEE Trans. Pattern Anal. Mach. Intell.*, vol. 31, no. 12, pp. 2290–2297, 2009.

[27] G. T. Toussaint, "The relative neighbourhood graph of a finite planar set," *Pattern Recogn.*, vol. 12, no. 4, pp. 261–268, 1980.

[28] O. Zienkiewicz, R. Taylor, and J. Zhu, *The Finite Element Method: Its Basis and Fundamentals*, 6th ed. Elsevier, 2005.

[29] J. E. Goodman and J. O'Rourke, *Handbook of Discrete and Computational Geometry*, 2nd ed. CRC Press LLC, 2004.

[30] M. A. Carreira-Perpinan and R. S. Zemel, "Proximity graphs for clustering and manifold learning," in *NIPS*, 2004.

[31] U. von Luxburg, "A tutorial on spectral clustering," *Stat. Comp.*, vol. 17, no. 4, pp. 395–416, 2007.

[32] J. A. Lee and M. Verleysen, *Nonlinear Dimensionality Reduction*. Springer, 2007.

2

Graph Cuts—Combinatorial Optimization in Vision

Hiroshi Ishikawa

Department of Computer Science and Engineering
Waseda University
Tokyo, Japan
Email: hfs@waseda.jp

CONTENTS

2.1 Introduction 26
    2.1.1 Chapter Summary 27
2.2 Markov Random Field 27
    2.2.1 Labeling Problem: A Simple Example 27
        2.2.1.1 Denoising Problem 28
        2.2.1.2 Image Similarity 29
        2.2.1.3 Denoising Criterion 30
    2.2.2 Markov Random Field 31
        2.2.2.1 Order of an MRF 32
        2.2.2.2 Energy of an MRF 33
        2.2.2.3 Bayesian Inference 34
2.3 Basic Graph Cuts: Binary Labels 35
    2.3.1 Minimum-Cut Algorithms 35
    2.3.2 The Graph Cuts 36
    2.3.3 Submodularity 39
    2.3.4 Nonsubmodular Case: The BHS Algorithm 40
    2.3.5 Reduction of Higher-Order Energies 43
2.4 Multi-Label Minimization 44
    2.4.1 Globally Minimizable Case: Energies with Convex Prior 45
    2.4.2 Move-Making Algorithms 49
        2.4.2.1 α-β Swap and α-Expansion 49
        2.4.2.2 Fusion Moves 51
        2.4.2.3 α-β Range Moves 52
        2.4.2.4 Higher-Order Graph Cuts 53
2.5 Examples 55
2.6 Conclusion 56
    Bibliography 57

2.1 Introduction

Many problems in computer vision, image processing, and computer graphics can be put into labeling problems [1]. In such a problem, an undirected graph is given as an abstraction of locations and their neighborhood structure, along with a set of labels. Then, the solutions to the problem is identified with *labelings*, or assignments of a label to each vertex in the graph. The problem is then to find the best labeling according to the criteria in the problem's requirements. An energy is a translation of the criteria into a function that evaluates how good the given labeling is, so that smaller energy for a labeling means a better corresponding solution to the problem. Thus, the problem becomes an "energy minimization problem". This separates the problem and the technique to solve it in a useful way by formulating the problem as an energy, it tends to make the problem more clearly defined, and also, once the problem is translated into an energy minimization problem, it can be solved using general algorithms.

In this chapter, we describe a class of combinatorial optimization methods for energy minimization called the *graph cuts*. It has become very popular in computer vision, image processing, and computer graphics, used for various tasks such as segmentation [2, 3, 4, 5, 6, 7, 8, 9, 10, 11, 12, 13], motion segmentation [14, 15, 16, 17], stereo [18, 19, 20, 21, 22, 23, 24], texture and image synthesis [25, 26], image stitching [27, 28, 29, 30], object recognition [31, 32, 33], computing geodesics and modeling gradient flows of contours and surfaces [34, 35], computational photography [36], and motion magnification [37].

Minimization of such energies in general is known to be NP-hard [38]. Stochastic approximation methods, such as the iterated conditional modes (ICM) [39] and simulated annealing [40] have been used, but the former is prone to falling into local minima, while the latter, although the convergence to global minima is theoretically guaranteed, it is very slow in practice.

Graph cuts are methods utilizing the s-t mincut algorithms. Known in operations research for a long time [41], it was first introduced into image processing in the late 1980s [42, 43], when it was shown capable of exactly minimizing an energy devised to denoise binary images. At the time, the exact solution was compared to the results by annealing, which turned out to oversmooth the images [44, 43].

In late 1990s, graph cuts were introduced to the vision community [19, 45, 6, 46, 23]. This time the method was used for multilabel cases, and it was shown [19, 6] that energies with real numbers as labels can be exactly minimized if the prior is linear in terms of the absolute value of the difference between the labels at neighboring sites. This was later generalized to the construction for the case when the prior is any convex function of the difference [47]. Also, approximation algorithms for the energies with for arbitrary label sets and Potts priors were introduced [19, 38], which lead to the current popularity.

At first, there was some duplication of efforts in the vision community. The condition for the binary energy to be exactly solved, long known in the OR community as the submodularity condition, was rediscovered [48]. The "linear" case mentioned above was also devised without the knowledge that a similar method had been used for the problem of task assignment to multiple processors [49]. Now, methods long known in the OR community have been introduced to vision, such as the BHS (a.k.a. QPBO / roof duality) algorithm [50, 51, 52, 53] that proved very useful.

Other new minimization methods such as belief propagation (BP) [54, 55, 56] and the tree-reweighted message passing (TRW) [57, 58]) have also been introduced to vision. An experimental comparison [59] of these methods with graph cuts were conducted using stereo and segmentation as examples. It showed that these new methods are much superior to ICM, which was used as the representative of the older methods. When the neighborhood structure was relatively sparse, such as the 4-neighbors, TRW performed better than graph cuts. However, according to another study [60], when the connectivity is larger, as in the case of stereo models with occlusion, graph cuts considerably outperformed both BP and TRW in both the minimum energy and the error compared to the ground truth.

2.1.1 Chapter Summary

We first introduce the energy minimization formalisms used in the image-related areas (§2.2). Next (§2.3), we describe the graph cut algorithm in the case of binary labels, which is the most fundamental case where the energy-minimization problem directly translates to the minimum-cut problem, and global minima can be computed in a polynomial time under certain condition called *submodularity*. For the binary case, we also describe the BHS algorithm, which can partially minimize even the energies that do not satisfy the submodularity condition, as well as the reduction of higher-order energies to first order. Finally (§2.4), we describe graph cuts to minimize energies when there are more than two labels. One approach to this is to translate multilabel energies into binary ones. In some cases, the resulting problem is globally optimizable. Another way is the approximate algorithms to solve the multilabel problems by iteratively solving binary problems with graph cuts.

2.2 Markov Random Field

2.2.1 Labeling Problem: A Simple Example

As the simplest example, we consider the denoising problem of binary (black-and-white) images. Such an image consists of pixels that can be black or white, arranged in a rectangle. This is an example of *labeling*. Each pixel in the image has a color;

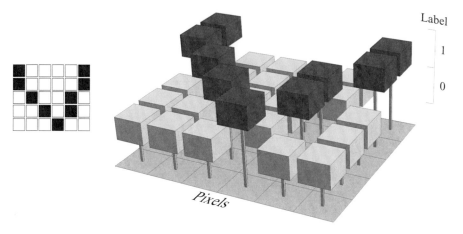

FIGURE 2.1
A labeling is an assignment of a label to each pixel. A binary image is a labeling with
two labels. The image on the left is a labeling, as shown on the right.

we consider this as assigning a label— 0 for white and 1 for black— to each pixel.
A labeling is thus a map from the set of pixels to the set of labels. Therefore, every
image is a labeling. Thus, the denoising problem is that of finding the "least noisy"
labeling.

More formally, let us denote the set of pixels by $\mathcal{P}$ and the set of labels by $\mathcal{L}$. For
example, in the case of the binary image denoising, $\mathcal{P} = \{(i,j) \mid i = 1, \cdots, n; \; j = 1, \cdots, m\}$ and $\mathcal{L} = \{0,1\}$, assuming that the image has $n \times m$ pixels. Then a
labeling is a map

$$X : \mathcal{P} \to \mathcal{L}. \tag{2.1}$$

Since the actual coordinates are not needed in the discussion, in the following we
denote the pixels by single letters such as $p, q \in \mathcal{P}$. The label that is assigned to pixel
p by a labeling X is denoted by X_p:

$$X : \mathcal{P} \ni p \mapsto X_p \in \mathcal{L}. \tag{2.2}$$

Figure 2.1 depicts an example of binary image and the labeling representing it.

For a set $\mathcal{P}$ of pixels and a set $\mathcal{L}$ of labels, we denote the set of all labelings by $\mathcal{L}^{\mathcal{P}}$.
There are $|\mathcal{L}|^{|\mathcal{P}|}$ possible such labelings; thus, the number of possible solutions to
the denoising problem is exponential in terms of the number of pixels. For a labeling
$X \in \mathcal{L}^{\mathcal{P}}$ and any subset $\mathcal{Q} \subset \mathcal{P}$, the labeling on $\mathcal{Q}$ defined by restricting X to $\mathcal{Q}$ is
denoted by $X_{\mathcal{Q}} \in \mathcal{L}^{\mathcal{Q}}$.

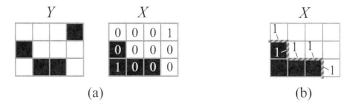

(a)   (b)

FIGURE 2.2
(a) Counting the number of pixels that has different colors for X and Y. (b) Counting neighboring pairs of pixels in X that have different colors.

2.2.1.1 Denoising Problem

In the denoising problem, we are given one labeling, the noisy image, and somehow find another labeling, the denoised image. The input and the output are related as follows:

1. They are the "same" image.

2. The output has less noise than the input.

To denoise an image by energy minimization, we need to know what these mean in terms of the labeling. We translate these notions into an *energy*

$$E : \mathcal{L}^\mathcal{P} \to \mathbb{R}, \tag{2.3}$$

which is a function on the set $\mathcal{L}^\mathcal{P}$ of labelings that gives a real number to each labeling. The real number $E(X)$ that is given a labeling X is called its *energy*. The energy evaluates how well the conditions above are met by a labeling. According to the problem at hand, we define the energy so that it gives a smaller value to a labeling when it is better according to some criteria that we consider reasonable or that we have observed to produce good results. Then we use an algorithm, such as graph cuts, to *minimize* the energy, i.e., the algorithm finds a labeling with as small an energy as possible. With graph cuts, sometimes the global minimum of the energy can be found; that is, we can actually find the labeling that gives the smallest energy among all the exponential number of possible labelings in $\mathcal{L}^\mathcal{P}$, or at least one of such minima.

2.2.1.2 Image Similarity

Suppose we are given a noisy image, or a labeling, Y, and that we would like to find the denoised version X. First, we consider the condition (1) above. That is, the noisy image Y and the denoised image X are supposed to be the same image. Since we cannot take this as meaning they are exactly the same (then we cannot reduce the noise), we quantify the difference between the two images and assume the condition means that the difference is small. How do we quantify the difference between two images? A popular method is to take the square of the L^2 norm:

$$\sum_{p \in \mathcal{P}} (X_p - Y_p)^2. \tag{2.4}$$

Another oft-used measure of the difference between two images is the sum of absolute difference:

$$\sum_{p \in \mathcal{P}} |X_p - Y_p|, \tag{2.5}$$

In our case, the two measures coincide, since the label is either 0 or 1. They both mean the number of the pixels that have different labels between X and Y, as illustrated in Figure 2.2(a).

2.2.1.3   Denoising Criterion

If we want to find the labeling X that minimize (2.4) or (2.5), it is easy: we take $X = Y$. This is natural, since we do not have any reason to arrive at anything else so far. To make X the denoised version of Y, we need to have some criterion to say some images are noisier than others. If we have the original image before noise was introduced, we can say that the closer the image is to the original, the less noisier it is. However, if we had the original, we don't need to denoise it but that is not usually the case. So we need a way to tell the level of noise in an image without comparing it with the original. If it is in the sense above, the closeness to the original, that is clearly impossible, since the original might be very "noisy" to start with. Nevertheless, we have a sense that some images are noisy and others are not, which presumably comes from the statistics of images that we encounter in life. Thus, knowing the statistics of images helps. Also, there are different kinds of noises that are likely to be introduced in practice whose statistics we know. If we know what caused the noise in the image, we can use the information to reduce the noise.

Here, however, we are using the denoising problem only as an example; so we simply assume that the noisier the image is, the rougher it is, that is, there are more changes in color between neighboring pixels.

Accordingly, to evaluate the roughness of image X, we count the number of changes in color between neighboring pixels (see Figure 2.2(b)) as follows:

$$\sum_{(p,q) \in \mathcal{N}} (X_p - X_q)^2. \tag{2.6}$$

Here, $\mathcal{N} \subset \mathcal{P} \times \mathcal{P}$ is the set of neighboring pairs of pixels. We assume that $\mathcal{N}$ is symmetric, i.e., $(q, p) \in \mathcal{N}$ if $(p, q) \in \mathcal{N}$. We also assume that $(p, p) \notin \mathcal{N}$ for any $p \in \mathcal{P}$.

The image that satisfies the "less noise" condition (2) the most, which minimizes

FIGURE 2.3

Denoising a binary image. Given the noisy image Y on the left, find the image X that is close to Y and has as few change as possible of labels between neighboring pixels. Since the two conditions usually conflict, we need to find a trade off. We encode it as the minimization of the energy (2.7).

the quantity (2.6), is a constant image, either all black or all white. On the other hand, the condition (1) is perfectly satisfied if we set $X = Y$, as mentioned above. Thus, the two conditions conflict unless Y is a constant image, leading to a need of a tradeoff. The heart of the energy minimization methods is that we attempt to treat such a tradeoff in a principled way by defining it as minimization of an energy. That is, we translate various factors in the tradeoff as terms which are added up to form an energy. The relative importance of each factor can be controlled by multiplying it by a weight.

In our case, we define the energy as a weighted sum of (2.4) and (2.6):

$$E(X) = \sum_{p \in \mathcal{P}} \lambda (X_p - Y_p)^2 + \sum_{(p,q) \in \mathcal{N}} \kappa (X_p - X_q)^2 \qquad (2.7)$$

We consider the energy a function of X, supposing Y is fixed. Here, λ and κ are positive weights. Although the two weights are fixed globally in this example, it can be varied from pixel to pixel and pair to pair, depending on the data Y and other factors; for instance, the smoothing factor κ is often changed according to the contrast in Y between the neighboring pixels. Figure 2.3 illustrates the tradeoff represented by the energy.

Assume that X minimizes $E(X)$. For each pixel p, $X_p = Y_p$ makes the first sum smaller. On the other hand, a smoother X without too many changes between neighboring pixels leads to a lesser second sum. Let us assume that $\lambda = \kappa$. If, for instance, there is an "isolated" pixel p such that Y assigns all its neighbors the opposite value from Y_p, the total energy would be smaller when the value X_p at p is the same with the neighbors, rather than choosing $X_p = Y_p$. Thus, minimizing $E(X)$ represents an effort to make X as close as possible to Y while fulfilling the condition that the values between neighboring pixels do not change too much. The tradeoff between the two conditions are controlled by the relative magnitude of the parameters λ and κ.

2.2.2 Markov Random Field

The energy minimization problem as above can be considered a maximum a posteriori (MAP) estimation problem of a Markov random field (MRF).

A Markov random field is a stochastic model with a number of random variables that are arranged in an undirected graph, so that each node has one variable. Let $\mathcal{P}$ and $\mathcal{N}$ be the sets of pixels and neighbors, as before. We will call the elements of $\mathcal{P}$ "sites"; they represent the notion of location in the problem. Since we are assuming $\mathcal{N}$ is symmetric, we can consider $(\mathcal{P}, \mathcal{N})$ an undirected graph. The sites most commonly correspond to the pixels in the images, especially in low-level problems such as the denoising example above. However, they sometimes represent other objects such as blocks in the images and segments (so-called superpixels) obtained by a quick oversegmentation of images.

A set of random variables $X = (X_p)_{p \in \mathcal{P}}$ is a *Markov random field* if its probability density function can be written as:

$$P(X) = \prod_{C \in \mathcal{C}} \varphi_C(X_C), \tag{2.8}$$

where $\mathcal{C}$ is the set of cliques of the undirected graph $(\mathcal{P}, \mathcal{N})$, and each function φ_C depends only on the vector of variables $X_C = (X_p)_{p \in C}$.

Assume that the random variables in X all take values in the same set $\mathcal{L}$. Then, we can think the set of random variables to be a random variable taking a value in the set $\mathcal{L}^{\mathcal{P}}$ of labelings. Thus, an MRF can be thought of as a random variable with values in the set of labelings whose probability density function (2.8) satisfies the condition above.

2.2.2.1 Order of an MRF

When the probability density function $P(X)$ in (2.8) can be written as

$$P(X) = \prod_{C \in \mathcal{C}, |C| \le k+1} \varphi_C(X_C), \tag{2.9}$$

with only the functions depending on the variables for cliques with at most $k + 1$ sites, it is called a k'th order MRF or an MRF of order k. For instance, an MRF of first order can be written as:

$$P(X) = \prod_{p \in \mathcal{P}} \varphi_p(X_p) \prod_{(p,q) \in \mathcal{N}} \varphi_{\{p,q\}}(X_{\{p,q\}}), \tag{2.10}$$

that depends only on the functions for cliques of size 1 and 2, i.e., sites and neighboring pairs of sites. An MRF of second order can be written as:

$$P(X) = \prod_{p \in \mathcal{P}} \varphi_p(X_p) \prod_{(p,q) \in \mathcal{N}} \varphi_{\{p,q\}}(X_{\{p,q\}}) \prod_{\{p,q,r\} \text{ clique}} \varphi_{\{p,q,r\}}(X_{\{p,q,r\}}). \tag{2.11}$$

Remember that the notation $X_{\{p,q,r\}}$ denotes the vector (X_p, X_q, X_r) of variables.

Note that there is an ambiguity in the factorization of the density function, since lower-order functions can be a part of a higher-order function. We can reduce this ambiguity somewhat by restricting $\mathcal{C}$ to be the set of maximal cliques. However, in practice we define the functions concretely; so there is no problem of ambiguity, and we would like to define them in a natural way. Therefore, we leave $\mathcal{C}$ to mean the set of all cliques.

2.2.2.2 Energy of an MRF

If we define in (2.8)

$$f_C(X_C) = -\log \varphi_C(X_C), \tag{2.12}$$

then maximizing $P(X)$ in (2.8) is equivalent to minimizing

$$E(X) = \sum_{C \in \mathcal{C}} f_C(X_C), \tag{2.13}$$

This function that assigns a real number $E(X)$ to a labeling X is called the *energy* of the MRF. An MRF is often defined with an energy instead of a density function. Given the energy (2.13), the density function (2.8) is obtained as

$$p(X) = e^{-E(X)} = \prod_{C \in \mathcal{C}} e^{-f_C(X_C)}. \tag{2.14}$$

This form of probability density function is called the *Gibbs distribution*.

The first-order MRF in (2.10) has the energy

$$E(X) = \sum_{p \in \mathcal{P}} g_p(X_p) + \sum_{(p,q) \in \mathcal{N}} h_{pq}(X_p, X_q), \tag{2.15}$$

where

$$g_p(X_p) = -\log \varphi_p(X_p), \tag{2.16}$$
$$h_{pq}(X_p, X_q) = h_{qp}(X_q, X_p) = -\log \varphi_{\{p,q\}}(X_{\{p,q\}}). \tag{2.17}$$

Most of the use of graph cuts have been for minimizing the energy of the form (2.15). Each term $g_p(X_p)$ in the first sum in (2.15) depends only on the label assigned to each site. The first sum is called the *data term*, because the most direct influence of data such as the given image, in deciding the label to assign to each site, often manifests itself through $g_p(X_p)$. For instance, in the case of the denoising problem in the previous section, this term influences the labeling X in the direction of making it closer to Y through the definition:

$$g_p(X_p) = \lambda(X_p - Y_p)^2. \tag{2.18}$$

Thus, the first term of (2.15) represents the data influence on the optimization process.

The second sum, on the other hand, reflects the prior assumption on the wanted outcome, such as less noise. In particular, it dictates the desirable property of the labels given to neighboring sites. Because this often means that there is less label change between neighboring sites, this term is called the *smoothing term*. In the case of the previous section, it is defined by:

$$h_{pq}(X_p, X_q) = \kappa(X_p - X_q)^2. \tag{2.19}$$

Sometimes, it is also called the *prior term*. The reason is that it often encodes the prior probability distribution, as will be explained next.

2.2.2.3 Bayesian Inference

Assume that we wish to estimate an unobserved parameter X, on the basis of observed data Y. The *maximum a posteriori* (MAP) estimate of X is its value that maximizes the posterior probability $P(X|Y)$, which is the conditional probability of X given Y. It is often used in the area of statistical inference, where some hidden variable is inferred from an observation. For instance, the denoising problem in the previous section can be thought of as such an inference.

The posterior probability $P(X|Y)$ is often derived from the prior probability $P(X)$ and the likelihood $P(Y|X)$. The prior $P(X)$ is the probability of X when there is no other condition or information. The likelihood $P(Y|X)$ is the conditional probability that gives the data probability under the condition that the unobserved variable in fact has value X. From these, the posterior is obtained by Bayes' rule:

$$P(X|Y) = \frac{P(Y|X)P(X)}{P(Y)}. \tag{2.20}$$

Suppose that we have an undirected graph $(\mathcal{P}, \mathcal{N})$ and a set $\mathcal{L}$ of labels. Also suppose that the unobserved parameter X takes as value a labeling, and we have a simple probabilistic model given by the following:

$$P(Y|X) = \frac{1}{Z_1(Y)} \prod_{p \in \mathcal{P}} \varphi_p^Y(X_p), \tag{2.21}$$

$$P(X) = \frac{1}{Z_2} \prod_{(p,q) \in \mathcal{N}} \varphi_{pq}(X_p, X_q), \tag{2.22}$$

where

$$Z_1(Y) = \sum_{X \in \mathcal{L}^{\mathcal{P}}} \prod_{p \in \mathcal{P}} \varphi_p^Y(X_p), \tag{2.23}$$

$$Z_2 = \sum_{X \in \mathcal{L}^{\mathcal{P}}} \prod_{(p,q) \in \mathcal{N}} \varphi_{pq}(X_p, X_q). \tag{2.24}$$

That is, the likelihood $P(Y|X)$ of the data Y given the parameter X is given in the form of the product of independent local distributions $\varphi_p^Y(X_p)$ for each site p that depends only on the label X_p, while the prior probability $P(X)$ is defined as a product of local joint distributions $\varphi_{pq}(X_p, X_q)$ depending on the labels X_p and X_q that the neighboring sites p and q are given.

By (2.20), it follows

$$P(X|Y) = \frac{1}{Z_1(Y)Z_2} \prod_{p\in\mathcal{P}} \varphi_p^Y(X_p) \prod_{(p,q)\in\mathcal{N}} \varphi_{pq}(X_p, X_q). \qquad (2.25)$$

If we fix the observed data Y, this makes it a first-order MRF. Thus, maximizing the posterior probability $P(X|Y)$ with a fixed Y is equivalent to the energy minimization problem of an MRF. For instance, our example energy minimization formulation of the denoising problem can be considered a MAP estimation problem with the likelihood and the prior:

$$P(Y|X) = \frac{1}{Z_1(Y)} \prod_{p\in\mathcal{P}} e^{-\lambda(X_p-Y_p)^2}, \qquad (2.26)$$

$$P(X) = \frac{1}{Z_2} \prod_{(p,q)\in\mathcal{N}} e^{-\kappa(X_p-X_q)^2}, \qquad (2.27)$$

which is consistent with a Gaussian noise model.

2.3 Basic Graph Cuts: Binary Labels

In this section, we discuss the basics of graph-cut algorithms, namely the case where the set of labels is binary: $\mathcal{L} = \mathbb{B} = \{0, 1\}$. The multiple label graph cuts depend on this basic case. As it is based on the algorithms to solve the s-t minimum cut problem, we start with a discussion on them. For details on the minimum cut and maximum flow algorithms, see textbooks such as [61, 62, 63].

2.3.1 Minimum-Cut Algorithms

Consider a directed graph $\mathcal{G} = (\mathcal{V}, \mathcal{E})$. Each edge $e_{i\to j}$ from v_i to v_j has weight w_{ij}. For the sake of notational simplicity, in the following we assume every pair of nodes $v_i, v_j \in \mathcal{V}$ has weight w_{ij} but $w_{ij} = 0$ if there is no edge $e_{i\to j}$.

Let us choose two nodes $s, t \in \mathcal{V}$; a *cut* of $\mathcal{G}$ with respect to the pair (s, t) is a partition of $\mathcal{V}$ into two sets $\mathcal{S} \subset \mathcal{V}$ and $\mathcal{T} = \mathcal{V} \setminus \mathcal{S}$ such that $s \in \mathcal{S}, t \in \mathcal{T}$ (Figure 2.4).

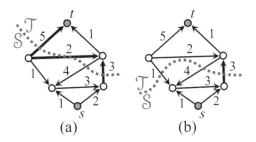

(a)   (b)

FIGURE 2.4
(a) A cut of $\mathcal{G}$ with respect to (s, t) divides the nodes into two sides. The cost of a cut is the total sum of the weights of the directed edges going from the s side ($\mathcal{S}$) to the t side ($\mathcal{T}$). The cost of the depicted cut is 10, the sum of the weights of the edges *in the cut*, indicated by thick arrows. (b) Another example: the cost of the cut is 3.

In the following, we denote this cut by $(\mathcal{S}, \mathcal{T})$. Then, the *cost* $c(\mathcal{S}, \mathcal{T})$ of cut $(\mathcal{S}, \mathcal{T})$ is the sum of the weights of the edges that go from $\mathcal{S}$ to $\mathcal{T}$:

$$c(\mathcal{S}, \mathcal{T}) = \sum_{v_i \in \mathcal{S}, v_j \in \mathcal{T}} w_{ij}. \tag{2.28}$$

Note that the directed edges going the other direction, from $\mathcal{T}$ to $\mathcal{S}$, do not count. An edge that contributes to the cost is said to be *in the cut*. A *minimum cut* is a cut with the minimum cost, and the minimum cut problem seeks to find such a cut for given the triple $(\mathcal{G}, s, t)$.

The minimum cut problem is well known to be equivalent to the *maximum flow* problem [64, 65]. When all edge weights are nonnegative, these problems are known to be tractable, which is the source of the usefulness of graph cuts. There are three groups of algorithms known: the '*augmenting path*' algorithms [64, 66], the '*push-relabel*' algorithms [67, 68, 69], and the '*network simplex*' algorithms [70]. Although asymptotically the push-relabel algorithms are the fastest, for typical image-related applications, actual performance seems to indicate that the augmenting path algorithms are the fastest. Boykov and Kolmogorov [71] experimentally compared the Dinic algorithm [72], which is a kind of augmenting path algorithm, a push-relabel algorithm, and their own augmenting path algorithm (the BK algorithm) in image restoration, stereo, and image segmentation applications to find the BK algorithm to be the fastest. This is the maximum-flow algorithm most used for graph cuts in image-related applications. There is also a push-relabel algorithm that was specifically developed for large regular grid graphs [73] used in higher-dimensional image processing problems, as well as algorithms to speed up the repeated minimization of an energy with small modifications [74, 75, 76], as commonly happens in movie segmentation.

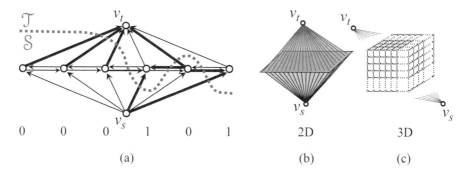

0   0   0   1   0   1   2D   3D

(a)   (b)   (c)

FIGURE 2.5
(a) The graph for binary MRF minimization. The edges in the cut are depicted as thick arrows. Each node other than v_s and v_t corresponds to a site. If a cut $(\mathcal{S}, \mathcal{T})$ places a node in $\mathcal{S}$, the corresponding site is labeled 0; if it is in $\mathcal{T}$, the site is labeled 1. The 0's and 1's at the bottom indicate the label each site is assigned. Here, the sites are arranged in 1D; but according to the neighborhood structure this can be any dimension as shown in (b) and (c).

2.3.2 The Graph Cuts

The graph-cut algorithms utilize the minimum cut (maximum flow) algorithms to minimize MRF energies. We construct a graph with special nodes s and t, as well as a node for each site. Then for a cut $(\mathcal{S}, \mathcal{T})$ of the graph with respect to the pair (s, t), each node must be either in $\mathcal{S}$ or $\mathcal{T}$; we interpret this as the site being labeled 0 or 1. That is, we define a one to one correspondence between the set of all labeling and that of all cuts.

Consider the first-order MRF energy (2.15), shown here again:

$$E(X) = \sum_{p \in \mathcal{P}} g_p(X_p) + \sum_{(p,q) \in \mathcal{N}} h_{pq}(X_p, X_q) \tag{2.29}$$

We construct a directed graph $\mathcal{G} = (\mathcal{V}, \mathcal{E})$ corresponding to this energy (Figure 2.5). The node set $\mathcal{V}$ contains the special nodes v_s and v_t (for consistency with the notation for graph edges in this book, we denote what we called s and t by v_s and v_t, respectively), as well as a node corresponding to each site p in $\mathcal{P}$:

$$\mathcal{V} = \{v_s, v_t\} \cup \{v_p \mid p \in \mathcal{P}\}. \tag{2.30}$$

From and to each node v_p corresponding to a site $p \in \mathcal{P}$, we add two edges, and also edges between nodes corresponding to neighboring sites:

$$\mathcal{E} = \{e_{s \to p} \mid p \in \mathcal{P}\} \cup \{e_{p \to t} \mid p \in \mathcal{P}\} \cup \{e_{p \to q} \mid (p, q) \in \mathcal{N}\}. \tag{2.31}$$

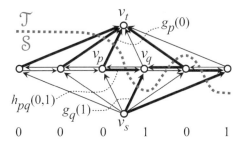

FIGURE 2.6
Weights for binary MRF minimization for the energy in the example in Section 2.2.1.
The numbers at the bottom indicates the labels assigned to the sites above. The thick
arrows indicate the edges in the cut. The cut separates the nodes v_p and v_t; thus,
the site p is assigned the label 0. The weight is exactly $g_p(0)$, which is added to the
cost of the cut. Similarly, the site q is assigned 1 and the cost is added $g_q(1)$. When
two neighboring sites are assigned different labels, an edge between the two is cut.
Here, the labels differ between p and q; thus, the weight $h_{pq}(0,1)$ of the edge $e_{p\to q}$
is added to the cost of the cut.

Now, we define a correspondence between a cut $(\mathcal{S}, \mathcal{T})$ of $\mathcal{G}$ with respect to
(v_s, v_t) and a labeling $X \in \mathcal{L}^{\mathcal{P}}$ as follows:

$$X_p = \begin{cases} 0 & (\text{if } v_p \in \mathcal{S}) \\ 1 & (\text{if } v_p \in \mathcal{T}). \end{cases} \tag{2.32}$$

That is X assigns 0 to site p if the node v_p corresponding to p is on the v_s side, and
1 if it is on the v_t side. This way, there is a one-to-one correspondence between cuts
of $\mathcal{G}$ with respect to (v_s, v_t) and labelings in $\mathcal{L}^{\mathcal{P}}$.

The minimum-cut algorithm can find a cut with the minimum cost efficiently;
thus, in order to find a labeling that minimizes the energy (2.15), we want to define a
weight of $\mathcal{G}$ so that the minimum cut corresponds to the minimum energy.

We consider the example in Section 2.2.1, with the energy terms (2.18) and
(2.19), and define:

$$w_{sp} = g_p(1), \tag{2.33}$$
$$w_{pt} = g_p(0), \tag{2.34}$$
$$w_{pq} = h_{pq}(0,1). \tag{2.35}$$

Consider the labeling X corresponding to a cut $(\mathcal{S}, \mathcal{T})$ of $\mathcal{G}$. First, note the data term
that depends only on one site. If $v_p \in \mathcal{S}$, then by (2.32) $X_p = 0$ and, as the edge
$e_{p\to t}$ is in the cut, by (2.34) $w_{pt} = g_p(0) = g_p(X_p)$ is added to the cost of the cut. If
$v_p \in \mathcal{T}$, then $X_p = 1$ and, as the edge $e_{s\to p}$ is in the cut, by (2.33) $w_{sp} = g_p(1) = g_p(X_p)$ is added to the cost. Either way, $g_p(X_p)$ is added to the cost of the cut.

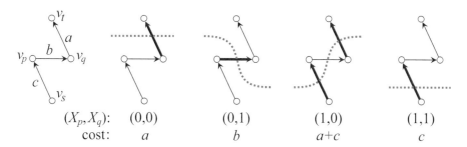

(X_p, X_q): (0,0) (0,1) (1,0) (1,1)
cost: a b $a+c$ c

FIGURE 2.7
Smoothing term and cost.

Next, we consider the smoothing term depending on two neighboring sites. For a neighboring pair of sites p and q, if $v_p \in \mathcal{S}, v_q \in \mathcal{T}$, then the edge $e_{p \to q}$ is in the cut. Thus, by (2.35) $h_{pq}(0, 1) = \kappa$ is added to the cut cost. By (2.32), it corresponds to $(X_p, X_q) = (0, 1)$ and the added weight to the cost coincides with the smoothing term that should be added to the energy in such combination of assignment. When both v_p and v_q belong to either of $\mathcal{S}$ and $\mathcal{T}$, the increase in cost is zero, corresponding to the energy $h_{pq}(0, 0) = h_{pq}(1, 1) = 0$.

This way, the energy for the labeling X precisely coincides with the cost of the corresponding cut. This construction is due to [43].

2.3.3 Submodularity

Thus, by finding the minimum-cut, we can find the global minimum of the energy (2.7) for our image denoising example, as well as the labeling that attains it. Can we generalize this to more general form of energy? The limiting factor is that the edge weights must be nonnegative in order to benefit from the polynomial-time minimum-cut algorithms: with the above construction, $g_p(X_p)$ and $h_{pq}(X_p, X_q)$ must all be nonnegative in (2.15). Also, it assumes $h_{pq}(0, 0) = h_{pq}(1, 1) = 0$.

However, there is still a little leeway in choosing edge weights. For instance, each v_p must belong to either $\mathcal{S}$ or $\mathcal{T}$; so one of the edges $e_{s \to p}$ and $e_{p \to t}$ is always in the cut. Therefore, adding the same value to w_{sp} and w_{pt} does not affect the choice of which edge to cut. This means that $g_p(X_p)$ can take any number because, if we have negative weight after defining w_{sp} and w_{pt} according to (2.33) and (2.34), we can add $- \min\{w_{sp}, w_{pt}\}$ to w_{sp} and w_{pt} to make both nonnegative. In fact, one of w_{sp} and w_{pt} can always be set to zero this way.

So the data term is arbitrary. How about the smoothing term? For a pair p, q of sites, let us define

$$w_{qt} = a, \quad w_{pq} = b, \quad w_{sp} = c, \tag{2.36}$$

as shown in Figure 2.7. Then the cut costs for the four possible combinations of (X_p, X_q) are:

$$h_{pq}(0,0) = a, \quad h_{pq}(0,1) = b, \quad h_{pq}(1,0) = a+c, \quad h_{pq}(1,1) = c. \tag{2.37}$$

Then we have

$$h_{pq}(0,1) + h_{pq}(1,0) - h_{pq}(0,0) - h_{pq}(1,1) = b. \tag{2.38}$$

Since the right-hand side is an edge weight, it must be nonnegative for the polynomial-time algorithms to be applicable. This is known as the *submodularity* condition:

$$h_{pq}(0,1) + h_{pq}(1,0) - h_{pq}(0,0) - h_{pq}(1,1) \geq 0. \tag{2.39}$$

Conversely, if this condition is met, we can set

$$a = h_{pq}(1,0) - h_{pq}(1,1), \tag{2.40}$$
$$b = h_{pq}(0,1) + h_{pq}(1,0) - h_{pq}(0,0) - h_{pq}(1,1), \tag{2.41}$$
$$c = h_{pq}(1,0) - h_{pq}(0,0). \tag{2.42}$$

Then we have the cut cost for each combination as follows:

$$(X_p, X_q) = (0,0): \quad h_{pq}(1,0) - h_{pq}(1,1), \tag{2.43}$$
$$(X_p, X_q) = (0,1): \quad h_{pq}(0,1) + h_{pq}(1,0) - h_{pq}(0,0) - h_{pq}(1,1), \tag{2.44}$$
$$(X_p, X_q) = (1,0): \quad h_{pq}(1,0) - h_{pq}(1,1) + h_{pq}(1,0) - h_{pq}(0,0), \tag{2.45}$$
$$(X_p, X_q) = (1,1): \quad h_{pq}(1,0) - h_{pq}(0,0). \tag{2.46}$$

If we add $h_{pq}(0,0) + h_{pq}(1,1) - h_{pq}(1,0)$ to the four cost values, we see each equals $h_{pq}(X_p, X_q)$. Since we are adding the same value to the four possible outcome of the pair, it does not affect the relative advantage among the alternatives. Thus, the minimum cut still corresponds to the labeling with the minimum energy.

We can do this for each neighboring pair. If all such pair p, q satisfies (2.39), we define:

$$w_{sq} = g_q(1) + \sum_{(p,q) \in \mathcal{N}} \{h_{pq}(1,0) - h_{pq}(0,0)\}, \tag{2.47}$$

$$w_{qt} = g_q(0) + \sum_{(p,q) \in \mathcal{N}} \{h_{pq}(1,0) - h_{pq}(1,1)\}, \tag{2.48}$$

$$w_{pq} = h_{pq}(0,1) + h_{pq}(1,0) - h_{pq}(0,0) - h_{pq}(1,1). \tag{2.49}$$

After that, if any edge weight w_{sq} or w_{qt} is negative, we can make it nonnegative as explained above.

In this way, we can make all the edge weights nonnegative. And from the construction, the minimum cut of this graph corresponds, by (2.32), to a labeling that minimizes the energy (2.15). The basic construction above for submodular first-order case has been known for at least 45 years [41].

2.3.4 Nonsubmodular Case: The BHS Algorithm

When the energy is not submodular, there are also known [53, 50] algorithms that give a partial solution, which were introduced to the image-related community, variously called as the QPBO algorithm [77] or roof duality [78]. Here, we describe the more efficient of the two algorithms, the BHS algorithm [50].

The solution of the algorithms has the following property:

1. The solution is a partial labeling, i.e., a labeling $X \in \mathbb{B}^Q$ on a subset Q of $\mathcal{P}$.

2. Given an arbitrary labeling $Y \in \mathbb{B}^{\mathcal{P}}$, the output labeling $X \in \mathbb{B}^Q$ has the property that, if we "overwrite" Y with X, the energy for the resulting labeling is not higher than that for the original labeling. Let us denote by $Y \triangleleft X$ the labeling obtained by overwriting Y by X:

$$(Y \triangleleft X)_p = \begin{cases} X_p & \text{if } p \in Q \\ Y_p & \text{otherwise.} \end{cases} \tag{2.50}$$

Then, this *autarky property* means $E(Y \triangleleft X) \le E(Y)$. If we take as Y a global minimum, then $Y \triangleleft X$ is also a global minimum. Thus, we see that the partial labeling X is always a part of a global minimum labeling, i.e., we can assign 0 or 1 to each site outside of Q to make it into the global minimum labeling $Y \triangleleft X$. It also means that if $Q = \mathcal{P}$, X is a global minimum labeling.

3. If the energy is submodular, all sites are labeled, giving a global minimum solution.

How many of the sites are labeled when the energy is not submodular is up to the individual problem.

Here, we give the graph construction for the BHS algorithm, following [77]. First, we *reparametrize* the given energy, i.e., modify it without changing the minimization problem, into the *normal form*. The energy (2.15) is said to be in normal form if

1. For each site p, $\min\{g_p(0), g_p(1)\} = 0$,

2. For each $(p, q) \in \mathcal{N}$ and $x = 0, 1$, $\min\{h_{pq}(0, x), h_{pq}(1, x)\} = 0$.

Note that the normal form is not unique. Beginning with any energy, we can reparametrize it into the normal form with the following algorithm:

1. While there is any combination of $(p, q) \in \mathcal{N}$ and $x \in \{0, 1\}$ that does not satisfy the second condition, repeatedly redefine

$$h_{pq}(0, x) \leftarrow h_{pq}(0, x) - \delta, \tag{2.51}$$
$$h_{pq}(1, x) \leftarrow h_{pq}(1, x) - \delta, \tag{2.52}$$
$$g_q(x) \leftarrow g_q(x) + \delta, \tag{2.53}$$

where $\delta = \min\{h_{pq}(0, x), h_{pq}(1, x)\}$.

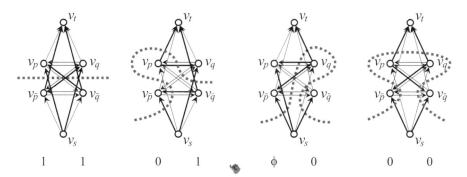

FIGURE 2.8
The graph construction for the BHS algorithm. The thick arrows indicate the edges in the cut. The graph has, in addition to v_s and v_t, two nodes v_p and $v_{\bar{p}}$ for each site p. We define a correspondence between the cuts of this graph and the output labeling as in (2.61). The assignment is shown at the bottom.

2. For each $p \in \mathcal{P}$, redefine

$$g_p(0) \leftarrow g_p(0) - \delta, \tag{2.54}$$
$$g_p(1) \leftarrow g_p(1) - \delta, \tag{2.55}$$

where $\delta = \min\{g_p(0), g_p(1)\}$.

Next, we build a graph $\mathcal{G} = (\mathcal{V}, \mathcal{E})$ as follows (Figure 2.8). Define two nodes for each site plus the two special nodes:

$$\mathcal{V} = \{v_s, v_t\} \cup \{v_p, v_{\bar{p}} \mid p \in \mathcal{P}\}. \tag{2.56}$$

There are four edges for each site, and four edges for each neighboring pair of sites:

$$\mathcal{E} = \{e_{s \to p}, e_{s \to \bar{p}}, e_{p \to t}, e_{\bar{p} \to t} \mid p \in \mathcal{P}\} \cup \{e_{p \to q}, e_{p \to \bar{q}}, e_{\bar{p} \to q}, e_{\bar{p} \to \bar{q}} \mid (p,q) \in \mathcal{N}\}. \tag{2.57}$$

Edge weights are defined as follows:

$$w_{sp} = w_{\bar{p}t} = \frac{1}{2}g_p(1), \qquad w_{pt} = w_{s\bar{p}} = \frac{1}{2}g_p(0), \tag{2.58}$$

$$w_{pq} = w_{\bar{q}\bar{p}} = \frac{1}{2}h_{pq}(0,1), \qquad w_{qp} = w_{\bar{p}\bar{q}} = \frac{1}{2}h_{pq}(1,0), \tag{2.59}$$

$$w_{p\bar{q}} = w_{q\bar{p}} = \frac{1}{2}h_{pq}(0,0), \qquad w_{\bar{q}p} = w_{\bar{p}q} = \frac{1}{2}h_{pq}(1,1). \tag{2.60}$$

For this graph, we find a minimum cut $(\mathcal{S}, \mathcal{T})$ and interpret it as follows:

$$\begin{cases} p \in \mathcal{Q} \text{ and } X_p = 0 & \text{if } v_p \in \mathcal{S}, v_{\bar{p}} \in \mathcal{T} \\ p \in \mathcal{Q} \text{ and } X_p = 1 & \text{if } v_p \in \mathcal{T}, v_{\bar{p}} \in \mathcal{S}, \\ p \notin \mathcal{Q} & \text{otherwise}. \end{cases} \tag{2.61}$$

As mentioned above, it depends on the actual problem to what extent this algorithm is effective. However, it is worth trying when dealing with nonsubmodular energies, as the construction is straightforward as above. There are also some techniques dealing with the sites that are unlabeled after the use of the algorithm [78].

2.3.5 Reduction of Higher-Order Energies

So far, the energies we dealt with have been of first order (2.15). Also in the vision literature, until recently most problems have been represented in terms of first-order energies, with a few exceptions that consider second-order ones [79, 48, 80]. Limiting the order to one restricts the representational power of the models, the rich statistics of natural scenes cannot be captured by such limited potentials. *Higher-order energies* can model more complex interactions and reflect the natural statistics better. There are also other reasons, such as enforcing connectivity [81] or histogram [82] in segmentation, for the need of optimizing higher-order energies. This has long been realized [83, 84, 85], and recent developments [86, 87, 88, 89, 90, 91], including various useful special cases of energies are discussed in the next chapter.

Here, we discuss the general binary case and introduce a method [92, 93] to transform general binary MRF minimization problems to the first-order case, for which the preceding methods can be used, at the cost of increasing the number of the binary variables. It is a generalization of earlier methods [48, 94] that allowed the reduction of second-order energies to first order.

An MRF energy that has labels in $\mathbb{B} = \{0, 1\}$ is a function of binary variables

$$f : \mathbb{B}^n \to \mathbb{R}, \tag{2.62}$$

where $n = |\mathcal{P}|$ is the number of sites/variables. Such functions are called *pseudo-Boolean functions* (PBFs.) Any PBF can be uniquely represented as a polynomial of the form

$$f(x_1, \ldots, x_n) = \sum_{S \subset \mathcal{I}} c_S \prod_{i \in S} x_i, \tag{2.63}$$

where $\mathcal{I} = \{1, \ldots, n\}$ and $c_S \in \mathbb{R}$. We refer the readers to [95, 96] for proof. Combined with the definition of the order, this implies that any binary-labeled MRF of order $d - 1$ can be represented as a polynomial of degree d. Thus, the problem of reducing higher-order binary-labeled MRFs to first order is equivalent to that of reducing general pseudo-Boolean functions to quadratic ones.

Consider a cubic PBF of variables x, y, z, with a coefficient $a \in \mathbb{R}$:

$$f(x, y, z) = axyz. \tag{2.64}$$

The reduction by [48, 94] is based on the following identity:

$$xyz = \max_{w \in \mathbb{B}} w(x + y + z - 2). \tag{2.65}$$

If $a < 0$,

$$axyz = \min_{w \in \mathbb{B}} aw(x + y + z - 2). \tag{2.66}$$

Thus, whenever $axyz$ appears in a minimization problem with $a < 0$, it can be replaced by $aw(x + y + z - 2)$.

If $a > 0$, we flip the variables (i.e., replace x by $1 - x$, y by $1 - y$, and z by $1 - z$) of (2.65) and consider

$$(1 - x)(1 - y)(1 - z) = \max_{w \in \mathbb{B}} w(1 - x + 1 - y + 1 - z - 2). \tag{2.67}$$

This is simplified to

$$xyz = \min_{w \in \mathbb{B}} w(x + y + z - 1) + (xy + yz + zx) - (x + y + z) + 1. \tag{2.68}$$

Therefore, if $axyz$ appears in a minimization problem with $a > 0$, it can be replaced by

$$a\{w(x + y + z - 1) + (xy + yz + zx) - (x + y + z) + 1\}. \tag{2.69}$$

Thus, either case, the cubic term can be replaced by quadratic terms. As any binary MRF of second order can be written as a cubic polynomial, each cubic monomial in the expanded polynomial can be converted to a quadratic polynomial using one of the formulae above, making the whole energy quadratic, i.e., of the first order. Of course, it does not come free; as we can see, the number of the variables increases.

This reduction works in the same way for higher order if $a < 0$:

$$ax_1 \ldots x_d = \min_{w \in \mathbb{B}} aw \{x_1 + \cdots + x_d - (d - 1)\}. \tag{2.70}$$

If $a > 0$, a different formula can be used [92, 93]:

$$ax_1 \cdots x_d = a \min_{w_1, \ldots, w_{n_d} \in \mathbb{B}} \sum_{i=1}^{n_d} w_i \left(c_{i,d}(-S_1 + 2i) - 1\right) + aS_2, \tag{2.71}$$

where

$$S_1 = \sum_{i=1}^{d} x_i, \qquad S_2 = \sum_{i=1}^{d-1} \sum_{j=i+1}^{d} x_i x_j = \frac{S_1(S_1 - 1)}{2}. \tag{2.72}$$

are the elementary symmetric polynomials in these variables and

$$n_d = \left\lfloor \frac{d - 1}{2} \right\rfloor, \qquad c_{i,d} = \begin{cases} 1 & \text{if } d \text{ is odd and } i = n_d, \\ 2 & \text{otherwise.} \end{cases} \tag{2.73}$$

Also, both formulas (2.70) and (2.71) can be modified by "flipping" some of the variables before and after transforming the higher-order term to a minimum or maximum [93]. For instance, another reduction for the quartic term can be obtained using the technique: if we define $\bar{x} = 1 - x$, we have $xyzt = (1 - \bar{x})yzt = yzt - \bar{x}yzt$. The right-hand side consists of a cubic term and a quartic term with a negative coefficient, which can be reduced using (2.70). This generalization gives rise to an exponential number (in terms of the occurrences of the variables in the function) of possible reductions.

2.4 Multi-Label Minimization

In this section, we deal with the case where there are more than two labels. First we describe the case when global optimization is possible, after which we introduce the move-making algorithms that can approximately minimize more general class of energies.

2.4.1 Globally Minimizable Case: Energies with Convex Prior

Assume that the labels have a linear order:

$$\mathcal{L} = \{l_0, \dots, l_k\}, \quad k \geq 2. \tag{2.74}$$

Also assume that, in the energy (2.15), the pairwise term $h_{pq}(X_p, X_q)$ is a *convex function* in terms of the difference of the indices of the labels, i.e., it can be written

$$h_{pq}(l_i, l_j) = \tilde{h}_{pq}(i - j) \tag{2.75}$$

with some function $\tilde{h}_{pq}(x)$ that satisfies

$$\tilde{h}_{pq}(i + 1) - 2\tilde{h}_{pq}(i) + \tilde{h}_{pq}(i - 1) \geq 0, \quad i = 1, \dots, k - 1. \tag{2.76}$$

In this case, the energy can be minimized globally using the minimum-cut algorithm by constructing a graph similar to the binary case we described in the preceding section [47].

As before, we construct a graph $\mathcal{G} = (\mathcal{V}, \mathcal{E})$ with special nodes v_s and v_t (Figure 2.9). For each site $p \in \mathcal{P}$, we define a series of nodes $\{v_p^1, \dots, v_p^k\}$:

$$\mathcal{V} = \{v_s, v_t\} \cup \{v_p^i \mid p \in \mathcal{P}, i = 1, \cdots, k\}. \tag{2.77}$$

For simplicity of notation, let us mean v_s by v_p^0 and v_t by v_p^{k+1} for any p.

We denote by $e_{(p,i) \to (q,j)}$ the directed edge going from v_p^i to v_q^j and by $w_{(p,i)(q,j)}$ its weight. Now we connect the nodes for the same site as follows. The first kind,

$$\mathcal{E}_1 = \{e_{(p,i) \to (p,i+1)} \mid p \in \mathcal{P}, i = 0, \cdots, k\} \tag{2.78}$$

starts with $v_s = v_p^0$ and goes $v_p^1, v_p^2, \dots$ in turn and ends at $v_t = v_p^{k+1}$.

To the opposite direction, the second kind of edges

$$\mathcal{E}_2 = \{e_{(p,i+1) \to (p,i)} \mid p \in \mathcal{P}, i = 1, \cdots, k - 1\} \tag{2.79}$$

have infinite weights. In Figure 2.9, they are depicted as dotted arrows. None of them

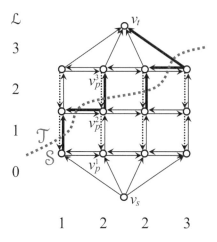

FIGURE 2.9

The graph for multilabel MRF minimization. There is a column of nodes and edges for each site. The thick arrows indicate the edges in the cut. We give an infinite weight to the downward dotted arrows, so that exactly one upward edge is in the cut for each column. The label assigned to each site is determined by which one of the upward edge is in the cut (the number on the left). The numbers at the bottom show the labels assigned to each site (column).

can be in any cut $(\mathcal{S}, \mathcal{T})$ with finite cost, as in the case shown in the figure; no dotted arrow goes from the $\mathcal{S}$ side to the $\mathcal{T}$ side. In practice, the weight can be some large-enough value. Because of these edges, among the column of edges

$$v_s = v_p^0 \longrightarrow v_p^1 \longrightarrow \dots \longrightarrow v_p^k \longrightarrow v_p^{k+1} = v_t \qquad (2.80)$$

in $\mathcal{E}_1$ over each site p, exactly one falls in the cut. This can be seen as follows: (1) the first node $v_s = v_p^0$ must be in $\mathcal{S}$; (2) the last node $v_t = v_p^{k+1}$ must be in $\mathcal{T}$; (3) so there must be at least one edge in the column going from $\mathcal{S}$ to $\mathcal{T}$; (4) but there cannot be two, since to go from $\mathcal{S}$ to $\mathcal{T}$ twice, there must be an edge in the column going from $\mathcal{T}$ to $\mathcal{S}$, but then the opposite edge with infinite weight would be in the cut.

Because of this, we can make a one-to-one correspondence between the cuts and the labelings as follows:

$$X_p = l_i \quad \Longleftrightarrow \quad v_p^i \in \mathcal{S}, v_p^{i+1} \in \mathcal{T} \qquad (2.81)$$

Therefore, we can transfer the data terms into cut cost in the similar way as the binary case by defining the weight as follows:

$$w_{(p,i)(p,i+1)} = g_p(l_i), \quad i = 0, \dots, k. \qquad (2.82)$$

As in the binary case, g_p can be any function, since we can add the same value to all the edges in the column if there is any with negative weight.

For the pairwise term $h_{pq}(X_p, X_q)$, we first examine the case shown in Figure 2.9. Here, there are horizontal edges between v_p^i and v_q^i, with the same "height" i. With these edges

$$\mathcal{E}_3 = \{e_{(p,i)\to(q,i)} \mid (p,q) \in \mathcal{N}, i = 1, \ldots, k\}, \tag{2.83}$$

we complete the construction of the graph by defining the set of edges $\mathcal{E} = \mathcal{E}_1 \cup \mathcal{E}_2 \cup \mathcal{E}_3$.

With this graph, if we give a constant positive number κ as weight to edges in $\mathcal{E}_3$, we have the smoothing term proportional to the difference between i and j:

$$h_{pq}(l_i, l_j) = \kappa |i - j|. \tag{2.84}$$

Now, assume that (2.75) and (2.76) are satisfied. For simplicity, we assume that $l_i = i, (i = 0, \cdots, k)$. Then the energy can be rewritten:

$$E(X) = \sum_{p \in \mathcal{P}} g_p(X_p) + \sum_{(p,q) \in \mathcal{N}} \tilde{h}_{pq}(X_p - X_q). \tag{2.85}$$

To minimize more general energies than (2.84), we need edges that are not horizontal (Figure 2.10). That is, for neighboring sites p and q, $\mathcal{E}_3$ would in general include edges $e_{(p,i)\to(q,j)}$ for all combination of i and j. We give them the weight

$$w_{(p,i)(q,j)} = \frac{1}{2} \left(\tilde{h}_{pq}(i - j + 1) - 2\tilde{h}_{pq}(i - j) + \tilde{h}_{pq}(i - j - 1) \right). \tag{2.86}$$

Because of the convexity condition (2.76), this is nonnegative. Without loss of generality, we can assume $\tilde{h}_{pq}(x) = \tilde{h}_{qp}(-x)$, because in the energy (2.85), $\tilde{h}_{pq}(X_p - X_q)$ and $\tilde{h}_{qp}(X_q - X_p)$ appear in pairs and thus we can redefine

$$\tilde{h}'_{pq}(x) = \frac{\tilde{h}_{pq}(x) + \tilde{h}_{qp}(-x)}{2} \tag{2.87}$$

if necessary.

Now, suppose edges $e_{(p,i)\to(p,i+1)}$ and $e_{(q,j)\to(q,j+1)}$ are in the cut $(\mathcal{S}, \mathcal{T})$, i.e., that $v_p^i, v_q^j \in \mathcal{S}$ and $v_p^{i+1}, v_q^{j+1} \in \mathcal{T}$. According to (2.81), this means $X_p = l_i = i, X_q = l_j = j$. Then we see that exactly the edges

$$\{e_{(p,l)\to(q,m)} \mid 0 \le l \le i,\ j + 1 \le m \le k + 1\} \tag{2.88}$$
$$\{e_{(q,l)\to(p,m)} \mid 0 \le l \le j,\ i + 1 \le m \le k + 1\} \tag{2.89}$$

are in the cut. We calculate the sum of the weights of these edges. First, summing up the weights of the edges in (2.88) (thick arrows in Figure 2.10) according to (2.86),

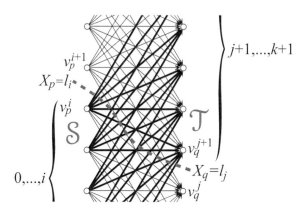

FIGURE 2.10
General smoothing edges for multilabel MRF minimization

we see most terms cancel out and

$$\sum_{l=0}^{i} \sum_{m=j+1}^{k+1} w_{(p,l)(q,m)}$$

$$= \frac{1}{2} \sum_{l=0}^{i} \sum_{m=j+1}^{k+1} \left(\tilde{h}_{pq}(l-m+1) - 2\tilde{h}_{pq}(l-m) + \tilde{h}_{pq}(l-m-1) \right)$$

$$= \frac{1}{2} \sum_{l=0}^{i} \sum_{n=l-k-1}^{l-j-1} \left(\tilde{h}_{pq}(n+1) - 2\tilde{h}_{pq}(n) + \tilde{h}_{pq}(n-1) \right)$$

$$= \frac{1}{2} \sum_{l=0}^{i} \left[\sum_{n=l-k}^{l-j} \left(\tilde{h}_{pq}(n) - \tilde{h}_{pq}(n-1) \right) - \sum_{n=l-k-1}^{l-j-1} \left(\tilde{h}_{pq}(n) - \tilde{h}_{pq}(n-1) \right) \right]$$

$$= \frac{1}{2} \sum_{l=0}^{i} \left(\tilde{h}_{pq}(l-j) - \tilde{h}_{pq}(l-k-1) - \tilde{h}_{pq}(l-j-1) + \tilde{h}_{pq}(l-k-2) \right)$$

$$= \frac{1}{2} \left(\tilde{h}_{pq}(i-j) - \tilde{h}_{pq}(-j-1) - \tilde{h}_{pq}(i-k-1) + \tilde{h}_{pq}(-k-2) \right).$$

$$(2.90)$$

Similarly, the edges in (2.89) add up to:

$$\frac{1}{2} \left(\tilde{h}_{qp}(j-i) - \tilde{h}_{qp}(-i-1) - \tilde{h}_{qp}(j-k-1) + \tilde{h}_{qp}(-k-2) \right). \quad (2.91)$$

Adding the two and using $\tilde{h}_{pq}(x) = \tilde{h}_{qp}(-x)$, we have the total sum of

$$\tilde{h}_{pq}(i-j) + r_{pq}(i) + r_{qp}(j), \quad (2.92)$$

where

$$r_{pq}(i) = \frac{\tilde{h}_{pq}(k+2) - \tilde{h}_{pq}(i-k-1) - \tilde{h}_{pq}(i+1)}{2}. \tag{2.93}$$

Since $r_{pq}(i)$ and $r_{qp}(j)$ depend only on $X_p = i$ and $X_q = j$, respectively, we can charge them to the data term. That is, instead of (2.82), we use

$$w_{(p,i)(p,i+1)} = g_p(i) - \sum_{(p,q)\in\mathcal{N}} r_{pq}(i). \tag{2.94}$$

and we have the sum above exactly $\tilde{h}_{pq}(i-j)$ which is $h_{pq}(l_i, l_j)$. Thus the cost of the cut is exactly the same as the energy of the corresponding labeling up to a constant; and we can use the minimum-cut algorithm to obtain the global minimum labeling.

This construction, since it uses the minimum-cut algorithms, in effect encodes the multi-label energy into a binary one and the convexity condition is the result of the submodularity condition. Generalizations have been made in this respect [97, 98, 99, 100], as well as a spatially continuous formulation [101, 102], which improves metrication error and efficiency. There has also been other improvement in time and memory efficiency [103, 104, 105].

2.4.2 Move-Making Algorithms

The above construction requires that the labels have a linear order, as well as that the energy pairwise term be convex. Since many problems do not satisfy the condition, iterative approximation algorithms are used more often in practice.

Move-making algorithms are iterative algorithms that repeatedly moves in the space of labelings so that the energy decreases. In general, the move can go to anywhere in the space, i.e., each site can change its label to any label; so, the best move would be to a global minimum. Since finding the global minimum in the general multi-label case is known to be NP-hard [106], we must restrict the range of possible moves so that finding the best move among them can be done in a polynomial time. The graph-cut optimization in each iteration of move-making algorithms can be thought of as prohibiting some of the labels at each site [107]. Depending on the restriction, this makes the optimization submodular or at least makes the graph much smaller than the full graph in the exact case.

Below, we introduce some of the approximation methods. There are also other algorithms that generalize move-making algorithms [108, 109].

2.4.2.1 α-β Swap and α-Expansion

Here, we first describe α-β *swap* and α-*expansion* moves, the two most popular algorithms in vision literature (Figure 2.11).

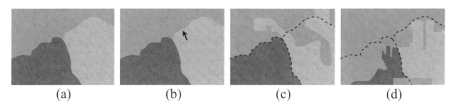

(a)   (b)   (c)   (d)

FIGURE 2.11
Three moves from the original labeling (a). (b) In the conventional one-pixel move
(ICM, annealing), one pixel at a time is changed. (c) In α-β swap, two labels α and
β are fixed and the sites currently labeled with one of the two are allowed to swap
the current label to the other. (d) In α-expansion, one label α is fixed and each site is
allowed to change its label to α.

For $\alpha, \beta \in \mathcal{L}$, an α-β swap from a labeling X is a move that allows only those
sites labeled α or β by X to change their labels to α or β. That is, a move $X \to X'$
is an α-β swap if

$$X_p \neq X'_p \quad \Rightarrow \quad X_p, X'_p \in \{\alpha, \beta\} \tag{2.95}$$

is satisfied. Similarly, an α-expansion is a move that allows each site to either switch
to label α or remain unchanged, i.e., a move $X \to X'$ is an α-expansion if

$$X_p \neq X'_p \quad \Rightarrow \quad X'_p = \alpha \tag{2.96}$$

is satisfied.

What these two moves have in common is that they both can be parametrized by
a binary labeling. To see this, define label functions $\mathcal{L} \times \{0, 1\} \to \mathcal{L}$ by:

$$x^{\alpha\beta}(l, b) = \begin{cases} l & \text{if } l \notin \{\alpha, \beta\} \\ \alpha & \text{if } l \in \{\alpha, \beta\} \text{ and } b = 0 \\ \beta & \text{if } l \in \{\alpha, \beta\} \text{ and } b = 1, \end{cases} \tag{2.97}$$

$$x^{\alpha}(l, b) = \begin{cases} l & \text{if } b = 0 \\ \alpha & \text{if } b = 1. \end{cases} \tag{2.98}$$

Then the α-β swap and α-expansion moves are defined by site-wise combination of
the current labeling X and a binary labeling Y. Fixing X, we evaluate the energy
(2.15) after the move:

$$E^{\alpha\beta}(Y) = \sum_{p \in \mathcal{P}} g_p(x^{\alpha\beta}(X_p, Y_p)) + \sum_{(p,q) \in \mathcal{N}} h_{pq}(x^{\alpha\beta}(X_p, Y_p), x^{\alpha\beta}(X_q, Y_q))$$

$$\tag{2.99}$$

$$E^{\alpha}(Y) = \sum_{p \in \mathcal{P}} g_p(x^{\alpha}(X_p, Y_p)) + \sum_{(p,q) \in \mathcal{N}} h_{pq}(x^{\alpha}(X_p, Y_p), x^{\alpha}(X_q, Y_q)) \tag{2.100}$$

What makes these moves important is that these energies can be shown to be sub-modular under simple criteria. Let us check the submodularity condition (2.39) for energy (2.99). We only have to check the case both X_p and X_q are in $\{\alpha, \beta\}$, as otherwise there is no real pairwise term depending on both Y_p and Y_q. Assuming $X_p, X_q \in \{\alpha, \beta\}$, the condition simply translates to:

$$h_{pq}(\alpha, \beta) + h_{pq}(\beta, \alpha) - h_{pq}(\alpha, \alpha) - h_{pq}(\beta, \beta) \geq 0. \qquad (2.101)$$

As for energy (2.100), let us write $\beta = X_p, \gamma = X_q$; then the condition is:

$$h_{pq}(\beta, \alpha) + h_{pq}(\alpha, \gamma) - h_{pq}(\beta, \gamma) - h_{pq}(\alpha, \alpha) \geq 0. \qquad (2.102)$$

Thus, if the energy satisfies (2.101) for all combination of $\alpha, \beta \in \mathcal{L}$, the optimal α-β swap move can always be found using the basic graph-cut algorithm. The algorithm iterate rotating through combinations of α and β in some order, until the energy stops lowering. Similarly, if the energy satisfies (2.102) for all combinations of $\alpha, \beta, \gamma \in \mathcal{L}$, the optimal α-expansion move can always be found. The algorithm rotates through α till the energy ceases to improve.

An important special case of both (2.101) and (2.102) is when h_{pq} is a metric, i.e., when conditions

$$h_{pq}(\alpha, \beta) \geq 0, \qquad (2.103)$$

$$h_{pq}(\alpha, \beta) = 0 \quad \Leftrightarrow \quad \alpha = \beta, \qquad (2.104)$$

$$h_{pq}(\alpha, \beta) + h_{pq}(\beta, \gamma) \geq h_{pq}(\alpha, \gamma), \qquad (2.105)$$

$$h_{pq}(\alpha, \beta) = h_{pq}(\beta, \alpha) \qquad (2.106)$$

are satisfied. In fact, only the first two are needed to satisfy (2.101); and only the first three would imply (2.102), when it is called a *semimetric*.

2.4.2.2 Fusion Moves

The *fusion move* [110, 111] is the most general form of move-making algorithm using binary MRF optimization. In each iteration, define the binary problem as the pixelwise choice between two arbitrary labelings, instead of between the current label and the fixed label α as in the case of α-expansion.

Let X, P be two multilabel labelings; then, the fusion of the two according to a binary labeling Y is the labeling defined by

$$F(X, P; Y)_p = \begin{cases} X_p & \text{if } Y_p = 0 \\ P_p & \text{if } Y_p = 1, \end{cases} \quad p \in \mathcal{P}. \qquad (2.107)$$

In each iteration, a binary-labeled MRF energy is defined as a function of Y, with X and P fixed:

$$\hat{E}(Y) = E(F(X, P; Y)). \qquad (2.108)$$

The Y that minimizes this energy is used to find the optimal fusion. As a move-making algorithm, the resulting fusion is often used as X in the next iteration; thus X is treated as the optimization variable, whereas P, called the *proposal*, is generated in some problem-specific way in each iteration.

Binary labeling problem defined this way is rarely submodular. So it became really useful only after the BHS (QPBO/roof duality) algorithms (Section 2.3.4), which give a partial labeling $Y \in \mathbb{B}^Q (Q \subset \mathcal{P})$ to minimization problems of a nonsubmodular energy, was introduced. For fusion, partial solution must be filled to make a full binary labeling Y. The autarky property guarantees that the energy of the fusion labeling $F(X, P; \mathbf{0} \lhd Y)$ is not more than that of X, i.e.,

$$E(F(X, P; \mathbf{0} \lhd Y)) = \hat{E}(\mathbf{0} \lhd Y) \leq \hat{E}(\mathbf{0}) = E(X). \qquad (2.109)$$

where $\mathbf{0} \in \mathbb{B}^{\mathcal{P}}$ is a labeling that assigns 0 to all sites. In other words, if we leave the labels unchanged at the sites where Y does not assign any label, the energy does not increase.

In fusion moves, it is crucial for the success and efficiency of the optimization to provide proposals P that fits the energies being optimized. Presumably, α expansion worked so well with the Potts energy because its proposal is the constant labeling. Using the outputs of other algorithms as proposals allows a principled integration of various existing algorithms [80]. In [112], a simple technique based on the gradient of the energy is proposed for generating proposal labelings that makes the algorithm much more efficient.

It is also worth noting that the set of labels can be very large, or even infinite, in fusion move, since the actual algorithm just chooses between two possible labels for each site.

2.4.2.3 α-β **Range Moves**

The exact minimization algorithm we described in Section 2.4.1 can be used for a class of move-making algorithms [113, 114], called α-β *range move* algorithms. When the set of labels has a linear order

$$\mathcal{L} = \{l_0, \ldots, l_k\}, \qquad (2.110)$$

they can efficiently minimize first-order energies of the form (2.15) with truncated-convex priors:

$$h_{pq}(l_i, l_j) = \min(\tilde{h}_{pq}(i - j), \theta), \qquad (2.111)$$

where $\tilde{h}_{pq}$ is a convex function, which satisfies (2.76), and θ a constant. This kind of prior limits the penalty for very large gaps in labels, thereby encouraging such gaps compared to the convex priors, which is sometimes desirable in segmenting images into objects. Here, we only describe the basic algorithm introduced in [113]; for other variations, see [114].

For two labels $\alpha = l_i, \beta = l_j, i < j$, the algorithm use the exact algorithm in Section 2.4.1 to find the optimal move among those that allow changing the labels in the range between the two labels at once. A move $X \to X'$ is called an α-β range move if it satisfies

$$X_p \neq X'_p \quad \Rightarrow \quad X_p, X'_p \in \{l_i, \ldots, l_j\}. \tag{2.112}$$

Let d be the maximum integer that satisfies $\tilde{h}_{pq}(d) \leq \theta$. The algorithm iterates α-β range moves with $\alpha = l_i, \beta = l_j, j - i = d$. Within this range, the prior (2.111) is convex, allowing the use of the exact algorithm. Let X be the current labeling and define

$$\mathcal{P}_{\alpha\text{-}\beta} = \{p \in \mathcal{P} \mid X_p \in \{l_i, \ldots, l_j\}\}. \tag{2.113}$$

Define a graph $\mathcal{G}_{\alpha\text{-}\beta} = (\mathcal{V}_{\alpha\text{-}\beta}, \mathcal{E}_{\alpha\text{-}\beta})$ as follows. The node set $\mathcal{V}_{\alpha\text{-}\beta}$ contains two special nodes v_s, v_t as well as a series of nodes $\{v_p^1, \ldots, v_p^d\}$ for each site p in $\mathcal{P}_{\alpha\text{-}\beta}$. As in Section 2.4.1, let v_p^0 be an alias for v_s, and v_p^{d+1} for v_t and denote a directed edge from v_p^n to v_q^m by $e_{(p,n)\to(q,m)}$. The edge set $\mathcal{E}_{\alpha\text{-}\beta}$ contains a directed edge $e_{(p,n)\to(p,n+1)}$ for each pair of $p \in \mathcal{P}_{\alpha\text{-}\beta}$ and $n = 0, 1, \ldots, d$, as well as the opposite edge $e_{(p,n+1)\to(p,n)}$ with an infinite weight for $n = 1, \ldots, d - 1$. This, as before, guarantees that exactly one edge among the series

$$v_s = v_p^0 \to v_p^1 \to \cdots \to v_p^d \to v_p^{d+1} = v_t \tag{2.114}$$

over each site $p \in \mathcal{P}_{\alpha\text{-}\beta}$ is in any cut with finite cost, establishing a one-to-one correspondence between the cut $(\mathcal{S}, \mathcal{T})$ and the new labeling:

$$X'_p = \begin{cases} l_{i+n} & (\text{if } p \in \mathcal{P}_{\alpha\text{-}\beta}, v_p^n \in \mathcal{S}, v_p^{n+1} \in \mathcal{T}) \\ X_p & (\text{if } p \notin \mathcal{P}_{\alpha\text{-}\beta}). \end{cases} \tag{2.115}$$

Thus, by defining the weights by

$$w_{(p,n)(p,n+1)} = g_p(l_{i+n}) \tag{2.116}$$

the data term is exactly reflected in the cut cost.

As for the smoothing term, any convex $\tilde{h}_{pq}$ can be realized by defining edges as in Section 2.4.1. However, that is only if both p and q belong to $\mathcal{P}_{\alpha\text{-}\beta}$. Although it can be ignored if neither belongs to $\mathcal{P}_{\alpha\text{-}\beta}$, it cannot be if only one does. This is, however, a unary term, as one of the pair has a fixed label. Thus, adding the effect to (2.116), we redefine

$$w_{(p,n)(p,n+1)} = g_p(l_{i+n}) + \sum_{\substack{(p,q)\in\mathcal{N} \\ q\in\mathcal{P}\backslash\mathcal{P}_{\alpha\text{-}\beta}}} \min(\tilde{h}_{pq}(l_{i+n} - X_q), \theta). \tag{2.117}$$

This way, the cost of a cut of $(\mathcal{G}_{\alpha\text{-}\beta}, v_s, v_t)$ is the energy of the corresponding labeling by (2.115) up to a constant. Thus, the minimum cut gives the optimal α-β range move.

2.4.2.4 Higher-Order Graph Cuts

Combining the fusion move and the reduction of general higher-order binary energy explained in Section 2.3.5, multiple-label higher-order energies can be approximately minimized [92, 93].

The energy is of the general form (2.13), shown here again:

$$E(X) = \sum_{C \in \mathcal{C}} f_C(X_C), \tag{2.118}$$

where $\mathcal{C}$ is a set of cliques (subsets) of the set $\mathcal{P}$ of sites, and X_C is the restriction of X to C.

The algorithm maintains the current labeling $X \in \mathcal{L}^{\mathcal{P}}$. In each iteration, the algorithm fuses X and a proposed labeling $P \in \mathcal{L}^{\mathcal{P}}$ by minimizing a binary energy. How P is prepared is problem specific, as is how X is initialized at the beginning of the algorithm.

The result of the merge according to a binary labeling $Y \in \mathbb{B}^{\mathcal{P}}$ is defined by (2.107) and its energy by (2.108). For the energy $E(X)$ of the form (2.13), the polynomial expression of $\hat{E}(Y)$ is

$$\hat{E}(Y) = \sum_{C \in \mathcal{C}} \sum_{\gamma \in \mathbb{B}^C} f_C(F(X_C, P_C; \gamma)) \, \theta_C^\gamma(Y_C), \tag{2.119}$$

where $F(X_C, P_C; \gamma) \in \mathcal{L}^C$ is the fusion labeling on the clique C determined by a binary vector $\gamma \in \mathbb{B}^C$:

$$F(X_C, P_C; \gamma)_p = \begin{cases} X_p & \text{if } \gamma_p = 0 \\ P_p & \text{if } \gamma_p = 1, \end{cases} \quad p \in C, \tag{2.120}$$

and $\theta_C^\gamma(Y_C)$ is a polynomial of degree $|C|$ defined by

$$\theta_C^\gamma(Y_C) = \prod_{p \in C} Y_p^{(\gamma)}, \quad Y_p^{(\gamma)} = \begin{cases} Y_p & \text{if } \gamma_p = 1 \\ 1 - Y_p & \text{if } \gamma_p = 0 \end{cases} \tag{2.121}$$

which is 1 if $Y_C = \gamma$ and 0 otherwise.

The polynomial $\hat{E}(Y)$ is then reduced into a quadratic polynomial using the technique described in Section 2.3.5. The result is a quadratic polynomial $\tilde{E}(\tilde{Y})$ in terms of the labeling $\tilde{Y} \in \mathbb{B}^{\mathcal{P}'}$, where $\mathcal{P}' \supset \mathcal{P}$. We use the BHS algorithm to minimize $\tilde{E}(\tilde{Y})$ and obtain a partial labeling $Z \in \mathbb{B}^{\mathcal{Q}'}$ on a subset $\mathcal{Q}'$ of $\mathcal{P}'$. Overwriting the zero labeling $\mathbf{0}$ by the restriction $Z_{\mathcal{P} \cap \mathcal{Q}'}$ of Z to the original sites, we obtain $\mathbf{0} \lhd Z_{\mathcal{P} \cap \mathcal{Q}'}$, which we denote by Y_Z. Then we have

$$E(F(X, P; Y_Z)) = \hat{E}(Y_Z) \le \hat{E}(\mathbf{0}) = E(X). \tag{2.122}$$

To see why, let us define

$$\mathcal{Y}_0 = \{\tilde{Y} \in \mathbb{B}^{\mathcal{P}'} \mid \tilde{Y}_{\mathcal{P}} = \mathbf{0}\}, \qquad \tilde{Y}_0^{\min} = \arg\min_{\tilde{Y} \in \mathcal{Y}_0} \tilde{E}(\tilde{Y}), \qquad (2.123)$$

$$\mathcal{Y}_Z = \{\tilde{Y} \in \mathbb{B}^{\mathcal{P}'} \mid \tilde{Y}_{\mathcal{P}} = Y_Z\}, \qquad \tilde{Y}_Z^{\min} = \arg\min_{\tilde{Y} \in \mathcal{Y}_Z} \tilde{E}(\tilde{Y}). \qquad (2.124)$$

Then, by the property of the reduction, we have $\hat{E}(\mathbf{0}) = \tilde{E}(\tilde{Y}_0^{\min})$ and $\hat{E}(Y_Z) = \tilde{E}(\tilde{Y}_Z^{\min})$. Also, if we define $\tilde{Z} = \tilde{Y}_0^{\min} \lhd Z$, we have $\tilde{E}(\tilde{Z}) \leq \tilde{E}(\tilde{Y}_0^{\min})$ by the autarky property. Since $\tilde{Z} \in \mathcal{Y}_Z$, we also have $\tilde{E}(\tilde{Y}_Z^{\min}) \leq \tilde{E}(\tilde{Z})$. Thus, $\hat{E}(Y_Z) = \tilde{E}(\tilde{Y}_Z^{\min}) \leq \tilde{E}(\tilde{Z}) \leq \tilde{E}(\tilde{Y}_0^{\min}) = \hat{E}(\mathbf{0})$.

Therefore, we update X to $F(X, P; Y_Z)$ and (often) decrease the energy. We iterate the process until some convergence criterion is met.

2.5 Examples

Here, we show denoising examples using the multilabel energy minimization techniques described in the previous section. Figure 2.12 shows the results of denoising an image with graph cuts. The original 8-bit grayscale image (a) was added an i. i. d. Gaussian additive noise with standard deviation $\sigma = 50$, resulting in the noisy image Y shown in (b).

The result of denoising it by finding X that globally minimizes the "total variation" energy:

$$E(X) = \sum_{p \in \mathcal{P}} (Y_p - X_p)^2 + \sum_{(p,q) \in \mathcal{N}} \kappa |X_p - X_q| \qquad (2.125)$$

with $\kappa = 8.75$, using the technique described in Section 2.4.1 and [47], is shown in (c).

The result of denoising Y with the higher-order graph cuts described in Section 2.4.2.4 is shown in (d). For the higher-order energy, we used the recent image statistical model called the *fields of experts* (FoE) [85], which represents the prior probability of an image X as the product of several Student's t-distributions:

$$P(X) \propto \prod_C \prod_{i=1}^{K} \left(1 + \frac{1}{2}(J_i \cdot X_C)^2\right)^{-\alpha_i}, \qquad (2.126)$$

where C runs over the set of all $n \times n$ patches in the image, and J_i is an $n \times n$ filter. The parameters J_i and α_i are learned from a database of natural images. Here, we used the energy

$$E(X) = \sum_{C \in \mathcal{C}} f_C(X_C), \qquad (2.127)$$

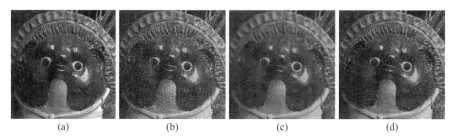

(a)   (b)   (c)   (d)

FIGURE 2.12
Denoising example. (a) Original image, (b) noise-added image ($\sigma = 50$), (c) denoised using the total variation energy, (d) denoised using the higher-order FoE energy.

where the set $\mathcal{C}$ of cliques consists of those cliques formed by singleton pixels and 2×2 patches. For the two kinds of cliques, the function f_C is defined by

$$f_C(X_C) = \frac{(Y_p - X_p)^2}{2\sigma^2}, \qquad\qquad (C = \{p\}) \qquad\qquad (2.128)$$

$$f_C(X_C) = \sum_{i=1}^{3} \alpha_i \log\left(1 + \frac{1}{2}(J_i \cdot X_C)^2\right). \qquad (C \text{ is a } 2 \times 2 \text{ patch}) \qquad (2.129)$$

The FoE parameters J_i and α_i for the experiment were kindly provided by Stefan Roth. The energy was approximately minimized using the higher-order graph cuts. We initialized X by Y and then used the following two proposals in alternating iterations: (i) a uniform random image created each iteration, and (ii) a blurred image, which is made every 30 iterations by blurring the current image X with a Gaussian kernel ($\sigma = 0.5625$).

2.6 Conclusion

In this chapter, we have provided an overview of the main ideas behind graph cuts and introduced the basic constructions that are most widely used as well as the ones the author is most familiar with.

Further Reading

The research on graph cuts has become extensive and diverse. There are also algorithms other than graph cuts [115, 116, 117] with similar purpose and sometimes superior performance. The interested reader may also refer to other books [118, 119], as well as other chapters of this book and the original papers in the bibliography.

Bibliography

[1] A. Blake and A. Zisserman, *Visual Reconstruction.* MIT Press, London, 1987.

[2] A. Blake, C. Rother, M. Brown, P. Perez, and P. Torr, "Interactive image segmentation using an adaptive GMMRF model," in *Proc. European Conference on Computer Vision (ECCV2004)*, 2004, pp. 428–441.

[3] Y. Boykov and G. Funka-Lea, "Graph cuts and efficient n-d image segmentation," *International Journal of Computer Vision*, vol. 70, no. 2, pp. 109–131, 2006.

[4] Y. Boykov and M.-P. Jolly, "Interactive graph cuts for optimal boundary & region segmentation of objects in N-D images," in *Proc. IEEE International Conference on Computer Vision (ICCV2001)*, vol. 1, 2001, pp. 105–112.

[5] M. Bray, P. Kohli, and P. H. A. Torr, "Posecut: Simultaneous segmentation and 3d pose estimation of humans using dynamic graph-cuts," in *Proc. European Conference on Computer Vision (ECCV2006)*, vol. 2, 2006, pp. 642–655.

[6] H. Ishikawa and D. Geiger, "Segmentation by grouping junctions," in *Proc. IEEE Computer Society Conference on Computer Vision and Pattern Recognition (CVPR'98)*, 1998, pp. 125–131.

[7] ——, "Higher-dimensional segmentation by minimum-cut algorithm," in *Ninth IAPR Conference on Machine Vision Applications (MVA 2005)*, 2005, pp. 488–491.

[8] Y. Li, J. Sun, and H.-Y. Shum, "Video object cut and paste." *ACM Trans. Graphics (Proc. SIGGRAPH2005)*, vol. 24, no. 3, pp. 595–600, 2005.

[9] Y. Li, J. Sun, C.-K. Tang, and H.-Y. Shum, "Lazy snapping," *ACM Trans. Graphicss (Proc. SIGGRAPH2004)*, vol. 23, no. 3, pp. 303–308, 2004.

[10] K. Li, X. Wu, D. Z. Chen, and M. Sonka, "Optimal surface segmentation in volumetric images—a graph-theoretic approach," *IEEE Trans. Pattern Analysis and Machine Intelligence*, vol. 28, no. 1, pp. 119–134, 2006.

[11] C. Rother, V. Kolmogorov, and A. Blake, ""grabcut": Interactive foreground extraction using iterated graph cuts," *ACM Trans. Graphics (Proc. SIGGRAPH2004)*, vol. 23, no. 3, pp. 309–314, 2004.

[12] J. Wang, P. Bhat, R. A. Colburn, M. Agrawala, and M. F. Cohen, "Interactive video cutout," *ACM Trans. Graphics (Proc. SIGGRAPH2005)*, vol. 24, no. 3, pp. 585–594, 2005.

[13] N. Xu, R. Bansal, and N. Ahuja, "Object segmentation using graph cuts based active contours," in *Proc. IEEE Computer Society Conference on Computer Vision and Pattern Recognition (CVPR2003)*, vol. 2, 2003, pp. 46–53.

[14] M. P. Kumar, P. H. Torr, and A. Zisserman, "Learning layered motion segmentations of video," *International Journal of Computer Vision*, vol. 76, no. 3, pp. 301–319, 2008.

[15] S. Roy and V. Govindu, "MRF solutions for probabilistic optical flow formulations," in *Proc International Conference on Pattern Recognition (ICPR2000)*, vol. 3, 2000, pp. 7053–7059.

[16] J. Wills, S. Agarwal, and S. Belongie, "What went where," in *Proc. IEEE Computer Society Conference on Computer Vision and Pattern Recognition (CVPR2003)*, vol. 1, 2003, pp. 37–44.

[17] J. Xiao and M. Shah, "Motion layer extraction in the presence of occlusion using graph cuts," *IEEE Trans. Pattern Analysis and Machine Intelligence*, vol. 27, no. 10, pp. 1644–1659, 2005.

[18] S. Birchfield and C. Tomasi, "Multiway cut for stereo and motion with slanted surfaces," in *Proc. IEEE International Conference on Computer Vision (ICCV'99)*, vol. 1, 1999, pp. 489–495.

[19] Y. Boykov, O. Veksler, and R. Zabih, "Markov random fields with efficient approximations," in *Proc. IEEE Computer Society Conference on Computer Vision and Pattern Recognition (CVPR'98)*, 1998, pp. 648–655.

[20] H. Ishikawa, "Multi-scale feature selection in stereo," in *Proc. IEEE Computer Society Conference on Computer Vision and Pattern Recognition (CVPR'99)*, vol. 1, 1999, pp. 1132–1137.

[21] H. Ishikawa and D. Geiger, "Local feature selection and global energy optimization in stereo," in *Scene Reconstruction, Pose Estimation and Tracking*, R. Stolkin, Ed.   Vienna, Austria: I-Tech Education and Publishing, 2007, pp. 411–430.

[22] V. Kolmogorov, A. Criminisi, A. Blake, G. Cross, and C. Rother, "Probabilistic fusion of stereo with color and contrast for bilayer segmentation," *IEEE Trans. Pattern Analysis and Machine Intelligence*, vol. 28, no. 9, pp. 1480–1492, 2006.

[23] S. Roy and I. Cox, "Maximum-flow formulation of the n-camera stereo correspondence problem," in *Proc. IEEE International Conference on Computer Vision (ICCV'98)*, 1998, pp. 492–499.

[24] S. Roy, "Stereo without epipolar lines : A maximum-flow formulation," *International Journal of Computer Vision*, vol. 34, pp. 147–162, 1999.

[25] V. Kwatra, A. Schödl, I. Essa, G. Turk, and A. Bobick, "Graphcut textures: Image and video synthesis using graph cuts," *ACM Trans. Graphics (Proc. SIGGRAPH2003)*, vol. 22, no. 3, pp. 277–286, 2003.

[26] M. H. Nguyen, J.-F. Lalonde, A. A. Efros, and F. de la Torre, "Image based shaving," *Computer Graphics Forum Journal (Eurographics 2008)*, vol. 27, no. 2, pp. 627–635, 2008.

[27] A. Agarwala, M. Dontcheva, M. Agrawala, S. Drucker, A. Colburn, B. Curless, D. Salesin, and M. Cohen, "Interactive digital photomontage," *ACM Trans. Graphics (Proc. SIGGRAPH2004)*, vol. 23, no. 3, pp. 294–302, 2004.

[28] A. Agarwala, M. Agrawala, M. Cohen, D. Salesin, and R. Szeliski, "Photographing long scenes with multi-viewpoint panoramas," *ACM Trans. Graphics (Proc. SIGGRAPH2006)*, vol. 25, no. 3, pp. 853–861, 2006.

[29] J. Hays and A. A. Efros, "Scene completion using millions of photographs," *ACM Trans. Graphics (Proc. SIGGRAPH2007)*, vol. 26, no. 3, 2007, Article# 4.

[30] C. Rother, S. Kumar, V. Kolmogorov, and A. Blake, "Digital tapestry," in *Proc. IEEE Computer Society Conference on Computer Vision and Pattern Recognition (CVPR2005)*, vol. 1, 2005, pp. 589–596.

[31] M. P. Kumar, P. H. S. Torr, and A. Zisserman, "Obj cut," in *Proc. IEEE Computer Society Conference on Computer Vision and Pattern Recognition (CVPR2005)*, vol. 1, 2005, pp. 18–25.

[32] D. Hoiem, C. Rother, and J. Winn, "3D layout CRF for multi-view object class recognition and segmentation," in *Proc. IEEE Computer Society Conference on Computer Vision and Pattern Recognition (CVPR2007)*, 2007.

[33] J. Winn and J. Shotton, "The layout consistent random field for recognizing and segmenting partially occluded objects," in *Proc. IEEE Computer Society Conference on Computer Vision and Pattern Recognition (CVPR2006)*, 2006, pp. 37–44.

[34] Y. Boykov and V. Kolmogorov, "Computing geodesics and minimal surfaces via graph cuts," in *Proc. IEEE International Conference on Computer Vision (ICCV2003)*, vol. 1, 2003, pp. 26–33.

[35] Y. Boykov, V. Kolmogorov, D. Cremers, and A. Delong, "An integral solution to surface evolution PDEs via geo-cuts," in *Proc. European Conference on Computer Vision (ECCV2006)*, vol. 3, 2006, pp. 409–422.

[36] A. Levin, R. Fergus, F. Durand, and W. T. Freeman, "Image and depth from a conventional camera with a coded aperture," *ACM Trans. Graphics (Proc. SIGGRAPH2007)*, vol. 26, no. 3, 2007, Article# 70.

[37] C. Liu, A. B. Torralba, W. T. Freeman, F. Durand, and E. H. Adelson, "Motion magnification," *ACM Trans. Graphics (Proc. SIGGRAPH2005)*, vol. 24, no. 3, pp. 519–526, 2005.

[38] Y. Boykov, O. Veksler, and R. Zabih, "Fast approximate energy minimization via graph cuts," *IEEE Trans. Pattern Analysis and Machine Intelligence*, vol. 23, no. 11, pp. 1222–1239, 2001.

[39] J. Besag, "On the statistical analysis of dirty pictures," *J. Royal Stat. Soc., Series B*, vol. 48, pp. 259–302, 1986.

[40] S. Geman and D. Geman, "Stochastic relaxation, Gibbs distributions, and the Bayesian restoration of images," *IEEE Trans. Pattern Analysis and Machine Intelligence*, vol. 6, pp. 721–741, 1984.

[41] P. L. Hammer, "Some network flow problems solved with pseudo-boolean programming," *Operations Res.*, vol. 13, pp. 388–399, 1965.

[42] D. M. Greig, B. T. Porteous, and A. H. Seheult, "Discussion of: On the statistical analysis of dirty pictures (by J. E. Besag)," *J. Royal Stat. Soc., Series B*, vol. 48, pp. 282–284, 1986.

[43] ——, "Exact maximum a posteriori estimation for binary images," *J. Royal Stat. Soc., Series B*, vol. 51, pp. 271–279, 1989.

[44] A. Blake, "Comparison of the efficiency of deterministic and stochastic algorithms for visual reconstruction," *IEEE Trans. Pattern Analysis and Machine Intelligence*, vol. 11, no. 1, pp. 2–12, 1989.

[45] H. Ishikawa and D. Geiger, "Occlusions, discontinuities, and epipolar lines in stereo," in *Proc. European Conference on Computer Vision (ECCV'98)*, 1998, pp. 232–248.

[46] ——, "Mapping image restoration to a graph problem," in *IEEE-EURASIP Workshop on Nonlinear Signal and Image Processing (NSIP'99)*, 1999, pp. 189–193.

[47] H. Ishikawa, "Exact optimization for Markov random fields with convex priors," *IEEE Trans. Pattern Analysis and Machine Intelligence*, vol. 25, no. 10, pp. 1333–1336, 2003.

[48] V. Kolmogorov and R. Zabih, "What energy functions can be minimized via graph cuts?" *IEEE Trans. Pattern Analysis and Machine Intelligence*, vol. 26, no. 2, pp. 147–159, 2004.

[49] C.-H. Lee, D. Lee, and M. Kim, "Optimal task assignment in linear array networks," *IEEE Trans. Computers*, vol. 41, no. 7, pp. 877–880, 1992.

[50] E. Boros, P. L. Hammer, and X. Sun, "Network flows and minimization of quadratic pseudo-boolean functions, Tech. Rep. RUTCOR Research Report RRR 17-1991, May 1991.

[51] E. Boros, P. L. Hammer, R. Sun, and G. Tavares, "A max-flow approach to improved lower bounds for quadratic unconstrained binary optimization (qubo)," *Discrete Optimization*, vol. 5, no. 2, pp. 501–529, 2008.

[52] E. Boros, P. L. Hammer, and G. Tavares, "Preprocessing of unconstrained quadratic binary optimization, Tech. Rep. RUTCOR Research Report RRR 10-2006, April 2006.

[53] P. L. Hammer, P. Hansen, and B. Simeone, "Roof duality, complementation and persistency in quadratic 0-1 optimization," *Math. Programming*, vol. 28, pp. 121–155, 1984.

[54] J. Pearl, *Probabilistic Reasoning in Intelligent Systems: Networks of Plausible Inference*. Morgan Kaufmann, San Francisco, 1998.

[55] P. Felzenszwalb and D. Huttenlocher, "Efficient belief propagation for early vision," *Int. J. Comput. Vis.*, vol. 70, pp. 41–54, 2006.

[56] T. Meltzer, C. Yanover, and Y. Weiss, "Globally optimal solutions for energy minimization in stereo vision using reweighted belief propagation," in *Proc. IEEE International Conference on Computer Vision (ICCV2005)*, 2005, pp. 428–435.

[57] M. J. Wainwright, T. S. Jaakkola, and A. S. Willsky, "Tree-based reparameterization framework for analysis of sum-product and related algorithms," *IEEE Trans. Information Theory*, vol. 49, no. 5, pp. 1120–1146., 2003.

[58] V. Kolmogorov, "Convergent tree-reweighted message passing for energy minimization," *IEEE Trans. Pattern Analysis and Machine Intelligence*, vol. 28, no. 10, pp. 1568–1583, 2006.

[59] R. Szeliski, R. Zabih, D. Scharstein, O. Veksler, V. Kolmogorov, A. Agarwala, M. Tappen, and C. Rother, "A comparative study of energy minimization methods for Markov random fields with smoothness-based priors," *IEEE Trans. Pattern Analysis and Machine Intelligence*, vol. 30, no. 7, pp. 1068–1080, 2008.

[60] V. Kolmogorov and C. Rother, "Comparison of energy minimization algorithms for highly connected graphs," in *Proc. European Conference on Computer Vision (ECCV2006)*, vol. 2, 2006, pp. 1–15.

[61] R. K. Ahuja, T. L. Magnanti, and J. B. Orlin, *Network Flows: Theory, Algorithms and Applications*. Prentice Hall, 1993.

[62] W. J. Cook, W. H. Cunningham, W. R. Pulleyblank, and A. Schrijver, *Combinatorial Optimization*. John Wiley & Sons, New-York, 1998.

[63] T. H. Cormen, C. E. Leiserson, R. L. Rivest, and C. Stein, *Introduction to Algorithms (Third Edition)*. MIT Press, 2009.

[64] L. R. Ford and D. R. Fulkerson, "Maximal flow through a network," *Can. J. Math.*, vol. 8, pp. 399–404, 1956.

[65] P. Elias, A. Feinstein, and C. E. Shannon, "A note on the maximum flow through a network," *IEEE Trans. Information Theory*, vol. 2, no. 4, pp. 117–119, 1956.

[66] L. Ford and D. Fulkerson, *Flows in Networks*. Princeton University Press, 1962.

[67] F. Alizadeh and A. V. Goldberg, "Implementing the push-relabel method for the maximum flow problem on a connection machine," *DIMACS Series in Discrete Mathematics and Theoretical Computer Science*, vol. 12, pp. 65–95, 1993.

[68] B. V. Cherkassky and A. V. Goldberg, "On implementing push-relabel method for the maximum flow problem," in *Proc. 4th International Programming and Combinatorial Optimization Conference*, 1995, pp. 157–171.

[69] A. V. Goldberg and R. E. Tarjan, "A new approach to the maximum-flow problem," *J. ACM*, vol. 35, pp. 921–940, 1988.

[70] A. V. Goldberg, "Efficient graph algorithms for sequential and parallel computers," Ph.D. dissertation, Massachussetts Institute of Technology, 1987.

[71] Y. Boykov and V. Kolmogorov, "An experimental comparison of min-cut/max-flow algorithms for energy minimization in vision," *IEEE Trans. Pattern Analysis and Machine Intelligence*, vol. 26, no. 9, pp. 1124–1137, 2004.

[72] E. A. Dinic, "Algorithm for solution of a problem of maximum flow in networks with power estimation," *Soviet Math. Dokl.*, vol. 11, pp. 1277–1280, 1970.

[73] A. Delong and Y. Boykov, "A scalable graph-cut algorithm for N-D grids," in *Proc. IEEE Computer Society Conference on Computer Vision and Pattern Recognition (CVPR2008)*, 2008.

[74] O. Juan and Y. Boykov, "Active graph cuts," in *Proc. IEEE Computer Society Conference on Computer Vision and Pattern Recognition (CVPR2006)*, vol. 1, 2006, pp. 1023–1029.

[75] P. Kohli and P. H. S. Torr, "Effciently solving dynamic Markov random fields using graph cuts," in *Proc. IEEE Computer Society Conference on Computer Vision and Pattern Recognition (CVPR2005)*, vol. 2, 2005, pp. 922–929.

[76] ——, "Measuring uncertainty in graph cut solutions–efficiently computing minmarginal energies using dynamic graph cuts," in *Proc. European Conference on Computer Vision (ECCV2006)*, vol. 2, 2006, pp. 20–43.

[77] V. Kolmogorov and C. Rother, "Minimizing non-submodular functions with graph cuts — a review," *IEEE Trans. Pattern Analysis and Machine Intelligence*, vol. 9, no. 7, pp. 1274–1279, 2007.

[78] C. Rother, V. Kolmogorov, V. Lempitsky, and M. Szummer, "Optimizing binary MRFs via extended roof duality," in *Proc. IEEE Computer Society Conference on Computer Vision and Pattern Recognition (CVPR2007)*, 2007.

[79] D. Cremers and L. Grady, "Statistical priors for efficient combinatorial optimization via graph cuts," in *Proc. European Conference on Computer Vision (ECCV2006)*, vol. 3, 2006, pp. 263–274.

[80] O. J. Woodford, P. H. S. Torr, I. D. Reid, and A. W. Fitzgibbon, "Global stereo reconstruction under second order smoothness priors," in *Proc. IEEE Computer Society Conference on Computer Vision and Pattern Recognition (CVPR2008)*, 2008.

[81] S. Vicente, V. Kolmogorov, and C. Rother, "Graph cut based image segmentation with connectivity priors," in *Proc. IEEE Computer Society Conference on Computer Vision and Pattern Recognition (CVPR2008)*, 2008.

[82] C. Rother, T. Minka, A. Blake, and V. Kolmogorov, "Cosegmentation of image pairs by histogram matching - incorporating a global constraint into MRFs," in *Proc. IEEE Computer Society Conference on Computer Vision and Pattern Recognition (CVPR2006)*, vol. 1, 2006, pp. 993–1000.

[83] H. Ishikawa and D. Geiger, "Rethinking the prior model for stereo," in *Proc. European Conference on Computer Vision (ECCV2006)*, vol. 3, 2006, pp. 526–537.

[84] G. L. Nemhauser, L. A. Wolsey, and M. L. Fisher, "Texture synthesis via a noncausal nonparametric multiscale markov random field," *IEEE Trans. Image Processing*, vol. 7, no. 6, pp. 925–931, 1998.

[85] S. Roth and M. J. Black, "Fields of experts: A framework for learning image priors," in *Proc. IEEE Computer Society Conference on Computer Vision and Pattern Recognition (CVPR2005)*, vol. 2, 2005, pp. 860–867.

[86] P. Kohli, M. P. Kumar, and P. H. S. Torr, "$\mathcal{P}^3$ & beyond: Move making algorithms for solving higher order functions," *IEEE Trans. Pattern Analysis and Machine Intelligence*, vol. 31, no. 9, pp. 1645–1656, 2009.

[87] P. Kohli, L. Ladicky, and P. H. S. Torr, "Robust higher order potentials for enforcing label consistency," *Int. J. Comput. Vis.*, vol. 82, no. 3, pp. 303–324, 2009.

[88] X. Lan, S. Roth, D. P. Huttenlocher, and M. J. Black, "Efficient belief propagation with learned higher-order markov random fields," in *Proc. European Conference on Computer Vision (ECCV2006)*, vol. 2, 2006, pp. 269–282.

[89] B. Potetz, "Efficient belief propagation for vision using linear constraint nodes," in *Proc. IEEE Computer Society Conference on Computer Vision and Pattern Recognition (CVPR2007)*, 2007.

[90] C. Rother, P. Kohli, W. Feng, and J. Jia, "Minimizing sparse higher order energy functions of discrete variables," in *Proc. IEEE Computer Society Conference on Computer Vision and Pattern Recognition (CVPR2009)*, 2009, pp. 1382–1389.

[91] N. Komodakis and N. Paragios, "Beyond pairwise energies: Efficient optimization for higher-order MRFs," in *Proc. IEEE Computer Society Conference on Computer Vision and Pattern Recognition (CVPR2009)*, 2009, pp. 2985–2992.

[92] H. Ishikawa, "Higher-order clique reduction in binary graph cut," in *Proc. IEEE Computer Society Conference on Computer Vision and Pattern Recognition (CVPR2009)*, 2009, pp. 2993–3000.

[93] ——, "Transformation of general binary MRF minimization to the first order case," *IEEE Trans. Pattern Analysis and Machine Intelligence*, 2011.

[94] D. Freedman and P. Drineas, "Energy minimization via graph cuts: Settling what is possible," in *Proc. IEEE Computer Society Conference on Computer Vision and Pattern Recognition (CVPR2005)*, vol. 2, 2005, pp. 939–946.

[95] E. Boros and P. L. Hammer, "Pseudo-boolean optimization," *Discrete Appl. Math.*, vol. 123, pp. 155–225, November 2002.

[96] P. L. Hammer and S. Rudeanu, *Boolean Methods in Operations Research and Related Areas*. Berlin, Heidelberg, New York: Springer-Verlag, 1968.

[97] S. Ramalingam, P. Kohli, K. Alahari, and P. H. S. Torr, "Exact inference in multi-label CRFs with higher order cliques," in *Proc. IEEE Computer Society Conference on Computer Vision and Pattern Recognition (CVPR2008)*, 2008.

[98] D. Schlesinger and B. Flach, "Transforming an arbitrary min-sum problem into a binary one," Dresden University of Technology, Tech. Rep. TUD-FI06-01, 2006.

[99]   D. Schlesinger, "Exact solution of permuted submodular MinSum problems," in *Proc. Int. Conference on Energy Minimization Methods in Computer Vision and Pattern Recognition (EMMCVPR2007)*, 2007, pp. 28–38.

[100] J. Darbon, "Global optimization for first order Markov random fields with submodular priors," in *12th Int. Workshop on Combinatorial Image Analysis*, 2008.

[101] T. Pock, T. Schoenemann, G. Graber, H. Bischof, and D. Cremers, "A convex formulation of continuous multi-label problems," in *Proc. European Conference on Computer Vision (ECCV2008)*, vol. 3, 2008, pp. 792–805.

[102] C. Zach, M. Niethammer, and J.-M. Frahm, "Continuous maximal flows and Wulff shapes: Application to MRFs," in *Proc. IEEE Computer Society Conference on Computer Vision and Pattern Recognition (CVPR2009)*, 2009, pp. 1382–1389.

[103] V. Kolmogorov and A. Shioura, "New algorithms for convex cost tension problem with application to computer vision," *Discrete Optim.*, vol. 6, no. 4, pp. 378–393, 2009.

[104] A. Chambolle, "Total variation minimization and a class of binary MRF models," in *Proc. Int. Workshop on Energy Minimization Methods in Computer Vision and Pattern Recognition (EMMCVPR2005)*, 2005, pp. 136–152.

[105] J. Darbon and M. Sigelle, "Image restoration with discrete constrained total variation part I: Fast and exact optimization," *J. Math. Imaging Vis.*, vol. 26, no. 3, 2006.

[106] V. Kolmogorov and R. Zabih, "Computing visual correspondence with occlusions using graph cuts," in *Proc. IEEE International Conference on Computer Vision (ICCV2001)*, vol. 2, 2001, pp. 508–515.

[107] P. Carr and R. Hartley, "Solving multilabel graph cut problems with multilabel swap," in *Digital Image Computing: Techniques and Applications*, 2009.

[108] M. Kumar and P. H. S. Torr, "Improved moves for truncated convex models," in *Proc. Neural Information Processing Systems (NIPS2008)*, 2008, pp. 889–896.

[109] S. Gould, F. Amat, and D. Koller, "Alphabet soup: A framework for approximate energy minimization," in *Proc. IEEE Computer Society Conference on Computer Vision and Pattern Recognition (CVPR2009)*, 2009, pp. 903–910.

[110] V. Lempitsky, C. Rother, and A. Blake, "Logcut — efficient graph cut optimization for markov random fields," in *Proc. IEEE International Conference on Computer Vision (ICCV2007)*, 2007.

[111] V. Lempitsky, S. Roth, and C. Rother, "Fusionflow: Discrete-continuous optimization for optical flow estimation," in *Proc. IEEE Computer Society Conference on Computer Vision and Pattern Recognition (CVPR2008)*, 2008.

[112] H. Ishikawa, "Higher-order gradient descent by fusion-move graph cut," in *Proc. IEEE International Conference on Computer Vision (ICCV2009)*, 2009, pp. 568–574.

[113] O. Veksler, "Graph cut based optimization for MRFs with truncated convex priors," in *Proc. IEEE Computer Society Conference on Computer Vision and Pattern Recognition (CVPR2007)*, 2007.

[114] ——, "Multi-label moves for MRFs with truncated convex priors," in *Proc. International Conference on Energy Minimization Methods in Computer Vision and Pattern Recognition (EMMCVPR2009)*, 2009, pp. 1–13.

[115] N. Komodakis and G. Tziritas, "A new framework for approximate labeling via graph-cuts," in *Proc. IEEE International Conference on Computer Vision (ICCV2005)*, vol. 2, 2005, pp. 1018–1025.

[116] ——, "Approximate labeling via graph-cuts based on linear programming," *IEEE Trans. Pattern Analysis and Machine Intelligence*, vol. 29, no. 8, pp. 1436–1453, 2007.

[117] N. Komodakis, G. Tziritas, and N. Paragios, "Fast, approximately optimal solutions for single and dynamic MRFs," in *Proc. IEEE Computer Society Conference on Computer Vision and Pattern Recognition (CVPR2007)*, 2007.

[118] Y. Boykov and O. Veksler, "Graph cuts in vision and graphics: Theories and applications," in *Handbook of Mathematical Models in Computer Vision*, N. Paragios, Y. Chen, and O. Faugeras, Eds. Springer-Verlag, 2006, pp. 79–96.

[119] A. Blake, P. Kohli, and C. Rother, Eds., *Advances in Markov Random Fields for Vision and Image Processing*. MIT Press, 2011.

3

Higher-Order Models in Computer Vision

Pushmeet Kohli

Machine Learning and Perception
Microsoft Research
Cambridge, UK
Email: pkohli@microsoft.com

Carsten Rother

Machine Learning and Perception
Microsoft Research
Cambridge, UK
Email: carrot@microsoft.com

CONTENTS

3.1 Introduction 65
    3.1.1 Outline 67
3.2 Higher-Order Random Fields 67
3.3 Patch and Region-Based Potentials 68
    3.3.1 Label Consistency in a Set of Variables 69
    3.3.2 Pattern-Based Potentials 71
3.4 Relating Appearance Models and Region-Based Potentials 76
3.5 Global Potentials 79
    3.5.1 Connectivity Constraint 79
    3.5.2 Constraints and Priors on Label Statistics 81
3.6 Maximum a Posteriori Inference 84
    3.6.1 Transformation-Based Methods 85
    3.6.2 Dual Decomposition 88
3.7 Conclusions and Discussion 89
    Bibliography 90

3.1 Introduction

Many computer vision problems such as object segmentation, disparity estimation, and 3D reconstruction can be formulated as pixel or voxel labeling problems. The conventional methods for solving these problems use pairwise conditional and markov random field (CRF/MRF) formulations [1], which allow for the exact or approximate inference of maximum a posteriori (MAP) solutions. MAP inference is performed using extremely efficient algorithms such as combinatorial methods (e.g., graph-cut [2, 3, 4] or the BHS-algorithm [5, 6]), or message passing-based techniques (e.g., belief propagation (BP) [7, 8, 9] or tree-reweighted (TRW) message passing [10, 11]).

The classical formulations for image-labeling problems represent all output elements using random variables. An example is the problem of interactive object cutout where each pixel is represented using a random variable which can take two possible labels: foreground or background. The conventionally used *pairwise random field* models introduce a statistical relationship between pairs of random variables, often only among the immediate 4 or 8 neighboring pixels. Although such models permit efficient inference, they have restricted expressive power. In particular, they are unable to enforce the high-level structural dependencies between pixels that have been shown to be extremely powerful for image-labeling problems. For instance, while segmenting an object in 2D or 3D, we might know that all its pixels (or parts) are connected. Standard pairwise MRFs or CRFs are not able to guarantee that their solutions satisfy such a constraint. To overcome this problem, a global potential function is needed which assigns all such invalid solutions a zero probability or an infinite energy.

Despite substantial work from several communities, pairwise MRF and CRF models for computer vision problems have not been able to solve image-labeling problems such as object segmentation fully. This has led researchers to question the richness of these classical pairwise energy-function-based formulations, which in turn has motivated the development of more sophisticated models. Along these lines, many have turned to the use of higher-order models that are more expressive, thereby enabling the capture of statistics of natural images more closely.

The last few years have seen the successful application of higher-order CRFs and MRFs to some low-level vision problems such as image restoration, disparity estimation and object segmentation [12, 13, 14, 15, 16, 17, 18, 19, 20, 21, 22, 23]. Researchers have used models composed of new families of higher-order potentials, i.e., potentials defined over multiple variables, which have higher modeling power and lead to more accurate models of the problem. Researchers have also investigated incorporation of constraints such as connectivity of the segmentation in CRF and MRF models. This is done by including higher-order or global potentials[1] that assign zero probability (infinite cost) to all label configurations that do not satisfy these constraints.

One of the key challenges with respect to higher-order models is the question of efficiently inferring the Maximum a Posterior (MAP) solution. Since, inference in pairwise models is very well studied, one popular technique is to transform the problem back to a pairwise random field. Interestingly, any higher-order function can be converted to a pairwise one, by introducing additional auxiliary random variables [5, 24]. Unfortunately, the number of auxiliary variables grows exponentially with the arity of the higher-order function, hence in practice only higher-order function with a few variables can be handled efficiently. However, if the higher-order

---

[1]Potentials defined over all variables in the problem are referred to as *global potentials*.

function contains some inherent "structure", then it is indeed possible to practically perform MAP inference in a higher-order random field where each higher-order function may act on thousands of variables [25, 15, 26, 21]. We will review various examples of such potential functions in this chapter.

There is a close relationship between higher-order random fields and random field models containing latent variables [27, 19]. In fact, as we will see later in the chapter, any higher-order model can be written as a pairwise model with auxiliary latent variables and vice versa [27]. Such transformations enable the use of powerful optimization algorithms and even result in global optimally solutions for some problem instances. We will explain the connection between higher-order models and models containing latent variables, using the problem of interactive foreground/background image segmentation as an example [28, 29].

3.1.1 Outline

This chapter deals with higher-order graphical models and their applications. We discuss a number of recently proposed higher-order random field models and the associated algorithms that have been developed to perform MAP inference in them. The structure of the chapter is as follows.

We start with a brief introduction of higher-order models in Section 3.2. In Section 3.3, we introduce a class of higher-order functions which encode interactions between pixels belonging to image patches or regions. In Section 3.4 we relate the conventional latent variable CRF model for interactive image segmentation [28] to a random field model with region-based higher-order functions. Section 3.5 discusses models which encode image-wide (global) constraints. In particular, we discuss the problem of image segmentation under a connectivity constraint and solving labeling problems under constraints on label-statistics. In Section 3.6, we discuss algorithms that have been used to perform MAP inference in such models. We concentrate on two categories of techniques: the transformation approach, and the problem (dual) decomposition approach. We also give pointers to many other inference techniques for higher-order random fields such as message passing [18, 21]. For topics on higher-order model that are not discussed in this chapter, we refer the reader to [30, 31].

3.2 Higher-Order Random Fields

Before proceeding further, we provide the basic notation and definitions used in this chapter. A random field is defined over a set of random variables $\mathbf{x} = \{x_i | i \in \mathcal{V}\}$. These variables are typically arranged on a lattice $\mathcal{V} = \{1, 2, ..., n\}$ and represent scene elements, such as pixels or voxels. Each random variable takes a value from the label set $\mathcal{L} = \{l_1, l_2, \ldots, l_k\}$. For example, in scene segmentation, the labels can represent semantic classes such as a building, tree, or a person. Any possible assignment of labels to the random variables will be called a *labeling* (also denoted by **x**). Clearly, in the above scenario the total number of labelings **x** is k^n.

An MRF or CRF model enforces a particular factorization of the posterior distribution $P(\mathbf{x}|\mathbf{d})$, where **d** is the observed data (e.g., RGB input image). It is common to define an MRF or CRF model through its so-called Gibbs energy function $E(\mathbf{x})$, which is the negative log of the posterior distribution of the random field, i.e.,

$$E(\mathbf{x}; \mathbf{d}) = -\log P(\mathbf{x}|\mathbf{d}) + \text{constant.} \qquad (3.1)$$

The energy (cost) of a labeling **x** is represented as a sum of potential functions, each of which depends on a subset of random variables. In its most general form, the energy function can be defines as:

$$E(\mathbf{x}; \mathbf{d}) = \sum_{c \in \mathcal{C}} \psi_c(\mathbf{x}_c). \qquad (3.2)$$

Here, c is called a *clique* which is a set of random variables $\mathbf{x}_c$ which are conditionally dependent on each other. The term $\psi_c(\mathbf{x}_c)$ denotes the value of the clique potential corresponding to the labeling $\mathbf{x}_c \subseteq \mathbf{x}$ for the clique c, and $\mathcal{C}$ is the set of all cliques. The degree of the potential $\psi_c(\cdot)$ is the size of the corresponding clique c (denoted by $|c|$). For example, a pairwise potential has $|c| = 2$.

For the well-studied special case of pairwise MRFs, the energy only consists of potentials of degree one and two, that is,

$$E(\mathbf{x}; \mathbf{d}) = \sum_{i \in \mathcal{V}} \psi_i(x_i; \mathbf{d}) + \sum_{(i,j) \in \mathcal{E}} \psi_{ij}(x_i, x_j). \qquad (3.3)$$

Here, $\mathcal{E}$ represents the set of pairs of variables which interact with each other. In the case of image segmentation, $\mathcal{E}$ may encode a 4-connected neighborhood system on the pixel-lattice.

Observe that the pairwise potential $\psi_{ij}(x_i, x_j)$ in Equation 3.3 does not depend on the image data. If we condition the pairwise potentials on the data, then we obtain a pairwise CRF model which is defined as:

$$E(\mathbf{x}; \mathbf{d}) = \sum_{i \in \mathcal{V}} \psi_i(x_i; \mathbf{d}) + \sum_{(i,j) \in \mathcal{E}} \psi_{ij}(x_i, x_j; \mathbf{d}). \qquad (3.4)$$

3.3   Patch and Region-Based Potentials

In general, it is computationally infeasible to exactly represent a general higher order potential function defined over many variables[2]. Some researchers have proposed higher-order models for vision problems which use potentials defined over a relatively small number of variables. Examples of such models include the work of Woodford et al. [23] on disparity estimation using a third-order smoothness potential, and El-Zehiry and Grady [12] on image segmentation with a curvature potential[3]. In such cases, it is also feasible to transform the higher-order energy into an equivalent pairwise energy function with the addition of a relatively small number of auxiliary variables, and minimize the resulting pairwise energy using conventional energy minimization algorithms.

Although the above-mentioned approach has been shown to produce good results, it is not able to deal with higher-order potentials defined over a very large number (hundreds or thousands) of variables. In the following, we present two categories of higher-order potentials that can be represented compactly and minimized efficiently. The first category encodes the property that pixels belonging to certain groups take the same label. While this is a powerful concept in several application domains, e.g., pixel-level object recognition, it is not always applicable, e.g., image denoising. The second category generalizes this idea by allowing groups of pixels to take arbitrary labelings, as long as the set of different labelings is small.

3.3.1   Label Consistency in a Set of Variables

A common method to solve various image-labeling problems like object segmentation, stereo and single view reconstruction is to formulate them using image segments (so-called superpixels [33]) obtained from unsupervised segmentation algorithms. Researchers working with these methods have made the observation that all pixels constituting the segments often have the same label, that is, they might belong to the same object or might have the same depth.

Standard super-pixel based methods use label consistency in super-pixels as a hard constraint. Kohli et al. [25] proposed a higher-order CRF model for image labeling that used label consistency in superpixels as a *soft constraint*. This was done by using higher-order potentials defined on the image segments generated using unsupervised segmentation algorithms. Specifically, they extend the standard pairwise

---

[2]Representation of a general m order potential function of k-state discrete variables requires k^m parameter values

[3]El-Zehiry and Grady [12] used potentials defined over 2×2 patches to enforce smoothness. Shekhovtsov et al. [32] have recently proposed a higher order model for encouraging smooth, low curvature image segmentations that uses potentials defined over much large sets of variables and was learnt using training data.

CRF model often used for object segmentation by incorporating higher-order potentials defined on sets or regions of pixels. In particular, they extend the pairwise CRF which is used in TextonBoost [34][4].

The Gibbs energy of the higher-order CRF of [25] can be written as:

$$E(\mathbf{x}) = \sum_{i \in \mathcal{V}} \psi_i(x_i) + \sum_{(i,j) \in \mathcal{E}} \psi_{ij}(x_i, x_j) + \sum_{c \in \mathcal{S}} \psi_c(\mathbf{x}_c), \qquad (3.5)$$

where $\mathcal{E}$ represents the set of all edges in a 4- or 8-connecting neighborhood system, $\mathcal{S}$ refers to a set of image segments (or superpixels), and ψ_c are higher-order *label consistency potentials* defined on them. In [25], the set $\mathcal{S}$ consisted of all segments of multiple segmentations of an image obtained using an unsupervised image segmentation algorithm such as mean-shift [35]. The labels constituting the label set $\mathcal{L}$ of the CRF represent the different objects. Every possible assignment of the random variables $\mathbf{x}$ (or configuration of the CRF) defines a segmentation.

The label consistency potential used in [25] is similar to the smoothness prior present in pairwise CRFs [36]. It favors all pixels belonging to a segment taking the same label. It takes the form of a P^n Potts model [15]:

$$\psi_c(\mathbf{x}_c) = \begin{cases} 0 & \text{if} \quad x_i = l_k, \forall i \in c, \\ \theta_1 |c|^{\theta_\alpha} & \text{otherwise.} \end{cases} \qquad (3.6)$$

where $|c|$ is the cardinality of the pixel set c[5], and θ_1 and θ_α are parameters of the model. The expression $\theta_1 |c|^{\theta_\alpha}$ gives the label inconsistency cost, i.e., the cost added to the energy of a labeling in which different labels have been assigned to the pixels constituting the segment. Figure 3.1(left) visualizes a P^n Potts potential.

The P^n Potts model enforces label consistency rigidly. For instance, if all but one of the pixels in a superpixel take the same label then the same penalty is incurred as if they were all to take different labels. Due to this strict penalty, the potential might not be able to deal with inaccurate super-pixels or resolve conflicts between overlapping regions of pixels. Kohli *et al.* [25] resolved this problem by using the *Robust* higher-order potentials defined as:

$$\psi_c(\mathbf{x}_c) = \begin{cases} N_i(\mathbf{x}_c)\frac{1}{Q}\gamma_{\max} & \text{if } N_i(\mathbf{x}_c) \le Q \\ \gamma_{\max} & \text{otherwise.} \end{cases} \qquad (3.7)$$

where $N_i(\mathbf{x}_c)$ denotes the number of variables in the clique c not taking the dominant label, i.e., $N_i(\mathbf{x}_c) = \min_k(|c| - n_k(\mathbf{x}_c))$, $\gamma_{\max} = |c|^{\theta_\alpha}(\theta_1 + \theta_2 G(c))$, where $G(c)$ is the measure of the quality of the superpixel c, and Q is the truncation parameter, which controls the rigidity of the higher-order clique potential. Figure 3.1(right) visualizes a robust P^n Potts potential.

---

[4]Kohli et al. ignore a part of the TextonBoost [34] energy that represents a global appearance model for each object-class. In Section 3.4 we will revisit this issue and show that in fact this global, appearance model is closely related to the higher-order potentials defined in [25].

[5]For the problem of [25], this is the number of pixels constituting super-pixel c.

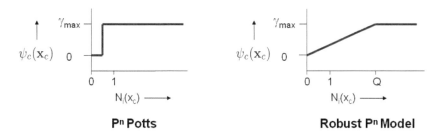

FIGURE 3.1
Behavior of the rigid P^n Potts potential (left) and the Robust P^n model potential (right). The figure shows how the cost enforced by the two higher-order potentials changes with the number of variables in the clique not taking the dominant label, i.e., $N_i(\mathbf{x}_c) = \min_k(|c| - n_k(\mathbf{x}_c))$, where $n_k(.)$ returns the number of variables x_i in $\mathbf{x}_c$ that take the label k. Q is the truncation parameter used in the definition of the higher-order potential (see equation 3.7).

Unlike the standard P^n Potts model, this potential function gives rise to a cost that is a linear truncated function of the number of inconsistent variables (see Figure 3.1). This enables the robust potential to allow some variables in the clique to take different labels. Figure 3.2 shows results for different models.

Lower-Envelope Representation of Higher-Order Functions

Kohli and Kumar [24] showed that many types of higher-order potentials including the Robust P^n model can be represented as lower envelopes of linear functions. They also showed that the minimization of such potentials can be transformed to the problem of minimizing a pairwise energy function with the addition of a small number of auxiliary variables which take values from a small label set.

It can be easily seen that the Robust P^n model (3.7) can be written as a lower envelope potential using $h + 1$ linear functions. The functions $f^q, q \in \mathcal{Q} = \{1, 2, \ldots, h + 1\}$ are defined using

$$\mu^q = \begin{cases} \gamma_a & \text{if } q = a \in \mathcal{L}, \\ \gamma_{\max} & \text{otherwise,} \end{cases}$$

$$w^q_{ia} = \begin{cases} 0 & \text{if } q = h + 1 \text{ or } a = q \in \mathcal{L}, \\ \alpha_a & \text{otherwise.} \end{cases}$$

The above formulation is illustrated in Figure 3.3 for the case of binary variables.

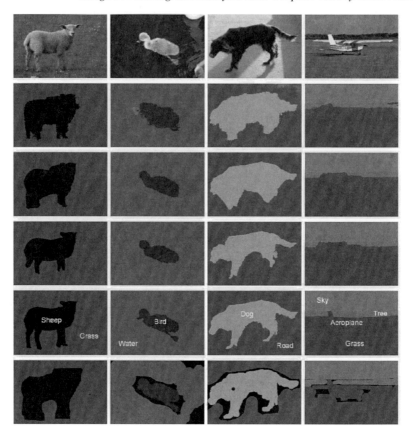

FIGURE 3.2

Some qualitative results. Please view in colour. First row: Original image. Second row: Unary likelihood labeling from TextonBoost [34]. Third row: Result obtained using a pairwise contrast preserving smoothness potential as described in [34]. Fourth row: Result obtained using the P^n Potts model potential [15]. Fifth row: Results using the Robust P^n model potential (3.7) with truncation parameter $Q = 0.1|c|$, where $|c|$ is equal to the size of the super-pixel over which the Robust P^n higher-order potential is defined. Sixth row: Hand-labeled segmentations. The ground truth segmentation are not perfect and many pixels (marked black) are un-labelled. Observe that the Robust P^n model gives best results. For instance, the leg of the sheep and bird have been accurately labeled, which was missing in the other results.

3.3.2   Pattern-Based Potentials

The potentials in the previous section were motivated by the fact that often a group of pixels have the same labeling. While this is true for a group of pixels which is inside

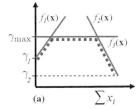 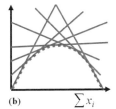

FIGURE 3.3

(a) Robust P^n model for binary variables. The linear functions f_1 and f_2 represents the penalty for variables not taking the labels 0 and 1, respectively. The function f_3 represents the robust truncation factor. (b) The general concave form of the robust P^n model defined using a larger number of linear functions.

an object, it is violated for a group which encodes a transitions between objects. Furthermore, the label consistency assumption is also not useful when the labeling represents, e.g., natural textures. In the following we will generalize the label-consistency potentials to so-called pattern-based potentials, which can model arbitrary labelings. Unfortunately, this generalization also implies that the underlying optimization will become harder (see Section 3.6).

Suppose we had a dictionary containing all possible 10×10 patches that are present in natural real-world images. One could use this dictionary to define a higher-order prior for the image restoration problem which can be incorporated in the standard MRF formulation. This higher-order potential is defined over sets of variables, where each set corresponds to a 10×10 image patch. It enforces that patches in the restored image come from the set of natural image patches. In other words, the potential function assigns a low cost (or energy) to the labelings that appear in the dictionary of natural patches. The rest of the labelings are given a high (almost constant) cost.

It is well known that only a small fraction of all possible labelings of a 10×10 patch actually appear in natural images. Rother *et al.* [26] used this sparsity property to compactly represent a higher-order potential prior for binary texture denoising by storing only the labelings that need to be assigned a low cost, and assigning a (constant) high cost to all other labelings.

They parameterize higher-order potentials by a list of possible labelings (also called patterns [37]) $\mathcal{X} = \{\mathbf{X}_1, \mathbf{X}_2, ..., \mathbf{X}_t\}$ of the clique variables $\mathbf{x}_c$, and their corresponding costs $\theta = \{\theta_1, \theta_2, ..., \theta_t\}$. They also include a high constant cost $\theta_{\max}$ for all other labelings. Formally, the potential functions can be defined as:

$$\psi_c(\mathbf{x}_c) = \begin{cases} \theta_q & \text{if } \mathbf{x}_c = \mathbf{X}_q \in \mathcal{X} \\ \theta_{\max} & \text{otherwise}, \end{cases} \tag{3.8}$$

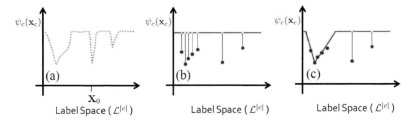

FIGURE 3.4
Different parameterizations of higher-order potentials. (a) The original higher-order potential function. (b) Approximating pattern-based potential which requires the definition of 7 labelings. (c) The compact representation of the higher-order function using the functional form defined in Equation (3.9). This representation (3.9) requires only $t = 3$ deviation functions.

where $\theta_q \leq \theta_{max}, \forall \theta_q \in \theta$. The higher-order potential is illustrated in Figure 3.4(b). This representation was concurrently proposed by Komodakis et al. [37].

Soft Pattern Potentials

The pattern-based potential is compactly represented and allows efficient inference. However, the computation cost is still quite high for potentials which assign a low cost to many labelings. Notice that the pattern-based representation requires one pattern per low-cost labeling. This representation cannot be used for higher-order potentials where a large number of labelings of the clique variables are assigned low weights ($< \theta_{max}$).

Rother et al. [26] observed that many low-cost label assignments tend to be close to each other in terms of the difference between labelings of pixels. For instance, consider the image segmentation task, which has two labels, foreground (f) and background (b). It is conceivable that the cost of a segmentation labeling ($fffb$) for 4 adjacent pixels on a line would be close to the cost of the labeling ($ffbb$). This motivated them to try to encode the cost of such groups of *similar* labelings in the higher-order potential in such a way that their transformation to quadratic functions does not require increasing the number of states of the switching variable z (see details in Section 3.6). The differences of the representations are illustrated in Figure 3.4(b) and (c).

They parameterized the compact higher-order potentials by a list of labeling deviation cost functions $\mathcal{D} = \{d_1, d_2, ..., d_t\}$, and a list of associated costs $\theta = \{\theta_1, \theta_2, ..., \theta_t\}$. They also maintain a parameter for the maximum cost θ_{max} that the potential can assign to any labeling. The deviation cost functions encode how the cost changes as the labeling moves away from some desired labeling. Formally, the

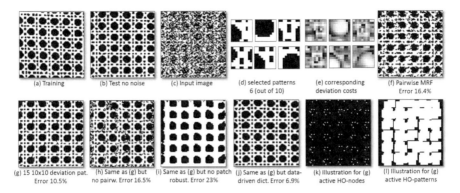

(a) Training   (b) Test no noise   (c) Input image   (d) selected patterns 6 (out of 10)   (e) corresponding deviation costs   (f) Pairwise MRF Error 16.4%

(g) 15 10x10 deviation pat. Error 10.5%   (h) Same as (g) but no pairw. Error 16.5%   (i) Same as (g) but no patch robust. Error 23%   (j) Same as (g) but data-driven dict. Error 6.9%   (k) Illustration for (g) active HO-nodes   (l) Illustration for (g) active HO-patterns

FIGURE 3.5

Binary texture restoration for Brodatz texture D101. (a) Training image (86×86 pixels). (b) Test image. (c) Test image with 60% noise, used as input. (d) 6 (out of 10) selected patterns of size 10×10 pixels. (e) Their corresponding deviation cost function. (f–j) Results of various different models (see text for details).

potential functions can be defined as:

$$\psi_c(\mathbf{x}_c) = \min\{\min_{q \in \{1,2,\ldots,t\}} \theta_q + d_q(\mathbf{x}_c), \theta_{\max}\}, \qquad (3.9)$$

where deviation functions $d_q : \mathcal{L}^{|c|} \to \mathbb{R}$ are defined as: $d_q(\mathbf{x}_c) = \sum_{i \in c; l \in \mathcal{L}} w_{il}^q \delta(x_i = l)$, where w_{il}^q is the cost added to the deviation function if variable x_i of the clique c is assigned label l. The function $\delta(x_i = l)$ is the Kronecker delta function that returns value 1 if $x_i = l$ and returns 0 for all assignments of x_i. This higher-order potential is illustrated in Figure 3.4(c). It should be noted that the higher-order potential (3.9) is a generalization of the pattern-based potential defined in Equation (3.8) and in [37]. Setting weights w_{il}^q as:

$$w_{il}^q = \begin{cases} 0 & \text{if } \mathbf{X}_q(i) = l \\ \theta_{\max} & \text{otherwise} \end{cases} \qquad (3.10)$$

makes potential (3.9) equivalent to Equation (3.8).

Note, that the above pattern-based potentials can also be used to model arbitrary higher-order potentials, as done in, e.g., [38], as long as the size of the clique is small.

Pattern-Based Higher-Order Potentials for Binary Texture Denoising

Pattern-based potentials are especially important for computer vision since many image labeling problems in vision are dependent on good prior models of patch labelings. In existing systems, such as new view synthesis, e.g., [13], or super-resolution, e.g., [39], patch-based priors are used in approximate ways and do not directly solve the underlying higher-order random field.

Rother et al. [26] demonstrated the power of the pattern-based potentials for the toy task of denoising a specific type of binary texture, i.e., Brodatz texture D101[6]. Given a training image, Figure 3.5(a), their goal was to denoise the input image (c) to achieve ideally (b). To derive the higher-order potentials, they selected a few patterns of size 10×10 pixels, which occur frequently in the training image (a) and are as different as possible in terms of their Hamming distance. They achieve this by k-means clustering over all training patches. Figure 3.5(d) depicts 6 (out of $k = 10$) such patterns.

To compute the deviation function for each particular pattern, they considered all patterns that belong to the same cluster. For each position within the patch, they record the frequency of having the same value. Figure 3.5(e) shows the associate deviation costs, where a bright value means low frequency (i.e., high cost). As expected, lower costs are at the edge of the pattern. Note, the scale and truncation of the deviation functions, as well as the weight of the higher-order function with respect to unary and pairwise terms, are set by hand in order to achieve best performance. The results for various models are shown in Figure 3.5(f–l). (Please refer to [26] for a detailed description of each model.)

Figure 3.5(f) shows the result with a learned pairwise MRF. It is apparent that the structure on the patch-level is not preserved. In contrast, the result in Figure 3.5(g), which uses the soft higher-order potentials and pairwise function, is clearly superior. Figure 3.5(h) shows the result with the same model as in (g) but where pairwise terms are switched off. The result is not as good since those pixels which are not covered by a patch are unconstrained and hence take the optimal noisy labeling. Figure 3.5(i) shows the importance of having patch robustness, i.e., that $\theta_{\max}$ in Equation (3.9) is not infinite, which is missing in the classical patch-based approaches (see [26]). Finally, 3.5(j) shows a result with the same model as in (g) but with a different dictionary. In this case the 15 representative patches are different for each position in the image. To achieve this, they used the noisy input image and hence have a CRF model instead of an MRF model (see details in [26]).

Figure 3.5(k-l) visualizes the energy for the result in (g). In particular, 3.5(k) illustrates in black those pixels where the maximum (robustness) patch cost $\theta_{\max}$ is paid. It can be observed that only a few pixels do not utilize the maximum cost. Figure 3.5(l) illustrates all 10×10 patches that are utilized, i.e., each white dot in 3.5(k) relates to a patch. Note that there is no area in 3.5(l) where a patch could be used that does not overlap with any other patch. Also, note that many patches do overlap.

---

[6]This specific denoising problem has also been addressed previously in [40].

| (a) input | (b) EM procedure | (c) DD procedure (global opt.) |

FIGURE 3.6

Interactive image segmentation using the interactive segmentation method proposed in [29]. The user places a yellow rectangle around the object (a). The result (b) is achieved with the iterative EM-style procedure proposed [29]. The result (c) is the global optimum of the function, which is achieved by transforming the energy to a higher-order random field and applying a dual-decomposition (DD) optimization technique [19]. Note that the globally optimal result is visually superior.

3.4 Relating Appearance Models and Region-Based Potentials

As mentioned in the previous Section 3.3.1, there is a connection between robust P^n Potts potentials and the TextonBoost model [34] which contains variables that encode the appearance of the foreground and background regions in the image. In the following, we will analyze this connection, which was presented in the work of Vicente et al. [19].

TextonBoost [34] has an energy term that models for each object-class segmentation an additional parametric appearance model. The appearance model is derived at test-time for each image individually. For simplicity, let us consider the interactive binary segmentation scenario, where we know beforehand that only two classes (fore- and background) are present. Figure 3.6 explains the application scenario. It has been shown in many works that having an additional appearance model for both fore- and background give improved results [28, 29]. The energy of this model takes the form:

$$E(\mathbf{x}, \theta^0, \theta^1) = \sum_{i \in \mathcal{V}} \psi_i(x_i, \theta^0, \theta^1, \mathbf{d}_i) + \sum_{(i,j) \in \mathcal{E}} w_{ij}|x_i - x_j|. \qquad (3.11)$$

Here, $\mathcal{E}$ is the set of 4-connected neighboring pixels, and $x_i \in \{0, 1\}$ is the segmentation label of pixel i (where 0 corresponds to the background and 1 to the foreground). The first term of Equation 3.11 is the likelihood term, where $\mathbf{d}_i$ is the RGB color at site i and θ^0 and θ^1 are, respectively, the background and foreground color models. Note that the color models θ^0, θ^1 act globally on all pixels in the respective segment. The second term is the standard contrast-sensitive edge term, see [28, 29] for details.

The goal is to minimize the energy (3.11) jointly for $\mathbf{x}$, θ^0 and θ^1. In [29] this optimization was done in an iterative, EM-style fashion. It works by iterating the following steps: (i) Fix color models θ^0, θ^1 and minimize energy (3.11) over segmentation $\mathbf{x}$. (ii) Fix segmentation $\mathbf{x}$, minimize energy (3.11) over color models θ^0, θ^1. The first step is solved via a maxflow algorithm, and the second one via standard machine-learning techniques for fitting a model to data. Each step is guaranteed not to increase the energy, but, of course, the procedure may get stuck in a local minimum, as shown in Figure 3.6(b).

In the following we show that the global variables can be eliminated by introducing global region-based potentials in the energy. This, then, allows for more powerful optimization techniques, in particular the dual-decomposition procedure. This procedure provides empirically a global optimum in about 60% of cases: see one example in Figure 3.6(c).

In [19] the color models were expressed in the form of histograms. We assume that the histogram has K bins indexed by $k = 1, ..., K$. The bin in which pixel i falls is denoted as b_i, and $\mathcal{V}_k \subseteq \mathcal{V}$ denotes the set of pixels assigned to bin k. The vectors θ^0 and θ^1 in $[0, 1]^K$ represent the distribution over fore- and background, respectively, and sum to 1. The likelihood model is then given by

$$\psi_i(x_i, \theta^0, \theta^1, d_i) = \sum_i - \log \theta_{b_i}^{x_i}, \tag{3.12}$$

where $\theta_{b_i}^{x_i}$ represents the likelihood of observing a pixel belonging to bin b_i which takes label x_i.

Rewriting the Energy via High-Order Cliques

Let us denote n_k^s to be the number of pixels i that fall into bin k and have label s, i.e., $n_k^s = \sum_{i \in \mathcal{V}_k} \delta(x_i - s)$. All these pixels contribute the same cost $- \log \theta_k^s$ to the term $\psi_i(x_i, \theta^0, \theta^1, d_i)$, therefore we can rewrite it as

$$\psi_i(x_i, \theta^0, \theta^1, d_i) = \sum_s \sum_k -n_k^s \log \theta_k^s. \tag{3.13}$$

It is well-known that for a given segmentation, $\mathbf{x}$ distributions θ^0 and θ^1 that minimize $\psi_i(x_i, \theta^0, \theta^1, d_i)$ are simply the empirical histograms computed over the appropriate segments: $\theta_k^s = n_k^s/n^s$ where n^s is the number of pixels with label $s : n^s = \sum_{i \in \mathcal{V}} \delta(x_i - s)$. Plugging optimal θ^0 and θ^1 into the energy (3.11) gives the following expression:

$$E(\mathbf{x}) = \min_{\theta^0, \theta^1} E(x, \theta^0, \theta^1) \tag{3.14}$$

$$= \sum_k h_k(n_k^1) + h(n^1) + \sum_{(i,j) \in \mathcal{E}} w_{ij}|x_i - x_j|, \tag{3.15}$$

$$h_k(n_k^1) = -g(n_k^1) - g(n_k - n_k^1) \tag{3.16}$$

$$h(n^1) = g(n^1) + g(n - n^1), \tag{3.17}$$

where $g(z) = z \, \log(z), n_k = |\mathcal{V}_k|$ is the number of pixels in bin k, and $n = |\mathcal{V}|$ is the total number of pixels.

It is easy to see that functions $h_k(\cdot)$ are concave and symmetric about $n_k/2$, and function $h(\cdot)$ is convex and symmetric about $n/2$. Unfortunately, as we will see in Section 3.6, the convex part makes the energy hard to be optimized. The form of Equation(3.15) allows an intuitive interpretation of this model. The first term (sum of concave functions) has a preference towards assigning all pixels in $\mathcal{V}_k$ to the same segment. The convex part prefers balanced segmentations, i.e., segmentations in which the background and the foreground have the same number of pixels.

Relationship to Robust P^n Model for Binary Variables

The concave functions $h_k(\cdot)$, i.e., Equation (3.16) have the form of a robust P^n Potts model for binary variables as illustrated in Figure 3.3(b). There are two main differences between the model of [25] and [19]. Firstly, the energy of [19] has a balancing term (Equation 3.17). Secondly, the underlying super-pixel segmentation is different. In [19], all pixels in the image that have the same colour are deemed to belong a single superpixel, whereas in [25] superpixels are spatially coherent. An interesting future work is to perform an empirical comparison of these different models. In particular, the balancing term (Equation 3.17) may be weighted differently, which can lead to improved results (see examples in [19]).

3.5 Global Potentials

In this section we discuss higher-order potential functions, which act on all variables in the model. For image-labeling problems, this implies a potential whose cost is affected by the labeling of every pixel. In particular, we will consider two types of higher-order functions: ones which enforce topological constraints on the labeling, such as connectivity of all foreground pixels, and those whose cost depends on the frequency of assigned labels.

3.5.1 Connectivity Constraint

Enforcing connectivity of a segmentation is a very powerful global constraint. Consider Figure 3.7 where the concept of connectivity is used to build an interactive segmentation tool. To enforce connectivity we can simply write the energy as

$$E(\mathbf{x}) = \sum_{i \in \mathcal{V}} \psi_i(x_i) + \sum_{(i,j) \in \mathcal{E}} w_{ij} |x_i - x_j| \quad s.t. \ \mathbf{x} \ being \ connected, \quad (3.18)$$

| (a) User input | (b) Graph cut | (c) Graph cut reduced coherency | (d) Additional user input | (e) DijkstraGC |

FIGURE 3.7

Illustrating the connectivity prior from [41]. (a) Image with user-scribbles (green, foreground; blue, background). Image segmentation using graph cut with standard (b) and reduced coherency (c). None of the results are perfect. By enforcing that the foreground object is 4-connected, a perfect result can be achieved (e). Note: this result is obtained by starting with the segmentation in (b) and then adding the 5 foreground user-clicks (red crosses) in (d).

where connectivity can, for instance, be defined on the standard 4-neighborhood grid. Apart from the connectivity constraint, the energy is a standard pairwise energy for segmentation, as in Equation (3.11). In [20] a modified version of this energy is solved with each user interaction. Consider the result in Figure 3.7(b) that is obtained with the input in 3.7(a). Given this result, the user places one red cross, e.g., at the tip of the fly's leg (Figure 3.7(d)), to indicate another foreground pixel. The algorithm in [20] then has to solve the subproblem of finding a segmentation where both islands (body of the fly and red cross) are connected. For this a new method called DijkstraGC was developed, which combines the shortest-path Dijkstra algorithm and graph cut. In [20] it is also shown that for some practical instances DijkstraGC is globally optimal. It is worth commenting that the connectivity constraint enforces a different from of regularization compared to standard pairwise terms. Hence, in practice the strength of the pairwise terms may be chosen differently when the connectivity constraint potential is used.

The problem of minimizing the energy (3.18) directly has been addressed in Nowozin et al. [42], using a constraint-generation technique. They have shown that enforcing connectivity does help object recognition systems. Very recently the idea of connectivity was used for 3D reconstruction, i.e., to enforce that objects are connected in 3D, see details in [41].

Bounding Box Constraint

Building on the work [42], Lempitsky et al. [43] extended the connectivity constraint to the so-called bounding box prior. Figure 3.8 gives an example where the bounding box prior helps to achieve a good segmentation result.

The bounding box prior is formalized with the following energy

$$E(\mathbf{x}) = \sum_{i \in \mathcal{V}} \psi_i(x_i) + \sum_{(i,j) \in \mathcal{E}} w_{ij}|x_i - x_j| \quad s.t. \ \forall C \in \Gamma \ \sum_{i \in C} x_i \geq 1, \quad (3.19)$$

FIGURE 3.8
Bounding box prior. (Left) Typical result of the image segmentation method proposed in [29] where the user places the yellow bounding box around the object, which results in the blue segmentation. The result is expected since in the absence of additional prior knowledge the head of the person is more likely background, due to the dark colors outside the box. (Right) Improved result after applying the bounding box constraint. It enforces that the segmentation is spatially "close" to the four sides of the bounding box and also that the segmentation is connected.

where Γ is the set of all 4-connected "crossing" paths. A crossing path C is a path that goes from the top to the bottom side of the box, or from the left to the right side. Hence, the constraint in (3.19) forces that along each path C, there is at least one foreground pixel. This constraint makes sure that there exist a segmentation that touches all 4 sides of the bounding box and that is also 4-connected. As in [42], the problem is solved by first relaxing it to continuous labels, i.e., $x_i \in [0, 1]$, and then applying a constraint-generation technique, where each constraint is a crossing path which violates the constraint in Equation (3.19). The resulting solution is then converted back to an integer solution, i.e., $x_i \in \{0, 1\}$, using a rounding schema called *pin-pointing*; see details in [43].

3.5.2 Constraints and Priors on Label Statistics

A simple and useful global potential is a cost based on the number of labels that are present in the final output. In its most simple form, the corresponding energy function has the form:

$$E(\mathbf{x}) = \sum_{i \in \mathcal{V}} \psi_i(x_i) + \sum_{(i,j) \in \mathcal{E}} \psi_{ij}(x_i, x_j) + \sum_{l \in L} c_l [\exists i : x_i = l], \qquad (3.20)$$

where L is the set of all labels, c_l is the cost for each label, and $[arg]$ is 1 if arg is true, and 0 otherwise. The above defined energy prefers a simpler solution over a more complex one.

The label cost potential has been used successfully in various domains, such as stereo [44], motion estimation [45], object segmentation [14], and object-instance recognition [46]. For instance, in [44] a 3D scene is reconstructed by a set of surfaces (planes or B-splines). A reconstruction with fewer surfaces is preferred due to the

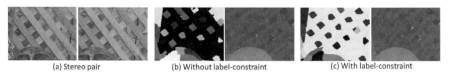

(a) Stereo pair   (b) Without label-constraint   (c) With label-constraint

FIGURE 3.9

Illustrating the label-cost prior. (a) Crop of a stereo image ("cones" image from Mid-dlebury database). (b) Result without a label-cost prior. In the left image each color represents a different surface, where gray-scale colors mark planes and nongray-scale colors, B-splines. The right image shows the resulting depth map. (c) Corre-sponding result with label-cost prior. The main improvement over (b) is that large parts of the green background (visible through the fence) are assigned to the same planar surface.

label-cost prior. One example where this prior helps is shown in Figure 3.9, where a plane, which is visible through a fence, is recovered as one plane instead of many planar fragments.

Various method have been proposed for minimizing the energy (3.20), including alpha-expansion (see details in [14, 45]). Extension to the energy defined in 3.20 have also been proposed. For instance, Ladicky et al. [14] have addressed the problem of minimizing energy functions containing a term whose cost is an arbitrary function of the set of labels present in the labeling.

Higher-Order Potentials Enforcing Label Counts

In many computer vision problems such as object segmentation or reconstruction, which are formulated in terms of labeling a set of pixels, we may know the number of pixels or voxels that can be assigned to a particular label. For instance, in the reconstruction problem, we may know the size of the object to be reconstructed. Such label count constraints are extremely powerful and have recently been shown to result in good solutions for many vision problems.

Werner [22] were one of the first to introduce constraints on label counts in en-ergy minimization. They proposed a n-ary maxsum diffusion algorithm for solving these problems, and demonstrated its performance on the binary image denoising problem. Their algorithm, however, could only produce solutions for some label counts. It was not able to guarantee an output for any arbitrary label count desired by the user. Kolmogorov et al. [47] showed that for submodular energy functions, the parametric maxflow algorithm [48] can be used for energy minimization with label counting constraints. This algorithm outputs optimal solutions for only few la-bel counts. Lim et al.[49] extended this work by developing a variant of the above algorithm they called *decomposed parametric maxflow*. Their algorithm is able to produce solutions corresponding to many more label counts.

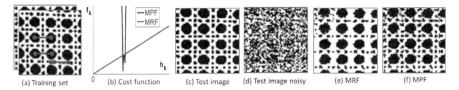

| (a) Training set | (b) Cost function | (c) Test image | (d) Test image noisy | (e) MRF | (f) MPF |

FIGURE 3.10

Illustrating the advantage of using an marginal probability field (MPF) over an MRF. (a) Set of training images for binary texture denoising. Superimposed is a pairwise term (translationally invariant with shift (15; 0); 3 exemplars in red). Consider the labeling $k = (1, 1)$ of this pairwise term ϕ. Each training image has a certain number $h_{(1,1)}$ of $(1, 1)$ labels, i.e. $h_{(1,1)} = \sum_i [\phi_i = (1, 1)]$. The negative logarithm of the statistics $\{h_{(1,1)}\}$ over all training images is illustrated in blue (b). It shows that all training images have roughly the same number of pairwise terms with label $(1, 1)$. The MPF uses the convex function f_k (blue) as cost function. It is apparent that the linear cost function of an MRF is a bad fit. (c) A test image and (d) a noisy input image. The result with an MRF (e) is inferior to the MPF model (f). Here the MPF uses a global potential on unary and pairwise terms.

| (a) Original | (b) Noisy Input | (c) MRF result | (d) MPF result | (e) Derivative histogram |

FIGURE 3.11

Results for image denoising using a pairwise MRF (c) and an MPF model (d). The MPF forces the derivatives of the solution to follow a mean distribution, which was derived from a large dataset. (f) Shows derivative histograms (discretized into the 11 bins). Here black is the target mean statistic, blue is for the original image (a), yellow for noisy input (b), green for the MRF result (c), and red for the MPF result (d). Note, the MPF result is visually superior and does also match the target distribution better. The runtime for the MRF (c) is 1096s and for MPF (d) 2446s.

Marginal Probability Fields

Finally, let us review the marginal probability field introduced in [50], which uses a global potential to overcome some fundamental limitations of maximum a posterior (MAP) estimation in Markov random fields (MRFs).

The prior model of a Markov random field suffers from a major drawback: the marginal statistics of the most likely solution (MAP) under the model generally do not match the marginal statistics used to create the model. Note, we refer to the marginal statistics of the cliques used in the model, which generally equates to those statistics deemed important. For instance, the marginal statistics of a single clique for a binary MRF are the number of 0s and 1s of the output labeling.

To give an example, given a corpus of binary training images that each contain 55% white and 45% black pixels (with no other significant statistic), a learned MRF prior will give each output pixel an independent probability of 0.55 of being white. Since the most likely value for each pixel is white, the most likely image under the model has 100% white pixels, which compares unfavorably with the input statistic of only 55%. When combined with data likelihoods, this model will therefore incorrectly bias the MAP solution towards being all white, the more so the greater the noise and hence data uncertainty.

The marginal probability field (MPF) overcomes this limitation. Formally, the MPF is defined as

$$E(\mathbf{x}) = \sum_k f_k \left(\sum_i [\phi_i(\mathbf{x}) = k] \right) , \qquad (3.21)$$

where $[arg]$ is defined as above, $\phi_i(x)$ returns the labeling of a factor at position i, k is an n-d vector, and f_k is the MPF cost kernel $\mathbb{R} \rightarrow \mathbb{R}^+$. For example, a pairwise factor of a binary random field has $|k| = 4$ possible states, i.e. $k \in \{(0,0), (0,1), (1,0), (1,1)\}$.

The key advantage of an MPF over an MRF is that the cost kernel f_k is arbitrary. In particular, by choosing a convex form for the kernel any arbitrary marginal statistics can be enforced. Figure 3.10 gives an example for binary texture denoising. Unfortunately, the underlying optimization problem is rather challenging, see details in [50]. Note that linear and concave kernels result in tractable optimization problems, e.g. for unary factors this has been described in Section 3.3.1 (see Figure 3.3(b) for a concave kernel for $\sum_i [x_i = 1]$).

The MPF can be used in many applications, such as denoising, tracking, segmentation, and image synthesis (see [50]). Figure 3.11 illustrates an example for image denoising.

3.6 Maximum a Posteriori Inference

Given an MRF, the problem of estimating the maximum a posteriori (MAP) solution can be formulated as finding the labeling $\mathbf{x}$ that has the lowest energy. Formally, this procedure (also referred to as energy minimization) involves solving the following problem:

$$\mathbf{x}^* = \arg \min_{\mathbf{x}} E(\mathbf{x}; \mathbf{d}). \qquad (3.22)$$

The problem of minimizing a general energy function is NP-hard in general, and remains hard even if we restrict the arity of potentials to 2 (pairwise energy functions). A number of polynomial time algorithms have been proposed in the literature for minimizing pairwise energy functions. These algorithms are able to find either exact solutions for certain families of energy functions or approximate solutions for general functions. These approaches can broadly be classified into two categories: message passing and move making. Message passing algorithms attempt to minimize approximations of the free energy associated with the MRF [10, 51, 52].

Move making approaches refer to iterative algorithms that *move* from one labeling to the other while ensuring that the energy of the labeling never increases. The move space (that is, the search space for the new labeling) is restricted to a subspace of the original search space that can be explored efficiently [2]. Many of the above approaches (both message passing [10, 51, 52] and move making [2]) have been shown to be closely related to the standard LP relaxation for the pairwise energy minimization problem [53].

Although there has been work on applying message passing algorithms for minimizing certain classes of higher-order energy functions [21, 18], the general problem has been relatively ignored. Traditional methods for minimizing higher-order functions involve either (1) converting them to a pairwise form by addition of auxiliary variables, followed by minimization using one of the standard algorithms for pairwise functions (such as those mentioned above) [5, 38, 37, 26], or (2) using dual-decomposition which works by decomposing the energy functions into different parts, solving them independently, and then merging the solution of the different parts [20, 50].

3.6.1 Transformation-Based Methods

As mentioned before, any higher-order function can be converted to a pairwise one by introducing additional auxiliary random variables [5, 24]. This enables the use of conventional inference algorithms such as belief propagation, tree-reweighted message passing, and graph cuts for such models. However, this approach suffers from the problem of combinatorial explosion. Specifically, a naive transformation can result in an exponential number of auxiliary variables (in the size of the corresponding clique) even for higher-order potentials with special structure [38, 37, 26].

In order to avoid the undesirable scenario presented by the naive transformation, researchers have recently started focusing on higher-order potentials that afford efficient algorithms [25, 15, 22, 23, 50]. Most of the efforts in this direction have been towards identifying useful families of higher-order potentials and designing algorithms specific to them. While this approach has led to improved results, its long-term impact on the field is limited by the restrictions placed on the form of the potentials. To address this issue, some recent works [24, 14, 45, 38, 37, 26] have attempted to characterize the higher-order potentials that are amenable to optimization. These works have successfully been able to exploit the *sparsity* of potentials and provide a convenient parameterization of tractable potentials.

Transforming Higher-Order Pseudo-Boolean Functions

The problem of transforming a general submodular higher-order function to a second order one has been well studied. Kolmogorov and Zabih [3] showed that all submodular functions of order three can be transformed to one of order two, and thus can be solved using graph cuts. Freedman and Drineas [54] showed how certain submodular higher-order functions can be transformed to submodular second order functions. However, their method, in the worst case, needed to add an exponential number of auxiliary binary variables to make the energy function second order.

The special form of the Robust P^n model (3.7) allows it to be transformed to a pairwise function with the addition of only two binary variables per higher-order potential. More formally, Kohli et al. [25] showed that higher-order pseudo-boolean functions of the form:

$$f(\mathbf{x}_c) = \min\left(\theta_0 + \sum_{i \in c} w_i^0(1 - x_i), \theta_1 + \sum_{i \in c} w_i^1 x_i, \theta_{\max}\right) \qquad (3.23)$$

can be transformed to submodular quadratic pseudo-boolean functions, and hence can be minimized using graph cuts. Here, $x_i \in \{0, 1\}$ are binary random variables, c is a clique of random variables, $\mathbf{x}_c \in \{0, 1\}^{|c|}$ denotes the labelling of the variables involved in the clique, and $w_i^0 \geq 0$, $w_i^1 \geq 0$, θ_0, θ_1, $\theta_{\max}$ are parameters of the potential satisfying the constraints $\theta_{\max} \geq \theta_0, \theta_1$, and

$$\left(\left(\theta_{\max} \leq \theta_0 + \sum_{i \in c} w_i^0(1 - x_i)\right) \vee \left(\theta_{\max} \leq \theta_1 + \sum_{i \in c} w_i^1 x_i\right)\right) = 1 \quad \forall \mathbf{x} \in \{0, 1\}^{|c|}$$
$$(3.24)$$

where $\vee$ is a boolean OR operator. The transformation to a quadratic pseudo-boolean function requires the addition of only two binary auxiliary variables, making it computationally efficient.

Theorem The higher-order pseudo-boolean function

$$f(\mathbf{x}_c) = \min\left(\theta_0 + \sum_{i \in c} w_i^0(1 - x_i), \theta_1 + \sum_{i \in c} w_i^1 x_i, \theta_{\max}\right) \qquad (3.25)$$

can be transformed to the submodular quadratic pseudo-boolean function

$$f(\mathbf{x}_c) = \min_{m_0, m_1}\left(r_0(1 - m_0) + m_0 \sum_{i \in c} w_i^0(1 - x_i) + r_1 m_1 + (1 - m_1) \sum_{i \in c} w_i^1 x_i - K\right)$$
$$(3.26)$$

by the addition of binary auxiliary variables m_0 and m_1. Here, $r_0 = \theta_{\max} - \theta_0$, $r_1 = \theta_{\max} - \theta_1$ and $K = \theta_{\max} - \theta_0 - \theta_1$. (See proof in [55]).

Multiple higher-order potentials of the form (3.23) can be summed together to obtain higher-order potentials of the more general form

$$f(\mathbf{x}_c) = F_c\left(\sum_{i \in c} x_i\right) \tag{3.27}$$

where $F_c : \mathbb{R} \to \mathbb{R}$ is any concave function. However, if the function F_c is convex (as discussed in Section 3.4) then this transformation scheme does not apply. Kohli and Kumar [24] have shown how the minimization of energy function containing higher-order potentials of the form 3.27 with convex functions F_c can be transformed to a compact max-min problem. However, this problem is computationally hard and does not lend itself to conventional maxflow based algorithms.

Transforming Pattern-Based Higher-Order Potentials

We will now describe the method used in [26] to transform the minimization of an arbitrary higher-order potential functions to the minimization of an equivalent quadratic function. We start with a simple example to motivate our transformation.

Consider a higher-order potential function that assigns a cost θ_0 if the variables $\mathbf{x}_c$ take a particular labeling $\mathbf{X}_0 \in \mathcal{L}^{|c|}$, and θ_1 otherwise. More formally,

$$\psi_c(\mathbf{x}_c) = \begin{cases} \theta_0 & \text{if} \quad \mathbf{x}_c = \mathbf{X}_0 \\ \theta_1 & \text{otherwise.} \end{cases} \tag{3.28}$$

where $\theta_0 \leq \theta_1$, and $\mathbf{X}_0$ denotes a particular labeling of the variables $\mathbf{x}_c$. The minimization of this higher-order function can be transformed to the minimization of a quadratic function using one additional *switching* variable z as:

$$\min_{\mathbf{x}_c} \psi_c(\mathbf{x}_c) = \min_{\mathbf{x}_c, z \in \{0,1\}} f(z) + \sum_{i \in c} g_i(z, x_i) \tag{3.29}$$

where the *selection* function f is defined as: $f(0) = \theta_0$ and $f(1) = \theta_1$, while the *consistency* function g_i is defined as:

$$g_i(z, x_i) = \begin{cases} 0 & \text{if} \quad z = 1 \\ 0 & \text{if} \quad z = 0 \text{ and } x_i = \mathbf{X}_0(i) \\ \text{inf} & \text{otherwise.} \end{cases} \tag{3.30}$$

where $\mathbf{X}_0(i)$ denotes the label of variable x_i in labeling $\mathbf{X}_0$.

Transforming Pattern-Based Higher-Order Potentials with Deviations

The minimization of a pattern-based potential with deviation functions (as defined in Section 3.3.2) can be transformed to the minimization of a pairwise function using a $(t + 1)$-state switching variable as:

$$\min_{\mathbf{x}_c} \psi_c(\mathbf{x}_c) = \min_{\mathbf{x}_c, z \in \{1,2,\dots,t+1\}} f(z) + \sum_{i \in c} g(z, x_i) \tag{3.31}$$

$$\text{where} \quad f(z) = \begin{cases} \theta_q & \text{if} \quad z = q \in \{1, .., t\} \\ \theta_{\max} & \text{if} \quad z = t + 1, \end{cases} \tag{3.32}$$

$$g_i(z, x_i) = \begin{cases} w_{il}^q & \text{if} \quad z = q \text{ and } x_i = l \in \mathcal{L} \\ 0 & \text{if} \quad z = t + 1. \end{cases} \tag{3.33}$$

The role of the switching variable in the above mentioned transformation can be seen as that of finding which *deviation* function will assign the lowest cost to any particular labeling. The reader should observe that the last, i.e., $(t + 1)^{th}$ state of the switching variable z does not penalize any labeling of the clique variables x_c. It should also be noted that the transformation method described above can be used to transform any general higher-order potential. However, in the worst case, the addition of a switching variable with $|\mathcal{L}|^{|c|}$ states is required, which makes minimization of even moderate order functions infeasible. Furthermore, in general the pairwise function resulting from this transformation is NP-hard.

3.6.2 Dual Decomposition

Dual decomposition has been successfully used for minimizing energy functions containing higher-order potentials. The approach works by decomposing the energy functions into different parts, solving them independently, and then merging the solution of the different parts. Since the merging step provides a lower bound on the original function, the process is repeated until the lower bound is optimal. For a particular task at hand the main question is on how to decompose the given problem into parts. This decomposition can have a major effect on the quality of the solution.

Let us explain the optimization procedure using the higher-order energy (3.15) for image segmentation. The function has the form

$$E(x) = \underbrace{\sum_k h_k(n_k^1) + \sum_{(i,j) \in \mathcal{E}} w_{ij}|x_i - x_j|}_{E^1(\mathbf{x})} + \underbrace{h(n^1)}_{E^2(\mathbf{x})}, \tag{3.34}$$

where $h_k(\cdot)$ are concave functions and $h(\cdot)$ is a convex function. Recall that n_k^1 and n^1 are functions of the segmentation: $n_k^1 = \sum_{i \in V_k} x_i$ and $n^1 = \sum_{i \in V} x_i$. It can be seen that the energy function (3.34) is composed of a submodular part ($E^1(\mathbf{x})$) and a supermodular ($E^2(\mathbf{x})$) part. As shown in [19] minimizing function (3.34) is an NP-hard problem.

We now apply the dual-decomposition technique to this problem. Let us rewrite the energy as

$$E(\mathbf{x}) = [E^1(\mathbf{x}) - \langle \mathbf{y}, \mathbf{x} \rangle] + [E^2(\mathbf{x}) + \langle \mathbf{y}, \mathbf{x} \rangle], \tag{3.35}$$

where $\mathbf{y}$ is a vector in $\mathbb{R}^n$, $n = |V|$, and $\langle \mathbf{y}, \mathbf{x} \rangle$ denotes the dot product between two vectors. In other words, we added unary terms to one subproblem and subtracted them from the other one. This is a standard use of the dual-decomposition approach for MRF optimization [56]. Taking the minimum of each term in (3.35), over **x** gives a lower bound on $E(\mathbf{x})$:

$$\phi(\mathbf{y}) = \underbrace{\min_{\mathbf{x}}[E^1(\mathbf{x}) - \langle \mathbf{y}, \mathbf{x}\rangle]}_{\phi^1(\mathbf{y})} + \underbrace{\min_{\mathbf{x}}[E^2(\mathbf{x}) + \langle \mathbf{y}, \mathbf{x}\rangle]}_{\phi^2(\mathbf{y})} \leq \min_{\mathbf{x}} E(\mathbf{x}) . \qquad (3.36)$$

Note that both minima, i.e., for $\phi^1(\mathbf{y})$ and $\phi^2(\mathbf{y})$, can be computed efficiently. In particular, the first term can be optimized via a reduction to an min s-t cut problem [25].

To get the tightest possible bound we need to maximize $\phi(\mathbf{y})$ over **y**. Function $\phi(\cdot)$ is concave, therefore, one could use some standard concave maximization technique, such as a subgradient method, which is guaranteed to converge to an optimal bound. In [19] it is shown that in this case the tightest bound can be computed in polynomial time using a parametric maxflow technique [47].

3.7 Conclusions and Discussion

In this chapter we reviewed a number of higher-order models for computer vision problems. We showed how the ability of higher-order models to encode complex statistics between pixels makes them an ideal candidate for image labelling problems. The focus of the chapter has been on models based on discrete variables. It has not covered some families of higher-order models such as fields of experts [17] and product of experts [57] that have been shown to lead to excellent results for problem such as image denoising.

We also addressed the inherent difficulty in representing higher-order models and in performing inference in them. Learning of higher-order models involving discrete variables has seen relatively little work, and should attract more research in the future.

Another family of models that are able to encode complex relationships between pixels are hierarchical models which contain latent variables. Typical examples of such models include deep belief nets (DBN) and restricted boltzmann machines (RBM). There are a number of interesting relationships between these models and higher-order random fields [27]. We believe the investigation of these relationships is a promising direction for future work. We believe this would lead to better understanding of the modeling power of both families of models, as well as lead to new insights which may help in the development of better inference and learning techniques.

Bibliography

[1] R. Szeliski, R. Zabih, D. Scharstein, O. Veksler, V. Kolmogorov, A. Agarwala, M. Tappen, and C. Rother, "A comparative study of energy minimization methods for Markov random fields." in *ECCV*, 2006, pp. 16–29.

[2] Y. Boykov, O. Veksler, and R. Zabih, "Fast approximate energy minimization via graph cuts." *PAMI*, vol. 23, no. 11, pp. 1222–1239, 2001.

[3] V. Kolmogorov and R. Zabih, "What energy functions can be minimized via graph cuts?." *PAMI*, vol. 26, no. 2, pp. 147–159, 2004.

[4] N. Komodakis, G. Tziritas, and N. Paragios, "Fast, approximately optimal solutions for single and dynamic MRFs," in *CVPR*, 2007, pp. 1–8.

[5] E. Boros and P. Hammer, "Pseudo-boolean optimization." *Discrete Appl. Math.*, vol. 123, no. 1-3, pp. 155–225, 2002.

[6] E. Boros, P. Hammer, and G. Tavares, "Local search heuristics for quadratic unconstrained binary optimization (QUBO)," *J. Heuristics*, vol. 13, no. 2, pp. 99–132, 2007.

[7] P. Felzenszwalb and D. Huttenlocher, "Efficient Belief Propagation for Early Vision," in *Proc. CVPR*, vol. 1, 2004, pp. 261–268.

[8] J. Pearl, "Fusion, propagation, and structuring in belief networks," *Artif. Intell.*, vol. 29, no. 3, pp. 241–288, 1986.

[9] Y. Weiss and W. Freeman, "On the optimality of solutions of the max-product belief-propagation algorithm in arbitrary graphs." *Trans. Inf. Theory*, 2001.

[10] V. Kolmogorov, "Convergent tree-reweighted message passing for energy minimization." *IEEE Trans. Pattern Anal. Mach. Intell.*, vol. 28, no. 10, pp. 1568–1583, 2006.

[11] M. Wainwright, T. Jaakkola, and A. Willsky, "Map estimation via agreement on trees: message-passing and linear programming." *IEEE Trans. Inf. Theory*, vol. 51, no. 11, pp. 3697–3717, 2005.

[12] N. Y. El-Zehiry and L. Grady, "Fast global optimization of curvature," in *CVPR*, 2010, pp. 3257–3264.

[13] A. Fitzgibbon, Y. Wexler, and A. Zisserman, "Image-based rendering using image-based priors." in *ICCV*, 2003, pp. 1176–1183.

[14] L. Ladicky, C. Russell, P. Kohli, and P. H. S. Torr, "Graph cut based inference with co-occurrence statistics," in *ECCV*, 2010, pp. 239–253.

[15] P. Kohli, M. Kumar, and P. Torr, "P^3 and beyond: Solving energies with higher order cliques," in *CVPR*, 2007.

[16] X. Lan, S. Roth, D. Huttenlocher, and M. Black, "Efficient belief propagation with learned higher-order markov random fields." in *ECCV*, 2006, pp. 269–282.

[17] S. Roth and M. Black, "Fields of experts: A framework for learning image priors." in *CVPR*, 2005, pp. 860–867.

[18] B. Potetz, "Efficient belief propagation for vision using linear constraint nodes," in *CVPR*, 2007.

[19] S. Vicente, V. Kolmogorov, and C. Rother, "Joint optimization of segmentation and appearance models," in *ICCV*, 2009, pp. 755–762.

[20] ——, "Graph cut based image segmentation with connectivity priors," in *CVPR*, 2008.

[21] D. Tarlow, I. E. Givoni, and R. S. Zemel, "Hop-map: Efficient message passing with high order potentials," *JMLR - Proceedings Track*, vol. 9, pp. 812–819, 2010.

[22] T. Werner, "High-arity interactions, polyhedral relaxations, and cutting plane algorithm for soft constraint optimisation (MAP-MRF)," in *CVPR*, 2009.

[23] O. Woodford, P. Torr, I. Reid, and A. Fitzgibbon, "Global stereo reconstruction under second order smoothness priors," in *CVPR*, 2008.

[24] P. Kohli and M. P. Kumar, "Energy minimization for linear envelope MRFs," in *CVPR*, 2010, pp. 1863–1870.

[25] P. Kohli, L. Ladicky, and P. Torr, "Robust higher order potentials for enforcing label consistency," in *CVPR*, 2008.

[26] C. Rother, P. Kohli, W. Feng, and J. Jia, "Minimizing sparse higher order energy functions of discrete variables," in *CVPR*, 2009, pp. 1382–1389.

[27] C. Russell, L. Ladicky, P. Kohli, and P. H. S. Torr, "Exact and approximate inference in associative hierarchical random fields using graph-cuts," in *UAI*, 2010.

[28] A. Blake, C. Rother, M. Brown, P. Perez, and P. Torr, "Interactive image segmentation using an adaptive GMMRF model," in *ECCV*, 2004, pp. I: 428–441.

[29] C. Rother, V. Kolmogorov, and A. Blake, "Grabcut: interactive foreground extraction using iterated graph cuts," in *SIGGRAPH*, 2004, pp. 309–314.

[30] A. Blake, P. Kohli, and C. Rother, *Advances in Markov Random Fields.* MIT Press, 2011.

[31] S. Nowozin and C. Lampert, *Structured Learning and Prediction in Computer Vision.* NOW Publishers, 2011.

[32] A. Shekhovtsov, P. Kohli, and C. Rother, "Curvature prior for MRF-based segmentation and shape inpainting," Center for Machine Perception, K13133 FEE Czech Technical University, Prague, Czech Republic, Research Report CTU–CMP–2011–11, September 2011.

[33] X. Ren and J. Malik, "Learning a classification model for segmentation." in *ICCV*, 2003, pp. 10–17.

[34] J. Shotton, J. Winn, C. Rother, and A. Criminisi, "*TextonBoost*: Joint appearance, shape and context modeling for multi-class object recognition and segmentation." in *ECCV (1)*, 2006, pp. 1–15.

[35] D. Comaniciu and P. Meer, "Mean shift: A robust approach toward feature space analysis." *IEEE Trans. Pattern Anal. Mach. Intell.*, vol. 24, no. 5, pp. 603–619, 2002.

[36] Y. Boykov and M. Jolly, "Interactive graph cuts for optimal boundary and region segmentation of objects in N-D images," in *ICCV*, 2001, pp. I: 105–112.

[37] N. Komodakis and N. Paragios, "Beyond pairwise energies: Efficient optimization for higher-order MRFs," in *CVPR*, 2009, pp. 2985–2992.

[38] H. Ishikawa, "Higher-order clique reduction in binary graph cut," in *CVPR*, 2009, pp. 2993–3000.

[39] W. T. Freeman, E. C. Pasztor, and O. T. Carmichael, "Learning low-level vision," *IJCV*, vol. 40, no. 1, pp. 25–47, 2000.

[40] D. Cremers and L. Grady, "Statistical priors for efficient combinatorial optimization via graph cuts," in *ECCV*, 2006, pp. 263–274.

[41] M. Bleyer, C. Rother, P. Kohli, D. Scharstein, and S. Sinha, "Object stereo: Joint stereo matching and object segmentation," in *CVPR*, 2011, pp. 3081–3088.

[42] S. Nowozin and C. H. Lampert, "Global connectivity potentials for random field models," in *CVPR*, 2009, pp. 818–825.

[43] V. S. Lempitsky, P. Kohli, C. Rother, and T. Sharp, "Image segmentation with a bounding box prior," in *ICCV*, 2009, pp. 277–284.

[44] M. Bleyer, C. Rother, and P. Kohli, "Surface stereo with soft segmentation," in *CVPR*, 2010, pp. 1570–1577.

[45] A. Delong, A. Osokin, H. N. Isack, and Y. Boykov, "Fast approximate energy minimization with label costs," in *CVPR*, 2010, pp. 2173–2180.

[46] D. Hoiem, C. Rother, and J. M. Winn, "3D layoutcrf for multi-view object class recognition and segmentation," in *CVPR*, 2007.

[47] V. Kolmogorov, Y. Boykov, and C. Rother, "Applications of parametric maxflow in computer vision," in *ICCV*, 2007, pp. 1–8.

[48] G. Gallo, M. Grigoriadis, and R. Tarjan, "A fast parametric maximum flow algorithm and applications," *SIAM J. on Comput.*, vol. 18, pp. 30–55, 1989.

[49] Y. Lim, K. Jung, and P. Kohli, "Energy minimization under constraints on label counts," in *ECCV*, 2010, pp. 535–551.

[50] O. Woodford, C. Rother, and V. Kolmogorov, "A global perspective on map inference for low-level vision," in *ICCV*, 2009, pp. 2319–2326.

[51] D. Sontag, T. Meltzer, A. Globerson, T. Jaakkola, and Y. . Weiss, "Tightening lp relaxations for map using message passing," in *UAI*, 2008.

[52] J. Yedidia, W. Freeman, and Y. Weiss, "Generalized belief propagation." in *NIPS*, 2000, pp. 689–695.

[53] C. Chekuri, S. Khanna, J. Naor, and L. Zosin, "A linear programming formulation and approximation algorithms for the metric labeling problem," *SIAM J. Discrete Math.*, vol. 18, no. 3, pp. 608–625, 2005.

[54] D. Freedman and P. Drineas, "Energy minimization via graph cuts: Settling what is possible." in *CVPR*, 2005, pp. 939–946.

[55] P. Kohli, L. Ladicky, and P. H. S. Torr, "Robust higher order potentials for enforcing label consistency," *IJCV*, vol. 82, no. 3, pp. 302–324, 2009.

[56] L. Torresani, V. Kolmogorov, and C. Rother, "Feature correspondence via graph matching: Models and global optimization," in *ECCV*, 2008, pp. 596–609.

[57] G. E. Hinton, "Training products of experts by minimizing contrastive divergence," *Neural Comput.*, vol. 14, no. 8, pp. 1771–1800, 2002.

4

A Parametric Maximum Flow Approach for Discrete Total Variation Regularization

Antonin Chambolle

CMAP, Ecole Polytechnique, CNRS
F-91128 Palaiseau, France
Email: antonin.chambolle@cmap.polytechnique.fr

Jérôme Darbon

CMLA, ENS Cachan, CNRS, PRES UniverSud
F-94235 Cachan, France
Email: jerome.darbon@cmla.ens-cachan.fr

CONTENTS

4.1 Introduction 93
4.2 Idea of the approach 95
4.3 Numerical Computations 97
    4.3.1 Binary Optimization 98
    4.3.2 Discrete Total Variations optimization 100
4.4 Applications 104
    Bibliography 106

4.1 Introduction

In this chapter, we consider the general reconstruction problem with total variation (TV) that takes the following form

$$E(\mathbf{u}) = F(\mathbf{u}) + \lambda J(\mathbf{u})$$

with $m \leq \mathbf{u} \leq M$, and where F and J, respectively, correspond to the data fidelity and the discrete total variation terms (to be defined later). We shall consider the three following cases for the data fidelity term F, which are widely used in image processing:

$$F(\mathbf{u}) = \begin{cases} \frac{1}{2}\|\mathbf{A}\mathbf{u} - \mathbf{f}\|^2 \\ \frac{1}{2}\|\mathbf{u} - \mathbf{f}\|^2 \\ \frac{1}{2}\|\mathbf{u} - \mathbf{f}\|_1 \end{cases} \tag{4.1}$$

where **f** is the observed data.

The most widely used approaches nowadays for performing nonsmooth optimization rely on variants of the proximal point algorithm [1, 2] or proximal splitting techniques [3, 4, 5, 6]. It is, therefore, an interesting problem to compute the proximity operator [1] of common regularizers for inverse problems in imaging. In this chapter we address the problem of computing the proximity operator of the discrete TV and we describe a few elementary imaging applications. Note that the techniques we are considering here, which rely on an ordering of the values at each pixel, work only for scalar-valued (grey) images and cannot be extended to vector-valued (color) data.

It was firstly observed by Picard and Ratliff in [7] that Boolean Markovian energies with second-order cliques can be mapped to a network flow on a oriented graph with a source and a sink (and minimizing the energy corresponds to computing the s,t-minimum-cut or solving maximum-flow by duality). This technique was used by Greig et al. in [8] to estimate the quality of the maximum a posteriori estimator of ferromagnetic Ising models for binary image restoration. Then, it was observed that solving a series of the above problems where source-node capacities are gradually increasing can be optimized with the time complexity of a single maximum-flow using a parametric maximum-flow approach [9, 10] (provided the number parameter is less than the number of pixels). A further improvement has been proposed by [11] who shows how to compute efficiently an exact solution, provided that the data fidelity term is separable and that the space of solutions is a discrete space.

This last approach has been rediscovered, and in some sense extended (in particular, to more general submodular interaction energies) by the authors of [12, 13, 14, 15, 16, 17, 18]. In particular, they describe in [15] a method which solves "exactly" (up to machine precision) the proximity operator of some discrete total variation energies. In the chapter we will describe with further details this technique along with some basic image processing applications.

Notations

In this paper we shall consider that an image is defined on a discrete grid Ω. The gray-level value of the image $\mathbf{u} \in \mathbb{R}^{\Omega}$ at the site $i \in \Omega$ is referred to as $u_i \in \mathbb{R}$.

A very general "discrete total variation" problem we can solve with a network-flow-based approach has the following form:

$$J(\mathbf{u}) = \sum_{i \neq j} w_{ij}(u_j - u_i)^+, \qquad (4.2)$$

where $x^+ = \max(x, 0)$ is the nonnegative part of $x \in \mathbb{R}$. Here, w_{ij} are nonnegative values that weight each interaction. For the sake of clarity we shall assume that $w_{ij} = w_{ji}$, and thus we consider discrete TV of the form

$$J(\mathbf{u}) = \sum_{\{i,j\}} w_{ij}|u_j - u_i|, \qquad (4.3)$$

where each interaction (i, j) appears only once in the sum. In practice, the grid is endowed with a neighborhood system and the weights w_{ij} are nonnegative only when i and j are neighbors. Two neighboring sites i and j are denoted by $i \sim j$.

For instance, one common example is to consider a regular 2D grid endowed with the 4-connectivity (i.e., 4 nearest neighbors) where the weights are all set to 1. In that case J is an approximation of the l_1-anisotropic total variation that formally writes as $\int (|\partial_x u| + |\partial_y u|)$. Less anisotropic versions can be obtained by considering more neighbors (with appropriate weights). Observe that we cannot tackle with our techniques the minimization of standard approximations of the isotropic TV written formally as $\int \sqrt{|\partial_x u|^2 + |\partial_y u|^2}$, as for instance in [19, 20]. Finally, note that one can extend the proposed approach to some higher-order interaction terms without too much effort by adding some internal nodes; We refer the reader to [15] for more details.

In the next section, we describe the parametric maximum approach for computing the proximity operator of the discrete TV. We follow with more details our description in [15] of the approach essentially due to Hochbaum [11]. Then we describe some applications.

4.2 Idea of the approach

Our goal is to minimize the energy $J(\mathbf{u}) + \frac{1}{2}\|\mathbf{u} - \mathbf{g}\|^2 = J(\mathbf{u}) + \frac{1}{2}\sum_{i \in \Omega}(u_i - g_i)^2$, or similar energies, where J has the form (4.3) and $\mathbf{g} \in \mathbb{R}^\Omega$ is given.

Here, the general form we are interested in is $J(\mathbf{u}) + \sum_i \psi(u_i)$, where ψ_i is a convex function (in our case, $\psi_i(z) = \frac{1}{2}(z - g_i)^2$ with $i \in \Omega$). The essential point here is that the derivative of ψ_i is nondecreasing. Nonseparable data terms (such as $\|\mathbf{A}\mathbf{u} - \mathbf{g}\|^2$) will be considered in Section 4.4.

In what follows, we will denote by $\psi_i'(z)$ the left derivative of ψ_i at z. All that follows would work in the same way if we considered the right-derivative, or any monotonous selection of the subgradient: the important point here is that $\psi_i'(z) \leq \psi_i'(z')$ for all $z \leq z'$.

For each $z \in \mathbb{R}$, we consider the minimization problem

$$\min_{\theta \in \{0,1\}^\Omega} J(\theta) + \sum_{i \in \Omega} \psi_i'(z)\theta_i \qquad (P_z)$$

The following comparison lemma is classical:

Lemma 4.2.1
Let $z > z'$, $i \in \Omega$ and assume that $\psi_i'(z) > \psi_i'(z')$ (which holds for instance if ψ_i is strictly convex): then, $\theta_i^z \leq \theta_i^{z'}$.

An easy proof is found for instance in [15]: the idea is to compare the energy of θ^z in (P_z) with the energy of $\theta^z \wedge \theta^{z'}$, and the energy of $\theta^{z'}$ in $(P_{z'})$ with the energy of $\theta^z \vee \theta^{z'}$, and sum both inequalities. Then, we get rid of the terms involving J thanks to its *submodularity*:

$$J(\theta^z \wedge \theta^{z'}) + J(\theta^z \vee \theta^{z'}) \leq J(\theta^z) + J(\theta^{z'}) \tag{4.4}$$

which is the essential property upon which is based most of the analysis of this chapter [21, 15, 22].

Remark 4.2.2

Now, if the ψ_i are not strictly convex, it easily follows from Lemma 4.2.1 that the conclusion remains true provided either θ^z is the minimal solution of (P_z), or $\theta^{z'}$ the maximal solution of $(P_{z'})$. Note that these smallest and largest solutions are well-defined since the energies of (P_z) are submodular.

To check this, we introduce for $\varepsilon > 0$ small a solution $\theta^{z,\varepsilon}$ to

$$\min_{\theta \in \{0,1\}^\Omega} J(\theta) + \sum_{i \in \Omega}(\psi_i'(z) + \varepsilon)\theta_i .$$

Then, the Lemma guarantees that $\theta_i^{z,\varepsilon} \leq \theta_i^{z'}$ for all i, as soon as $z' \leq z$. In the limit, we easily check that $\underline{\theta}_i^z = \lim_{\varepsilon \to 0} \theta_i^{z,\varepsilon} = \sup_{\varepsilon \to 0} \theta_i^{z,\varepsilon}$ is a solution of (P_z), still satisfying $\underline{\theta}_i^z \leq \theta_i^{z'}$ for any $i \in \Omega$ and $z' \leq z$. This shows both that $\underline{\theta}^z$ is the smallest solution to (P_z), and that it is below any solution to $(P_{z'})$ with $z' < z$, which is the claim of Remark 4.2.2.

We will also denote by $\overline{\theta}^z$ the maximal solution of (P_z). This allows us to define two vectors $\underline{u}$ and $\overline{u}$ as the following for any $i \in \Omega$

$$\begin{cases} \underline{u}_i = \sup\{z, \underline{\theta}_i^z = 1\} , \\ \overline{u}_i = \sup\{z, \overline{\theta}_i^z = 1\} . \end{cases} \tag{4.5}$$

Let us now show the following result:

Proposition 4.2.3

We have that $\underline{u}$ and $\overline{u}$ are, respectively, the minimal and maximal solutions of

$$\min_{\mathbf{u}} \left(J(\mathbf{u}) + \sum_{i \in \Omega} \psi_i(u_i) \right) . \tag{4.6}$$

Proof: The proof relies on the (discrete, and straightforward) "co-area formula", which states that

$$J(\mathbf{u}) = \int_{-\infty}^{\infty} J(\chi_{\{\mathbf{u}>z\}})dz , \tag{4.7}$$

and simply follows from the remark that for any $a, b \in \mathbb{R}$, $|a - b| = \int_{\mathbb{R}} |\chi_{\{a>s\}} - \chi_{\{b>s\}}| \, ds$.

Let us first choose for any z a solution θ^z of (P_z), and let

$$u_i = \sup\{z, \theta_i^z = 1\}. \tag{4.8}$$

We first observe that for any z, $\{i : u_i > z\} \subseteq \{i : \theta_i^z = 1\} \subseteq \{i : u_i > s\}$, so that, letting $\mathcal{U} = \{u_i : i \in \Omega\}$ we have that for any $z \in \mathbb{R} \setminus \mathcal{U}$,

$$\{i : u_i > z\} = \{i : \theta_i^z = 1\} = \{i : u_i > z\} \tag{4.9}$$

so that almost all level sets of $\mathbf{u}$ are minimizers of (P_z). Then, if $\mathbf{v} \in \mathbb{R}^\Omega$ and $m \le \min\{u_i, v_i : i \in \Omega\}$, we find that

$$
\begin{aligned}
J(\mathbf{u}) + \sum_{i \in \Omega} \psi_i(u_i) &= \int_m^\infty J(\chi_{\{\mathbf{u}>z\}}) \, dz + \sum_{i \in \Omega} \int_m^\infty \psi_i'(z)\chi_{\{u_i>z\}} \, dz \\
&= \int_m^\infty \left(J(\theta^z) + \sum_{i \in \Omega} \psi_i'(z)\theta_i^z \right) dz \le \int_m^\infty \left(J(\chi_{\{\mathbf{v}>s\}}) + \sum_{i \in \Omega} \psi_i'(z)\chi_{\{v_i>z\}} \right) dz \\
&= J(\mathbf{v}) + \sum_{i \in \Omega} \psi_i(v_i) \quad (4.10)
\end{aligned}
$$

and it follows that $\mathbf{u}$ is a minimizer of (4.6).

Now, conversely, if we take for $\mathbf{v}$ another minimizer in (4.10), we deduce that

$$0 = \int_m^\infty \left(\left(J(\chi_{\{\mathbf{v}>s\}}) + \sum_{i \in \Omega} \psi_i'(z)\chi_{\{v_i>z\}} \right) - \left(J(\theta^z) + \sum_{i \in \Omega} \psi_i'(z)\theta_i^z \right) \right) dz$$

and as the term under the integral is nonnegative, we deduce that it is zero for almost all z: in other words, $\chi_{\{\mathbf{v}\ge z\}}$ is a minimizer of (P_z) for almost all z. In particular, $\chi_{\{\mathbf{v}\ge z\}} \supseteq \underline{\theta}^z$ (as the latter is the minimal solution), from which we deduce that $\mathbf{v} \ge \underline{\mathbf{u}}$. We deduce that $\underline{\mathbf{u}}$ is the minimal solution of (4.6). In the same way we can obviously show that $\bar{\mathbf{u}}$ is the maximal solution. This proves the result. $\square$

4.3 Numerical Computations

This section describes a parametric network flow based approach for minimizing discrete total variations with separable convex data fidelity terms. In general, the computed solution will be an ϵ solution with respect to the l_∞-norm in finite time.

For some particular cases, an exact solution can be obtained. The signal is made of N components. For the sake of clarity, it is assumed that the solution lives between m and M with $m < M$. If these bounds are too tight, then the algorithm will output an optimal solution with the constraint $m \leq \mathbf{u} \leq M$. We first present a network-flow-based approach originally proposed by Picard and Ratliff [7] for minimizing binary problems (P_z) (see also [8]). Then, we describe the parametric maximum-flow approach, which extends this approach to solve directly a series of such problems for varying z, and therefore can produce a solution to the discrete total variation minimization problems.

4.3.1   Binary Optimization

In their seminal paper [7], Picard and Ratliff propose an approach for reducing the problem of minimizing an Ising-like binary problem

$$\min_{\theta \in \{0,1\}^\Omega} J(\theta) - \sum_i \beta_i \theta_i \,, \tag{4.11}$$

to a minimum-cut problem or by linear duality a maximum-flow problem. The idea is to associate a graph to this energy such that its minimum-cut yields an optimal solution of the original energy. The graph has $(n + 2)$ nodes; there is one node associated to each variable θ_i besides two special nodes called the *source* and the *sink* that are, respectively, referred to as s and t. For the sake of clarity, we shall use i for either denoting a node of the graph or the index to access a component of the solution vector. We consider oriented edges, to which we associate capacities (i.e., intervals) defined as follows, for any nodes i and j:

$$c(i,j) \;=\; \begin{cases} [0, (\beta_j)^+] & \text{for } i = s, j \in \Omega \\ [0, (\beta_i)^-] & \text{for } i \in \Omega, j = t \\ c(i,j) = [0, w_{ij}] & \text{for } i, j \in \Omega \end{cases} \tag{4.12}$$

where $(x)^+ = \max(0, x)$ and $(x)^- = \max(0, -x)$. In a slight abuse of notation, we will also use $c(i,j)$ to denote the maximal element of the interval $c(i,j)$ — positive whenever $c(i,j) \neq \{0\}$.

A flow in a network is a real-valued vector $(f_{ij})_{i,j \in \Omega \cup \{s,t\}}$, where each component is associated to an arc. A flow f is said feasible if and only if

- for any $(i,j) \in \Omega \cup \{s,t\}$, we have $f_{ij} \in c(i,j)$;

- for any $i \in \Omega$, $\sum_{j \neq i} f_{ij} = \sum_{j \neq i} f_{ji}$.

Formally speaking, the last condition expresses that "inner" nodes $i \in \Omega$ receive as much flow as they send, so that f can be seen as a "quantity" flowing from s (which can only send, as no edge of positive capacity points towards s) to t (which can only receive). In particular, we can define the total flow $F(f)$ through the graph as the quantity $\sum_{i \neq s} f_{si} = \sum_{i \neq t} f_{it}$.

A cut associated to this network is a partition of the nodes into two subsets, denoted by $(\mathcal{S}, \mathcal{T})$, such that $s \in \mathcal{S}$ and $t \in \mathcal{T}$. The capacity of a cut $C(\mathcal{S}, \mathcal{T})$ is the sum of the maximal capacities of the edges that have initial node in $\mathcal{S}$ and final node in $\mathcal{T}$. It is thus defined as follows

$$C(\mathcal{S}, \mathcal{T}) = \sum_{\{i,j\}, i \in \mathcal{S}, j \in \mathcal{T}} c(i,j), \tag{4.13}$$

Now, let $\theta \in \{0,1\}^\Omega$ be the characteristic function of the $\mathcal{S} \cap \Omega$. We get

$$C(\mathcal{S}, \mathcal{T}) = \sum_{i,j \in \Omega} (\theta_i - \theta_j)^+ c(i,j) + \sum_{i \in \Omega} (1 - \theta_i)^+ c(s,i) + \theta_i c(i,t) \tag{4.14}$$

$$= J(u) + \sum_{i \in \Omega} \theta_i \beta_i + \sum_{i \in \Omega} (-\beta_i)^-, \tag{4.15}$$

which is the binary energy to optimize up to a constant. In other words, minimizing a binary energy of the form of (4.11) can be performed by computing the minimum cut of its associated network. Such a cut can be computed in polynomial time. It turns out that the dual problem to the minimum cut problem is the so-called maximum flow problem of maximizing $F(f)$ among all feasible flows. Hence, the maximum flow problem consists in sending the maximum amount of flow from the source to the sink such that the flow is feasible and that the divergence is free at all nodes except the source and the sink. The best theoretical time complexity algorithms for computing a maximum-flow on a general graph are based on a "push-relabel" approach [23, 24] (more recent algorithms can compute faster approximate flows [25]). However, for graph with few incident arcs (as it is the generally case for image processing and computer vision problems) the augmenting path approach yields excellent practical time [26].

To describe this approach let us first introduce the notion of residual capacity. Given a flow f, the residual capacity $c_f(i,j)$ associated to the arc (i,j) corresponds to the maximum amount of flow that can still be "sent" through the arc (i,j) while remaining a feasible flow. If δ is sent through an arc (i,j), its residual capacity should be reduced by δ while the capacity of the reverse (j,i) should be increased by δ (indeed, since the flow can be sent back), and it follows:

$$c_f(i,j) = c(i,j) - f_{ij} + f_{ji} \tag{4.16}$$

The *residual graph* is the graph endowed with the residual capacities (and whose arcs, thus, are all edges of nonzero residual capacity).

The idea of an augmenting path approach for computing a maximum-flow consists in starting with the null flow (which is feasible) and in searching for a path from the source to the node that can send a positive amount of flow (with respect to the residual capacities). If such a path is found, then the maximal amount of flow is sent though it while the residual capacities are updated. This process is iterated until no augmenting path is found: it follows that all paths from s to t pass through at least an edge of residual capacity zero (which is said to be "saturated"), and one can check that the actual flow is the maximum-flow.

A minimum-cut $(\mathcal{S}, \mathcal{T})$ corresponds to any 2-partition such that all arcs (i, j) with $i \in \mathcal{S}$ and $j \in \mathcal{T}$ have a null residual capacities. An easy way to find one is to let $\mathcal{S}$ be the set of nodes that are connected to s in the residual graph (defined by keeping only the edges of positive residual capacity), and $\mathcal{T}$ its complement. Note that uniqueness of the cut is not guaranteed. The choice above, for instance, will select the smallest solution $\mathcal{S}$.

4.3.2 Discrete Total Variations optimization

The goal of this section is to extend the previous approach to optimize discrete total variation problems. First, let us note that the above binary optimization approach allows us to solve any problem (P_z), choosing $\beta_i = -\psi_i'(z)$ for each i, and find the smallest, or largest, solution. In other words, given a level z we can easily decide whether the optimal value $\underline{u}_i$ (or $\overline{u}_i$) is above or below z, by solving a simple maximal flow problem. Thus, a straightforward and naive way to reconstruct an ϵ-solution is to solve all the binary problems that correspond to levels $z = m + k\epsilon$ with $k \in \{0, \ldots \lfloor \frac{M-m}{\epsilon} \rfloor\}$.

We present several improvements that drastically reduce the amount of computations compared to this approach. A first improvement consists in using a dyadic approach. Assume that we have computed the binary solution θ^z for the level z. Now, suppose that we wish to compute the solution $\theta^{z'}$ for the level z' with $z > z'$. Since the monotone property holds we have that $\theta^z \leq \theta^{z'}$. Thus, if we have $\theta_i^z = 1$ for some i then we will get $\theta_i^{z'} = 1$, and, conversely, $\theta_i^{z'} = 0$ implies $\theta_i^z = 0$. In other words, we already know the value of the solution for some pixels. Thus, instead of considering the original energy, we consider its restriction defined by setting all known binary variables to their optimal values. It is easy to see that this new energy is still an Ising-like energy that allows its optimization using the network-flow approach (indeed, the only interesting case happens for terms $|u_j - u_i|$ when one optimal assignment for a variable is known while it is not for the other; it is easy to see that this original pairwise interaction boils down to a unary term after restriction). The search of the optimal value can be performed in a dyadic way that is similar to some binary search algorithms. This consists in splitting the search space in two equal parts so that each variable is involved in $O(\log_2 \frac{M-m}{\epsilon})$ optimizations. Such an approach has been considered in [13, 17].

The above arguments for the dyadic approach also yields another improvement. Indeed, recall that the reduction operation removes interactions between variables. One can consider the connected components of variables that are in interaction with each other and solve independently each of the optimizations associated with each of these connected components. It is easy to see that each of these problems are independent of each other and thus, that an optimal solution is still computed. Using connected components tends to reduce the size of the problems in which a variable is involved with although it does not change the total number of times. Besides, note that this reduces the worst case theoretical time complexity (as the theoretical worst-case time complexity for computing a maximum-flow is much larger than linear; see [24] for instance). Such an approach is described in more details in [13] and [15]. Also, note that the dyadic approach and the connected component point of view correspond to a divide-and-conquer approach (see [13]) that could also be used for improving performance through parallelization.

The main improvement for computing an ϵ-solution consists in reusing the optimal flow computed at some stage for the next one. This is the main idea of the "parametric maximum-flow" [10, 9, 11]. Due to the convexity of the data-fidelity terms, the capacities $c(s, i)$ and $c(i, t)$ are, respectively, nonincreasing and nondecreasing with respect to z, while the other capacities remain unchanged. Under these assumptions, we have that the set of nodes S connected to the source is growing as the level z is decreasing (this is nothing else than another formulation of Lemma 4.2.1). This inclusion property allows the authors of [27] to design an algorithm that solves the maximum-flow problem for a series of z with the time complexity of a single maximum-flow algorithm. Their analysis relies on the use of a preflow-based algorithm [23, 24]. A similar analysis can be done for augmenting path algorithms but in this case, to our knowledge, a sharp estimation of the time complexity remains unknown.

Assume we have computed the solution for a level z. We can replace the original capacities with the residual capacities (since we can reconstruct the original capacities from the flow and the residual capacities); the result is a graph where all paths from the source to the sink are saturated somewhere. Then, we increase the residual capacities of all arcs $c(i, t)$[1] from an internal node to the sink by an amount $\delta > 0$ (this is to simplify; of course, this quantity could also depend on i). We rerun the augmenting-path algorithm from the current flow with the new capacities. Now, if a node i was already connected to $\mathcal{T}$ (hence, disconnected from s), then it will remain in $\mathcal{T}$ since it cannot receive any flow, as all paths starting from s to i are saturated. (From a practical point of view, it is, in fact, not even necessary to update these capacities.) Consider, on the other hand, the case of a node i that was in S, hence, connected to the source. Since the capacity of the arc (i, t) is increased, it gets connected also to the sink: there is now at least one path from s to t passing through i along which the flow can be further augmented. Then, i may fall back again into S, or switch to $\mathcal{T}$. Eventually, as we increase z (and, hence, gradually increase the residual capacities of (i, t) and run the max-flow algorithm again), The new flow will saturate arcs that are closer to the sink, and at some point the node i will switch to $\mathcal{T}$ (for some large enough value of z).

This is the idea of the parametric maximum-flow . In practice, we start from a minimal level and, through positive increments, eventually reaches the maximal

---

[1] In the sequel, $c(\cdot, \cdot)$ will always denote the residual capacities.

level. These increments can be arbitrary, i.e., not uniform. Choosing increments of $\delta = \epsilon$ naturally yields an ϵ−approximation in the quadratic case, where $\psi_i'(z) = z - g_i$ for every $i \in \Omega$.

The final improvement consists in merging all these points of view into a single one. Let us first consider the merging of the dyadic and connected components point of views with the parametric maximum-flow. Instead of starting with the minimal level for z, we can start with the median. After we run the maximum-flow, we know for each pixel if $u_i \geq z$ or $u_i \leq z$. We can separate the support of the image into its connected components. Let us denote this partition by $\{\mathcal{S}_k\}$ and $\{\mathcal{T}_k\}$, which corresponds to all connected components of internal nodes that are connected to the source and to the sink, respectively. For each connected component we can associate the minimal and maximal possible values in which the optimal value lives in. Every time, the length of this interval is divided by two by running the maximum-flow with the level z that corresponds to the median of this set. Instead of rerunning the maximum-flow approach for each problem associated to each connected component (which would require to rebuild the graph), we apply the maximum-flow approach. This is performed as follows. We set to zero all capacities from arcs (j, i) with $j \in \mathcal{T}_k$ and $i \in \mathcal{S}_{k'}$ for any k, k'. This means that we remove the arcs between i and j (indeed, since the residual capacity of the reverse arc (i, j) was already zero after the maximum-flow). For any node, we update the capacity $c(i, t)$ such that it corresponds to the capacity associated to the median the set of the possible optimal values: for nodes $i \in \mathcal{S}_k$ this means increasing $c(i, t)$ while it means increasing $c(s, i)$ for nodes $i \in \mathcal{T}_k$. The essential improvement (without the connected component point of view) we get is that the number of times a node is involved in a maximum-flow optimization is of the order by the logarithm (in base 2) of the range of the solution (i.e., the maximal z minus the minimal z). The improvement due to the connected component version states that the number of maximum-flows per node is *at most* the above quantity. There is a trade-off between considering the connected component version or not that essentially depends on the original data.

Finally, the dyadic approach can be replaced by a more general approach, which computes an exact solution (at least in theory and in practice up to machine precision) for some particular cases of separable data fidelity terms (e.g., quadratic, piecewise affine, piecewise quadratic, ...). The idea is essentially due to D. Hochbaum [11]. We consider the case where $\psi_i(z) = (z - g_i)^2/2$, hence $\psi_i'(z) = z - g_i$.

The idea follows from the remark that the optimal energy in (P_z),

$$z \mapsto \min_{\theta \in \{0,1\}^\Omega} J(\theta) + \sum_{i \in \Omega} (z - g_i)\theta_i \tag{4.17}$$

is piecewise affine. To see this, we let, as before, θ^z be the solution for each z, and $\mathcal{U} = \{u_i : i \in \Omega\}$ be the set of values of the (unique, as in that case the energy for $\mathbf{u}$ is strictly convex) solution u. Then, for $z \notin \mathcal{U}$, $\theta^z = \chi_{\{\mathbf{u}>z\}}$ is unique and constant in a neighborhood of z, while if $z \in \mathcal{U}$, there are at least two solutions

$$\overline{\theta}^z = \chi_{\{\mathbf{u}\geq z\}} = \lim_{\varepsilon \downarrow 0} \theta^{z-\varepsilon} \text{ and } \underline{\theta}^z = \chi_{\{\mathbf{u}>z\}} = \lim_{\varepsilon \downarrow 0} \theta^{z+\varepsilon}. \tag{4.18}$$

It easily follows that the optimal energy (4.17) is affine in $\mathbb{R} \setminus \mathcal{U}$, and for $z^0 \in \mathcal{U}$ (which are called the "breakpoints" in [11]), it is precisely

$$J(\overline{\theta}^{z^0}) - \sum_{i \in \Omega} g_i \overline{\theta}_i^{z^0} + z \left| \{ i \, : \, u_i \geq z^0 \} \right| \qquad (4.19)$$

for $z \leq z^0$ close enough to z^0, and

$$J(\underline{\theta}^{z^0}) - \sum_{i \in \Omega} g_i \underline{\theta}_i^{z^0} + z \left| \{ i \, : \, u_i > z^0 \} \right| \qquad (4.20)$$

for $z \geq z^0$ close enough to z^0 (where $|\cdot|$ denotes the cardinality of a set). Moreover, at the breakpoint z^0, the slope of this affine function decreases (the difference is $|\{ \mathbf{u} = z^0 \}|$), hence, this piecewise affine function is also concave.

An idea for identifying these breakpoints easily follows. Assume we have solved the problem for two values $z < z'$, finding two solutions $\theta^z, \theta^{z'}$. Then, we know that there exists $z^* \in [z, z']$ such that

$$\mathcal{E}^* := J(\theta^z) - \sum_{i \in \Omega} g_i \theta_i^z + z^* \sum_{i \in \Omega} \theta_i^z = J(\theta^{z'}) - \sum_{i \in \Omega} g_i \theta_i^{z'} + z^* \sum_{i \in \Omega} \theta_i^{z'}. \qquad (4.21)$$

Let us perform a new minimization (by the maximal flow algorithm) at the level z^* (to solve (P_{z^*})). Then, there are two possibilities:

- The optimal energy at z^* is $\mathcal{E}^*$, and in particular both θ^z and $\theta^{z'}$ are solutions (and, in fact, we will actually find that the solution is $\theta^{z'}$ if we take care to always select, from the graph-cut algorithm, the minimal solution to the problem); in this case, we have identified a breakpoint.

- The optimal energy at z^* is below $\mathcal{E}^*$: this means that there is, at least, one breakpoint in $[z, z^*]$ and one other in $[z^*, z']$ that we can try to identify by the same procedure.

In practice, we first pick a minimal level $z_0 = m$, and a maximal level $z_1 = M$. If $m \leq \min_i g_i$ and $M \geq \max_i g_i$, then $\theta_i^{z_0} \equiv 1$ while $\theta_i^{z_1} \equiv 0$. It is easy to check that the value $z_2 = z^*$, which would be computed by (4.21) (with $z = z_0$, $z' = z_1$), is the average $(\sum_i g_i)/|\Omega|$. But an obvious reason is that if it were a breakpoint, then $\mathbf{u}$ would be a constant, minimizing $\sum_i (u - g_i)^2$: hence, the average of the g_i's.

This observation allows to compute efficiently the values "z^*" at each stage, without having to evaluate the quantities in (4.21). Indeed, if z_2 is not a breakpoint, we need to find two new potential breakpoints $z_3 \in [z_0, z_2]$ and $z_4 \in [z_2, z_1]$. After having performed the cut with the capacities associated to z_2, we are left with a residual graph, a set $\mathcal{S}$ of nodes connected to s, and its complement $\mathcal{T}$, and we know that $u_i \leq z_2$ if $i \in \mathcal{T}$, and $u_i \geq z_2$ else. First, we separate again $\mathcal{S}$ and $\mathcal{T}$ by setting to zero the capacities $c(i, j)$ for $i \in \mathcal{T}$ and $j \in \mathcal{S}$ (in that case, $c(j, i)$ is already

0). Now, in $\mathcal{T}$, the cut at the value z_3 will decide which nodes i have $u_i \geq z_3$ and which have $u_i \leq z_3$. The residual graph encodes a new problem of the form $\min_{\mathbf{u}} \tilde{J}(\mathbf{u}) + \|\mathbf{u} - \tilde{\mathbf{g}}\|^2$, defined on the nodes of $\mathcal{T} \setminus \{t\}$, for some "total-variation" like function $\tilde{J}$ and a data term $\tilde{\mathbf{g}}$, which is defined by the residual capacities $c(i, t)$. Hence, again, if the solution $\tilde{u}$ were constant in $\mathcal{T}$, it would assume the value $\tilde{z}^*$ given by the average of these capacities, $\tilde{z}^* = \sum_{i \in \mathcal{T}} c(i, t) / |\mathcal{T} \setminus \{t\}|$. This means that the new cut has to be performed after having set to $\tilde{z}^*$ the capacities $c(s, i)$, $i \in \mathcal{T} \setminus \{t\}$, and it corresponds to the level $z_3 = z_2 - \tilde{z}^*$.

Symmetrically, in $\mathcal{S}$, we must set to $\hat{z}^* = \sum_{i \in \mathcal{S}} c(s, i) / |\mathcal{S} \setminus \{s\}|$ all the capacities $c(i, t)$, which corresponds to choosing the level $z_4 = z_2 + \hat{z}^*$ as a potential new breakpoint.

This method can be easily run, in a dichotomic-like way, until all the breakpoints have been identified and the exact solution $\mathbf{u}$ computed. An improvement, as before, will consist in choosing the new breakpoint values not globally in each region that has to be separated, but on each connected component (in Ω) of these regions (it means that, after having identified the connected component of these regions, we set in each connected component that is connected to t all the values of $c(s, i)$ to the average of the values $c(i, t)$, and perform symmetrically in the connected components linked to s).

4.4   Applications

This section is devoted to the presentation of some numerical experiments in image processing. We consider anistropic total variations defined on the 8 nearest neighbors: the weights on the 4 nearest ones (i.e., left, right, top, and bottom) are set to 1, while the weights for the 4 diagonal interactions are set to $1/\sqrt{2}$.

We first consider the case of denoising. Figure 4.1 depicts an original image and a version corrupted by an additive Gaussian noise with standard deviation $\sigma = 12$ and null mean. The denoised version obtained by minimizing exactly the discrete total variation with a separable quadratic data fidelity term is depicted in Figure 4.2.

Next, we consider the optimization of DTV energies with nonseparable data fidelities. We assume that the latter are differentiable with continuous derivatives; this is the case for instance for terms of the form $\frac{1}{2}\|\mathbf{Au} - \mathbf{f}\|^2$ with $\|\mathbf{A}\| < \infty$ that are used in image restoration for deconvolution purposes. One popular way to optimize these energies relies on proximal splitting techniques, which consist of alternating an explicit descent on the smooth part of the energy with an implicit step (or proximal step) for the nonsmooth part. This is given by the proximity operator , defined by:

$$\text{Prox}_{\lambda J}(\mathbf{f}) = \underset{\mathbf{u}}{\text{argmin}} \; \frac{1}{2}\|\mathbf{u} - \mathbf{f}\|^2 + \lambda J(\mathbf{u}) = (I + \lambda J)^{-1}(\mathbf{f}). \qquad (4.22)$$

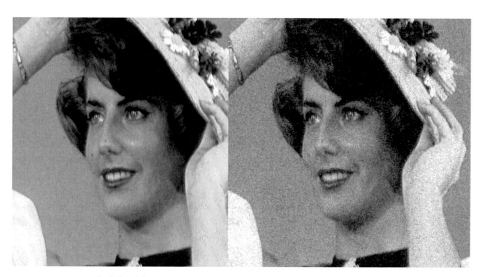

Original image   Noisy image

FIGURE 4.1
Original image on the left and corrupted version on the right.

FIGURE 4.2
Denoised image by minimizing DTV with a separable quadratic term.

---

1. Step 0: Set $\mathbf{v}^1 \leftarrow \mathbf{u}^0$ and $t_1 \leftarrow 1$

2. Step k: for $(k \geq 1)$ compute

    (a) $\mathbf{u}^k = \text{Prox}_{\lambda J}(\mathbf{v}^k)$

    (b) $t^{k+1} = \frac{1+\sqrt{1+4t_k^2}}{2}$

    (c) $\mathbf{v}^{k+1} = \mathbf{u}_k + \frac{t_k-1}{t_{k+1}}(\mathbf{u}^k - \mathbf{u}^{k-1})$

FIGURE 4.3
Algorithm for fast proximal iterations.

This operator is computed exactly by the approach described in this chapter. The general proximal forward–backward iteration [5, 6, 28] takes the following form

$$\mathbf{u}^{n+1} = \text{Prox}_{\lambda J}(\mathbf{u}^n - \frac{1}{L}\nabla F(\mathbf{u}^n)) \tag{4.23}$$

with $L = |\nabla F|_\infty$, and this scheme converges toward a solution of $F + \lambda J$, provided such a solution does exist. We illustrate this approach with $F(\mathbf{u}) = \frac{1}{2}\|\mathbf{Au} - \mathbf{f}\|^2$ for image restoration of blurred and noisy images. We consider a blurring operator $\mathbf{A}$ that corresponds to a uniform kernel of size 7×7. The original image of Figure 4.1 is blurred using this kernel and then corrupted by an additive Gaussian noise of standard variation $\sigma = 3$ and null mean to yield to the corrupted image depicted in Figure 4.4 (left). The reconstructed image obtained by iterating the splitting method is depicted in Figure 4.4 (right).

The speed of the proximal splitting algorithm can be improved using acceleration techniques [28, 29, 30]. Compared to the standard proximal iterations, the fast iterates are computed using particular linear combinations of two iterative sequences. Such an algorithm is presented in Figure 4.3. To show the improvement of such an approach we depict in Figure 4.5 the squared distance of the current estimate from the global minimizer (estimated as the solution obtained 1200 iterations) for both the standard and the fast proximal splitting in function of the iterations.

Finally, note that the inversion of $\mathbf{A}^*\mathbf{A}$ could also be performed using Douglas-Rachford splitting [4].

Bibliography

[1] J. Moreau, "Proximité et dualité dans un espace hilbertien," *Bulletin de la S.M.F.*, vol. 93, pp. 273–299, 1965.

[2] R. Rockafellar and R. Wets, *Variational Analysis*. Springer-Verlag, Heidelberg, 1998.

FIGURE 4.4
Blurred and noisy image on the left and reconstructed version on the right.

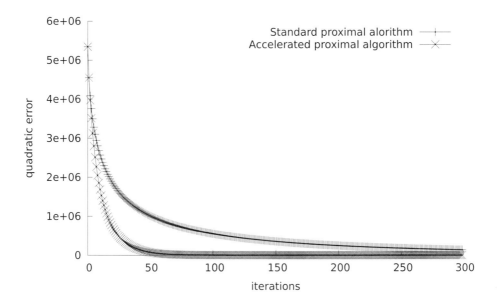

FIGURE 4.5
Plot of the squared distance of the estimate minimizer to the global minimizer with respect to the iterations.

[3] J. Eckstein and D. Bertsekas, "On the Douglas-Rachford splitting method and the proximal point algorithm for maximal monotone operators," *Math. Program.*, vol. 55, pp. 293–318, 1992.

[4] P. Lions and B. Mercier, "Splitting algorithms for the sum of two nonlinear operators," *SIAM J. on Numerical Analysis*, vol. 16, pp. 964–979, 1992.

[5] I. Daubechies, M. Defrise, and C. De Mol, "An iterative thresholding algorithm for linear inverse problems with a sparsity constraint," *Comm. Pure Appl. Math.*, vol. 57, no. 11, pp. 1413–1457, 2004. [Online]. Available: http://dx.doi.org/10.1002/cpa.20042

[6] P. L. Combettes and V. R. Wajs, "Signal recovery by proximal forward-backward splitting," *Multiscale Model. Simul.*, vol. 4, no. 4, pp. 1168–1200 (electronic), 2005. [Online]. Available: http://dx.doi.org/10.1137/050626090

[7] J. C. Picard and H. D. Ratliff, "Minimum cuts and related problems," *Networks*, vol. 5, no. 4, pp. 357–370, 1975.

[8] D. M. Greig, B. T. Porteous, and A. H. Seheult, "Exact maximum a posteriori estimation for binary images," *J. R. Statist. Soc. B*, vol. 51, pp. 271–279, 1989.

[9] G. Gallo, M. D. Grigoriadis, and R. E. Tarjan, "A fast parametric maximum flow algorithm and applications," *SIAM J. Comput.*, vol. 18, no. 1, pp. 30–55, 1989.

[10] M. Eisner and D. Severance, "Mathematical techniques for efficient record segmentation in large shared databases," *J. Assoc. Comput. Mach.*, vol. 23, no. 4, pp. 619–635, 1976.

[11] D. S. Hochbaum, "An efficient algorithm for image segmentation, Markov random fields and related problems," *J. ACM*, vol. 48, no. 4, pp. 686–701 (electronic), 2001.

[12] Y. Boykov, O. Veksler, and R. Zabih, "Fast approximate energy minimization via graph cuts," *IEEE Trans. Pattern Anal. Mach. Intell.*, vol. 23, no. 11, pp. 1222–1239, 2001.

[13] J. Darbon and M. Sigelle, "Image restoration with discrete constrained Total Variation part I: Fast and exact optimization," *J. Math. Imaging Vis.*, vol. 26, no. 3, pp. 261–276, December 2006.

[14] ——, "Image restoration with discrete constrained Total Variation part II: Levelable functions, convex priors and non-convex cases," *Journal of Mathematical Imaging and Vision*, vol. 26, no. 3, pp. 277–291, December 2006.

[15] A. Chambolle and J. Darbon, "On total variation minimization and surface evolution using parametric maximum flows," *International Journal of Computer Vision*, vol. 84, no. 3, pp. 288–307, 2009.

[16] J. Darbon, "Global optimization for first order Markov random fields with submodular priors," *Discrete Applied Mathematics*, June 2009. [Online]. Available: http://dx.doi.org/10.1016/j.dam.2009.02.026

[17] A. Chambolle, "Total variation minimization and a class of binary MRF models," in *Energy Minimization Methods in Computer Vision and Pattern Recognition*, ser. Lecture Notes in Computer Science, 2005, pp. 136–152.

[18] H. Ishikawa, "Exact optimization for Markov random fields with convex priors," *IEEE Trans. Pattern Analysis and Machine Intelligence*, vol. 25, no. 10, pp. 1333–1336, 2003.

[19] A. Chambolle, S. Levine, and B. Lucier, "An upwind finite-difference method for total variation–based image smoothing," *SIAM Journal on Imaging Sciences*, vol. 4, no. 1, pp. 277–299, 2011. [Online]. Available: http://link.aip.org/link/?SII/4/277/1

[20] A. Chambolle and T. Pock, "A first-order primal-dual algorithm for convex problems with applications to imaging," *Journal of Mathematical Imaging and Vision*, vol. 40, pp. 120–145, 2011. [Online]. Available: http://dx.doi.org/10.1007/s10851-010-0251-1

[21] F. Bach, "Convex analysis and optimization with submodular functions: a tutorial," INRIA, Tech. Rep. HAL 00527714, 2010.

[22] L. Lovász, "Submodular functions and convexity," in *Mathematical programming: the state of the art (Bonn, 1982)*. Berlin: Springer, 1983, pp. 235–257.

[23] R. K. Ahuja, T. L. Magnanti, and J. B. Orlin, *Network flows.* Englewood Cliffs, NJ: Prentice Hall Inc., 1993.

[24] T. H. Cormen, C. E. Leiserson, R. L. Rivest, and C. Stein, *Introduction to Algorithms.* The MIT Press, 2001.

[25] P. Christiano, J. A. Kelner, A. Madry, D. A. Spielman, and S.-H. Teng, "Electrical flows, laplacian systems, and faster approximation of maximum flow in undirected graphs," *CoRR*, vol. abs/1010.2921, 2010.

[26] Y. Boykov and V. Kolmogorov, "An experimental comparison of min-cut/max-flow algorithms for energy minimization in vision," *IEEE Trans. Pattern Analysis and Machine Intelligence*, vol. 26, no. 9, pp. 1124–1137, September 2004.

[27] A. V. Goldberg and R. E. Tarjan, "A new approach to the maximum flow problem," in *STOC '86: Proc. of the eighteenth annual ACM Symposium on Theory of Computing.* New York, NY, USA: ACM Press, 1986, pp. 136–146.

[28] A. Beck and M. Teboulle, "Fast gradient-based algorithms for constrained total variation image denoising and deblurring problems," *Trans. Img. Proc.*, vol. 18, pp. 2419–2434, November 2009. [Online]. Available: http://dx.doi.org/10.1109/TIP.2009.2028250

[29] Y. Nesterov, *Introductory Lectures on Convex Optimization: A Basic Course.* Kluwer, 2004.

[30] ——, "Gradient methods for minimizing composite objective function," Catholic University of Louvain, Tech. Rep. CORE Discussion Paper 2007/76, 2007.

5

Targeted Image Segmentation Using Graph Methods

Leo Grady

Department of Image Analytics and Informatics
Siemens Corporate Research
Princeton, NJ, USA
Email: Leo.Grady@siemens.com

CONTENTS

5.1	The Regularization of Targeted Image Segmentation	113
	5.1.1 Power Watershed Formalism	114
	5.1.2 Total Variation and Continuous Max-Flow	116
5.2	Target Specification	118
	5.2.1 Segmentation Targeting With User Interaction	119
	5.2.1.1 Seeds and Scribbles	119
	5.2.1.2 Object Boundary Specification	120
	5.2.1.3 Sub-Region Specification	120
	5.2.2 Segmentation Targeting with Appearance Information	121
	5.2.2.1 Prior Appearance Models	121
	5.2.2.2 Optimized Appearance Models	122
	5.2.3 Segmentation Targeting with Boundary Information	123
	5.2.3.1 Boundary Polarity	125
	5.2.4 Segmentation Targeting with Shape Information	125
	5.2.4.1 Implicit Shape Models	127
	5.2.4.2 Explicit Shape Models	129
	5.2.5 Segmentation Targeting with Object Feature Information	130
	5.2.6 Segmentation Targeting with Topology Information	131
	5.2.6.1 Connectivity Specification	131
	5.2.6.2 Genus Specification	132
	5.2.7 Segmentation Targeting by Relational Information	133
5.3	Conclusion	134
	Bibliography	135

Traditional image segmentation is the process of subdividing an image into smaller regions based on some notion of homogeneity or cohesiveness among groups of pixels. Several prominent traditional segmentation algorithms have been based on graph theoretic methods [1, 2, 3, 4, 5, 6]. However, these methods have been notoriously difficult to quantitatively evaluate since there is no formal definition of the segmentation problem (see [7, 8] for some approaches to evaluating traditional segmentation algorithms).

In contrast to the traditional image segmentation scenario, many real-world applications of image segmentation instead focus on identifying those pixels belonging to a *specific* object or objects (which we will call **targeted segmentation**) for which there are some known characteristics. In this chapter, we focus only on the extraction of a single object (i.e., labeling each pixel as **object** or **background**). The segmentation of a specific object from the background is *not* just a special case of the traditional image segmentation problem that is restricted to two labels. Instead, a targeted image segmentation algorithm must input the additional information that determines *which* object is being segmented. This additional information, which we will call **target specification**, can take many forms: user interaction, appearance models, pairwise pixel affinity models, contrast polarity, shape models, topology specification, relational information, and/or feature inclusion. The ideal targeted image segmentation algorithm could input any or all of the target specification information that is available about the target object and use it to produce a high-quality segmentation.

Graph-theoretic methods have provided a strong basis for approaching the targeted image segmentation problem. One aspect of graph-theoretic methods that has made them so appropriate for this problem is that it is fairly straightforward to incorporate different kinds of target specification (or, at least, to prove that they are NP-hard). In this chapter, we begin by showing how the image graph may be constructed to accommodate general target specification and review several of the most prominent models for targeted image segmentation, with an emphasis on their commonalities. Having established the general models, we show specifically how the various types of target specification may be used to identify the target object.

A targeted image segmentation algorithm consists of two parts: a *target specification* that identifies the desired object and a *regularization* that completes the segmentation from the target specification. The goal of the regularization algorithm is to incorporate as little or as much target specification that is available and still produce a meaningful segmentation of the target object under generic assumptions of object cohesion. We first treat different methods of performing the regularization in a general framework before discussing how targeting information can be combined with the regularization. In our opinion, the regularization is the core piece of a targeted segmentation algorithm and defines the primary difference between different algorithms. The target specification informs the algorithm what is *known* about the target object, while the regularization determines the segmentation with an underlying model of objects that describes what is *unknown* about the object. Put differently, the various varieties of target specification may be combined fairly easily with different types of regularization methods, but it is the regularization component that determines how the final output segmentation will be defined from that information.

The exposition will generally be in terms of *nodes* and *edges* rather than pixels or voxels. Although nodes are typically identified with pixels or voxels, we do not want to limit our exposition to 2D or 3D (unless otherwise specified), and we additionally want to allow for graphs representing images derived from superpixels or more general structures (see Chapter 1).

Before beginning, we note that targeted image segmentation is very similar to the alpha-matting problem in Chapter 9, and many of the methods strongly overlap. The main difference between targeted image segmentation and alpha-matting is that alpha-matting necessarily produces a real-valued blending coefficient at each pixel, while targeted segmentation ultimately seeks to identify which pixels belong to the object (although real-valued confidences can be quite useful in targeted segmentation problems). Moreover, the literature on alpha-matting and targeted segmentation is typically focused on different aspects of the problem. Alpha-matting algorithms are usually interactive with the assumption that most pixels are labeled except those at the boundary (a *trimap* input). The focus for alpha-matting algorithms is therefore on getting a meaningful blending coefficient for pixels that contain some translucency such as hair or fur. In contrast, the focus in targeted segmentation algorithms is on both interactive or automatic solutions in which there are few or no prelabeled pixels. Targeted image segmentation methods using graphs are also closely related to semi-supervised clustering algorithms. The primary differences between targeted image segmentation and semi-supervised clustering are: (1) The graph in image segmentation is typically a lattice rather than an arbitrary network, (2) The pixels in image segmentation are embedded in $\mathbb{R}^N$, while a semi-supervised clustering may not have an embedding, and (3) Pixels are associated with values (e.g., RGB) while points to be clustered may not be. The effect of these differences is that some ways of providing target specification in image segmentation may not apply to semi-supervised clustering, such as shape, appearance models, genus, or object features.

5.1 The Regularization of Targeted Image Segmentation

Many algorithms have been proposed for regularizing the targeted image segmentation problem. One class of algorithms were formulated originally in a continuous domain and optimized using active contours or level sets. More recently, a second set of algorithms have been formulated directly on a graph (or reformulated on a graph from the original continuous definition) and optimized using various methods in combinatorial and convex optimization. Algorithms in both classes ultimately take the form of finding a minimum of the energy

$$E(\mathbf{x}) = E_{\text{unary}}(\mathbf{x}) + \lambda E_{\text{binary}}(\mathbf{x}), \tag{5.1}$$

in which $x_i \in \mathbb{R}$, $0 \le x_i \le 1$; any pixels known to belong to the object (foreground) set, $v_i \in \mathcal{F}$, would be assigned $x_i = 1$, and any pixels known to belong to the background set, $v_i \in \mathcal{B}$, would be assigned $x_i = 0$. The parameter λ may be used to control the relative weighting of the two terms (i.e., to modulate the confidence of the information represented by each term). Any pixels with known labels are termed **seeds** or sometimes **scribbles** (since they are often provided interactively). The value of **x** at a pixel therefore represents the affinity of that pixel to belong to object (and may also be interpreted as the success probability of a Bernoulli distribution modeling the pixel label [9]). Given an **x** that minimizes (5.1), a hard segmentation may be produced by assigning a pixel with value $x_i > \theta = 0.5$ to the *object* and a pixel with value $x_i \leq \theta$ to the *background*. Since the explicit goal of a targeted segmentation algorithm is to find a hard (binary) segmentation, it may seem strange that the above formulation performs the optimization over a real-valued **x** and sets a threshold, since it is well known that *relaxing* a general binary optimization problem to real values and thresholding can produce a suboptimal solution to the binary optimization problem [10]. However, as we will see shortly, several targeted segmentation algorithms which are not derived from optimization problems may be computed by performing the above optimization over real values and then thresholding at 0.5. Consequently, for these algorithms, there is no "relaxation" since the algorithms are not defined by optimization problems except insofar as the computations to perform the algorithms may be viewed as a continuous-valued optimization. Furthermore, it has been shown that those segmentation algorithms reviewed below that are defined as binary optimization problems still produce minimal binary solutions, whether they are optimized over binary variables or whether the solution is produced by thresholding a real-valued solution.

The first term in (5.1) is called the *unary* term since it is formulated as a function of each node independently. Similarly, the *binary* term is formulated as a function of both nodes and edges. The *unary* term is sometimes called the **data** or **region** term and the *binary* term is sometimes called the **boundary** term. In this work we will prefer the terms *unary* and *binary* to describe these terms, since we will see that the unary and binary terms are used to represent many types of target specification beyond just data and boundaries.

All of the regularization methods discussed in this section are formulated on a *primal* graph in which each pixel is associated with a node to be labeled. This formulation has the longest history in the image segmentation literature, and it is now best understood how to perform target specification with different information for regularization algorithms with this formulation. Furthermore, primal methods are easy to apply in 3D since we can simply associate each voxel with a node. Recently, various authors have been looking into formulations of regularization and target specification on the *dual* graph (starting with [11, 12] and then extended by [13, 14, 15, 16]), but this thread is too recent to further review here. See [17, 12] for more information on primal and dual graphs in this context.

5.1.1 Power Watershed Formalism

Five prominent algorithms for targeted image segmentation were shown to unite under a single form of the unary and binary terms that differ only by a parameter choice.

Specifically, if we let the unary and binary terms equal (termed the **basic energy model** in [17])

$$E_{\text{unary}}(\mathbf{x}) = \sum_{v_i \in \mathcal{V}} w_{\mathcal{B}i}^q |x_i - 0|^P + w_{\mathcal{F}i}^q |x_i - 1|^P, \tag{5.2}$$

and

$$E_{\text{binary}}(\mathbf{x}) = \sum_{e_{ij} \in \mathcal{E}} w_{ij}^q |x_i - x_j|^P, \tag{5.3}$$

then Table 5.1 shows how different targeted segmentation algorithms are derived from differing values of p and q. The values and meaning of the unary object weight, $w_{\mathcal{F}i}$, unary background weight, $w_{\mathcal{B}i}$, and binary weights w_{ij} vary with respect to the target specification and are discussed more extensively in Section 5.2. Unless otherwise stated, we assume that $w_{\mathcal{F}i}, w_{\mathcal{B}i} > 0$, $\forall v_i \in \mathcal{V}$ and $w_{ij} > 0$, $\forall e_{ij} \in \mathcal{E}$. Specifically, it was shown in [18, 9, 19, 20] that the graph cuts algorithm [21, 22], the random walker algorithm [23], the shortest paths (geodesic) algorithm [24, 25], the conventional watershed algorithm [26] and the power watershed algorithm [27] could all be seen as providing a solution to the same energy (with different parameters), despite very different motivations and optimization methods. Specifically, the graph cuts algorithm optimizes the energy with a max-flow computation [28], the random walker algorithm optimizes the energy with a linear system solve, the shortest paths algorithm uses Dijkstra's algorithm to perform the optimization (or [29]), and the two watershed algorithms use speciality algorithms devised for this optimization, which are detailed in [26, 27]. It is important to note that the random walker and shortest paths algorithms are not motivated by optimization problems, but rather by determining the likelihood that a random walker first reaches a seed or by determining which pixel is connected by shortest path to the seeds. In fact, it is quite surprising that the random walker and shortest path algorithms, with motivations not derived from optimization, can still be viewed as the thresholded solutions to an optimization problem. However, it is important not to confuse the fact that random walker and shortest paths are equivalent to thresholding continuous optimization problems with a relaxation of a binary optimization problem. In fact, it is surprising that the optimization problems representing each algorithm (regardless of formulation) are so directly related to each other with simple parameter changes.

The generalized shortest path algorithm may also be considered to further subsume additional algorithms. In particular, the image foresting transform [24] showed that algorithms based on fuzzy connectedness [30, 31] could also be viewed as special cases of the shortest path algorithm (albeit with a different notion of distance). Furthermore, the GrowCut algorithm [32], derived in terms of cellular automata, was also shown to be equivalent to the shortest path based algorithm [33]. Consequently, the GrowCut algorithm can be seen as directly belonging to the framework described in Table 5.1, while the fuzzy connectedness algorithms are closely related (it has been suggested in [34] that fuzzy connectedness segmentation is obtained by a binary optimization of the case for $p \to \infty$ and $q \to \infty$). A recent treatment of the connection between these algorithms, fuzzy connectedness and a few other algorithms may be found in Miranda and Falcão [35].

p \ q	finite	∞
1	Graph cuts	Watershed $p = 1$
2	Random walker	Power watershed $p = 2$
∞	ℓ_1 norm Voronoi	Shortest paths (geodesics)

TABLE 5.1

Minimization of the general model presented in Equation (5.1) with Equations (5.2) and (5.3) leads to several well-known algorithms in the literature. This connection is particularly surprising since the random walker and shortest path algorithms were not originally formulated or motivated in the context of an energy optimization. Furthermore, graph cuts is often viewed as a binary optimization algorithm, even though optimization of Equation (5.1) for a real-valued variable leads to a cut of equal length (after thresholding the real-valued solution). Note that "∞" is used to indicate the solution obtained as the parameter approaches infinity (see [27] for details). The expression "ℓ_1 norm Voronoi" is intended to represent an intuitive understanding of this segmentation on a 4-connected (2D) or 6-connected (3D) lattice.

The optimization of fractional values of p was considered in [36, 9], in which this algorithm was termed *P-brush* and optimized using iterative reweighted least squares. Additionally, it was shown in [36, 9] that the optimal solutions are strongly continuous as p varies smoothly. Consequently, a choice of q finite and $p = 1.5$ would yield an algorithm which was a hybrid of graph cuts and random walker. It was shown in [18, 36, 9] that if q is finite, lower values of p create solutions with increased shrinking bias, while higher values of p create solutions with increased sensitivity to the locations of user-defined seeds. Furthermore, it was shown that the metrication artifacts are removed for $p = 2$ (due to the connection with the Laplace equation [9]), but are present for the $p = 1$ and $p = \infty$ solutions.

The regularization algorithms in the power watershed framework all assume that some object and background information is specified in the form of hard constraints or unary terms. In practice, it is possible that the object is specified using only object hard constraints or unary terms. In this case, it is possible to extend the models in the power watershed framework with a generic background term such as a balloon force [37, 38] or by using the isoperimetric algorithm [5, 39], as shown in [17]

5.1.2 Total Variation and Continuous Max-Flow

One concern about the $p = 1$ and finite q model above (graph cuts) is the metrication issues that can arise as a result of the segmentation boundary being measured by the edges cut, which creates a dependence on the edge structure of the image lattice [9]. This problem has been addressed in the literature [40, 41] by adding more edges, but this approach can sometimes create computational problems (such as memory usage), particularly in 3D image segmentation. Consequently, the total variation and continuous max-flow algorithms are sometimes preferred over graph cuts because they (approximately) preserve the geometrical description of the solution as minimizing boundary length (surface area), but also reduce metrication errors (without additional memory) by using a formulation at the *nodes* in order to preserve rotational invariance [42, 43, 44, 20].

The total variation (TV) functional is most often expressed using a continuous domain (first used in computer vision by [45, 46]). However, formulations on a graph have been considered [47, 48]. Although TV was initially considered for image denoising, it is natural to apply the same algorithm to targeted segmentation by using an appropriate unary term and interpretation of the variable [44, 17]. In the formalism given in (5.1), the segmentation algorithm by total variation has the same unary term as (5.2), but the binary term is given by [48]

$$E_{\text{binary}}^{\text{TV}}(\mathbf{x}) = \sum_{v_i \in \mathcal{V}} \left(\sum_{j, \forall e_{ij} \in \mathcal{E}} w_{ij} |x_i - x_j|^2 \right)^{\frac{p}{2}}. \tag{5.4}$$

However, note that some authors have preferred a node weighting for TV, which would instead give

$$E_{\text{binary}}^{\text{TV}}(\mathbf{x}) = \sum_{v_i \in \mathcal{V}} w_i \left(\sum_{j, \forall e_{ij} \in \mathcal{E}} |x_i - x_j|^2 \right)^{\frac{p}{2}}. \tag{5.5}$$

Since our discussions in Section 5.2 will treat node weights as belonging to the unary term and edge weights as belonging to the binary term, we will explicitly prefer the formulation given in Equation (5.4).

Optimization of the TV functional is generally more complicated than optimization of any of the algorithms in the power watershed framework. The TV functional is convex and may therefore be optimized with gradient descent [48], but this approach is typically too slow to be useful in many practical image segmentation situations. However, fast methods for performing TV optimization have appeared recently [49, 50, 51, 52], which makes these methods more competitive in speed compared to the algorithms in the power watershed framework described in Section 5.1.1.

The dual problem to minimizing the TV functional is often considered to be the continuous max-flow (CMF) problem [53, 54]. However, more recent studies have shown that this duality is fragile in the sense that weighted TV and weighted max-flow are not always dual in the continuous domain [55] and that standard discretizations of both TV and CMF are not strongly dual to each other (weighted or unweighted) [20]. Consequently, it seems best to consider the CMF problem to be a separate, but related, problem to TV. Although CMF was introduced into the image segmentation literature in its continuous form [42], subsequent work has shown how to write the CMF problem on an arbitrary graph [20]. This graph formulation of the continuous max-flow problem (termed **combinatorial continuous max-flow**) is *not* equivalent to the standard max-flow problem formulated on a graph by Ford and Fulkerson [56]. The difference between combinatorial continuous max-flow (CCMF) and conventional max-flow is that the capacity constraints for conventional max-flow are defined on edges, while the capacity constraints for CCMF are defined on nodes. As with TV, this formulation on nodes affords a certain rotational independence which leads to a reduction of metrication artifacts.

The native CMF and CCMF problems do not have explicit unary terms in the sense defined in Equation (5.1), although such terms could be added via the dual of these algorithms. The incorporation of unary terms into the CCMF model was addressed specifically in [20]. Furthermore, both the CMF and CCMF problems for image segmentation were formulated using node weights [42, 20] to represent unary terms as well as flow constraints (similar to the second TV represented in Equation (5.5)). However, as with the second TV, this use of node weights in both contexts does not fit the formalism adopted in Section 5.2. Although the standard formulations use node weights for the "binary" term (the capacity constraints), this could be easily modified to adopt edge weights in the manner formulated by [53] for the continuous formulation of CMF or adopted by [48] for the formulation of an edge-weighted TV. A fast optimization method for the CCMF problem was given by Couprie et al. in [20].

5.2 Target Specification

In the previous discussion we discussed the basic regularization models underlying a variety of targeted segmentation algorithms. However, before these regularization models can be applied for targeted object segmentation it is necessary to specify additional information. Specifically, the targeting information may be encoded on the graph by manipulating unary term weights, binary term weights, constraints, and/or extra terms to Equation (5.1). Therefore, some types of target specification will have little effect on our ability to find exact solutions for the algorithms in Section 5.1, while other kinds of target specification force us to employ approximate solutions only (e.g., the objective function becomes nonconvex).

In this section we will review a variety of ways to provide a target specification that have appeared in the literature. In general, multiple sources of target specification may be combined by simply including all constraints, or by adding together the weights defining the unary term and binary term specifications. However, note that *some* form of target specification is necessary to produce a nontrivial solution of the regularization algorithms in Section 5.1. While discussing different methods of target specification, we will deliberately be neutral to the choice of underlying regularization model that was presented in Section 5.1. The general form of all the models given in Equation (5.1) can accommodate any of the target specification methods

(a) Seeds/scribbles   (b) Bounding box   (c) Rough   boundary   (d) Bounding polygon
                                                        specification

FIGURE 5.1

Different user interaction modes that have been used to specify the segmentation target. These interaction modes may also be populated automatically (e.g., by finding a bounding box automatically).

that will be reviewed, since each model can utilize target information encoded in user interaction, additional terms, unary node weights, and/or binary edge weights.

5.2.1 Segmentation Targeting With User Interaction

An easy way of specifying the target object is to allow a user to specify the desired object. By building a system that can incorporate user interaction, it becomes possible to distribute a system that is capable of segmenting a variety of (a priori unknown) targets rather than a single known target. Present-day image editing software (e.g., Adobe Photoshop) always includes some sort of interactive targeted image segmentation tool.

There are essentially three methods in the literature to allow a user to specify the target object. The first method is to label some pixels inside/outside the object, the second method is to specify some indication of the object boundary and the third is to specify the sub-region within the image that contains the object. Figure 5.2.1 gives an example of these methods of user interaction.

5.2.1.1 Seeds and Scribbles

In this user interaction mode, the user specifies labels (typically with a mouse or touchscreen) for some pixels inside the target object and for some pixels outside the object. These user-labeled pixels are often called **seeds**, **scribbles**, or **hard constraints**. Assume that all pixels labeled object (foreground) belong to the set $\mathcal{F}$ and all pixels labeled background belong to set $\mathcal{B}$ where $\mathcal{F} \cap \mathcal{B} = \emptyset$. These seeds are easily incorporated into (5.1) by performing the optimization subject to the constraints that $x_i = 1$ if $v_i \in \mathcal{F}$ and $x_i = 0$ if $v_i \in \mathcal{B}$.

An equivalent way of writing these seeds is to cast them as unary terms, such that (in the context of (5.2)) $w_{\mathcal{F}i} = \infty$ if $v_i \in \mathcal{F}$ and $w_{\mathcal{B}i} = \infty$ if $v_i \in \mathcal{B}$. In this way, the unary terms effectively enforce the hard constraints. Enforcement of the seeds through unary terms has two advantages. The first advantage is that the grid structure of the image lattice is not disrupted by adding a unary term, which can be important for the efficient implementation of some algorithms (see [57]). The second advantage is that the unary formulation of the seeds allows one to incorporate some level of uncertainty in the seed locations by setting the weights to some finite value instead of infinity. In this setting, a smaller weight encodes less confidence in the location while a larger weight encodes greater confidence. Note that it is possible to incorporate some seeds as hard constraints and others as unary terms. In this way, one may build a system that requires only object seeds (set as hard constraints) in which the background seeds were set at every other pixel (i.e., $\mathcal{B} = \mathcal{V} - \mathcal{F}$) via low-weight unary terms. Consequently, the user would only need to provide object interactions to specify the target, which would reduce the amount of user time necessary to achieve the segmentation.

Other interfaces for user seeding have also been explored in the literature. For example, Heiler et al. [58] suggested allowing the user to specify that two pixels belong to different labels without specifying which pixel is object and which is background. This sort of interaction may be incorporated by introducing a negative edge weight to encode repulsion [59]. Specifically, if v_i and v_j are specified to have an opposite label, this constraint can be encoded by introducing a new edge e_{ij} with weight $w_{ij} = -\infty$, which forces the optimization in Equation (5.1) to place these nodes on opposite sides of the partition. Note, however, that the introduction of negative edge weights requires an explicit constraint on the optimization in Equation (5.1) to force $0 \leq x_i \leq 1$ in order to avoid the possibility of an unbounded solution.

5.2.1.2 Object Boundary Specification

Another natural user interface is to roughly specify part or all of the desired object boundary. This boundary specification has been encoded into the above algorithms in one of two ways by editing a polygon or by providing a soft brush. Boundary specification is substantially different in 2D and 3D segmentation, since 2D object boundaries are contours and 3D object boundaries are surfaces. In general, approaches that have a used boundary specification have operated mainly in 2D image segmentation.

A similar boundary-based approach to interactively specify the object boundary was proposed by Wang et al. [60], who asks the user to freehand paint the boundary with a large brush (circle) such that the true boundary lies inside the painted region. Given this painted region (which the authors assumed could be painted quickly), hard constraints for the object label were placed on the inside of the user-painted boundary and hard constraints for the background label were placed on the outside of the user-painted boundary. See Chapter 9 for more details on this user interface.

5.2.1.3 Sub-Region Specification

Sub-region specification was introduced to save the user time, as an alternative to placing seeds or marking the object boundary (even roughly). The most common approach to sub-region specification has been to ask the user to draw a box surrounding the target object [61, 62, 63, 64]. This interface is convenient for a user, but also has the advantage that it is easy to connect to an automatic preprocessing algorithm to produce a fully automated segmentation, that is, by using machine learning with a sliding window to find the correct bounding box (a Viola-Jones type detector [65]). We showed with a user study in [64] that a single box input was sufficient to fully specify a target object over a wide variety of object sizes, shapes, backgrounds and homogeneity levels, even when the images do not contain explicit semantic content.

Rother et al. [61] introduced one of the most well-known methods for making use of the bounding box interface for segmentation called GrabCut. In the GrabCut algorithm background constraints are placed on the outside of the user-supplied box, while allowing an appearance model to drive the object segmentation in an iterative manner (see Section 5.2.2). Lempitsky et al. [62] suggested constraining the segmentation such that the user-drawn box is tightly fit to the resulting segmentation, which they term the *tightness prior*. Grady et al. [64] adopted the same interface for the PIBS algorithm (probabilistic isoperimetric box segmentation), which used no unary term, but instead set directed edge weights based on an estimate of the object appearance probability for each pixel and applied the isoperimetric segmentation algorithm [5, 39] with the bounding box set as background seeds.

5.2.2 Segmentation Targeting with Appearance Information

The purpose of an appearance model is to use statistics about the distribution of intensity/color/texture etc., inside, and outside the object to specify the target object. In most cases, appearance models are incorporated as unary terms in the segmentation. Broadly speaking, there are two approaches to appearance modeling. In the first approach, the model is established once prior to performing segmentation. In the second approach, the appearance model is learned adaptively from the segmentation to produce a joint estimation of both the appearance model and the image segmentation.

5.2.2.1 Prior Appearance Models

In the earliest appearance of these targeted segmentation models, it was assumed that the probability was known that would match the image intensity/color at each pixel to an object or background label. Specifically, if g_i represents the intensity/color at pixel v_i, then it was assumed that $p(g_i|x_i = 1)$ and $p(g_i|x_i = 0)$ were known. The representation of $p(g_i|x_i)$ may be parametric or nonparametric. Presumably, these probabilities are based on training data or known a priori (as in the classical case of binary image segmentation). Given known values of the probabilities, $p(g_i|x_i = 1)$

and $p(g_i|x_i = 0)$, these probabilities are used to affect the unary terms in the above model by setting, for (5.2), $w_{\mathcal{F}i} = p(g_i|x_i = 1)$ and $w_{\mathcal{B}i} = p(g_i|x_i = 0)$. Note that $w_{\mathcal{F}i} = -\log(p(g_i|x_i = 1))$ and $w_{\mathcal{B}i} = -\log(p(g_i|x_i = 0))$ have also been used instead in conjunction with a derivation from an markov random field (see [21]).

If the probability models are not known a priori, then they may be estimated from user-drawn seeds [22]. Specifically, the seeds may be viewed as a *sample* of the object and background distributions. Consequently, these samples may be used with a kernel estimation method to estimate $p(g_i|x_i = 1)$ and $p(g_i|x_i = 0)$. When **g** represents multichannel image data (e.g., color), then a parametric method, such as a Gaussian mixture model, is typically preferred [66].

In addition to modeling $p(g_i|x_i)$ directly, several groups have used a transformed version of g_i to establish unary terms. The purpose of this transformation is to model objects with a more complex appearance, which may be better characterized by a texture than by a simple intensity/color distribution. Specifically, transformations which appear in the literature include the outputs of textons/filter banks [67], structure tensors [68, 69], tensor voting [70], or features learned on-the-fly from image patches [71]. Formally, we may view these methods as estimating $p(g_{\mathcal{W}_i}|x_i)$, where $\mathcal{W}_i$ is a window of pixels centered on pixel v_i.

Sometimes the target segmentation has a known shape that can be exploited to model $p(g_{\mathcal{W}_i}|x_i)$ by estimating the similarity of a window around pixel v_i to the known shape. This matching may be viewed as the application of a shape filter to produce a response that is interpreted as $p(g_{\mathcal{W}_i}|x_i)$. Shape filters of this nature have usually been applied to elongated or vessel-like objects. Vessel-like objects have frequently been detected using properties of the Hessian matrix associated with a window centered on a pixel. Frangi et al. [72] used the eigenvalues of the Hessian matrix to define a *vesselness measure* that was applied in Freiman et al. [73] as a unary term for defining the vesselness score. A similar approach was used by Esneault et al. [74] who used a cylinder matching function to estimate $p(g_{\mathcal{W}_i}|x_i)$ and gave a unary term for the segmentation of vessel-like objects. Using shape information in this way is fundamentally different from the shape models described in Section 5.2.4 since the use of shape information here is to directly estimate $p(g_{\mathcal{W}_i}|x_i)$ via a filter response, while the models in Section 5.2.4 use the fit of a top-down model to fit a unary term.

5.2.2.2 Optimized Appearance Models

Models in the previous section assumed that the appearance models were known a priori or could be inferred from a presentation of the image and/or user interaction. However, one may also view the appearance model as a variable, which may be inferred from the segmentation process. In this setting, we may rewrite Equation (5.1) as

$$E(P, \mathbf{x}) = E_{\text{unary}}(P, \mathbf{x}) + \lambda E_{\text{binary}}(P, \mathbf{x}), \tag{5.6}$$

in which P represents the appearance model, that is, P is shorthand for $w_{\mathcal{F}i}$ and $w_{\mathcal{B}i}$. Segmentation algorithms which take this form typically alternate between iterations in which (5.6) is optimized for $\mathbf{x}$ relative to a fixed P and then optimized for P relative to a fixed $\mathbf{x}$. This alternation is a result of the difficulty in finding a joint global optimization for both P and $\mathbf{x}$, and the fact that each iteration of the alternating optimization approach is guaranteed to produce a solution with lower or equal energy from the previous iteration.

The most classic approach with an optimized appearance model is the Mumford-Shah model [75, 76] in which $w_{\mathcal{F}i}$ and $w_{\mathcal{B}i}$ are a function of an idealized object and background image, $\mathbf{a}$ and $\mathbf{b}$. Specifically, we may follow [77] to write the piecewise smooth Mumford-Shah model as

$$E(P,\mathbf{x}) = \left(\mathbf{x}^T\left(\mathbf{g}-\mathbf{a}\right)^2\right) + \left(\mathbf{1}-\mathbf{x}\right)^T\left(\mathbf{g}-\mathbf{b}\right)^2\right) + \\ \mu\left(\mathbf{x}^T|\mathbf{A}|^T\left(\mathbf{A}\mathbf{a}\right)^2 + \left(\mathbf{1}-\mathbf{x}\right)^T|\mathbf{A}|^T\left(\mathbf{A}\mathbf{b}\right)^2\right) + \lambda E_{\text{binary}}(\mathbf{x}), \tag{5.7}$$

where $\mathbf{A}$ is the incidence matrix (see definition in Chapter 1). The μ and λ parameters are free parameters used to weight the relative importance of the terms. The form presented here represents a small generalization of the classic form, also in [77], in which $\mathbf{x}$ was assumed to be binary and $E_{\text{binary}}(\mathbf{x}) = \sum_{e_{ij}} w_{ij}|x_i - x_j|$, that is, that $p = q = 1$ in (5.3). The optimization of Equation (5.7) may be performed by alternating steps in which $\mathbf{a}, \mathbf{b}$ are fixed and $\mathbf{x}$ is optimized with steps in which $\mathbf{x}$ is fixed an $\mathbf{a}, \mathbf{b}$ are optimized. Since each alternating iteration of the optimization is convex, the energy of successive solutions will monotonically decrease.

A simplified version of the Mumford-Shah model is known as the piecewise constant model (also called the Chan-Vese model [78]) in which $\mathbf{a}$ and $\mathbf{b}$ are represented by constant functions (i.e., $a_i = k_1, \forall v_i \in \mathcal{V}$ and $b_i = k_2, \forall v_i \in \mathcal{V}$). We may write the piecewise constant Mumford-Shah model as

$$E(P,\mathbf{x}) = \left(\mathbf{x}^T\left(\mathbf{g}-k_1\mathbf{1}\right)^2 + \left(\mathbf{1}-\mathbf{x}\right)^T\left(\mathbf{g}-k_2\mathbf{1}\right)^2\right) + \lambda E_{\text{binary}}(\mathbf{x}). \tag{5.8}$$

Therefore, the piecewise constant Mumford-Shah model neglects the second term in (5.7). The same alternating optimization of Equation (5.8) is possible, except that the optimization for $\mathbf{a}$ and $\mathbf{b}$ is extremely simple, that is, $a_i = k_1 = \frac{\mathbf{x}^T\mathbf{g}}{\mathbf{x}^T\mathbf{1}}$ and $b_i = k_2 = \frac{(\mathbf{1}-\mathbf{x})^T\mathbf{g}}{(\mathbf{1}-\mathbf{x})^T\mathbf{1}}$.

A similar optimized appearance model was presented in the GrabCut algorithm [61] which was applied to color images. In this approach, the object and background color distributions were modeled as two Gaussian mixture models (GMM) with unknown parameters. Specifically, GrabCut optimizes the model

$$E(\theta,\mathbf{x}) = H(\theta_{\mathcal{F}},\mathbf{g},\mathbf{x}) + H(\theta_{\mathcal{B}},\mathbf{g},(\mathbf{1}-\mathbf{x})) + \lambda E_{\text{binary}}(\mathbf{x}), \tag{5.9}$$

in which θ represents the parameters for the GMM and $H(\theta_{\mathcal{F}}, g_i, x_i)$ represents the (negative log) probability that g_i belongs to the object GMM (see [61] for more details). As with the optimization of the Mumford-Shah models, the optimization of the GrabCut energy also proceeds by alternating between estimation of $\mathbf{x}$ (given a fixed $\theta_{\mathcal{F}}$ and $\theta_{\mathcal{B}}$) and estimation of $\theta_{\mathcal{F}}$ and $\theta_{\mathcal{B}}$ (given a fixed $\mathbf{x}$).

5.2.3 Segmentation Targeting with Boundary Information

A widespread method for specifying a target is to impose a model of the target object boundary appearance that may be incorporated into the edge weights of the binary term in Equation (5.3). Depending on the form of the boundary regularization, the edge weights represent either the *affinity* between pixels (i.e., small weight indicates a disconnect between pixels) or the *distance* between pixels (i.e., large weight indicates a disconnect between pixels). These descriptions are reciprocal to each other, as described in [17], i.e., $w_{\text{affinity}} = \frac{1}{w_{\text{distance}}}$. To simplify our discussion in this section, we assume that the edge weights represent *affinity* weights.

The most classical approach to incorporating boundary appearance into a targeted segmentation method is to assume that the object boundary is more likely to pass between pixels with large image contrast. The functions which are typically used to encode boundary contrast were originally developed in the context of non-convex descriptions of the image denoising problem [79, 80, 81] and later enshrined in the segmentation literature via the normalized cuts method [3] and the geodesic active contour method [82]. Since these weighting functions may be derived from corresponding M-estimators [83, 17] we use the term *Welsch function* to describe

$$w_{ij} = \exp\left(-\beta \|g_i - g_j\|_2^2\right), \tag{5.10}$$

and *Cauchy function* to describe

$$w_{ij} = \frac{1}{1 + \beta \|g_i - g_j\|_2^2}, \tag{5.11}$$

where g_i represents the intensity (grayscale) value or a color (or multispectral) vector associated with pixel v_i. In practice, these edge weighting functions are often slightly altered to avoid zero weights (which can disconnect the graph) and to accommodate the dynamic range in different images. To avoid zero edge weights, a small constant may be added (e.g., $\epsilon = 1e^{-6}$). The dynamic range of an image may be accommodated by dividing β by a constant ρ where $\rho = \max_{e_{ij}} \|g_i - g_j\|_2^2$ or $\rho = \text{mean}_{e_{ij}} \|g_i - g_j\|_2^2$. Another possibility for choosing ρ more robustly is to choose the edge difference corresponding to the 90th percentile of all edge weights (as done in [80]). In one study comparing the effectiveness of the Welsch and Cauchy functions, it was determined that the Cauchy function performed slightly better on average (and is also computationally less expensive to compute at runtime) [84].

One problem with the assumption that the object boundary passes between pixels with large image contrast is that the difference between the target object and its boundary may be more subtle than some contrasts present in the image. For example, soft tissue boundaries in a medical computed tomography image are often much lower contrast than the tissue/air boundaries present at the lung or external body interfaces. Another problem is that the contrast model is only appropriate for objects with a smooth appearance, but is less appropriate for textured objects. To account for low contrast or textured objects, the weighting functions in (5.10) and

(5.11) may be modified by replacing g_i with $H(g_i)$ where H is a function which maps raw pixel content to a transform space where the object appearance is relatively constant. For example, if the target object exhibits a striped texture, then H could represent the response of a "stripe filter" which causes the target object to produce a relatively uniform response. In fact, any of the appearance models described in Section 5.2.2 can be used to establish more sophisticated boundary models by adopting $H(g_i) = p(g_i|x_i = 1)$. This sort of model can be very effective even when the target object has a relatively smooth appearance, since it can be used to detect boundaries which are more subtle in the original image [84, 64].

5.2.3.1 Boundary Polarity

A problem with the boundary models in the previous section is that they describe the probability that the boundary passes between two pixels without accounting for the *direction* of the transition [85]. This problem is illustrated in Figure 5.2 which shows how a more subtle correct boundary may be overwhelmed by a strong boundary with the wrong transition. Alternately, we may say that the target specification may be ambiguous without a specification of the boundary polarity, as illustrated in Figure 5.2.

A model of the target object appearance allows us to specify that the target segmentation boundary transitions from the pixels which have the object appearance to pixels which do not [86]. Mechanically, this polarity model has been specified by formulating the graph models on a *directed* graph in which each undirected edge e_{ij} is replaced by two directed edges e_{ij} and e_{ji} such that w_{ij} and w_{ji} may or may not be equal [85, 86]. Given an appearance model from Section 5.2.2, it was proposed in [86] to formulate these directed edges via

$$w_{ij} = \begin{cases} \exp\left(-\beta_1 ||H(g_i) - H(g_j)||_2^2\right) & \text{if } H(g_i) > H(g_j), \\ \exp\left(-\beta_2 ||H(g_i) - H(g_j)||_2^2\right) & \text{else,} \end{cases} \qquad (5.12)$$

where $\beta_1 > \beta_2$. In other words, boundary transitions from pixels matching the target object appearance to pixels that do not are less costly (from the standpoint of energy minimization). Note that the same approach could be applied using the Cauchy weighting formula.

Edge weights have been shown to play the role on a graph that corresponds to the *metric* in conventional multivariate calculus [17]. In this respect, it was shown that the weights of a directed graph could be viewed specifically as representing a Finsler metric on a graph [87].

5.2.4 Segmentation Targeting with Shape Information

Object shape is a natural and semantic descriptor for specifying a target segmentation. However, even semantically, the description of object shape can take many

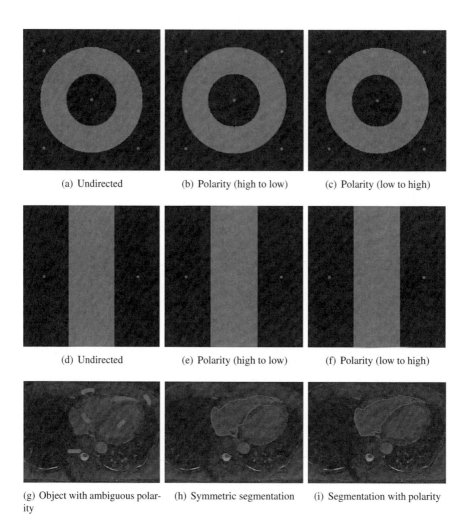

(a) Undirected   (b) Polarity (high to low)   (c) Polarity (low to high)

(d) Undirected   (e) Polarity (high to low)   (f) Polarity (low to high)

(g) Object with ambiguous polarity   (h) Symmetric segmentation   (i) Segmentation with polarity

FIGURE 5.2

(a-f) Ambiguous synthetic images where the use of polarity can be used to appropriately specify the target object. (g-i) A real image with ambiguous boundary polarity. It is unclear whether the endocardium (inside) or epicardium (outside) is desired. Using an undirected model chooses the endocardium (inside) in one part of the segmentation and the epicardium (outside) in the other part. However, polarity information can be used to specify the target as the endocardium (inside boundary).

forms, from general descriptions (e.g., "The object is elongated") to parametric descriptions (e.g., "The object is elliptical" or "The object looks like a leaf") and feature-driven descriptions (e.g., "The object has four long elongated parts and one short elongated part connected to a broad base"). Targeted segmentation algorithms using shape information have followed the same patterns of description. In this section we will cover both *implicit* (general) shape descriptions and *explicit* (parametric) shape descriptions. The discussion on feature-driven object descriptions will be postponed to Section 5.2.5.

It is important to distinguish between those shape models in which the graph is being used to regularize the *appearance* of a shape feature and those models in which the *regularization* provides the shape model. To illustrate this point, consider the segmentation of elongated shapes. The shape can be encoded by either a standard regularization of an elongation appearance measure (e.g., a vesselness measure from Section 5.2.2.1 with TV or random walker regularization) or by a regularization that biases the segmentation toward elongated solutions (e.g., a curvature regularization) or both. Put differently, we must distinguish between a data/unary term that measures elongation and a binary or higher order term that promotes elongation. Section 5.2.2.1 treated the appearance-based type, and this section treats the regularization type.

5.2.4.1 Implicit Shape Models

The targeted segmentation models described in Section 5.1 implicitly favor large, compact (blob-like) shapes. Specifically, for an object of fixed area with a single central seed, the targeted algorithms described in Section 5.1 favor a circle (up to metrication artifacts) when applied on a graph with weights derived from Euclidean distance of the edge length. This property may be seen in graph cuts and TV because a circle is the shape that minimizes the boundary length over all shapes with a fixed area (known classically from the isoperimetric problem). The shortest path algorithm also finds a circle because the segmentation returns all pixels within a fixed distance from the central seed (where the distance is fixed by the area constraint). Similarly, the random walker algorithm also favors a circle, since the segmentation includes all pixels within a fixed resistance distance (where the distance is fixed by the area constraint).

Although many target segmentation objects are compact in shape, it is not always the case. For an example in medical imaging, thin blood vessels are a common segmentation target, as are thin surfaces such as the grey matter in cerebral cortex. Segmenting these objects can be accomplished with the algorithms in Section 5.2 if the bias toward compact shapes can be overcome by a strong appearance term or a sharp boundary contrast (leading to an unambiguous binary term via the edge weighting). However, in many real images we cannot rely on strong appearance terms or boundary contrast. A variety of graph-based models have been introduced recently that measure a discrete notion of the *curvature* of a segmentation boundary [88, 89, 90, 14, 91]. Specifically, these models seek to optimize a discrete form of the elastica equation proposed by Mumford [92], which he showed to embody the Bayes optimal solution of a model of curve continuity. The continuous form of the elastica equation is

$$e(\mathcal{C}) = \int_{\mathcal{C}} (a + b\kappa^2)ds \quad a, b > 0, \tag{5.13}$$

where κ denotes scalar curvature, and ds is the arc length element. At the time of writing, there are several different ideas that have been published for writing an appropriate discretization of this model and performing the optimization. Since it is not yet clear if these or some other idea will become preferred for performing curvature optimization, we leave the details of the various methods to the references.

Another implicit shape model in the literature biases the segmentation toward rectilinear shapes [93]. Specifically, this model takes advantage of the work by Zunic and Rosin [94] on rectilinear shape measurement, where it was shown that the class of shapes which minimize the ratio

$$Q(P) = \frac{\mathrm{Per}_1(P)}{\mathrm{Per}_2(P)}, \tag{5.14}$$

are rectilinear shapes (with respect to the specified X and Y axis). Here, $\mathrm{Per}_1(P)$ denotes the ℓ_1 perimeter of shape P according to the specified X and Y axis. Similarly, $\mathrm{Per}_2(P)$ denotes the perimeter of shape P with the ℓ_2 metric. In other words,

$$\mathrm{Per}_1(P) = \int |u'(t)| + |v'(t)|dt, \tag{5.15}$$
$$\mathrm{Per}_2(P) = \int \sqrt{u'^2(t) + v'^2(t)}dt, \tag{5.16}$$

for some parametrization $u(t), v(t)$ of the shape boundary P. Note that (5.14) does not provide a rotationally invariant measure of rectilinearity (i.e., this approach assumes that the rectangular object roughly aligns with the image axes). Noting that the perimeter of the segmentation may be measured by an ℓ_1 or ℓ_2 metric, depending on the edge structure and weighting (see [93, 40]), it is possible to create two graphs to represent the ℓ_1 and ℓ_2 metrics, displayed in Figure 5.2.4.1. Since a rectilinear shape is small in ℓ_1 compared to ℓ_2, this method of implicit rectilinear shape segmentation was called the *opposing metrics method*. Given an ℓ_1 graph, represented by Laplacian matrix $\mathbf{L}_1$ and an ℓ_2 graph, represented by Laplacian matrix $\mathbf{L}_2$, a binary indicator vector $\mathbf{x}$ that minimizes the ratio

$$E_{\mathrm{rect}}(\mathbf{x}) = \frac{\mathrm{Per}_1(\Omega\mathbf{x})}{\mathrm{Per}_2(\Omega\mathbf{x})} \approx \frac{\mathbf{x}^T \mathbf{L}_1 \mathbf{x}}{\mathbf{x}^T \mathbf{L}_2 \mathbf{x}}, \tag{5.17}$$

will be a rectilinear shape. Consequently, the term in (5.17) may be used as a regularization to implicitly bias the segmentation toward a rectilinear shape. Note that the authors of [93] chose to relax the binary constraint on $\mathbf{x}$ to produce a generalized eigenvector problem, which may be solved efficiently for sparse matrices using established methods. A final observation of the energy in (5.17) is that the numerator fits precisely into the power watershed framework in Section 5.1, particularly (5.3).

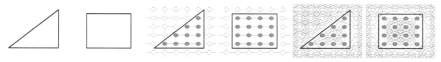

(a) 4-unit triangle. (b) 4-unit square. (c) Cut-measured (d) Cut-measured (e) Cut-measured (f) Cut-measured ℓ_1 perimeter: ℓ_1 perimeter: 16, perimeter on L_1 perimeter on L_1 perimeter on L_2 perimeter on L_2 16, ℓ_2 perimeter: ℓ_2 perimeter: 16 graph: 16 graph: 16 graph: 11.8317 graph: 13.9192 13.6569

FIGURE 5.3

A rectilinear shape is known to minimize the ratio of the perimeter measured with an ℓ_1 metric and the perimeter measured with an ℓ_2 metric [94]. It is possible to measure the perimeter of a object, expressed as a cut, with respect to various metrics with arbitrary accuracy by adjusting the graph topology and weighting [40]. The opposing metrics strategy for segmenting rectilinear shapes is to look for the segmentation that minimizes the cut with respect to one graph, L_1, while maximizing the cut with respect to a second graph, L_2.

Consequently, one interpretation of the opposing metrics method of [93] is that it relies on the same sort of binary regularization as before, except that it uses a second graph (represented by the denominator) to modify the implicit default shape from a circle to a rectilinear shape.

In this section it was shown how most traditional first-order graph-based segmentation algorithms implicitly bias the segmentation toward compact circular shapes. However, by modifying the functionals to include the higher-order term of curvature, it was possible to remove the bias toward compact shapes, meaning that the energy was better suited for elongated objects such as blood vessels. Additionally, by further modifying the functional to include a second graph (present in the denominator) it is possible to implicitly bias the segmentation toward rectilinear objects. One advantage of these implicit methods is that it has been frequently demonstrated that a global optimum of the energy functional is possible to achieve, meaning that the algorithms have no dependence on initialization. In the next section we consider those methods that add an explicit shape bias by using training data or a parameterized shape. While this approach allows for the targeting of more complex shapes, a global optimum becomes very hard to achieve, meaning that these methods rely on a good initialization.

5.2.4.2 Explicit Shape Models

One of the first approaches to incorporating an explicit shape model into graph-based segmentation was by Slabaugh and Unal [95] who constrained the segmentation to return an elliptical object. The idea was effectively to add a third term to the conventional energy,

$$E(\mathbf{x}, \theta) = E_{\text{data}}(\mathbf{x}) + E_{\text{binary}}(\mathbf{x}) + E_{\text{ellipse}}(\mathbf{x}, \theta), \qquad (5.18)$$

where θ represents the parameters of an ellipse, and

$$E_{\text{ellipse}}(\mathbf{x}, \theta) = \sum_{v_i \in \mathcal{V}} |M_i^\theta - x_i|, \tag{5.19}$$

where M^θ represents the mask corresponding to the ellipse parameterized by θ such that pixels inside the ellipse are marked '1' and pixels outside the mask are marked '0'.

Since finding a global optimum of (5.18) is challenging, Slabaugh and Unal [95] performed an alternating optimization of (5.18) in which $\mathbf{x}$ was first optimized for with a fixed θ (in [95] this optimization was performed only over a narrow band surrounding the current solution) and then θ was optimized for a fixed $\mathbf{x}$. Although this formulation was introduced for an elliptical shape, it is clear that it could be extended to any parameterized shape.

A problem with the above approach to targeting with an explicit shape model is that complex shapes may be difficult to effectively parameterize and fit to a fixed solution for $\mathbf{x}$. Consequently, an alternative approach is to use a shape template that can be fit to the current segmentation. This was the approach adopted by Freedman and Zhang [96], who used a distance function computed on the fit template to modify the edge weights to promote a segmentation matching the template shape.

Freedman and Zhang's method for incorporating shape information suffers from the fact that it does not account for any shape variability, but rather assumes that the target segmentation for a particular image closely matches the template. Shape variability of a template may be accounted for using a PCA decomposition of a set of training shapes, as described in the classical work of Tsai et al. [97] using a level set framework. A similar method of accounting for shape variability was formulated on a graph by Zhu-Jacquot and Zabih [98] who model the shape template, U, as

$$U = \bar{U} + \sum_{k=1}^{K} w_k U^k, \tag{5.20}$$

where $\bar{U}$ is the mean shape and U^k is the kth principal component of a set of training shapes. The weights w_k and the set of rigid transformations is optimized at each iteration of [98] with the current segmentation $\mathbf{x}$ fixed. Then, the segmentation $\mathbf{x}$ is determined using the fit shape template to define a unary term (as opposed to the binary term in [96]).

5.2.5 Segmentation Targeting with Object Feature Information

In some segmentation tasks, substantial training data is available to provide sufficient target specification to specify the desired object. Typically, we assume that the training data contains both images and segmentations of the desired object. From this training data, it may be possible to learn a set of characteristic geometric or appearance features to describe the target object and relationships between those features.

One of the first methods to adopt this approach in the context of targeted graph-based segmentation was the ObjCut method of Kumar et al. [99]. In this algorithm, the authors build on the work of pictorial structures [100] and assume that the target object may be subdivided into a series of parts that may vary in relationship to each other. The composition of the model into parts (and their relationships) is directed by the algorithm designer, but the appearance and variation of each part model is learned using the training data. The algorithm operates by fitting the orientation and scale parameters of the parts and then using the fitted pictorial structure model to define unary terms for the segmentation. Levin and Weiss build on this work to employ a conditional random field to perform joint training of the model using both the high-level module and the low-level image data [101].

5.2.6 Segmentation Targeting with Topology Information

If the target object has a known topology, then it is possible to use this information to find an appropriate segmentation. Unfortunately, specification of a global property of the segmentation, like topology, has been challenging. Topology specification has seen increasing interest recently, however, all of the work we are aware of has focused on this sort of specification in the context of the graph cuts segmentation algorithm, in which the problems have generally been shown to be NP-Hard. Two types of topology specification have been explored in the literature—Connectivity and genus.

5.2.6.1 Connectivity Specification

When the segmentation target is a single object, it is reasonable to assume that the object is connected (although this assumption is not always correct, as in the case of occlusion). A segmentation, $\mathcal{S}$, is considered connected if $\forall v_i \in \mathcal{S}, v_j \in \mathcal{S}, \exists \pi_{ij}$, s.t. if $v_k \in \pi_{ij}$, then $v_k \in \mathcal{S}$.

In general, any of the segmentation methods in Section 5.1 do not guarantee that the returned segmentation is connected. However, under certain conditions, it is known that any algorithm in the power watershed framework will guarantee connectivity. The conditions for guaranteed connectivity are: (1) No unary terms, (2) The set of object seeds is a connected set, and (3) For all edges, $w_{ij} > 0$. Given these conditions, any algorithm in the power watershed framework produces a connected object, since the **x** value for every nonseeded node is always between the **x** values of its neighbors, meaning that if x_i is above the 0.5 threshold (i.e., included in the segmentation $\mathcal{S}$), then it has at least one neighboring node that is also above threshold (i.e., included in the segmentation $\mathcal{S}$). More specifically, any solution **x** that minimizes

$$E(\mathbf{x}) = \sum_{e_{ij} \in E} w_{ij}^q |x_i - x_j|^p, \tag{5.21}$$

for some choice of $p \geq 0$ and $q \geq 0$ has the property for any nonseeded node that

$$x_i = \arg\min \sum_j w_{ij}^q |x_i - x_j|^p, \tag{5.22}$$

for established values of all x_j. There is some maximum $x_{\max}$ and minimum $x_{\min}$ among the neighbors of v_i. Specifically, we know that $x_{\min} \leq x_i \leq x_{\max}$ since any x_i lower than $x_{\min}$ would give a larger value in the argument of (5.22) than $x_{\min}$. The same argument applies to $x_{\max}$, meaning that the optimal x_i falls within the range of its neighbors, and consequently there exists a neighbor v_j for which $x_j \geq x_i$. Since $\mathcal{S}$ is defined as the set of all nodes for which $x_i > 0.5$, then any node in $\mathcal{S}$ contains at least one neighbor in $\mathcal{S}$, meaning that there is a connected path from any node in $\mathcal{S}$ to a object seed. By assumption, the set of object seeds is connected, so every node inside $\mathcal{S}$ is connected to every other node inside $\mathcal{S}$. Since this principle applies to any choice of p and q, then under the above conditions the object is guaranteed to be connected for graph cuts, random walker, shortest paths, watershed, power watershed, and fuzzy connectedness. A slight modification of the same argument also allows us to guarantee that TVSeg produces a connected object segmentation under very similar conditions.

In practice, the conditions that guarantee connectedness of the segmentation may be violated in those circumstances in which unary terms are used or where the set of input seeds is not connected. In these situations, it has been proven for graph cuts that the problem of finding the minimal solution while guaranteeing connectivity is NP-Hard [102]. To our knowledge, it is still unknown whether it is possible to guarantee connectivity for any of the other algorithms in the power watershed framework. Vincente et al. chose to approximate a related, but simpler, problem [103]. Specifically, Vincente et al. consider that two points in the image have been designated for connectivity (either interactively or automatically) and provide an algorithm that generates a segmentation that approximately optimizes the graph cut criterion while guaranteeing the connectivity of the designated points. Vincente et al. then demonstrate that their algorithm will sometimes find the global optimum, while guaranteeing connectivity of the designated points, on real-world images. Nowozin and Lampert adopted a different approach to the problem of enforcing connectivity for the graph cuts algorithm by instead solving a related optimization problem for the global optimum [104].

5.2.6.2 Genus Specification

Many target objects have a known genus, particularly in biomedical imaging. This genus information can help the algorithm avoid leaving internal voids (if the object is known to be simply connected) or falsely filling voids when the object is known to have internal holes. There has been little work to date on genus specification for targeted graph-based segmentation algorithms. The first work on this topic is by Zeng et al. [102] who proved that it is NP-hard to optimize the graph cuts criterion with a genus constraint (the other algorithms in the power watershed framework have not been examined with a genus constraint). However, Zeng et al. provide an algorithm that enforces a genus constraint (keeping the genus of an initial segmentation) while approximately optimizing the graph cuts criterion. Their algorithm is based on the topology-preserving level sets work of [105] which uses the *simple point* concept from digital topology [106]. Effectively, if an algorithm adds or subtracts only simple points from an initial segmentation, then the modified segmentation will have the same genus as the original. Zeng et al. show that their algorithm for genus preservation has the same asymptotic complexity as the graph cuts algorithm without genus constraints. The algorithm by Zeng et al. for genus specification was further simplified and improved by Danek and Maska [107]. Therefore, these genus specification algorithms can be used to enforce object connectivity by making the initial input segmentation connected.

5.2.7 Segmentation Targeting by Relational Information

Our focus in this chapter is on the specification of target information to define the segmentation of a single object. However, certain kinds of target specification can only be defined for an object in relation to other objects. In this section, we examine this relational target specification and so slightly expand our scope beyond single-object segmentation.

Similar to Chapter 10, the goal of using relational information (sometimes known as *geometric constraints*) is to enforce a certain geometrical relationship between two objects. Specifically, the following relationships between two regions (A and B) were studied by Delong and Boykov [108]:

1. *Containment*: Region B must be inside region A, perhaps with a repulsion force between boundaries.

2. *Exclusion*: Regions A and B cannot overlap at any pixel, perhaps with a repulsion force between boundaries.

3. *Attraction*: Penalize the area $A - B$, exterior to B, by some cost $\alpha > 0$ per unit area. Thus, A will prefer not to grow too far beyond the boundary of B.

The basic idea of [108] is to create two graphs, $\mathcal{G}^A = \{\mathcal{V}^A, \mathcal{E}^A\}$ and $\mathcal{G}^B = \{\mathcal{V}^B, \mathcal{E}^B\}$, each to represent the segmentation of A or B. The segmentation of A on $\mathcal{G}^A$ is represented by the indicator vectors $\mathbf{x}^A$ (and $\mathbf{x}^B$, respectively, for $\mathcal{G}^B$). Then, a new energy term is introduced

$$E_{\text{Geometric}} \left(\mathbf{x}^A, \mathbf{x}^B \right) = \sum_{v_i \in \mathcal{V}^A, v_j \in \mathcal{V}^B} w_{ij}(x_i, x_j), \tag{5.23}$$

meaning that the two graphs $\mathcal{G}^A$ and $\mathcal{G}^B$ are effectively fully connected with each other. The weights to encode the three geometric relationships are given in 5.2.7.

$\mathcal{A}$ contains $\mathcal{B}$				$\mathcal{A}$ excludes $\mathcal{B}$				$\mathcal{A}$ attracts $\mathcal{B}$		
x_i^A	x_i^B	w_{ij}		x_i^A	x_i^B	w_{ij}		x_i^A	x_i^B	w_{ij}
0	0	0		0	0	0		0	0	0
0	1	∞		0	1	0		0	1	0
1	0	0		1	0	0		1	0	α
1	1	0		1	1	∞		1	1	0

TABLE 5.2
Three types of relational information between two objects.

Delong and Boykov studied the optimization of (5.23) in the context of the graph cuts model and found that containment and attraction are easy to optimize since the energy is submodular. However, the energy for the exclusion relationship is non-submodular and therefore more challenging except under limited circumstances (see [108] for more details). Furthermore, encoding these geometrical relationships for more than two objects can lead to optimization problems for which a global optimum is not easy to obtain. Ulén et al. [109] applied this model to cardiac segmentation and further expanded the discussion of encoding these constraints for multiple objects. These geometric interactions have not been studied in the context of any of the other regularization models of Section 5.1, but they would be straightforward to incorporate.

5.3   Conclusion

Traditional image segmentation seeks to subdivide an image into a series of meaningful objects. In contrast, a targeted segmentation algorithm uses input target specification to determine a specific desired object. A targeted image segmentation algorithm consists of two parts: a *target specification* which specifies the desired object and a *regularization* that completes the segmentation from the target specification. The ideal regularization algorithm could incorporate as little or as much target specification that was available and still produce a meaningful segmentation of the target object.

Targeted image segmentation algorithms have matured in recent years. Today, there are many effective regularization methods available with efficient optimization algorithms. Additionally, many different types of target specification have been accommodated, including prior information about appearance, boundary, shape, user interaction, learned features, relational information, and topology.

Despite this increasing maturity, more remains to be done. The regularization methods reviewed in Section 5.1 are not as effective for segmenting elongated or sheet-like objects. Additionally, the use of prior information about higher-order statistics of the target object have just started to be explored (see Chapter 3 for more on this topic). Finally, the exploration has been piecemeal of the optimization of different types of target specification combined with different regularization methods (or combined with other types of target specification). The optimization of different regularization methods in the presence of different types of target specification has not always been treated in the literature. We believe that the future will see the development of new and exciting methods for regularization that may be optimized efficiently in the presence of one or more different types of target specification, which will bring the community to the point of effectively solving the targeted image segmentation problem.

Bibliography

[1] C. Zahn, "Graph theoretical methods for detecting and describing Gestalt clusters," *IEEE Transactions on Computation*, vol. 20, pp. 68–86, 1971.

[2] Z. Wu and R. Leahy, "An optimal graph theoretic approach to data clustering: Theory and its application to image segmentation," *IEEE Transactions on Pattern Analysis and Machine Intelligence*, vol. 15, no. 11, pp. 1101–1113, 1993.

[3] J. Shi and J. Malik, "Normalized cuts and image segmentation," *IEEE Transactions on Pattern Analysis and Machine Intelligence*, vol. 22, no. 8, pp. 888–905, Aug. 2000.

[4] P. F. Felzenszwalb and D. P. Huttenlocher, "Efficient graph-based image segmentation," *International Journal of Computer Vision*, vol. 59, no. 2, pp. 167–181, September 2004.

[5] L. Grady and E. L. Schwartz, "Isoperimetric graph partitioning for image segmentation," *IEEE Transactions on Pattern Analysis and Machine Intelligence*, vol. 28, no. 3, pp. 469–475, March 2006.

[6] J. Roerdink and A. Meijster, "The watershed transform: definitions, algorithms, and parallellization strategies," *Fund. Informaticae*, vol. 41, pp. 187–228, 2000.

[7] Y. Zhang, "A survey on evaluation methods for image segmentation," *Pattern Recognition*, vol. 29, no. 8, pp. 1335–1346, 1996.

[8] D. Martin, C. Fowlkes, D. Tal, and J. Malik, "A database of human segmented natural images and its application to evaluating segmentation algorithms and measuring ecological statistics," in *Proc. ICCV*, 2001.

[9] D. Singaraju, L. Grady, A. K. Sinop, and R. Vidal, "P-brush: A continuous valued MRF for image segmentation," in *Advances in Markov Random Fields for Vision and Image Processing*, A. Blake, P. Kohli, and C. Rother, Eds. MIT Press, 2010.

[10] G. L. Nemhauser and L. A. Wolsey, *Integer and Combinatorial Optimization*. Wiley-Interscience, 1999.

[11] L. Grady, "Computing exact discrete minimal surfaces: Extending and solving the shortest path problem in 3D with application to segmentation," in *Proc. of CVPR*, vol. 1, June 2006, pp. 69–78.

[12] ——, "Minimal surfaces extend shortest path segmentation methods to 3D," *IEEE Trans. on Pattern Analysis and Machine Intelligence*, vol. 32, no. 2, pp. 321–334, Feb. 2010.

[13] T. Schoenemann, F. Kahl, and D. Cremers, "Curvature regularity for region-based image segmentation and inpainting: A linear programming relaxation," in *Proc. of CVPR*. IEEE, 2009, pp. 17–23.

[14] P. Strandmark and F. Kahl, "Curvature regularization for curves and surfaces in a global optimization framework," in *Proc. of EMMCVPR*, 2011, pp. 205–218.

[15] F. Nicolls and P. Torr, "Discrete minimum ratio curves and surfaces," in *Proc. of CVPR*. IEEE, 2010, pp. 2133–2140.

[16] T. Windheuser, U. Schlickewei, F. Schmidt, and D. Cremers, "Geometrically consistent elastic matching of 3D shapes: A linear programming solution," in *Proc. of ICCV*, vol. 2, 2011.

[17] L. Grady and J. R. Polimeni, *Discrete Calculus: Applied Analysis on Graphs for Computational Science*. Springer, 2010.

[18] A. K. Sinop and L. Grady, "A seeded image segmentation framework unifying graph cuts and random walker which yields a new algorithm," in *Proc. of ICCV 2007*, IEEE Computer Society. IEEE, Oct. 2007.

[19] C. Allène, J.-Y. Audibert, M. Couprie, and R. Keriven, "Some links between extremum spanning forests, watersheds and min-cuts," *Image and Vision Computing*, vol. 28, no. 10, 2010.

[20] C. Couprie, L. Grady, L. Najman, and H. Talbot, "Combinatorial continuous max flow," *SIAM J. on Imaging Sciences*, vol. 4, no. 3, pp. 905–930, 2011.

[21] D. Greig, B. Porteous, and A. Seheult, "Exact maximum *a posteriori* estimation for binary images," *Journal of the Royal Statistical Society, Series B*, vol. 51, no. 2, pp. 271–279, 1989.

[22] Y. Boykov and M.-P. Jolly, "*Interactive graph cuts* for optimal boundary & region segmentation of objects in N-D images," in *Proc. of ICCV 2001*, 2001, pp. 105–112.

[23] L. Grady, "Random walks for image segmentation," *IEEE Trans. on Pattern Analysis and Machine Intelligence*, vol. 28, no. 11, pp. 1768–1783, Nov. 2006.

[24] A. X. Falcão, R. A. Lotufo, and G. Araujo, "The image foresting transformation," *IEEE Transactions on Pattern Analysis and Machine Intelligence*, vol. 26, no. 1, pp. 19–29, 2004.

[25] X. Bai and G. Sapiro, "A geodesic framework for fast interactive image and video segmentation and matting," in *ICCV*, 2007.

[26] S. Beucher and F. Meyer, "The morphological approach to segmentation: the watershed transformation," in *Mathematical Morphology in Image Processing*, E. R. Dougherty, Ed. Taylor & Francis, Inc., 1993, pp. 433–481.

[27] C. Couprie, L. Grady, L. Najman, and H. Talbot, "Power watershed: A unifying graph-based optimization framework," *IEEE Trans. on Pat. Anal. and Mach. Int.*, vol. 33, no. 7, pp. 1384–1399, July 2011.

[28] Y. Boykov and V. Kolmogorov, "An experimental comparison of min-cut/max-flow algorithms for energy minimization in vision," *IEEE Transactions on Pattern Analysis and Machine Intelligence*, pp. 1124–1137, 2004.

[29] L. Yatziv, A. Bartesaghi, and G. Sapiro, "A fast $O(N)$ implementation of the fast marching algorithm," *Journal of Computational Physics*, vol. 212, pp. 393–399, 2006.

[30] J. Udupa and S. Samarasekera, "Fuzzy connectedness and object definition: Theory, algorithms, and applications in image segmentation," *Graphical Models and Image Processing*, vol. 58, pp. 246–261, 1996.

[31] P. Saha and J. Udupa, "Relative fuzzy connectedness among multiple objects: Theory, algorithms, and applications in image segmentation," *Computer Vision and Image Understanding*, vol. 82, pp. 42–56, 2001.

[32] V. V. and K. V., "GrowCut — Interactive Multi-Label N-D Image Segmentation," in *Proc. of Graphicon*, 2005, pp. 150–156.

[33] A. Hamamci, G. Unal, N. Kucuk, and K. Engin, "Cellular automata segmentation of brain tumors on post contrast MR images," in *Proc. of MICCAI*. Springer, 2010, pp. 137–146.

[34] K. C. Ciesielski and J. K. Udupa, "Region-based segmentation: Fuzzy connectedness, graph cut and related algorithms," in *Biomedical Image Processing*, ser. Biological and Medical Physics, Biomedical Engineering, T. M. Deserno, Ed. Springer-Verlag, 2011, pp. 251–278.

[35] P. A. V. Miranda and A. X. Falcão, "Elucidating the relations among seeded image segmentation methods and their possible extensions," in *Proceedings of Sibgrapi*, 2011.

[36] L. G. Dheeraj Singaraju and R. Vidal, "P-brush: Continuous valued MRFs with normed pairwise distributions for image segmentation," in *Proc. of CVPR 2009*, IEEE Computer Society. IEEE, June 2009.

[37] L. D. Cohen, "On active contour models and balloons," *CVGIP: Image understanding*, vol. 53, no. 2, pp. 211–218, 1991.

[38] L. Cohen and I. Cohen, "Finite-element methods for active contour models and balloons for 2-D and 3-D images," *IEEE Transactions on Pattern Analysis and Machine Intelligence*, vol. 15, no. 11, pp. 1131–1147, 1993.

[39] L. Grady and E. L. Schwartz, "Isoperimetric partitioning: A new algorithm for graph partitioning," *SIAM Journal on Scientific Computing*, vol. 27, no. 6, pp. 1844–1866, June 2006.

[40] Y. Boykov and V. Kolmogorov, "Computing geodesics and minimal surfaces via graph cuts," in *Proceedings of International Conference on Computer Vision*, vol. 1, October 2003.

[41] O. Daněk and P. Matula, "An improved Riemannian metric approximation for graph cuts," *Discrete Geometry for Computer Imagery*, pp. 71–82, 2011.

[42] B. Appleton and H. Talbot, "Globally optimal surfaces by continuous maximal flows," *IEEE Transactions on Pattern Analysis and Machine Intelligence*, vol. 28, no. 1, pp. 106–118, Jan. 2006.

[43] M. Unger, T. Pock, and H. Bischof, "Interactive globally optimal image segmentation," Inst. for Computer Graphics and Vision, Graz University of Technology, Tech. Rep. 08/02, 2008.

[44] M. Unger, T. Pock, W. Trobin, D. Cremers, and H. Bischof, "TVSeg - Interactive total variation based image segmentation," in *Proc. of British Machine Vision Conference*, 2008.

[45] D. Shulman and J.-Y. Herve, "Regularization of discontinuous flow fields," in *Proc. of Visual Motion*, 1989, pp. 81–86.

[46] L. Rudin, S. Osher, and E. Fatemi, "Nonlinear total variation based noise removal algorithms," *Physica D*, vol. 60, no. 1-4, pp. 259–268, 1992.

[47] S. Osher and J. Shen, "Digitized PDE method for data restoration," in *In Analytical-Computational methods in Applied Mathematics*, E. G. A. Anastassiou, Ed. Chapman & Hall/CRC,, 2000, pp. 751–771.

[48] A. Elmoataz, O. Lézoray, and S. Bougleux, "Nonlocal discrete regularization on weighted graphs: A framework for image and manifold processing," *IEEE Transactions on Image Processing*, vol. 17, no. 7, pp. 1047–1060, 2008.

[49] A. Chambolle, "An algorithm for total variation minimization and applications," *J. Math. Imaging Vis.*, vol. 20, no. 1–2, pp. 89–97, 2004.

[50] J. Darbon and M. Sigelle, "Image restoration with discrete constrained total variation part I: Fast and exact optimization," *Journal of Mathematical Imaging and Vision*, vol. 26, no. 3, pp. 261–276, Dec. 2006.

[51] X. Bresson, S. Esedoglu, P. Vandergheynst, J.-P. Thiran, and S. Osher, "Fast global minimization of the active contour/snake model," *Journal of Mathematical Imaging and Vision*, vol. 28, no. 2, pp. 151–167, 2007.

[52] A. Chambolle and T. Pock, "A first-order primal-dual algorithm for convex problems with applications to imaging," *Journal of Mathematical Imaging and Vision*, vol. 40, no. 1, pp. 120–145, 2011.

[53] M. Iri, "Theory of flows in continua as approximation to flows in networks," *Survey of Mathematical Programming*, 1979.

[54] G. Strang, "Maximum flows through a domain," *Mathematical Programming*, vol. 26, pp. 123–143, 1983.

[55] R. Nozawa, "Examples of max-flow and min-cut problems with duality gaps in continuous networks," *Mathematical Programming*, vol. 63, no. 2, pp. 213–234, Jan. 1994.

[56] L. R. Ford and D. R. Fulkerson, "Maximal flow through a network," *Canadian Journal of Mathematics*, vol. 8, pp. 399–404, 1956.

[57] L. Grady, "A lattice-preserving multigrid method for solving the inhomogeneous poisson equations used in image analysis," in *Proc. of ECCV*, ser. LNCS, D. Forsyth, P. Torr, and A. Zisserman, Eds., vol. 5303. Springer, 2008, pp. 252–264.

[58] M. Heiler, J. Keuchel, and C. Schnörr, "Semidefinite clustering for image segmentation with a-priori knowledge," in *Proc. of DAGM*, 2005, pp. 309–317.

[59] S. X. Yu and J. Shi, "Understanding popout through repulsion," in *Proc. of CVPR*, vol. 2. IEEE Computer Society, 2001.

[60] J. Wang, M. Agrawala, and M. Cohen, "Soft scissors: An interactive tool for realtime high quality matting," in *Proc. of SIGGRAPH*, 2007.

[61] C. Rother, V. Kolmogorov, and A. Blake, ""GrabCut" — Interactive foreground extraction using iterated graph cuts," in *ACM Transactions on Graphics, Proceedings of ACM SIGGRAPH 2004*, vol. 23, no. 3. ACM, 2004, pp. 309–314.

[62] V. Lempitsky, P. Kohli, C. Rother, and T. Sharp, "Image segmentation with a bounding box prior," in *Proc. of ICCV*, 2009, pp. 277–284.

[63] H. il Koo and N. I. Cho, "Rectification of figures and photos in document images using bounding box interface," in *Proc. of CVPR*, 2010, pp. 3121–3128.

[64] L. Grady, M.-P. Jolly, and A. Seitz, "Segmentation from a box," in *Proc. of ICCV*, 2011.

[65] P. Viola and M. Jones, "Robust real-time face detection," *International Journal of Computer Vision*, vol. 57, no. 2, pp. 137–154, 2004.

[66] A. Blake, C. Rother, M. Brown, P. Perez, , and P. Torr, "Interactive image segmentation using an adaptive GMMRF model," in *Proc. of ECCV*, 2004.

[67] X. Huang, Z. Qian, R. Huang, and D. Metaxas, "Deformable-model based textured object segmentation," in *Proc. of EMMCVPR*, 2005, pp. 119–135.

[68] J. Malcolm, Y. Rathi, and A. Tannenbaum, "A graph cut approach to image segmentation in tensor space," in *Proc. Workshop Component Analysis Methods*, 2007, pp. 18–25.

[69] S. Han, W. Tao, D. Wang, X.-C. Tai, and X. Wu, "Image segmentation based on grabcut framework integrating multiscale nonlinear structure tensor," *IEEE Trans. on TIP*, vol. 18, no. 10, pp. 2289–2302, Oct. 2009.

[70] H. Koo and N. Cho, "Graph cuts using a Riemannian metric induced by tensor voting," in *Proc. of ICCV*, 2009, pp. 514–520.

[71] J. Santner, M. Unger, T. Pock, C. Leistner, A. Saffari, and H. Bischof, "Interactive texture segmentation using random forests and total variation," in *Proc. of BVMC*, 2009.

[72] A. F. Frangi, W. J. Niessen, K. L. Vincken, and M. A. Viergever, "Multiscale vessel enhancement filtering," in *Proc. of MICCAI*, ser. LNCS, 1998, pp. 130–137.

[73] M. Freiman, L. Joskowicz, and J. Sosna, "A variational method for vessels segmentation: Algorithm and application to liver vessels visualization," in *Proc. of SPIE*, vol. 7261, 2009.

[74] S. Esneault, C. Lafon, and J.-L. Dillenseger, "Liver vessels segmentation using a hybrid geometrical moments/graph cuts method," *IEEE Trans. on Bio. Eng.*, vol. 57, no. 2, pp. 276–283, Feb. 2010.

[75] D. Mumford and J. Shah, "Optimal approximations by piecewise smooth functions and associated variational problems," *Comm. Pure and Appl. Math.*, vol. 42, pp. 577–685, 1989.

[76] A. Tsai, A. Yezzi, and A. Willsky, "Curve evolution implementation of the Mumford-Shah functional for image segmentation, denoising, interpolation, and magnification," *IEEE Transactions on Image Processing*, vol. 10, no. 8, pp. 1169–1186, 2001.

[77] L. Grady and C. Alvino, "The piecewise smooth Mumford-Shah functional on an arbitrary graph," *IEEE Transactions on Image Processing*, vol. 18, no. 11, pp. 2547–2561, Nov. 2009.

[78] T. Chan and L. Vese, "Active contours without edges," *IEEE Transactions on Image Processing*, vol. 10, no. 2, pp. 266–277, 2001.

[79] S. Geman and D. McClure, "Statistical methods for tomographic image reconstruction," in *Proc. 46th Sess. Int. Stat. Inst. Bulletin ISI*, vol. 52, no. 4, Sept. 1987, pp. 4–21.

[80] P. Perona and J. Malik, "Scale-space and edge detection using anisotropic diffusion," *IEEE Transactions on Pattern Analysis and Machine Intelligence*, vol. 12, no. 7, pp. 629–639, July 1990.

[81] D. Geman and G. Reynolds, "Constrained restoration and the discovery of discontinuities," *IEEE Transactions on Pattern Analysis and Machine Intelligence*, vol. 14, no. 3, pp. 367–383, March 1992.

[82] V. Caselles, R. Kimmel, and G. Sapiro, "Geodesic active contours," *International journal of computer vision*, vol. 22, no. 1, pp. 61–79, 1997.

[83] M. J. Black, G. Sapiro, D. H. Marimont, and D. Heeger, "Robust anisotropic diffusion," *IEEE Transactions on Image Processing*, vol. 7, no. 3, pp. 421–432, March 1998.

[84] L. Grady and M.-P. Jolly, "Weights and topology: A study of the effects of graph construction on 3D image segmentation," in *Proc. of MICCAI 2008*, ser. LNCS, D. M. et al., Ed., vol. Part I, no. 5241. Springer-Verlag, 2008, pp. 153–161.

[85] Y. Boykov and G. Funka-Lea, "Graph cuts and efficient N-D image segmentation," *International Journal of Computer Vision*, vol. 70, no. 2, pp. 109–131, 2006.

[86] D. Singaraju, L. Grady, and R. Vidal, "Interactive image segmentation of quadratic energies on directed graphs," in *Proc. of CVPR 2008*, IEEE Computer Society. IEEE, June 2008.

[87] V. Kolmogorov and Y. Boykov, "What metrics can be approximated by geo-cuts, or global optimization of length/area and flux," in *Proc. of the International Conference on Computer Vision (ICCV)*, vol. 1, 2005, pp. 564–571.

[88] T. Schoenemann, F. Kahl, and D. Cremers, "Curvature regularity for region-based image segmentation and inpainting: A linear programming relaxation," in *Proc. of ICCV*, Kyoto, Japan, 2009.

[89] N. El-Zehiry and L. Grady, "Fast global optimization of curvature," in *Proc. of CVPR 2010*, June 2010.

[90] ——, "Contrast driven elastica for image segmentation," in *Submitted to CVPR 2012*, 2012.

[91] B. Goldluecke and D. Cremers, "Introducing total curvature for image processing," in *Proc. of ICCV*, 2011.

[92] D. Mumford, "Elastica and computer vision," *Algebraic Geometry and Its Applications*, pp. 491–506, 1994.

[93] A. K. Sinop and L. Grady, "Uninitialized, globally optimal, graph-based rectilinear shape segementation — The opposing metrics method," in *Proc. of ICCV 2007*, IEEE Computer Society. IEEE, Oct. 2007.

[94] J. Zunic and P. Rosin, "Rectilinearity measurements for polygons," *IEEE Trans. on Pat. Anal. and Mach. Int.*, vol. 25, no. 9, pp. 1193–1200, Sept. 2003.

[95] G. Slabaugh and G. Unal, "Graph cuts segmentation using an elliptical shape prior," in *Proc. of ICIP*, vol. 2, 2005.

[96] D. Freedman and T. Zhang, "Interactive graph cut based segmentation with shape priors," in *Proc. of CVPR*, 2005, pp. 755–762.

[97] A. Tsai, A. Yezzi, W. Wells, C. Tempany, D. Tucker, A. Fan, W. E. Grimson, and A. Willsky, "A shape-based approach to the segmentation of medical images using level sets," *IEEE TMI*, vol. 22, no. 2, pp. 137–154, 2003.

[98] J. Zhu-Jacquot and R. Zabih, "Graph cuts segmentation with statistical shape priors for medical images," in *Proc. of SITIBS*, 2007.

[99] M. Kumar, P. H. S. Torr, and A. Zisserman, "Obj cut," in *Proc. of CVPR*, vol. 1. IEEE, 2005, pp. 18–25.

[100] P. Felzenszwalb and D. Huttenlocher, "Efficient matching of pictorial structures," in *Proc. of CVPR*, vol. 2. IEEE, 2000, pp. 66–73.

[101] A. Levin and Y. Weiss, "Learning to combine bottom-up and top-down segmentation," in *Proc. of ECCV.* Springer, 2006, pp. 581–594.

[102] Y. Zeng, D. Samaras, W. Chen, and Q. Peng, "Topology cuts: A novel min-cut/max-flow algorithm for topology preserving segmentation in ND images," *Computer vision and image understanding*, vol. 112, no. 1, pp. 81–90, 2008.

[103] S. Vicente, V. Kolmogorov, and C. Rother, "Graph cut based image segmentation with connectivity priors," in *Proc. of CVPR*, 2008.

[104] S. Nowozin and C. Lampert, "Global interactions in random field models: A potential function ensuring connectedness," *SIAM Journal on Imaging Sciences*, vol. 3, p. 1048, 2010.

[105] X. Han, C. Xu, and J. Prince, "A topology preserving level set method for geometric deformable models," *IEEE Transactions on Pattern Analysis and Machine Intelligence*, pp. 755–768, 2003.

[106] G. Bertrand, "Simple points, topological numbers and geodesic neighborhoods in cubic grids," *Pattern Recognition letters*, vol. 15, no. 10, pp. 1003–1011, 1994.

[107] O. Danek and M. Maska, "A simple topology preserving max-flow algorithm for graph cut based image segmentation," in *Proc. of the Sixth Doctoral Workshop on Mathematical and Engineering Methods in Computer Science*, 2010, pp. 19–25.

[108] A. Delong and Y. Boykov, "Globally optimal segmentation of multi-region objects," in *Proc. of ICCV.* IEEE, 2009, pp. 285–292.

[109] J. Uln, P. Strandmark, and F. Kahl, "Optimization for multi-region segmentation of cardiac MRI," in *Proc. of the MICCAI Workshop on Statistical Atlases and Computational Models of the Heart: Imaging and Modelling Challenges*, 2011.

6

A Short Tour of Mathematical Morphology on Edge and Vertex Weighted Graphs

Laurent Najman

Université Paris-Est
Laboratoire d'Informatique Gaspard-Monge
Equipe A3SI - ESIEE Paris
France
Email: l.najman@esiee.fr

Fernand Meyer

Centre de Morphologie Mathématique
Ecole des Mines de Paris
Fontainebleau, France
Email: f.meyer@cmm.ensmp.fr

CONTENTS

6.1 Introduction 142
6.2 Graphs and lattices 143
    6.2.1 Lattice of Graphs 143
    6.2.2 Lattice of Weights 144
6.3 Neighborhood Operations on Graphs 145
    6.3.1 Adjunctions on Graphs 145
6.4 Filters 149
6.5 Connected Operators and Filtering with the Component Tree 152
6.6 Watershed Cuts 154
    6.6.1 The Drop of Water Principle 155
    6.6.2 Catchment Basins by a Steepest Descent Property 156
    6.6.3 Minimum Spanning Forests 157
    6.6.4 Illustrations to Segmentation 157
6.7 MSF Cut Hierarchy and Saliency Maps 161
    6.7.1 Uprootings and MSF Hierarchies 162
    6.7.2 Saliency Maps 163
    6.7.3 Application to Image Segmentation 163
6.8 Optimization and the Power Watershed 164
    6.8.1 Applications 168
6.9 Conclusion 169
    Bibliography 169

6.1 Introduction

Mathematical morphology is a discipline of image analysis that was introduced in the mid-1960s by two researchers at the École des Mines in Paris: Georges Matheron [1] and Jean Serra [2, 3]. Historically, it was the first consistent nonlinear image analysis theory, which from the very start included not only theoretical results but also many practical aspects. Due to the algebraic nature of morphology, the space on which the operators are defined can be either continuous or discrete. However, it was only in 1989 [4] that researchers from the CMM at the École des Mines began to study morphology on graphs, soon formalized in [5]. Fairly recent developments [6, 7, 8, 9, 10, 11, 12] in this direction has several motivations and beneficial consequences that we are going to review in this chapter. Rather than trying to cover every aspect of the theory, we choose to present a comprehensive subset based on a recent unifying graph theoretical framework developed by the A3SI team of the LIGM at Paris-Est University. The presentation is roughly divided in three parts, dealing respectively with basic operators (mainly based on [10]), hierarchical segmentation (mainly based on [13] and [14]), and optimization (mainly based on [15]). For the reader interested in a more complete presentation of morphology, we recommend [16] and the recent [17].

One of the fundamental ideas of mathematical morphology is to compare unkown objects with known ones. We begin by presenting the tools that make such an idea practicable: mathematical structures called *lattices* (Section 6.2), allowing us to compare weighted and nonweighted edges and vertices. We then present (Section 6.3) several dilations and erosions, which always come in pairs (they are called *adjunct operators*). From there, we can build (Section 6.4) some morphological filters, called *opening* or *closing*. During the presentation of those various operators, we often give an interpretation of classical graph operators in morphological terms. We conclude this part by presenting (Section 6.5) some connected operators based on pruning a tree representation of the image, illustrating their usage for image filtering and simplification.

The second part of the chapter deals with hierarchical segmentation. During the course of the chapter's first part, several times the minimum spanning tree appears. This tree is the oldest combinatorial optimization problem [18, 19]. Since the seminal work of Zahn [20], the minimum spanning tree has been extensively used in classification. Its first appearance in image processing dates from 1986, in a paper by Morris et al. [21]. Meyer was the first to explicitly use it in a morphological context [22]. A strong momentum to its usage in segmentation has been provided, thanks to a 2004 paper by Felzenswalb and Huttenlocher [23]. In the second part of the chapter, we revisit the watershed [24], and show that in the framework of edge-weighted graphs, the watershed has very strong links with the minimum spanning tree (Section 6.6). Thanks to such links, we can use the watershed for building hierarchies of segmentations (Section 6.7), relying on the filtering tools seen in the first part of the chapter.

The main principle of morphology, comparison, is rather different from the optimization paradigm. However, rather than opposing these two viewpoints, it is more fruitful to explore their connections. In the last part of the chapter (Section 6.8), we turn towards optimization and show that the watershed presented in the previous part can be extended to be used as an optimization tool.

6.2 Graphs and lattices

In mathematical morphology, we compare objects with respect to each other. The mathemathical structure that allows us to make such an operation effective is called a *lattice*. Recall that a (complete) lattice is a partially ordered set, that also has a least upper bound, called *supremum*, and a greatest lower bound, called *infimum*. More formally, a lattice [25] $(\mathcal{L}, \leq)$ is a set $\mathcal{L}$ (the space) endowed with an *ordering* relationship $\leq$, which is reflexive ($\forall x \in \mathcal{L}, x \leq x$), anti-symmetric ($x \leq y$ and $y \leq x \implies x = y$), and transitive ($x \leq y$ and $y \leq z \implies x \leq z$). This ordering is such that for all x and y, we can define both a larger element $x \vee y$ and a smaller element $x \wedge y$. Such a lattice is said to be *complete* if any subset $\mathcal{P}$ of $\mathcal{L}$ has a *supremum* $\bigvee \mathcal{P}$ and an *infimum* $\bigwedge \mathcal{P}$ that both belongs to $\mathcal{L}$. The supremum is formally the smallest amongst all elements of $\mathcal{L}$ that are greater than all the elements of $\mathcal{P}$, and, conversely, the infimum is the largest element of $\mathcal{L}$ that is smaller than all the elements of $\mathcal{P}$.

Most of the morphological theory can be presented and developed at this abstract level, without making references to the properties of the underlying space. However, studying what impact such properties have can indeed be interesting in some situations. In the sequel, we study some lattices that can be built from graph spaces, and what kind of (morphological) operators can be built from such lattices.

6.2.1 Lattice of Graphs

We define a *graph* as a pair $\mathcal{G} = (\mathcal{V}(\mathcal{G}), \mathcal{E}(\mathcal{G}))$ where $\mathcal{V}(\mathcal{G})$ is a set and $\mathcal{E}(\mathcal{G})$ is composed of unordered pairs of distinct elements in $\mathcal{V}(\mathcal{G})$, i.e., $\mathcal{E}(\mathcal{G})$ is a subset of $\{\{v_1, v_2\} \subseteq \mathcal{V}(\mathcal{G}) \mid v_1 \neq v_2\}$. Each element of $\mathcal{V}(\mathcal{G})$ is called a *vertex or a point (of $\mathcal{G}$)*, and each element of $\mathcal{E}(\mathcal{G})$ is called an *edge* (of $\mathcal{G}$). In the sequel, to simplify the notations, e_{ij} stands for the edge $\{v_i, v_j\} \in \mathcal{E}(\mathcal{G})$.

Let $\mathcal{G}_1$ and $\mathcal{G}_2$ be two graphs. If $\mathcal{V}(\mathcal{G}_2) \subseteq \mathcal{V}(\mathcal{G}_1)$ and $\mathcal{E}(\mathcal{G}_2) \subseteq \mathcal{E}(\mathcal{G}_1)$, then $\mathcal{G}_1$ and $\mathcal{G}_2$ are ordered and we write $\mathcal{G}_2 \sqsubseteq \mathcal{G}_1$. If $\mathcal{G}_2 \sqsubseteq \mathcal{G}_1$, we say that $\mathcal{G}_2$ is a *subgraph* of $\mathcal{G}_1$, or that $\mathcal{G}_2$ is *smaller* than $\mathcal{G}_1$ and that $\mathcal{G}_1$ is *greater* than $\mathcal{G}_2$.

Important remark. Hereafter, the workspace is a graph $\mathcal{G} = (\mathcal{V}(\mathcal{G}), \mathcal{E}(\mathcal{G}))$, and we consider the sets $\mathcal{V}(\mathbb{G})$, $\mathcal{E}(\mathbb{G})$ and $\mathbb{G}$ of, respectively, all subsets of $\mathcal{V}(\mathcal{G})$, all subsets of $\mathcal{E}(\mathcal{G})$ and all subgraphs of $\mathcal{G}$. We also use the classical notations $\mathcal{V} = \mathcal{V}(\mathcal{G})$ and $\mathcal{E} = \mathcal{E}(\mathcal{G})$.

Let $\mathcal{S}_0, \mathcal{S}_1 \subseteq \mathbb{G}$ be the sets of, respectively, the graphs made of a single vertex and the graphs made of a pair of vertices linked by an edge, i.e., $\mathcal{S}_0 = \{(\{v\}, \emptyset) \mid v \in \mathcal{V}(\mathcal{G})\}$ and $\mathcal{S}_1 = \{(\{v_i, v_j\}, \{e_{ij}\}) \mid e_{ij} \in \mathcal{E}(\mathcal{G})\}$. We set $\mathcal{S} = \mathcal{S}_0 \cup \mathcal{S}_1$. Any graph $\mathcal{G}^1 \in \mathbb{G}$ is *generated* by the family $\mathcal{F} = \{\mathcal{G}_1, \ldots, \mathcal{G}_\ell\}$ of all elements in $\mathcal{S}$ smaller than $\mathcal{G}^1$: $\mathcal{G}^1 = (\bigcup_{i \in [1,\ell]} \mathcal{V}(\mathcal{G}_i), \bigcup_{i \in [1,\ell]} \mathcal{E}(\mathcal{G}_i))$; we say that the elements of $\mathcal{F}$ are the *generators* of $\mathcal{G}^1$ [10]. Conversely, any family $\mathcal{F}$ of elements in $\mathcal{S}$ generates an element of $\mathbb{G}$. Hence, $\mathcal{S}$ *(sup-) generates* $\mathbb{G}$.

Clearly, the ordering $\sqsubseteq$ on graphs amount to having $\mathcal{G}_2 \sqsubseteq \mathcal{G}_1$ when all generators of $\mathcal{G}_2$ are also generators of $\mathcal{G}_1$. Therefore, ordering $\sqsubseteq$ provides a *lattice* structure on the set $\mathbb{G}$. Indeed, the largest graph smaller than a family $\mathcal{F} = \{\mathcal{G}_1, \ldots, \mathcal{G}_\ell\}$ of elements in $\mathbb{G}$ is the graph generated by the generators common to all $\mathcal{G}_i$, $i \in [1, \ell]$; this *infimum* is denoted by $\sqcap \mathcal{F}$. Similarly, the *supremum* $\sqcup \mathcal{F}$ is generated by the union of the families of generators of all $\mathcal{G}_i$, $i \in [1, \ell]$.

If $\mathcal{V}(\mathcal{G}_1) \subseteq \mathcal{V}(\mathcal{G})$ (respectively $\mathcal{E}(\mathcal{G}_2) \subseteq \mathcal{E}(\mathcal{G})$), we denote by $\overline{\mathcal{V}(\mathcal{G}_1)}$ (respectively $\overline{\mathcal{E}(\mathcal{G}_2)}$) the *complementary set of* $\mathcal{V}(\mathcal{G}_1)$ (respectively $\mathcal{E}(\mathcal{G}_2)$) *in* $\mathcal{V}(\mathcal{G})$ (respectively $\mathcal{E}(\mathcal{G})$), that is $\overline{\mathcal{V}(\mathcal{G}_1)} = \mathcal{V}(\mathcal{G}) \setminus \mathcal{V}(\mathcal{G}_1)$ (respectively $\overline{\mathcal{E}(\mathcal{G}_2)} = \mathcal{E}(\mathcal{G}) \setminus \mathcal{E}(\mathcal{G}_2)$). Observe that, if $\mathcal{G}_1$ is a subgraph of $\mathcal{G}$, then, except in some degenerated cases, the pair $(\overline{\mathcal{V}(\mathcal{G}_1)}, \overline{\mathcal{E}(\mathcal{G}_1)})$ is not a graph.

Property 1 ([10])

The set $\mathbb{G}$ of the subgraphs of $\mathcal{G}$ forms a complete lattice, sup-generated by the set $\mathcal{S} = \mathcal{S}_0 \cup \mathcal{S}_1$, but not complemented. The supremum and the infimum of any family $\mathcal{F} = \{\mathcal{G}_1, \ldots \mathcal{G}_\ell\}$ of elements in $\mathbb{G}$ are given by, respectively, $\sqcap \mathcal{F} = (\bigcap_{i \in [1,\ell]} \mathcal{V}(\mathcal{G}_i), \bigcap_{i \in [1,\ell]} \mathcal{E}(\mathcal{G}_i))$ and $\sqcup \mathcal{F} = (\bigcup_{i \in [1,\ell]} \mathcal{V}(\mathcal{G}_i), \bigcup_{i \in [1,\ell]} \mathcal{E}(\mathcal{G}_i))$.

6.2.2 Lattice of Weights

As a fixed grid is able to represent images by assigning grey tones to pixels, the graph $\mathcal{G}$ is able to generate a number of derived graphs by assigning weights to the vertices and edges of the graph $\mathcal{G}$. According to the applications, the weights can be real or integer, taking their values in $\mathbb{R}, \mathbb{R}^+, \mathbb{N}, [-n, +n], [0, +n]$. The weight of a vertex v_i is written w_i, while the one of an edge e_{ij} is written w_{ij}. The set of all (edges and vertices) weights is written w. The case where the weights are binary, belonging to $\{0, 1\}$ may be interpreted as presence/absence: $w_{ij} = 1$ and $w_k = 1$ express respectively the existence of an edge e_{ij} and of a vertex v_k. All edges and vertices of $\mathcal{G}$ with weight 0 do not exist in this weighted graph.

A possible lattice structure on the weights is given by the following. Let w^1 and w^2 be two sets of weights, we have $w^1 \prec w^2$ whenever $w_{ij}^1 < w_{ij}^2$ and $w_k^1 < w_k^2$. The supremum (*respectively* infimum) of a family of sets of weights is a set of weights where the weight of a given element is the greatest (*respectively* lowest) possible weight of the weights of the same element in the family.

Remark 2

In a graph, there may be isolated vertices, that is vertices which are not adjacent to an edge. On the contrary, each edge is adjacent to two vertices. For this reason not any binary distribution of weights on the vertices and edges of $\mathcal{G}$ represents a graph. It is the case if and only if each edge with weight 1 is adjacent to two vertices with weight 1. The same holds for any weight distributions. w correspond to a graph if each edge with weight w_{ij} is such that $w_{ij} \leq w_i$ and $w_{ij} \leq w_j$. In other words, this is the case if the extremities of an edge have weights which are not lower than the weight of the edge. Graphs may be used for modeling many different structures. In some cases vertex weight only have a physical meaning, and the edge weights simply serve for storing intermediate results in the computation of the weights of the vertices. In other cases, it is the converse. In such situations one does not care whether the weights represent a graph. In other situations, on the contrary, one cares to define operators transforming one graph into another graph.

6.3 Neighborhood Operations on Graphs

Morphology really starts if one considers the neighborhood relations between vertices and edges. We now define operators taking as arguments such neighborhoods. The construction is incremental, from the smallest neighborhood to larger neighborhoods.

6.3.1 Adjunctions on Graphs

In the graph $\mathcal{G}$, we can consider sets of points as well as sets of edges. Therefore, it is convenient to consider operators that go from one kind of sets to the other one. In this section, we investigate such operators and we study their morphological properties. Then, based on these operators, we propose several dilations and erosions, acting on the lattice of all subgraphs of $\mathcal{G}$.

Let $\mathcal{V}(\mathcal{G}_1)$ be a subset of $\mathcal{V}(\mathcal{G})$, we denote by $\mathbb{G}_{\mathcal{V}(\mathcal{G}_1)}$ the set of all subgraphs of $\mathcal{G}$ whose vertex set is $\mathcal{V}(\mathcal{G}_1)$. Let $\mathcal{E}(\mathcal{G}_2)$ be a subset of $\mathcal{E}(\mathcal{G})$. We denote by $\mathbb{G}_{\mathcal{E}(\mathcal{G}_2)}$ the set of all subgraphs of $\mathcal{G}$ whose edge set is $\mathcal{E}(\mathcal{G}_2)$.

Definition 3 (edge-vertex correspondences, [10])

We define the operators $\delta_{\mathcal{E}\mathcal{V}}$, $\varepsilon_{\mathcal{E}\mathcal{V}}$ from $\mathcal{E}(\mathbb{G})$ into $\mathcal{V}(\mathbb{G})$ and the operators $\varepsilon_{\mathcal{V}\mathcal{E}}$, $\delta_{\mathcal{V}\mathcal{E}}$

from $\mathcal{V}(\mathbb{G})$ into $\mathcal{E}(\mathbb{G})$ as follows:

	$\mathcal{E}(\mathbb{G}) \to \mathcal{V}(\mathbb{G})$			$\mathcal{V}(\mathbb{G}) \to \mathcal{E}(\mathbb{G})$		
Provide the object with a graph structure	$\mathcal{E}(\mathcal{G}_1)$ $\delta_{\mathcal{E}\mathcal{V}}(\mathcal{E}(\mathcal{G}_1))$ $(\delta_{\mathcal{E}\mathcal{V}}(\mathcal{E}(\mathcal{G}_1)), \mathcal{E}(\mathcal{G}_1))$ $\sqcap\mathbb{G}_{\mathcal{E}(\mathcal{G}_1)}$	*such* $=$	*that* $\to$	$\mathcal{V}(\mathcal{G}_1)$ $\varepsilon_{\mathcal{V}\mathcal{E}}(\mathcal{V}(\mathcal{G}_1))$ $(\mathcal{V}(\mathcal{G}_1), \varepsilon_{\mathcal{V}\mathcal{E}}(\mathcal{V}(\mathcal{G}_1)))$ $\sqcup\mathbb{G}_{\mathcal{V}(\mathcal{G}_1)}$	*such* $=$	*that* $\to$
Provide its complement with a graph structure	$\mathcal{E}(\mathcal{G}_1)$ $\varepsilon_{\mathcal{E}\mathcal{V}}(\mathcal{E}(\mathcal{G}_1))$ $(\varepsilon_{\mathcal{E}\mathcal{V}}(\mathcal{E}(\mathcal{G}_1)), \mathcal{E}(\mathcal{G}_1))$ $\sqcap\mathbb{G}_{\overline{\mathcal{E}(\mathcal{G}_1)}}$	*such* $=$	*that* $\to$	$\mathcal{V}(\mathcal{G}_1)$ $\delta_{\mathcal{V}\mathcal{E}}(\mathcal{V}(\mathcal{G}_1))$ $(\mathcal{V}(\mathcal{G}_1), \delta_{\mathcal{V}\mathcal{E}}(\mathcal{V}(\mathcal{G}_1)))$ $\sqcup\mathbb{G}_{\overline{\mathcal{V}(\mathcal{G}_1)}}$	*such* $=$	*that* $\to$

In other words, if $\mathcal{V}(\mathcal{G}_1) \subseteq \mathcal{V}(\mathcal{G})$ and $\mathcal{E}(\mathcal{G}_2) \subseteq \mathcal{E}(\mathcal{G})$, $(\delta_{\mathcal{E}\mathcal{V}}(\mathcal{E}(\mathcal{G}_2)), \mathcal{E}(\mathcal{G}_2))$ is the smallest subgraph of $\mathcal{G}$ whose edge set is $\mathcal{E}(\mathcal{G}_2)$, $(\mathcal{V}(\mathcal{G}_1), \varepsilon_{\mathcal{V}\mathcal{E}}(\mathcal{V}(\mathcal{G}_1)))$ is the largest subgraph of $\mathcal{G}$ whose vertex set is $\mathcal{V}(\mathcal{G}_1)$, $(\varepsilon_{\mathcal{E}\mathcal{V}}(\mathcal{E}(\mathcal{G}_2)), \mathcal{E}(\mathcal{G}_2))$ is the smallest subgraph of $\mathcal{G}$ whose edge set is $\overline{\mathcal{E}(\mathcal{G}_2)}$, and $(\mathcal{V}(\mathcal{G}_1), \delta_{\mathcal{V}\mathcal{E}}(\mathcal{V}(\mathcal{G}_1)))$ is the largest subgraph of $\mathcal{G}$ whose vertex set is $\overline{\mathcal{V}(\mathcal{G}_1)}$.

These operators are illustrated in Figures 6.1(a–f). The following property locally characterizes them.

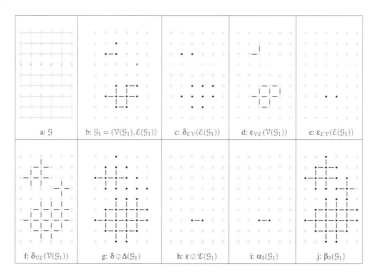

a: $\mathcal{G}$ b: $\mathcal{G}_1 = (\mathcal{V}(\mathcal{G}_1), \mathcal{E}(\mathcal{G}_1))$ c: $\delta_{\mathcal{E}\mathcal{V}}(\mathcal{E}(\mathcal{G}_1))$ d: $\varepsilon_{\mathcal{V}\mathcal{E}}(\mathcal{V}(\mathcal{G}_1))$ e: $\varepsilon_{\mathcal{E}\mathcal{V}}(\mathcal{E}(\mathcal{G}_1))$

f: $\delta_{\mathcal{V}\mathcal{E}}(\mathcal{V}(\mathcal{G}_1))$ g: $\delta \oslash \Delta(\mathcal{G}_1)$ h: $\varepsilon \oslash \pounds(\mathcal{G}_1)$ i: $\alpha_3(\mathcal{G}_1)$ j: $\beta_3(\mathcal{G}_1)$

FIGURE 6.1
Dilations and erosions.

Property 4 ([10])
For any $\mathcal{E}(\mathcal{G}_1) \subseteq \mathcal{E}(\mathcal{G})$ and $\mathcal{V}(\mathcal{G}_2) \subseteq \mathcal{V}(\mathcal{G})$:

1. $\delta_{\mathcal{E}\mathcal{V}} : \mathcal{E}(\mathcal{G}) \to \mathcal{V}(\mathcal{G})$ *is such that* $\delta_{\mathcal{E}\mathcal{V}}(\mathcal{E}(\mathcal{G}_1)) = \{v_i \in \mathcal{V}(\mathcal{G}) \mid \exists e_{ij} \in \mathcal{E}(\mathcal{G}_1)\}$;

2. $\varepsilon_{\mathcal{VE}}$: $\mathcal{V}(\mathcal{G}) \rightarrow \mathcal{E}(\mathcal{G})$ is such that $\varepsilon_{\mathcal{VE}}(\mathcal{V}(\mathcal{G}_2)) = \{e_{ij} \in \mathcal{E}(\mathcal{G}) \mid v_i \in \mathcal{V}(\mathcal{G}_2) \text{ and } v_j \in \mathcal{V}(\mathcal{G}_2)\}$;

3. $\varepsilon_{\mathcal{EV}}$: $\mathcal{E}(\mathcal{G}) \rightarrow \mathcal{V}(\mathcal{G})$ is such that $\varepsilon_{\mathcal{EV}}(\mathcal{E}(\mathcal{G}_1)) = \{v_i \in \mathcal{V}(\mathcal{G}) \mid \forall e_{ij} \in \mathcal{E}(\mathcal{G}), e_{ij} \in \mathcal{E}(\mathcal{G}_1)\}$;

4. $\delta_{\mathcal{VE}}$: $\mathcal{V}(\mathcal{G}) \rightarrow \mathcal{E}(\mathcal{G})$ is such that $\delta_{\mathcal{VE}}(\mathcal{V}(\mathcal{G}_2)) = \{e_{ij} \in \mathcal{E}(\mathcal{G}) \mid \text{ either } v_i \in \mathcal{V}(\mathcal{G}_2) \text{ or } v_j \in \mathcal{V}(\mathcal{G}_2)\}$.

In other words, $\delta_{\mathcal{EV}}(\mathcal{E}(\mathcal{G}_1))$ is the set of all vertices that belong to an edge of $\mathcal{E}(\mathcal{G}_1)$, $\varepsilon_{\mathcal{VE}}(\mathcal{V}(\mathcal{G}_2))$ is the set of all edges whose two extremities are in $\mathcal{V}(\mathcal{G}_2)$, $\varepsilon_{\mathcal{EV}}(\mathcal{E}(\mathcal{G}_1))$ is the set of all vertices which do not belong to any edge of $\overline{\mathcal{E}(\mathcal{G}_1)}$, and $\delta_{\mathcal{VE}}(\mathcal{V}(\mathcal{G}_2))$ is the set of all edges which have at least one extremity in $\mathcal{V}(\mathcal{G}_2)$.

From this characterization, we can recognize the general graph version of some operators introduced by Meyer and Angulo [8] (see also [7]) for the hexagonal grid. By translating supremum and infimum of the graph lattice to the corresponding supremum and infimum of the lattice of weigths, we obtain the weighted version of these operators. The four basic operators of Definition 3 translate to: $(\delta_{\mathcal{EV}}(w))_i = \max\{w_{ij} \mid e_{ij} \in \mathcal{E}\}$, $\varepsilon_{\mathcal{EV}}(w)_i = \min\{w_{ij} \mid e_{ij} \in \mathcal{E}\}$, $(\delta_{\mathcal{VE}}(w))_{ij} = \max\{w_i, w_j\}$, and $(\varepsilon_{\mathcal{VE}}(w))_{ij} = \min\{w_i, w_j\}$.

Before further analyzing the operators defined above, let us briefly recall some algebraic tools that are fundamental in mathematical morphology [26].

Given two lattices $\mathcal{L}_1$ and $\mathcal{L}_2$, an operator δ : $\mathcal{L}_1 \rightarrow \mathcal{L}_2$ is called a *dilation* when it preserves the supremum (i.e., $\forall \mathcal{X} \subseteq \mathcal{L}_1, \delta(\vee_1 \mathcal{X}) = \vee_2\{\delta(x) \mid x \in \mathcal{X}\}$, where $\vee_1$ is the supremum in $\mathcal{L}_1$ and $\vee_2$ the supremum in $\mathcal{L}_2$). Similarly, an operator which preserves the infimum is called an *erosion*.

Two operators ε : $\mathcal{L}_1 \rightarrow \mathcal{L}_2$ and δ : $\mathcal{L}_2 \rightarrow \mathcal{L}_1$ form an *adjunction* (ε, δ) when for any x in $\mathcal{L}_2$ and any y in $\mathcal{L}_1$, we have $\delta(x) \leq_1 y \Leftrightarrow x \leq_2 \varepsilon(y)$, where $\leq_1$ and $\leq_2$ denote the order relations on, respectively, $\mathcal{L}_1$ and $\mathcal{L}_2$. Given two operators ε and δ, if the pair (ε, δ) is an adjunction, then ε is an erosion and δ is a dilation.

Given two complemented lattices $\mathcal{L}_1$ and $\mathcal{L}_2$, two operators α and β from $\mathcal{L}_1$ into $\mathcal{L}_2$ are *dual (with respect to the complement) of each other* when, for any $x \in \mathcal{L}_1$, we have $\beta(x) = \overline{\alpha(\overline{x})}$. If α and β are dual of each other, then β is an erosion whenever α is a dilation.

Property 5 (dilation, erosion, adjunction, duality, [10])

1. *Both $(\varepsilon_{\mathcal{VE}}, \delta_{\mathcal{EV}})$ and $(\varepsilon_{\mathcal{EV}}, \delta_{\mathcal{VE}})$ are adjunctions.*

2. *Operators $\varepsilon_{\mathcal{VE}}$ and $\delta_{\mathcal{VE}}$ (respectively $\varepsilon_{\mathcal{EV}}$ and $\delta_{\mathcal{EV}}$) are dual of each other.*

3. *Operators $\delta_{\mathcal{EV}}$ and $\delta_{\mathcal{VE}}$ are dilations.*

4. *Operators $\varepsilon_{\mathcal{EV}}$ and $\varepsilon_{\mathcal{VE}}$ are erosions.*

Let us compose these dilations and erosions to act on $\mathcal{V}(\mathbb{G})$ and $\mathcal{E}(\mathbb{G})$.

Definition 6 (vertex-dilation, vertex-erosion)
We define δ and ε that act on $\mathcal{V}(\mathbb{G})$ (i.e., $\mathcal{V}(\mathbb{G}) \to \mathcal{V}(\mathbb{G})$) by $\delta = \delta_{\mathcal{E}\mathcal{V}} \circ \delta_{\mathcal{V}\mathcal{E}}$ and $\varepsilon = \varepsilon_{\mathcal{E}\mathcal{V}} \circ \varepsilon_{\mathcal{V}\mathcal{E}}$.

As compositions of, respectively, dilations and erosions, δ and ε are, respectively, a dilation and an erosion. Moreover, by composition of adjunctions and dual operators, δ and ε are dual, and (ε, δ) is an adjunction.

In fact, it can be shown that δ and ε correspond exactly to the usual notions of an erosion and of a dilation of a set of vertices in a graph [4, 5]. It means, in particular, that when $\mathcal{V}(\mathcal{G})$ is a subset of the grid points $\mathbb{Z}^d$ and when the edge set $\mathcal{E}(\mathcal{G})$ is obtained from a symmetrical structuring element, then the operators defined above are equivalent to the usual binary dilation and erosion by the considered structuring element. For instance, in Figure 6.1, $\mathcal{V}(\mathcal{G})$ is a rectangular subset of $\mathbb{Z}^2$ and $\mathcal{E}(\mathcal{G})$ corresponds to the basic "cross" structuring element. It can be verified that the vertex sets in Figure 6.1 (g) and (h), obtained by applying δ and ε to $\mathcal{V}(\mathcal{G}_1)$ (Figure 6.1(b), are the dilation and the erosion by a "cross" structuring element of $\mathcal{V}(\mathcal{G}_1)$.

We now consider a dual/adjunct pair of dilation and erosion acting on $\mathcal{E}(\mathbb{G})$.

Definition 7 (edge-dilation, edge-erosion, [10])
We define Δ and $\mathcal{E}$ that act on $\mathcal{E}(\mathbb{G})$ by $\Delta = \delta_{\mathcal{V}\mathcal{E}} \circ \delta_{\mathcal{E}\mathcal{V}}$ and $\mathcal{E} = \varepsilon_{\mathcal{V}\mathcal{E}} \circ \varepsilon_{\mathcal{E}\mathcal{V}}$.

Definition 8 ([10])
We define the operators $\delta \oslash \Delta$ and $\varepsilon \oslash \mathcal{E}$ by, respectively, $(\delta(\mathcal{V}(\mathcal{G}_1)), \Delta(\mathcal{E}(\mathcal{G}_1)))$ and $(\varepsilon(\mathcal{V}(\mathcal{G}_1)), \mathcal{E}(\mathcal{E}(\mathcal{G}_1)))$, for any $\mathcal{G}_1 \in \mathbb{G}$.

For instance, Figures 6.1(f) and 6.1(g) present the results obtained by applying the operator $\delta \oslash \Delta$ and the operator $\varepsilon \oslash \mathcal{E}$ to the subgraph $\mathcal{G}_1$ (Figure 6.1(b)) of $\mathcal{G}$ (Figure 6.1(a)).

Theorem 9 (graph-dilation, graph-erosion,[10])
The operators $\delta \oslash \Delta$ and $\varepsilon \oslash \mathcal{E}$ are respectively a dilation and an erosion acting on the lattice $(\mathbb{G}, \sqsubseteq)$. Furthermore, $(\varepsilon \oslash \mathcal{E}, \delta \oslash \Delta)$ is an adjunction.

Note that since lattice $\mathbb{G}$ is sup-generated by set $\mathcal{S}$, it suffices to know the dilation of the graphs in $\mathcal{S}$ for characterizing the dilation of the graphs in $\mathbb{G}$.

Compared to classical morphological operators on sets, the dilations and erosions introduced in this section furthermore convey some connectivity properties different than the ones which can be deduced from classical dilations and erosions. Observe, for instance, in Figure 6.1(g), that some 4-adjacent vertices of $\delta(\mathcal{V}(\mathcal{G}_1))$ are not linked by an edge in the graph $\delta \oslash \Delta(\mathcal{G}_1)$. These properties can be useful in further processing involving for instance connected operators [27, 28, 29, 30].

Thanks to the operators presented in Definition 3, other intersecting adjunctions (hence dilations/erosions) can be defined on $\mathbb{G}$:

1. (α_1, β_1) such that $\forall \mathcal{G}_1 \in \mathbb{G}$, $\alpha_1(\mathcal{G}_1) = (\mathcal{V}(\mathcal{G}), \mathcal{E}(\mathcal{G}_1))$ and $\beta_1(\mathcal{G}_1) = (\delta_{\mathcal{E}\mathcal{V}}(\mathcal{E}(\mathcal{G}_1)), \mathcal{E}(\mathcal{G}_1))$;

2. (α_2, β_2) such that $\forall \mathcal{G}_1 \in \mathbb{G}$, $\alpha_2(\mathcal{G}_1) = (\mathcal{V}(\mathcal{G}_1), \varepsilon_{\mathcal{V}\mathcal{E}}(\mathcal{V}(\mathcal{G}_1)))$ and $\beta_2(\mathcal{G}_1) = (\mathcal{V}(\mathcal{G}_1), \emptyset)$;

3. (α_3, β_3) such that $\forall \mathcal{G}_1 \in \mathbb{G}$, $\alpha_3(\mathcal{G}_1) = (\varepsilon_{\mathcal{E}\mathcal{V}}(\mathcal{E}(\mathcal{G}_1)), \varepsilon_{\mathcal{V}\mathcal{E}} \circ \varepsilon_{\mathcal{E}\mathcal{V}}(\mathcal{E}(\mathcal{G}_1)))$ and $\beta_3(\mathcal{G}_1) = (\delta_{\mathcal{E}\mathcal{V}} \circ \delta_{\mathcal{V}\mathcal{E}}(\mathcal{V}(\mathcal{G}_1)), \delta_{\mathcal{V}\mathcal{E}}(\mathcal{V}(\mathcal{G}_1)))$.

The adjunction (α_3, β_3) is illustrated in Figures 6.1(i) and 6.1(j). Note also that, using usual graph terminologies, β_1 (respectively α_2) can be defined as the operator which associates to a graph the graph induced by its edge set (respectively, vertex set).

A more general version of those operators can be obtained in the framework of complexes [12], allowing us to deal for example with meshes. There also exist other recent graph-based approaches to morphology, for example based on hypergraphs [11] or on discrete differential equations [9].

6.4 Filters

In mathematical morphology, a *filter* is an operator α acting on a lattice $\mathcal{L}$ that is increasing (i.e., $\forall x, y \in \mathcal{L}$, $\alpha(x) \leq \alpha(y)$ whenever $x \leq y$) and idempotent (i.e., $\forall x \in \mathcal{L}$, $\alpha(\alpha(x)) = \alpha(x)$). A filter α on $\mathcal{L}$ that is extensive (i.e., $\forall x \in \mathcal{L}$, $x \leq \alpha(x)$) is called a *closing on* $\mathcal{L}$, whereas a filter α on $\mathcal{L}$ that is anti-extensive (i.e., $\forall x \in \mathcal{L}$, $\alpha(x) \leq x$) is called an *opening on* $\mathcal{L}$. It is known that composing the two operators of an adjunction yields an opening or a closing, depending on the order in which the operators are composed [26]. In this section, the operators of Section 6.3 are composed to obtain filters on $\mathcal{V}(\mathbb{G})$, $\mathcal{E}(\mathbb{G})$ and $\mathbb{G}$.

Definition 10 (opening, closing, [10])
1. *We define γ_1 and ϕ_1, which act on $\mathcal{V}(\mathbb{G})$, by $\gamma_1 = \delta \circ \varepsilon$ and $\phi_1 = \varepsilon \circ \delta$.*

2. *We define Γ_1 and Φ_1, which act on $\mathcal{E}(\mathbb{G})$, by $\Gamma_1 = \Delta \circ \mathcal{E}$ and $\Phi_1 = \mathcal{E} \circ \Delta$.*

3. *We define the operators $\gamma \oslash \Gamma_1$ and $\phi \oslash \Phi_1$ by, respectively, $\gamma \oslash \Gamma_1(\mathcal{G}_1) = (\gamma_1(\mathcal{V}(\mathcal{G}_1)), \Gamma_1(\mathcal{E}(\mathcal{G}_1)))$ and $\phi \oslash \Phi_1(\mathcal{G}_1) = (\phi_1(\mathcal{V}(\mathcal{G}_1)), \Phi_1(\mathcal{E}(\mathcal{G}_1)))$ for any $\mathcal{G}_1 \in \mathbb{G}$.*

Figures 6.2(b) and 6.2(f) present the result of $\gamma \oslash \Gamma_1$ and $\phi \oslash \Phi_1$ for, respectively, the subgraph of Figure 6.2(a) and the one of Figure 6.2(e).

The opening γ_1 and the closing ϕ_1 correspond to the classical opening and closing on the vertices. The closing Γ_1 and the opening Φ_1 are the corresponding edge-version of these operators. By combination, we obtain $\gamma \oslash \Gamma_1$ and $\phi \oslash \Phi_1$.

In fact, by composing $\delta_{\mathcal{E}\mathcal{V}}$ with $\varepsilon_{\mathcal{V}\mathcal{E}}$ and $\delta_{\mathcal{E}\mathcal{V}}$ with $\varepsilon_{\mathcal{V}\mathcal{E}}$, we obtain smaller filters.

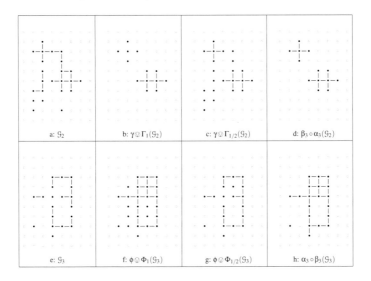

FIGURE 6.2
Openings and closings ($\mathcal{G}$ is induced by the 4-adjacency relation).

Definition 11 (half-opening, half-closing, [10])

1. *We define $\gamma_{1/2}$ and $\phi_{1/2}$, that act on $\mathcal{V}(\mathbb{G})$, by $\gamma_{1/2} = \delta_{\mathcal{E}\mathcal{V}} \circ \epsilon_{\mathcal{V}\mathcal{E}}$ and $\phi_{1/2} = \epsilon_{\mathcal{E}\mathcal{V}} \circ \delta_{\mathcal{V}\mathcal{E}}$.*

2. *We define $\Gamma_{1/2}$ and $\Phi_{1/2}$, that act on $\mathcal{E}(\mathbb{G})$ by $\Gamma_{1/2} = \delta_{\mathcal{V}\mathcal{E}} \circ \varepsilon_{\mathcal{E}\mathcal{V}}$ and $\Phi_{1/2} = \varepsilon_{\mathcal{V}\mathcal{E}} \circ \delta_{\mathcal{E}\mathcal{V}}$.*

3. *We define the operators $\gamma \oslash \Gamma_{1/2}$ and $\phi \oslash \Phi_{1/2}$ by, respectively, $\gamma \oslash \Gamma_{1/2}(\mathcal{G}_1) = (\gamma_{1/2}(\mathcal{V}(\mathcal{G}_1)), \Gamma_{1/2}(\mathcal{E}(\mathcal{G}_1)))$ and $\phi \oslash \Phi_{1/2}(\mathcal{G}_1) = (\phi_{1/2}(\mathcal{V}(\mathcal{G}_1)), \Phi_{1/2}(\mathcal{E}(\mathcal{G}_1)))$, for any $\mathcal{G}_1 \in \mathbb{G}$.*

Thanks to Property 4, the operators defined above can be locally characterized. Let $\mathcal{V}(\mathcal{G}_1) \subseteq \mathcal{V}(\mathcal{G})$ and $\mathcal{E}(\mathcal{G}_2) \subseteq \mathcal{E}(\mathcal{G})$, we have:

$$
\begin{aligned}
\gamma_{1/2}(\mathcal{V}(\mathcal{G}_1)) &= \{v_i \in \mathcal{V}(\mathcal{G}_1) \mid \exists e_{ij} \in \mathcal{E}(\mathcal{G}) \text{ with } v_j \in \mathcal{V}(\mathcal{G}_1)\} \\
&= \mathcal{V}(\mathcal{G}_1) \setminus \{v_i \in \mathcal{V}(\mathcal{G}_1) \mid \forall e_{ij} \in \mathcal{E}(\mathcal{G}), v_j \notin \mathcal{V}(\mathcal{G}_1)\} \\
\Gamma_{1/2}(\mathcal{E}(\mathcal{G}_2)) &= \{e_{ij} \in \mathcal{E}(\mathcal{G}) \mid \{e_{ik} \in \mathcal{E}(\mathcal{G})\} \subseteq \mathcal{E}(\mathcal{G}_2)\} \\
&= \mathcal{E}(G_2) \setminus \{e \in \mathcal{E}(Y) \mid \forall v_i \in e, \exists e_{ij} \in \mathcal{E}(\mathcal{G}) \text{ with } e_{ij} \notin \mathcal{E}(\mathcal{G}_2)\} \\
\phi_{1/2}(\mathcal{V}(\mathcal{G}_1)) &= \{v_i \in \mathcal{V}(\mathcal{G}) \mid \text{ either } v_i \in \mathcal{V}(\mathcal{G}_1) \text{ or } \forall e_{ij} \in \mathcal{E}(\mathcal{G}), v_j \in \mathcal{V}(\mathcal{G}_1)\} \\
&= \mathcal{V}(\mathcal{G}_1) \cup \{v_i \in \overline{\mathcal{V}(\mathcal{G}_1)} \mid \forall e_{ij} \in \mathcal{E}(\mathcal{G}), v_j \in \mathcal{V}(\mathcal{G}_1)\} \\
\Phi_{1/2}(\mathcal{E}(\mathcal{G}_2)) &= \{e_{ij} \in \mathcal{E}(\mathcal{G}) \mid \exists e_{ik} \in \mathcal{E}(\mathcal{G}_2) \text{ and } \exists e_{jl} \in \mathcal{E}(\mathcal{G}_2)\} \\
&= Y \cup \{e_{ij} \in \overline{\mathcal{E}(\mathcal{G}_2)} \mid v_i \in \delta_{\mathcal{E}\mathcal{V}}(\mathcal{E}(\mathcal{G}_2)) \text{ and } v_j \in \delta_{\mathcal{E}\mathcal{V}}(\mathcal{E}(\mathcal{G}_2))\}.
\end{aligned}
$$

Informally speaking, $\gamma_{1/2}$ removes from $\mathcal{G}_1$ its isolated vertices, whereas $\Gamma_{1/2}$ removes from $\mathcal{G}_2$ the edges that do not contain a vertex completely covered by edges in $\mathcal{G}_2$. It may be seen furthermore that $\gamma_{1/2}$ (respectively, $\Gamma_{1/2}$) is the dual of $\phi_{1/2}$ (respectively, $\Phi_{1/2}$). Thus, $\phi_{1/2}$ adds to $\mathcal{G}_1$ the vertices of $\overline{\mathcal{V}(\mathcal{G}_1)}$ completely surrounded by elements of $\mathcal{G}_1$ whereas $\Phi_{1/2}$ adds to $\mathcal{G}_2$ the edges of $\overline{\mathcal{E}(\mathcal{G}_2)}$ whose two extremities belong to at least one edge in $\mathcal{G}_2$ (see, for instance, Figure 6.2).

The family $\mathbb{G}$ is closed under the operators presented in Definition 10.3 since they are obtained by composition of operators also satisfying this property. Furthermore, it can be deduced from the local characterization of the operators $\gamma_{1/2}, \Gamma_{1/2}, \phi_{1/2}$ and $\Phi_{1/2}$ that the family $\mathbb{G}$ is also closed under the operators of Definition 11.3. Hence, thanks to the properties of adjunctions recalled in the introduction of this section, the following theorem can be established.

Theorem 12 (graph-openings, graph-closings,[10])

1. *The operators $\gamma_{1/2}$ and γ_1 (respectively, $\Gamma_{1/2}$ and Γ_1) are openings on $\mathcal{V}(\mathbb{G})$ (respectively, $\mathcal{E}(\mathbb{G})$) and $\phi_{1/2}$, and Φ_1 (respectively, $\Phi_{1/2}$ and ϕ_1) are closings on $\mathcal{V}(\mathbb{G})$ (respectively, $\mathcal{E}(\mathbb{G})$).*

2. *The family $\mathbb{G}$ is closed under $\gamma \oslash \Gamma_{1/2}, \phi \oslash \Phi_{1/2}, \gamma \oslash \Gamma_1$, and $\phi \oslash \Phi_1$.*

3. *The operators $\gamma \oslash \Gamma_{1/2}$ and $\gamma \oslash \Gamma_1$ are openings on $\mathbb{G}$ and $\phi \oslash \Phi_{1/2}$, and $\phi \oslash \Phi_1$ are closings on $\mathbb{G}$.*

Composing the operators of the adjunctions (α_i, β_i), defined at the end of Section 6.3, also yields remarkable openings and closings. Indeed, it can be easily seen that: $\alpha_1 \circ \beta_1 = \alpha_1, \alpha_2 \circ \beta_2 = \alpha_2, \beta_1 \circ \alpha_1 = \beta_1$, and $\beta_2 \circ \alpha_2 = \beta_2$. Thus, α_1 and α_2 are both a closing and an erosion, and β_1 and β_2 are both a dilation and an opening. This means, in particular, that α_1 and α_2 are idempotent extensive erosions and that β_1 and β_2 are idempotent anti-extensive dilations. The opening and the closing resulting from the adjunction (α_3, β_3) are illustrated in Figures 6.2d and 6.2h.

The weighted versions of those various operators also have an interpretation related to classical notions of graph theory. Consider, for example, the weighted version of the opening $\Gamma_{1/2} = \delta_{\mathcal{V}\mathcal{E}}\varepsilon_{\mathcal{E}\mathcal{V}}$. The dilation $\delta_{\mathcal{V}\mathcal{E}}$ assigns to each edge the highest weight of its adjacent vertices; but each adjacent vertex has been assigned by $\varepsilon_{\mathcal{E}\mathcal{V}}$ the weight of its lowest edge. So if w_{ij} is left unchanged by $\Gamma_{1/2}$, it means that the edge e_{ij} is (one of) the lowest edges of v_i or of v_j, say v_i. Its weight is higher or equal to the weight of the lowest edge of v_j. Thus, using some well-known properties of minimum spanning trees (see Section 6.6.3 for a precise definition of a minimum spanning tree), it can be shown that the graph induced by the set of edges that are left invariant by $\Gamma_{1/2}$ is the union of all mininimum spanning trees of this graph, closely related to the *Gabriel graph* [31].

It is possible to associate with any lattice $\mathcal{L}$, the lattice of all increasing operators on $\mathcal{L}$. In this context, two filters, φ_1 and φ_2, on the lattice $\mathcal{L}$ are said *ordered* if, for any $\in \mathcal{L}, \varphi_1(x) \leq \varphi_2(x)$ or if, for any $x \in \mathcal{L}, \varphi_2(x) \leq \varphi_1(x)$. A usual way to build

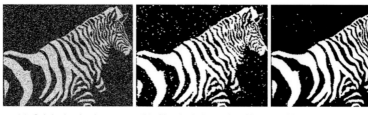

(a) Original noisy image   (b) Classical alternating filter   (c) Vertex-edge alternating filter

FIGURE 6.3
Illustration of classical versus vertex-edge alternating filter (see [10]).

a hierarchy of filters (i.e., an ordered family of filters) from an adjunct pair (α, β) of erosion and dilation consists of building the dilations and erosions obtained by iterating several times α and β. In general, composing these iterated versions of α and β leads to hierarchies of filters when the number of iterations increases. By combining and iterating the operators that we have defined in this section, we obtain [10] hierarchies of filters in the lattice $\mathbb{G}$ that perform better than the classical ones, in the sense that they are able to remove more noise (see Figure 6.3.)

6.5   Connected Operators and Filtering with the Component Tree

Gray-level connected operators [27] act by merging neighboring "flat" zones. They cannot create new contours and, as a result, they cannot introduce in the output image a structure that is not present in the input image. Furthermore, they cannot modify the position of existing boundaries between regions, and therefore have very good contour preservation properties.

To create "flat" zones, a simple operation on a weighted graph is thresholding, that produces a level set. For $\lambda \in \mathbb{R}^+$, the level-sets of an edge-weighted graph are denoted $w_{\mathcal{E}}^{<}[\lambda] = \{e_{ij} \in \mathcal{E} | w_{ij} \leq \lambda\}$. We define $\mathcal{C}_{\mathcal{E}}^{<}$ as the set composed of all the pairs $[\lambda, C]$, where $\lambda \in \mathbb{R}^+$ and C is a (connected) component of the graph $w_{\mathcal{E}}^{<}[\lambda]$.

We note that one can reconstruct w from $\mathcal{C}_{\mathcal{E}}^{<}$; more precisely, we have:

$$w_{ij} = \min\{\lambda \mid [\lambda, C] \in \mathcal{C}_{\mathcal{E}}^{<}, e_{ij} \in \mathcal{E}(C)\} \tag{6.1}$$

One can remark that two elements of $\mathcal{C}_{\mathcal{E}}^{<}$ are either disjoint or nested, thus it is easy to order the elements of $\mathcal{C}_{\mathcal{E}}^{<}$ in a tree. This tree is called the *(min) component tree*, and there exists fast algorithms to compute it [32, 33].

Using $\mathcal{C}_{\mathcal{E}}^{<}$ allows us to deal with specific components of w. For example, a minimum of w is a component C such that there exists λ with $[\lambda, C] \in \mathcal{C}_{\mathcal{E}}^{<}$, and it does

(a)　Component
tree

(b) Height

(c) Area

(d) Volume

(e) Area filtering

FIGURE 6.4

Illustration of, respectively, a component tree (dashed lines), the height, the area, and the volume of a component and an area filtering.

not exist $[\lambda_1, C_1] \in \mathcal{C}_{\mathcal{E}}^{\leq}$ with $C_1 \sqsubset C$ and $C_1 \neq C$. Observe that a minimum of w is a graph. The set of all minima of w is denoted $\mathfrak{M}$. Similarly, we can define $\mathcal{C}_{\mathcal{E}}^{\geq}$ allowing us to deal with maxima of edge-weighted graphs $\mathcal{C}_{\mathcal{V}}^{\leq}$ and $\mathcal{C}_{\mathcal{V}}^{\geq}$, allowing us to deal with node-weighted graphs, and $\mathcal{C}_{\mathcal{V}\mathcal{E}}^{\leq}$, and $\mathcal{C}_{\mathcal{V}\mathcal{E}}^{\leq}$ allowing us to deal with node- and edge- weighted graphs.

We mention that other trees are possible, for example, the binary partition tree [34], the tree of level lines, also known as the inclusion tree or the tree of shapes [35], etc. The interest of the tree of shapes is that it allows us to interact both with maxima and minima, in a self-dual manner. See [36] for a survey of the usage of those trees in image processing. Here, we illustrate the usage of the component tree for filtering.

Given Equation (6.1), filtering a graph can be seen as equivalent to removing some components of (say) $\mathcal{C}_{\mathcal{E}}^{\leq}$. Such a filtering is called *flooding*. One way to make this idea practicable is to design an attribute that tells us if a given component should be kept or not. Among the numerous attributes that can be computed, three are natural: the height, the area, and the volume (Figure 6.4). Let $[\lambda, C] \in \mathcal{C}_{\mathcal{E}}^{\leq}$. We define

$$\text{height}([\lambda, C]) = \max\{\lambda - w_{ij} | e_{ij} \in \mathcal{E}(C)\} \tag{6.2}$$

$$\text{area}([\lambda, C]) = |\mathcal{V}(C)| \tag{6.3}$$

$$\text{volume}([\lambda, C]) = \sum_{e_{ij} \in \mathcal{E}(C)} (\lambda - w_{ij}) \tag{6.4}$$

For example, removing all the components whose area is lower than a threshold is a closing of w (Figure 6.4.e). Note that a filtering with either the height or the volume is not a closing, as such a filtering is not idempotent.

Figure 6.5 illustrates this kind of filtering. Figure 6.5.a is an image of a cell in which we want to extract the ten bright lobes. If we consider that the brighter a pixel is, the higher its weight, then the tree we want to process is $\mathcal{C}_{\mathcal{V}}^{\geq}$. Figure 6.5.b shows that Figure 6.5(a) contains numerous maxima. Figure 6.5(c) is the filtered image obtained by a volume-based filtering, and Figure 6.5(d) shows the maxima of this filtered image.

As the components of the level sets can be ordered into a tree, they form a hierarchy of components. In the context of classification [37], hierarchies are widely

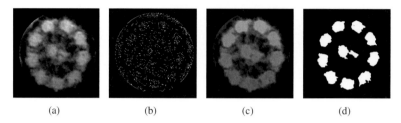

(a)   (b)   (c)   (d)

FIGURE 6.5
(a) Original image. (b) Maxima of image (a), in white. (c) Filtered image. (d) Maxima of image (c), which corresponds to the ten most significant lobes of the image (a).

studied, and the morphological framework can shed some light on the tools that are developed in that community. For example, let us look at the application Ψ that associates to any set of edge-weights w the map $\Psi(w)$ such that for any edge $e_{ij} \in \mathcal{E}$, $\Psi(w)(e_{ij}) = \min\{\lambda \mid [\lambda, C] \in \mathcal{C}_{\mathcal{E}}^{<}, v_i \in \mathcal{V}(C), v_j \in \mathcal{V}(C)\}$. Intuitively, and using a geographical metaphor, $\Psi(w)(e_{ij})$ is the lowest altitude to which one has to climb to go from v_i to v_j. It is straightforward to see that $\Psi(w) \leq w$, that $\Psi(\Psi(w)) = \Psi(w)$ and that if $w' \leq w$, $\Psi(w') \leq \Psi(w)$. Thus, Ψ is an opening [38]. We remark that the subset of strictly positive maps that are defined on the complete graph $(\mathcal{V}, \mathcal{V} \times \mathcal{V})$ and that are open with respect to Ψ is the set of ultrametric distances on $\mathcal{V}$. The mapping Ψ is known under several names, including "subdominant ultrametric" and "ultrametric opening". It is well known that Ψ is associated to the simplest method for hierarchical classification called *single linkage clustering* [39, 40], closely related to Kruskal's algorithm [41] for computing a minimum spanning tree.

Before concluding the section, let us mention another flooding that has been used as a preprocessing step for watershed segmentation with markers. It is called *geodesic reconstruction from the markers* [42, 43], where the markers is a subgraph. It consists in removing all the components of $\mathcal{C}_{\mathcal{E}}^{<}$ that are not marked (i.e., that do not contain a vertex of the markers) and is given as a function w_R such that, for every edge e, we set $w_R(e)$ to be equal to the level λ of the lowest component of w containing e and at least one vertex of the markers. Observe that any component of w_R indeed contains at least one vertex of the markers. Given some markers, the geodesic reconstruction is a closing. Such a closing is also useful as an efficient preprocessing for a fast power-watershed optimization algorithm (see [15] and Section 6.8).

6.6 Watershed Cuts

There exist many possible ways for defining a watershed [24, 44, 45, 46, 13, 47]. Intuitively, the watershed of a function (seen as a topographical surface) is composed of the locations from which a drop of water could flow towards different minima. The framework of edge-weighted graph allows the formalization of this principle and the proof of several remarkable properties [13] that we review in the sequel of the section. We first show that watershed cuts can be equivalently defined by their "catchment basins" (through a steepest descent property) or by the "dividing lines" separating these catchment basins (through the drop of water principle). As far as we know, in discrete frameworks, a similar property does not hold. The second property establishes the optimality of watershed-cuts: there is an equivalence between the watershed-cuts and the separations induced by a minimum spanning forest relative to the minima.

6.6.1 The Drop of Water Principle

Definition 13 (from Definition 12 in [46])
Let $\mathcal{G}_1$ and $\mathcal{G}_2$ be two non-empty subgraphs of $\mathcal{G}$. We say that $\mathcal{G}_2$ is an extension of $\mathcal{G}_1$ (in $\mathcal{G}$) if $\mathcal{G}_1 \sqsubseteq \mathcal{G}_2$ and if any component of $\mathcal{G}_2$ contains exactly one component of $\mathcal{G}_1$.

The notion of extension is very general. Many segmentation algorithms iteratively extend some seed components in a graph; they produce an extension of the seeds. Most of them terminate once they have reached an extension that cover all the vertices of the graph. The separation thus produced is called a *graph cut*. Let $\mathcal{S} \subseteq \mathcal{E}$. We denote by $\bar{\mathcal{S}}$ *the complementary set of $\mathcal{S}$ in $\mathcal{E}$*, i.e., $\bar{\mathcal{S}} = \mathcal{E} \setminus \mathcal{S}$. Recall that the graph induced by $\mathcal{S}$, given by the dilation β_1, is the graph whose edge set is $\mathcal{S}$ and whose vertex set is made of all points that belong to an edge in $\mathcal{S}$, i.e., ($\{v \in \mathcal{V} \mid \exists e \in \mathcal{S}, v \in e\}, \mathcal{S}$). In the following, the graph induced by $\mathcal{S}$ is also denoted by $\mathcal{S}$.

Definition 14
Let $\mathcal{G}_1 \sqsubseteq \mathcal{G}$ and $\mathcal{S} \subseteq \mathcal{E}$. We say that $\mathcal{S}$ is a (graph) cut for $\mathcal{G}_1$ if $\bar{\mathcal{S}}$ is an extension of $\mathcal{G}_1$ and if $\mathcal{S}$ is minimal for this property, i.e., if $\mathcal{T} \subseteq \mathcal{S}$ and $\bar{\mathcal{T}}$ is an extension of $\mathcal{G}_1$, then we have $\mathcal{T} = \mathcal{S}$.

We introduce the watershed cuts of an edge-weighted graph. To this end, we formalize the drop of water principle. Intuitively, the catchment basins constitute an extension of the minima, and they are separated by "lines" from which a drop of water can flow down towards distinct minima.

Let $\pi = (v_0, \cdots, v_n)$ be a path. The path π is *descending* if, for any $i \in [1, l-1]$, $w(v_{i-1}, v_i) \geq w(v_i, v_{i+1})$.

Definition 15 (drop of water principle, [13])
Let $\mathcal{S} \subseteq \mathcal{E}$. We say that $\mathcal{S}$ is a watershed cut (or simply a watershed) if $\bar{\mathcal{S}}$ is an extension of $\mathcal{M}$ and if for any $e_{ij} = \{v_i, v_j\} \in \mathcal{S}$, there exist $\pi_1 = (v_0^1 = v_i, \cdots, v_n^1)$ and $\pi_2 = (v_0^2 = v_j, \cdots, v_m^2)$, which are two descending paths in $\bar{\mathcal{S}}$, such that v_n^1 and v_m^2 are vertices of two distinct minima of w; $w_{ij} \geq w(\{v_0^1, v_1^1\})$ (respectively, $w_{ij} \geq w(\{v_0^2, v_1^2\})$), whenever π_1 (respectively, π_2) is not trivial.

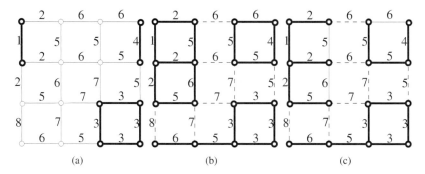

FIGURE 6.6
An edge-weighted graph $\mathcal{G}$ The edges and vertices drawn in bold are (a), the minima of w; (b), an extension of $\mathcal{M}$; (c), an MSF rooted in $\mathcal{M}$. In (b), the set of dashed edges is a watershed cut of w.

We illustrate the previous definition on the edge-weighted graph depicted in Figure 6.6. The weights w contain three minima (in bold in Figure6.6(a)). Let us denote by $\mathcal{S}$ the set of dashed edges depicted in Figure 6.6(b) and by $e = \{v_1, v_2\}$ the only edge whose altitude is 8. It may be seen that $\bar{\mathcal{S}}$ (in bold Figure 6.6(b)) is an extension of $\mathcal{M}$. We also remark that there exists π_1 (respectively π_2) a descending path in $\bar{\mathcal{S}}$ from v_1 (respectively v_2) to the minimum at altitude 1 (respectively, 3). The altitude of the first edge of π_1 (respectively, π_2) is lower than the altitude of e. It can be verified that the previous properties hold true for any edge in $\mathcal{S}$. Thus, $\mathcal{S}$ is a watershed of F.

6.6.2 Catchment Basins by a Steepest Descent Property

A popular alternative to Definition 15 consists of defining a watershed exclusively by its catchment basins and the paths of steepest descent of w [48, 44, 49, 45], and does not involve any property of the divide. The following theorem (17) establishes the consistency of watershed cuts in edge-weighted graphs: they can be equivalently defined by a steepest descent property on the catchment basins (regions) or by the drop of water principle on the cut (border) which separate them. As far as we know, there is no definition of watershed in vertex-weighted graphs that verifies a similar property. This theorem thus emphasizes that the framework of edge-weighted graphs is adapted for the study of discrete watersheds.

Let $\pi = (v_0, \cdots, v_l)$ be a path in $\mathcal{G}$. The path π is *a path of steepest descent (for w)* if, for any $i \in [1, l]$, $w(v_{i-1}, v_i) = (\varepsilon_\varepsilon\nu w)_{i-1} = \min\{w_{ij} | \forall j \text{ such that } e_{ij} \in \mathcal{E}\}$.

Definition 16 ([13])
Let $\mathcal{S} \subseteq \mathcal{E}$ be a cut for $\mathcal{M}$. We say that $\mathcal{S}$ is a basin cut (of w) if, from each point of V

to $\mathcal{M}$, there exists, in the graph induced by $\bar{\mathcal{S}}$, a path of steepest descent for w. If $\mathcal{S}$ is a basin cut of w, any component of $\bar{\mathcal{S}}$ is called a catchment basin (of w, for $\mathcal{S}$).

Theorem 17 (consistency, [13])
Let $\mathcal{S} \subseteq \mathcal{E}$ be a cut for $\mathcal{M}$. The set $\mathcal{S}$ is a watershed if and only if $\mathcal{S}$ is a basin cut of w.

6.6.3 Minimum Spanning Forests

We establish the optimality of watersheds. To this end, we introduce the notion of minimum spanning forests rooted in some subgraphs of $\mathcal{G}$. Each of these forests induces a unique graph cut. The main result of this study (Th. 19) states that a graph cut is induced by a minimum spanning rooted in the minima of a function if and only if it is a watershed of this function.

Definition 18 (rooted MSF, [14])
Let $\mathcal{G}_1$ and $\mathcal{G}_2$ be two nonempty subgraphs of $\mathcal{G}$. We say that $\mathcal{G}_2$ is rooted in $\mathcal{G}_1$ if $\mathcal{V}(\mathcal{G}_1) \subseteq \mathcal{V}(\mathcal{G}_2)$ and if the vertex set of any component of $\mathcal{G}_2$ contains the vertex set of exactly one component of $\mathcal{G}_1$. Recall that the weight of $\mathcal{G}_1$ is the sum of its weight, i.e., $\sum_{e_{ij} \in \mathcal{E}(G_1)} w_{ij}$

We say that $\mathcal{G}_2$ is a minimum spanning forest (MSF) rooted in $\mathcal{G}_1$ if (1) $\mathcal{G}_2$ is spanning for $\mathcal{V}$, (2) $\mathcal{G}_2$ is rooted in $\mathcal{G}_1$, and (3) the weight of $\mathcal{G}_2$ is less than or equal to the weight of any graph $\mathcal{G}_3$ satisfying (1) and (2) (i.e., $\mathcal{G}_3$ is both spanning for $\mathcal{V}$ and rooted in $\mathcal{G}_1$). The set of all minimum spanning forests rooted in $\mathcal{G}_1$ is denoted by $MSF(G_1)$.

The above definition of rooted MSFs allows the usual notions of graph theory based on trees and forests to be recovered. In particular, if v is a vertex of V, it can be seen that any element in $MSF(\{v\}, \emptyset)$ is a *minimum spanning tree of* (G, w), and that, conversely, any minimum spanning tree of (G, w) belongs to $MSF(\{v\}, \emptyset)$.

Let us consider the edge-weighted graph $\mathcal{G}$ and the subgraph $\mathcal{M}$ (in bold) of Figure 6.6(a). It can be verified that the graph $\mathcal{G}_1$ (in bold in Figure 6.6(c)) is an MSF rooted in $\mathcal{M}$.

We now have the mathematical tools to present the main result of this section (Th. 19), which establishes the optimality of watersheds.

Let $\mathcal{G}_1$ be a subgraph of $\mathcal{G}$ and let $\mathcal{G}_2$ be a spanning forest rooted in $\mathcal{G}_1$. There exists a unique cut for $\mathcal{G}_2$, and this cut is also a cut for $\mathcal{G}_1$. We say that this unique cut is the *cut induced by* $\mathcal{G}_2$.

Theorem 19 (optimality, [13])
Let $\mathcal{S} \subseteq \mathcal{E}$. The set $\mathcal{S}$ is a cut induced by a MSF rooted in $\mathcal{M}$ if and only if $\mathcal{S}$ is a watershed cut.

6.6.4   Illustrations to Segmentation

In this subsection, we illustrate the versatility of the proposed framework to perform segmentation on different kinds of geometric objects. Firstly, we show how to segment triangulated surfaces by watershed cuts, and, secondly, we apply the watershed cuts to the segmentation of diffusion tensors images, which are medical images associating a tensor to each voxel.

3D shape acquisition and digitizing have received more and more attention for a decade, leading to an increasing amount of 3D surface-models (or meshes) such as the one in Figure 6.7(d). In a recent work [50], a new search engine has been proposed for indexing and retrieving objects of interests in a database of meshes (EROS 3D) provided by the French Museum Center for Research. One key idea of this search engine is to use region descriptors rather than global shape descriptors. In order to produce such descriptors, it is then essential to obtain meaningful mesh segmentations.

Informally, a *mesh M* in the 3D Euclidean space is a set of *triangles*, *sides* of triangles, and *points* such that each side is included in exactly two triangles (see Figure 6.7(a)). In order to perform a watershed cut on such a mesh, we build a graph $\mathcal{G} = (\mathcal{V}, \mathcal{E})$ whose vertex set $\mathcal{V}$ is the set of all triangles in M and whose edge set $\mathcal{E}$ is composed by the pairs $e_{ij} = \{v_i, v_j\}$ such that v_i and v_j are two triangles of M that share a common side (see Figure 6.7(a)). The graph $\mathcal{G}$ is known under the term *2-dual of the surface mesh* [51].

To obtain a segmentation of the mesh M thanks to a watershed cut, we need to weight the edges of $\mathcal{G}$ (or equivalently the sides of M) by a map whose values are high around the boundaries of the regions that we want to separate. We have found that the interesting contours on the EROS 3D meshes are mostly located on concave zones. Therefore, we weight the edges of $\mathcal{G}$ by a weighting w, which behaves like the inverse of the mean curvature of the surface (see [50] for more details). Then, we can compute a watershed cut (in bold in Figure 6.7(b)) which leads to a natural and accurate mesh segmentation in the sense that the "borders" of the regions are made of sides of triangles (in bold in Figure 6.7c) of high curvature.

The direct application of this method on the mesh shown Figure 6.7(d) leads to a strong over-segmentation (Figure 6.7(e)) due to the huge number of local minima. By using the methodology introduced in Section 6.5 to remove from w all the minima which have a depth lower than a predefined threshold (here, 50) A watershed cut of the filtered w is depicted in Figure 6.7(f).

In the medical context, *Diffusion Tensor Images (DTIs)* [52] provide a unique insight into oriented structures within tissues. A *DTI T* maps the set of voxels $\mathcal{V} \subseteq \mathbb{Z}^3$ (i.e., $\mathcal{V}$ is a cuboid of $\mathbb{Z}^3$) into the set of 3×3 tensors (i.e., 3×3 symmetric positive definite matrices). The value $T(v)$ of a DTI T at a voxel $v \in \mathcal{V}$ describes the diffusion of water molecules at v. For instance, the first eigenvector of $T(v)$ (i.e., the one whose associated eigenvalue is maximal) provides the principal direction of water molecules diffusion at point x and its associated eigenvalue gives the magnitude of

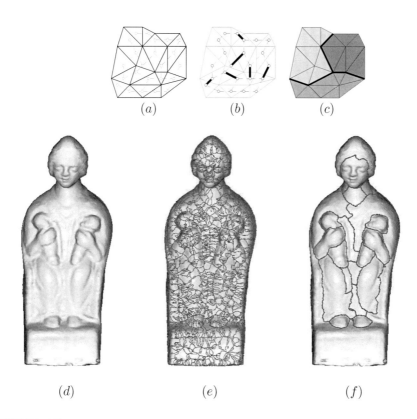

(a)   (b)   (c)

(d)   (e)   (f)

FIGURE 6.7
Surface segmentation by watershed cut. (a) A mesh in black and its associated graph in gray. (b) A cut on this graph (in bold); and (c), the corresponding segmentation of the mesh. (d) Rendering of the mesh of a sculpture. (e) A watershed (in red) of a map F which behaves like the inverse of the mean curvature and, in (f), a watershed of a filtered version of F. The mesh shown in (d) is provided by the French Museum Center for Research.

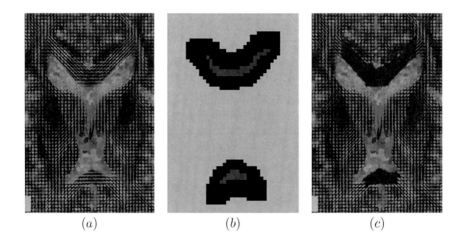

(a)   (b)   (c)

FIGURE 6.8

Diffusion tensor images segmentation. (a) A close-up on a cross-section of a 3D brain DTI. (b) Image representation of the markers (same cross-section as (a)), obtained from a statistical atlas, for the corpus callosum (dark gray) and for its background (light gray) (c) Segmentation of the corpus callosum by an MSF-cut for the markers. The tensors belonging to the component of the MSF that extends the marker labelled "corpus callosum" are removed from the initial DTI thus, the corresponding voxels appear black.

the diffusion along this direction. Since water molecules highly diffuse along fiber tracts and since the white matter of the brain is mainly composed of fiber tracts, DTIs are particularly adapted to the study of brain architecture. Figure 6.8a shows a representation of a cross-section of a brain DTI where the tensors are represented by ellipsoids. Indeed, the data of a tensor is equivalent to the one of an ellipsoid. In the brain, the corpus callosum is an important structure made of fiber tracts connecting homologous areas of each hemisphere. In order to track the fibers that pass through the corpus callosum, it is necessary to segment it first. The next paragraph briefly reviews how to reach this goal, thanks to watershed cuts [47].

We consider the graph $\mathcal{G} = (\mathcal{V}, \mathcal{E})$ induced by the 6-adjacency and defined by $e_{ij} \in \mathcal{E}$ iff $v_i \in \mathcal{V}, v_j \in \mathcal{V}$ and $\Sigma_{k \in \{1,2,3\}} |v_i^k - v_j^k| = 1$, where $v_i = (v_i^1, v_i^2, v_i^3)$ and $v_j = (v_j^1, v_j^2, v_j^3)$. In order to weight any edge e_{ij} of $\mathcal{G}$ by a dissimilarity measure between the tensors $T(v_i)$ and $T(v_j)$, we choose the Log-Euclidean distance, which is known to satisfy interesting properties [53]. Then, we associate to each edge $e_{ij} \in E$ the weight $w_{ij} = \| \log(T(v_i)) - \log(T(v_j)) \|$, where $\log$ denotes the matrix logarithm and $\|.\|$ the Euclidean (sometimes also called Frobenius) norm on matrices. To segment the corpus callosum in this graph, we extract (thanks to a statistical atlas), markers for both the corpus callosum and its background, and we compute an MSF-cut for these markers. An illustration of this procedure is shown in Figure 6.8.

To conclude this section, let us mention that spatio-temporal graphs are also feasible. For example, in [54], a 3D+t segmentation of the left-ventricle on cine-MR images was computed, showing an improvemnt with respect to the segmentation of each 3D volume separately.

6.7 MSF Cut Hierarchy and Saliency Maps

We now study some optimality properties of hierarchical segmentations (see [55, 56, 57, 58, 59, 60] for examples of hierarchical segmentations) in the framework of edge-weighted graphs, where the cost of an edge is given by a dissimilarity between two points of an image. Since the pioneering work of [39, 40] stating an equivalence between hierarchies and minimum spanning trees (MST), a large number of hierarchical schemes rely on the construction of such a tree (see [21] for one of the oldest). We formalize a fundamental operation called *uprooting* that allows us to merge a marked region with one of its neighbors with the cheapest cost. When applied sequentially on the weighted graph of neighboring regions, the uprooting builds a MST of this neighboring graph. Intuitively, one can see that, if one starts from a minimum spanning forest (MSF) rooted in the minima of the image (or, equivalently, from a watershed cut), then one builds a hierarchy of MSFs of the original image itself, the last uprooting step giving an MST of this original image. More surprisingly, Th. 23

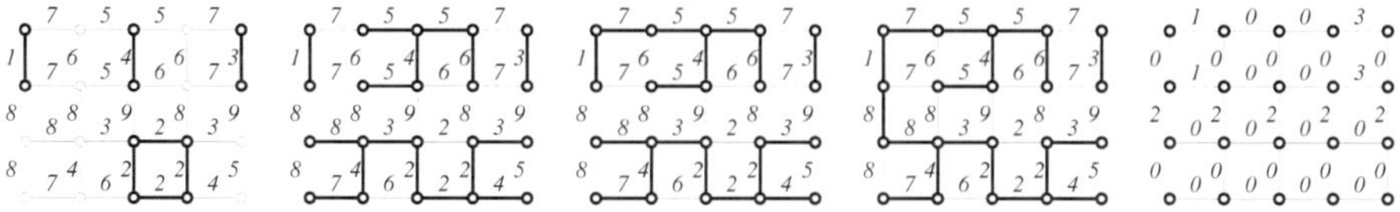

FIGURE 6.9
(a) An edge-weighted graph $\mathcal{G}$, the minima of w are depicted in bold. (b) A watershed cut represented by dashed edges; the watershed cut is equivalent to the graph $\mathcal{G}_0 \in MSF(\mathcal{M})$ represented in bold. (c,d) two bold graphs called, respectively, $\mathcal{G}_1$ and $\mathcal{G}_2$ such that $\mathcal{T} = \langle \mathcal{G}_0, \mathcal{G}_1, \mathcal{G}_2 \rangle$ is both an MSF hierarchy for and an uprooting by $\langle M_1, M_2 \rangle$ (where M_i is the minimum of w whose altitude is i); their associated cuts are represented by dashed edges. (e) The saliency map of the MSF hierarchy $\langle \mathcal{G}_0, \mathcal{G}_1, \mathcal{G}_2 \rangle$.

states that the two processes are equivalent: any MST of the original image can be built from an uprooting sequence on a watershed cut. Thus, watershed cuts are the only watershed that allows the building of hierarchical segmentations that are optimal with respect to the original image, in the sense that they "preserve" the MST of the original image.

Definition 20 (MSF hierarchy, [14])
Let $\mathcal{M} = \langle M_1, \ldots, M_\ell \rangle$ be a sequence of pairwise distinct minima of w and let $\mathcal{T} = \langle \mathcal{G}_0, \ldots \mathcal{G}_\ell \rangle$ be a sequence of subgraphs of $\mathcal{G}$. We say that $\mathcal{T}$ is an MSF hierarchy for S if for any $i \in [0, \ell]$, the graph $\mathcal{G}_i$ is an MSF rooted in $\sqcup[\mathcal{M} \setminus \{M_j \mid j \in [1, i]\}]$, and for any $i \in [1, \ell]$, we have $\mathcal{G}_{i-1} \sqsubseteq \mathcal{G}_i$.

Theorem 21 ([61, 14])
Any $\mathcal{G}_i$ of an MSF hierarchy is a watershed cut of the geodesic reconstruction w_R where the markers are the minima rooting the MSF $\mathcal{G}_i$.

6.7.1   Uprootings and MSF Hierarchies

In this section, we introduce a simple transformation, called *uprooting*, that allows a forest $\mathcal{G}_1$ rooted in a graph $\mathcal{G}_2$ to be incrementally transformed into a forest $\mathcal{G}_3$ rooted in a graph $\mathcal{G}_4$ obtained by removing some components of $\mathcal{G}_2$. Through an equivalence theorem, we establish an important link between the uprooting transform and the MSF hierarchies. This result opens the way toward efficient algorithms for computing MSF hierarchies.

Let $\mathcal{G}_1$ be a subgraph of $\mathcal{G}$ that is spanning for $\mathcal{V}$, and let $v \in \mathcal{V}$. We denote by $CC_v(\mathcal{G}_1)$ the component of $\mathcal{G}_1$ whose vertex set contains v. Let $\mathcal{V}' \subseteq \mathcal{V}$, we set $CC_{\mathcal{V}'}(\mathcal{G}_1) = \sqcup\{CC_v(\mathcal{G}_1) \mid v \in \mathcal{V}'\}$.

Let $\mathcal{G}_1 \sqsubseteq \mathcal{G}$, and let $e_{ij} = \{v_i, v_j\} \in \mathcal{E}$. The edge e_{ij} is *outgoing from* $\mathcal{G}_1$ if e_{ij} is made of a vertex in $\mathcal{V}(\mathcal{G}_1)$ and of a vertex in $\overline{\mathcal{V}(\mathcal{G}_1)}$. In the following, by abuse of notation, we write $\mathcal{G}_1 \sqcup \{e_{ij}\}$ for the supremum of $\mathcal{G}_1$, and the graph induced by $\{e_{ij}\}$: $\mathcal{G}_1 \sqcup \{e_{ij}\} = (\mathcal{V}(\mathcal{G}_1) \cup e_{ij}, \mathcal{E}(\mathcal{G}_1) \cup \{e_{ij}\})$.

Let $\mathcal{G}_1, \mathcal{G}_2$, and $\mathcal{G}_3$ be three subgraphs of $\mathcal{G}$ such that $\mathcal{G}_1$ is spanning for $\mathcal{V}$ and such that $\mathcal{G}_1 \neq \mathcal{G}_2$. We say that $\mathcal{G}_2$ is an *elementary uprooting of* $\mathcal{G}_1$ by $\mathcal{G}_3$ if there exists an edge e of minimum weight among the edges outgoing from $CC_{\mathcal{V}(\mathcal{G}_3)}(\mathcal{G}_1)$ such that $\mathcal{G}_2 = \mathcal{G}_1 \sqcup \{u\}$. We also say that $\mathcal{G}_2$ is an *elementary uprooting of* $\mathcal{G}_1$ by $\mathcal{G}_3$ if $\mathcal{G}_2 = \mathcal{G}_1$ and if there is no edge outgoing from $CC_{\mathcal{V}(\mathcal{G}_3)}(\mathcal{G}_1)$.

Definition 22 ([14])
Let $\mathcal{S} = \langle M_1, \ldots, M_\ell \rangle$ *be a sequence of pairwise distinct minima of* w. *An* uprooting *by* $\mathcal{S}$ *is a sequence* $\langle \mathcal{G}_0, \ldots, \mathcal{G}_\ell \rangle$ *of graphs such that* $\mathcal{G}_0 \in MSF(\mathcal{M})$ *and, for any* $i \in [1, \ell]$, $\mathcal{G}_i$ *is an elementary uprooting of* $\mathcal{G}_{i-1}$ *by* M_i.

Theorem 23 ([14])
Let $\mathcal{S} = \langle M_1, \ldots, M_\ell \rangle$ *be a sequence of pairwise distinct minima of* w. *Let* $\mathcal{T} = \langle \mathcal{G}_0, \ldots \mathcal{G}_\ell \rangle$ *be a sequence of subgraphs of* $\mathcal{G}$. *The sequence* $\mathcal{T}$ *is an MSF hierarchy for* $\mathcal{S}$ *if and only if the sequence* $\mathcal{T}$ *is an uprooting by* $\mathcal{S}$.

6.7.2 Saliency Maps

The cuts of the sequence of any MSF hierarchy can be stacked to form a weighting called the saliency map. The saliency map allows us to easily assess the quality and the robustness of a hierarchical segmentation. Furthermore, it is a weighted graph, and can be further processed if needed, for example, to remove small regions unwanted in the hierarchy.

Let us give a precise defition of the saliency map [56, 14]. We first need to define, for a graph $\mathcal{G}$, the map ϕ as $\phi(\mathcal{G}) = \sqcup\{\alpha_2(\mathcal{G}_i) \mid \mathcal{G}_i$ is a component of $\mathcal{G}\}$. In other words, $\phi(\mathcal{G})$ is the union of all the graphs induced by the connected components of $\mathcal{G}$. It is easy to see that ϕ is an opening.

Let $\mathcal{M} = \langle M_1, \ldots, M_\ell \rangle$ be a sequence of pairwise distinct minima of w and Let $\mathcal{T} = \langle \mathcal{G}_0, \ldots \mathcal{G}_\ell \rangle$ be a MSF hierarchy for $\mathcal{M}$. The *saliency map* s for $\mathcal{T}$ is the map such that for any $e_{ij} \in \mathcal{E}$, either $s_{ij} = k$ if there exists k such that k is the lowest number satisfying $e_{ij} \in \phi(\mathcal{G}_k)$ or $s_{ij} = l + 1$ if it does not exists any $\mathcal{G}_k$ such that $e_{ij} \in \phi(\mathcal{G}_k)$. Observe, in particular, that if $e_{ij} \in \mathcal{E}(\phi(\mathcal{G}_0))$, $s_{ij} = 0$, and that $s[\lambda] = \phi(\mathcal{G}_\lambda)$ for $\lambda \in \{0, \cdots, l\}$.

Under the term of *ultrametric watershed*, one can give a computable definition to saliency maps, allowing to show the equivalenve between the set of hierarchical segmentations and the set of saliency maps [60].

6.7.3 Application to Image Segmentation

To use the framework in practice, we have to design an order on the minima of w. Let μ be an attribute on $\mathcal{C}_{\mathcal{E}}^{\leq}$ (i.e., a function from $\mathcal{C}_{\mathcal{E}}^{\leq}$ to $\mathbb{R}$ that is increasing on $\mathcal{C}_{\mathcal{E}}^{\leq}$) such as the area, the volume, or the height. We compute an *extinction measure* [62, 63] μ_e to each minima of w by first definining a strict total order relation $\prec$ on the set of minima (based for example on the altitude of each minimum) such that $M_0 \prec M_1 \prec \cdots \prec M_l$; then we set $\mu_e(M_0) = \infty$ and

$$
\mu_e(M_i) = \min\{\mu([\lambda, C]) \mid [\lambda, C] \in \mathcal{C}_{\mathcal{E}}^{\leq}, \atop \text{there exists } M_j \text{ a minimum of } w, \ M_j \prec M_i\} \tag{6.5}
$$

The map μ_e will define the order $\prec_e$ of the sequence for the hierachy of MSF: $M_i \prec_e M_j$ whenever $\mu_e(M_i) < \mu_e(M_j)$.

Other choices for the ordering are possible: for example, waterfall [64], that consists in computing a sequence of watersheds of watersheds, each step leading to a drastic reduction in the number of basins. Another interesting ordering is done by optimization [57]). Let us consider a two-term-based energy function of the form $\lambda C + D$, where D is a goodness-of-fit term and C is a regularization term. Finding an optimum of this function is NP-hard in the general case. On the other hand, on a hierarchy (hence, on a watershed cut), when the goodness-of-fit term decreases with the fineness of the partitions, and, inversely, that the regularization term increases with this fineness, we can show that finding an optimum can be done in linear time, by dynamic programming. Such an optimization is an efficient way to control the flooding, which is stopped when the optimum is reached. Varying the λ parameter allows us to obtain a complete hierarchy of segmentations.

Figure 6.10(b), (c), and (d) illustrates some saliency maps on the image of Figure 6.10a. The underlying graph is the one induced by the 4-adjacency relation whose edges are weighted by a simple color gradient (maximum, over the RGB channels, of the absolute differences of pixel values). The minima are ordered, thanks to extinction values [62] related to depth, volume, and color consistency.

6.8 Optimization and the Power Watershed

In this section, we review the Power Watershed framework [15] (see also Leo Grady's chapter, this book). In the previous sections dealing with watershed cuts, the weights encode a dissimilarity such as a gradient. Classicaly, in the context of segmentation and clustering applications based on optimization, the weights encode affinity such that vertices connected by an edge with high weight are considered to be strongly connected and edges with a low weight represent nearly disconnected vertices. One

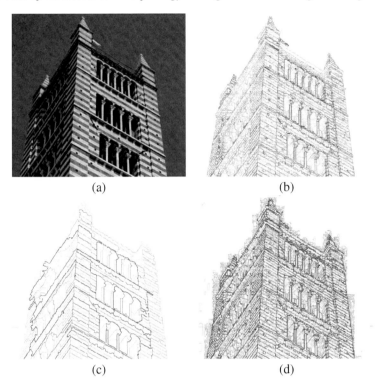

(a)　　　　　　　　　　(b)

(c)　　　　　　　　　　(d)

FIGURE 6.10
Illustration of saliencies of watershed cuts (original picture (a) from *koakoo*: http://blog.photos-libres.fr/).

common choice for generating weights from image intensities is to set a Gaussian weighting such as

$$w_{ij} = \exp(-\beta(\nabla I)^2), \qquad (6.6)$$

where ∇I is the normalized gradient of the image I. The gradient for a grey level image is $I_i - I_j$. As the weights are inverted, the maxima are considered instead of minima, and a thalweg is computed instead of watershed. A thalweg is the deepest continuous line along a valley. In the rest of the paper, we continue to use by convention the term "watershed" instead of "thalweg".

Given foreground F and background B seeds, and p, q two real positive values, the energy presented for binary segmentation in [15] is given by

$$x^{\star pq} = \arg\min_x \sum_{e_{ij} \in E} w_{ij}^p |x_i - x_j|^q \text{ s.t. } x(F) = 1, \ x(B) = 0. \qquad (6.7)$$

In this energy, w_{ij} can be interpreted as a weight enforcing a regularization of the

q \ p	0	finite	∞
1	Collapse to seeds	Max-Flow (Graph cuts)	Power watershed $q = 1$
2	ℓ_2 norm Voronoi	Random walker	Power watershed $q = 2$
∞	ℓ_1 norm Voronoi	ℓ_1 norm Voronoi	Shortest Path Forest

TABLE 6.1

The generalized scheme of [15] for image segmentation includes several popular segmentation algorithms as special cases of the parameters p and q. The power watershed may be optimized efficiently with a maximum spanning forest calculation.

contours, such that any (usually unwanted) high-frequency content is penalized in x. The energy defined in (6.7) essentially forces x to remain smooth within the object, while allowing it to vary quickly close to point clusters near the boundary of the object. The data constraints enforce fidelity of x to a specified configuration, taking the values zero and one as the reconstructed object indicator. Observe that the values of x may not necessarily be binary when the value of q is strictly greater than one.

The different values of p and q lead to different algorithms for optimizing the energy. Those algorithms form the underpinning for many of the advanced image segmentation methods in the literature. Table 6.1 gives a reference for the different algorithms generated by various values of p and q. The limit cases are the limit of the minimizers of Equation (6.7), e.g., the case $p \to \infty$ reads

$$x^{\star q} = \lim_{p \to \infty} x^{\star pq}. \tag{6.8}$$

Let us highlight the main choices for those parameters.

- When the power on the weight, p, is finite, and the exponent $q = 1$, we recover [61] the Max-Flow (Graph Cuts) energy which can be optimized by a max flow algorithm.

- When $q = 2$, we obtain a combinatorial Dirichlet problem also known as the random walker problem [65].

- When q and p converge toward infinity with the same speed, then [66] a solution to (6.8) can be computed by the shortest path (geodesics) algorithm [6, 67].

- As described in [15], by raising the power p toward infinity, and varying the power q, we obtain a family of segmentation models, which we refer to as **power watershed**, and that we detailed below.

A primary advantage of power watershed with varying q is that the main computational burden of these algorithms depends on an MSF computation, which is extremely efficient [68]. For example, interpreted from the standpoint of the Gaussian weighting function in (6.6), it is clear that we may associate $\beta = p$ to understand that the watershed equivalence comes from operating the weighting function in a particular parameter range. In the case $q = 1$, an important insight from this connection is that *above some value of β we can replace the expensive max-flow computation with an efficient maximum spanning forest computation.*

Let us review some important theoretical results. First, we highlight that the cut obtained when minimizing the energy (6.8) is a watershed cut [13], and thus a maximum spanning forests [13] (MSF). Let $x^\star$ be a solution to (6.8) and s be the segmentation result defined as $s_i = 1$ if $x_i^\star \geq \frac{1}{2}$, 0 if $x_i^\star < \frac{1}{2}$. The set of edges e_{ij} such that $s_i \neq s_j$ is a q-cut.

Theorem 24 ([15])
For $q \geq 1$, any q-cut is a watershed cut of the (geodesically) reconstructed weights.

Furthermore, if every connected component of $\mathbb{M}$ contains at least a vertex of $B \cup F$, and $q \geq 1$, then any q-cut when $p \to \infty$ is an MSF cut for w.

Algorithm 1: Power watersheds algorithm $p \to \infty, q > 1$

Data: A weighted graph $G(V, E)$ comprising known labels $x(B), x(F)$.
Result: A **potential function** $\bar{x}$ (solution of (6.8) thanks to Th. 25.)
while *any node has an unknown label* **do**
    Find a maximal subgraph $S \in G$ composed of edges of maximal weight;
    if *S contains any nodes with known x* **then**
        find x_S minimizing (6.7) for q on the subset S;
        consider all x_S values produced by this operation as known;
    else
        merge all of the nodes in S into a single node, such that when the value of x for this merged node becomes known, all merged nodes are assigned the same value of x and considered known;

An algorithm is presented in Algorithm 1 to optimize the (unique) watershed that optimizes the energy for $q > 1$ and $p \to \infty$. The case of parameter $p \to \infty$ is particularly interesting.

1. The power watershed algorithm in the case $q > 1$ has a well-defined behavior in the absence or lack of weight information (presence of plateaus); indeed, as the energy is strictly convex, the solution to (6.8) is unique.

2. The worst-case complexity of the power watershed algorithm in the case $p \to \infty$ is given by the cost of optimizing (6.7) for the given q. In the best-case scenario (all weights have unique values), the power watershed algorithm has the same asymptotic complexity as the algorithm used for a MSF computation (quasi-linear). In practical applications where the plateaus have a size less than some

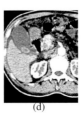
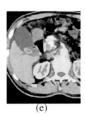

(a)   (b)   (c)   (d)   (e)

FIGURE 6.11
Two applications of power watershed. (a-c) Shape fitting using power watershed [69]. (a) reconstructed bunny. (b) A close-up on one of the ears. (c) A close-up on the set of original scanned noisy points measurements. (d-e) Filtering of a liver image by anisotropic diffusion driven by power watershed [70]. In (e) noise and small vessels are both removed from (d), leading to a result which may be used as a first step before segmentation.

fixed value K, then the complexity of the power watershed algorithm matches the quasi-linear complexity of the standard watershed-cut algorithm.

The main properties of Algorithm 1 (with $q \geq 1$) are summarized in the following theorem. This theorem states that the energy found by the algorithm is the correct one, *i.e.* $x^{*q} = \bar{x}$, and that furthermore, the computed cut is a MSF cut.

Theorem 25 ([15])
If $q > 1$, the potential x^{*} obtained by minimizing the energy of (6.7) subject to the boundary constraints converges toward the potential $\bar{x}$ obtained by Algorithm 1 as $p \to \infty$.

Furthermore, for any $q \geq 1$, the cut C defined by the segmentation s computed by Algorithm 1 is an MSF cut for w.

6.8.1 Applications

Minimizing exactly the energy $E_{1,1}$ is possible by using the graph cuts (i.e., $q = 1$) algorithm in the case of two labels, but is NP-hard if constraints impose more than two different labels. However, the other algorithms presented in the framework (and, in particular, the watershed cuts and the power watershed) can efficiently perform seeded segmentation with as many labels as desired [15]. As several examples of segmentation have already been shown in this chapter, we rather highlight two other applications that take advantage the unique optimization characteristics of the power watershed.

The first example is *shape fitting*. Surface reconstruction from a set of noisy point measurements has been a well studied problem over the last decades. Recently, variational [71, 72] and discrete optimization [73] approaches have been applied to solve it, demonstrating good robustness to outliers thanks to a global energy minimization scheme. In [69], Couprie *et al.* use the power watershed framework to derive a specific watershed algorithm for surface reconstruction. The proposed algorithm is fast, robust to marker placement, and produces smooth surfaces. Figure 6.11(a–c) shows a surface reconstructed from noisy scanned dot sets using the power watershed algorithm.

The second example deals with filtering. In [70], Couprie et al. reformulate the problem of anisotropic diffusion as an L_0 optimization problem, and show that power watersheds are able to optimize this energy quickly and effectively. This study paves the way for using the power watershed as a useful general-purpose minimizer in many different computer vision contexts. An example of such an L_0 optimization is presented in Figures 6.11(d–e).

6.9 Conclusion

We have presented in this paper some applications of mathematical morphology to weighted graphs. The translation of the abstract framework of lattices to graphs allows us to obtain morphological operators thinner (in the sense that they process smaller details) than the usual ones commonly defined only on the vertices. We have shown how to filter an image using connected filters based on a tree representation of the image. We have exhibited the links between watersheds and minimum spanning trees, allowing us to make the equivalence between hierarchical segmentation based on watershed cuts and minimum spanning trees. We conclude the chapter by exploring links between the morphological and the optimization approaches.

For practical applications, we want to stress the importance of having a framework with an open-source generic implementation of existing algorithms, not limited to the pixel framework, but also able to deal transparently with edges, or, more generally, with graphs and complexes [74].

Bibliography

[1] G. Matheron, *Random sets and integral geometry*. New York: John Wiley & Sons, 1975.

[2] J. Serra, *Image analysis and mathematical morphology*. London, U.K.: Academic Press, 1982.

[3] J. Serra, Ed., *Image analysis and mathematical morphology. Volume 2: Theoretical advances*. London, U.K.: Academic Press, 1988.

[4] L. Vincent, "Graphs and mathematical morphology," *Signal Processing*, vol. 16, pp. 365–388, 1989.

[5] H. Heijmans and L. Vincent, "Graph morphology in image analysis," in *Mathematical morphology in image processing*, E. Dougherty, Ed.   New York: Marcel Dekker, 1993, vol. 34, ch. 6, pp. 171–203.

[6] A. X. Falcão, R. A. Lotufo, and G. Araujo, "The image foresting transformation," *IEEE PAMI*, vol. 26, no. 1, pp. 19–29, 2004.

[7] F. Meyer and R. Lerallut, "Morphological operators for flooding, leveling and filtering images using graphs," in *Graph-based Representations in Pattern Recognition (GbRPR'07)*, vol. LNCS 4538, 2007, pp. 158–167.

[8] F. Meyer and J. Angulo, "Micro-viscous morphological operators," in *Mathematical Morphology and its Application to Signal and Image Processing (ISMM 2007)*, 2007, pp. 165–176.

[9] V.-T. Ta, A. Elmoataz, and O. Lézoray, "Partial difference equations over graphs: Morphological processing of arbitrary discrete data," in *ECCV 2008*, ser. Lecture Notes in Computer Science, D. Forsyth, P. Torr, and A. Zisserman, Eds.   Springer Berlin / Heidelberg, 2008, vol. 5304, pp. 668–680.

[10] J. Cousty, L. Najman, and J. Serra, "Some morphological operators in graph spaces," in *International Symposium on Mathematical Morphology 2009*, ser. LNCS, vol. 5720.   Springer Verlag, Aug. 2009, pp. 149–160.

[11] I. Bloch and A. Bretto, "Mathematical morphology on hypergraphs: Preliminary definitions and results," in *Discrete Geometry for Computer Imagery*, ser. Lecture Notes in Computer Science, I. Debled-Rennesson, E. Domenjoud, B. Kerautret, and P. Even, Eds.   Springer Berlin / Heidelberg, 2011, vol. 6607, pp. 429–440.

[12] F. Dias, J. Cousty, and L. Najman, "Some morphological operators on simplicial complexes spaces," in *DGCI 2011*, ser. LNCS, no. 6607.   Springer, 2011, pp. 441–452.

[13] J. Cousty, G. Bertrand, L. Najman, and M. Couprie, "Watershed Cuts: Minimum Spanning Forests and the Drop of Water Principle," *IEEE PAMI*, vol. 31, no. 8, pp. 1362–1374, Aug. 2009.

[14] J. Cousty and L. Najman, "Incremental algorithm for hierarchical minimum spanning forests and saliency of watershed cuts," in *ISMM 2011*, ser. LNCS, no. 6671, 2011, pp. 272–283.

[15] C. Couprie, L. Grady, L. Najman, and H. Talbot, "Power Watersheds: A Unifying Graph Based Optimization Framework," *IEEE Transactions on Pattern Analysis and Machine Intelligence*, vol. 33, no. 7, pp. 1384–1399, July 2011, http://doi.ieeecomputersociety.org/10.1109/TPAMI.2010.200. [Online]. Available: http://powerwatershed.sourceforge.net/

[16] P. Soille, *Morphological Image Analysis*.   Springer-Verlag, 1999.

[17] L. Najman and H. Talbot, Eds., *Mathematical morphology: from theory to applications*.   ISTE-Wiley, June 2010.

[18] Nesetril, Milkova, and Nesetrilova, "Otakar Boruvka on Minimum Spanning Tree problem: Translation of both the 1926 papers, comments, history," *DMATH: Discrete Mathematics*, vol. 233, 2001.

[19] G. Choquet, "étude de certains réseaux de routes," *Comptes-rendus de l'Acad. des Sciences*, pp. 310–313, 1938.

[20] C. Zahn, "Graph-theoretical methods for detecting and desciding Gestalt clusters," *IEEE Transactions on Computers*, vol. C-20, no. 1, pp. 99–112, 1971.

[21] O. J. Morris, M. d. J. Lee, and A. G. Constantinides, "Graph theory for image analysis: an approach based on the shortest spanning tree," *IEE Proc. on Communications, Radar and Signal*, vol. 133, no. 2, pp. 146–152, 1986.

[22] F. Meyer, "Minimum spanning forests for morphological segmentation," in *Procs. of the second international conference on Mathematical Morphology and its Applications to Image Processing*, September 1994, pp. 77–84.

[23] P. Felzenszwalb and D. Huttenlocher, "Efficient graph-based image segmentation," *International Journal of Computer Vision*, vol. 59, pp. 167–181, 2004.

[24] L. Vincent and P. Soille, "Watersheds in digital spaces: An efficient algorithm based on immersion simulations," *IEEE PAMI*, vol. 13, no. 6, pp. 583–598, June 1991.

[25] G. Birkhoff, *Lattice Theory*, ser. American Mathematical Society Colloquium Publications. American Mathematical Society, 1995, vol. 25.

[26] C. Ronse and J. Serra, "Algebric foundations of morphology," in *Mathematical Morphology*, L. Najman and H. Talbot, Eds. ISTE-Wiley, 2010, pp. 35–79.

[27] P. Salembier and J. Serra, "Flat zones filtering, connected operators, and filters by reconstruction," *IEEE TIP*, vol. 4, no. 8, pp. 1153–1160, Aug 1995.

[28] C. Ronse, "Set-theoretical algebraic approaches to connectivity in continuous or digital spaces," *JMIV*, vol. 8, no. 1, pp. 41–58, 1998.

[29] U. Braga-Neto and J. Goutsias, "Connectivity on complete lattices: new results," *Comput. Vis. Image Underst.*, vol. 85, no. 1, pp. 22–53, 2002.

[30] G. K. Ouzounis and M. H. Wilkinson, "Mask-based second-generation connectivity and attribute filters," *IEEE PAMI*, vol. 29, no. 6, pp. 990–1004, 2007.

[31] R. K. Gabriel and R. R. Sokal, "A new statistical approach to geographic variation analysis," *Systematic Zoology*, vol. 18, no. 3, pp. 259–278, Sep. 1969.

[32] P. Salembier, A. Oliveras, and L. Garrido, "Anti-extensive connected operators for image and sequence processing," *IEEE TIP*, vol. 7, no. 4, pp. 555–570, April 1998.

[33] L. Najman and M. Couprie, "Building the component tree in quasi-linear time," *IEEE TIP*, vol. 15, no. 11, pp. 3531–3539, 2006.

[34] P. Salembier and L. Garrido, "Binary partition tree as an efficient representation for image processing, segmentation and information retrieval," *IEEE Transactions on Image Processing*, vol. 9, no. 4, pp. 561–576, Apr. 2000.

[35] V. Caselles and P. Monasse, *Geometric Description of Images as Topographic Maps*, ser. Lecture Notes in Computer Science. Springer, 2010, vol. 1984.

[36] P. Salembier, "Connected operators based on tree pruning strategies," in *Mathematical morphology:from theory to applications*, L. Najman and H. Talbot, Eds. ISTE-Wiley, 2010, pp. 179–198.

[37] J. Benzécri, *L'Analyse des données: la Taxinomie*. Dunod, 1973, vol. 1.

[38] B. Leclerc, "Description combinatoire des ultramétriques," *Mathématique et sciences humaines*, vol. 73, pp. 5–37, 1981.

[39] N. Jardine and R. Sibson, *Mathematical taxonomy*. Wiley, 1971.

[40] J. Gower and G. Ross, "Minimum spanning tree and single linkage cluster analysis," *Appl. Stats.*, vol. 18, pp. 54–64, 1969.

[41] J. B. Kruskal, "On the shortest spanning subtree of a graph and the traveling salesman problem," *Proceedings of the American Mathematical Society*, vol. 7, pp. 48–50, February 1956.

[42] F. Meyer and S. Beucher, "Morphological segmentation," *JVCIR*, vol. 1, no. 1, pp. 21–46, Sept. 1990.

[43] S. Beucher and F. Meyer, "The morphological approach to segmentation: the watershed transformation," in *Mathematical morphology in image processing*, ser. Optical Engineering, E. Dougherty, Ed. New York: Marcel Dekker, 1993, vol. 34, ch. 12, pp. 433–481.

[44] L. Najman and M. Schmitt, "Watershed of a continuous function," *Signal Processing*, vol. 38, no. 1, pp. 99–112, 1994.

[45] J. B. T. M. Roerdink and A. Meijster, "The watershed transform: Definitions, algorithms and parallelization strategies," *Fundamenta Informaticae*, vol. 41, no. 1-2, pp. 187–228, 2001.

[46] G. Bertrand, "On topological watersheds," *JMIV*, vol. 22, no. 2-3, pp. 217–230, May 2005.

[47] J. Cousty, G. Bertrand, L. Najman, and M. Couprie, "Watershed cuts: thinnings, shortest-path forests and topological watersheds," *IEEE PAMI*, vol. 32, no. 5, pp. 925–939, 2010.

[48] F. Meyer, "Topographic distance and watershed lines," *Signal Processing*, vol. 38, no. 1, pp. 113–125, July 1994.

[49] P. Soille and C. Gratin, "An efficient algorithm for drainage networks extraction on DEMs," *Journal of Visual Communication and Image Representation*, vol. 5, no. 2, pp. 181–189, June 1994.

[50] S. Philipp-Foliguet, M. Jordan, L. Najman, and J. Cousty, "Artwork 3D Model Database Indexing and Classification," *Pattern Recogn.*, vol. 44, no. 3, pp. 588–597, Mar. 2011.

[51] L. J. Grady and J. R. Polimeni, *Discrete Calculus: Applied Analysis on Graphs for Computational Science*, 1st ed. Springer, Aug. 2010.

[52] P. J. Basser, J. Mattiello, and D. LeBihan, "MR diffusion tensor spectroscopy and imaging." *Biophys. J.*, vol. 66, no. 1, pp. 259–267, 1994.

[53] V. Arsigny, P. Fillard, X. Pennec, and N. Ayache, "Log-Euclidean metrics for fast and simple calculus on diffusion tensors," *Magnetic Resonance in Medicine*, vol. 56, no. 2, pp. 411–421, August 2006.

[54] J. Cousty, L. Najman, M. Couprie, S. Clément-Guinaudeau, T. Goissen, and J. Garot, "Segmentation of 4D cardiac MRI: automated method based on spatio-temporal watershed cuts," *IVC*, vol. 28, no. 8, pp. 1229–1243, Aug. 2010.

[55] F. Meyer, "The dynamics of minima and contours," in *Mathematical Morphology and its Applications to Image and Signal Processing*, P. Maragos, R. Schafer, and M. Butt, Eds. Boston: Kluwer, 1996, pp. 329–336.

[56] L. Najman and M. Schmitt, "Geodesic saliency of watershed contours and hierarchical segmentation," *IEEE PAMI*, vol. 18, no. 12, pp. 1163–1173, December 1996.

[57] L. Guigues, J. P. Cocquerez, and H. L. Men, "Scale-sets image analysis," *IJCV*, vol. 68, no. 3, pp. 289–317, 2006.

[58] P. A. Arbeláez and L. D. Cohen, "A metric approach to vector-valued image segmentation," *IJCV*, vol. 69, no. 1, pp. 119–126, 2006.

[59] F. Meyer and L. Najman, "Segmentation, minimum spanning tree and hierarchies," in *Mathematical Morphology: from theory to application*, L. Najman and H. Talbot, Eds. London: ISTE-Wiley, 2010, ch. 9, pp. 229–261.

[60] L. Najman, "On the equivalence between hierarchical segmentations and ultrametric watersheds," *Journal of Mathematical Imaging and Vision*, vol. 40, no. 3, pp. 231–247, July 2011, arXiv:1002.1887v2. [Online]. Available: http://www.laurentnajman.org

[61] C. Allène, J.-Y. Audibert, M. Couprie, and R. Keriven, "Some links between extremum spanning forests, watersheds and min-cuts," *IVC*, vol. 28, no. 10, pp. 1460–1471, Oct. 2010.

[62] C. Vachier and F. Meyer, "Extinction value: a new measurement of persistence," in *IEEE Workshop on Nonlinear Signal and Image Processing*, 1995, pp. 254–257.

[63] G. Bertrand, "On the dynamics," *IVC*, vol. 25, no. 4, pp. 447–454, 2007.

[64] S. Beucher, "Watershed, hierarchical segmentation and waterfall algorithm," in *Mathematical Morphology and its Applications to Image Processing*, J. Serra and P. Soille, Eds. Kluwer Academic Publishers, 1994, pp. 69–76.

[65] L. Grady, "Random walks for image segmentation," *IEEE PAMI*, vol. 28, no. 11, pp. 1768–1783, 2006.

[66] A. K. Sinop and L. Grady, "A seeded image segmentation framework unifying graph cuts and random walker which yields a new algorithm," in *Proc. of ICCV'07*, 2007.

[67] G. Peyre, M. Pechaud, R. Keriven, and L. Cohen, "Geodesic methods in computer vision and graphics," *Foundations and Trends in Computer Graphics and Vision*, vol. 5, no. 3-4, pp. 197–397, 2010. [Online]. Available: http://hal.archives-ouvertes.fr/hal-00528999/

[68] B. Chazelle, "A minimum spanning tree algorithm with inverse-Ackermann type complexity," *Journal of the ACM*, vol. 47, pp. 1028–1047, 2000.

[69] C. Couprie, X. Bresson, L. Najman, H. Talbot, and L. Grady, "Surface reconstruction using power watershed," in *ISMM 2011*, ser. LNCS, no. 6671, 2011, pp. 381–392.

[70] C. Couprie, L. Grady, L. Najman, and H. Talbot, "Anisotropic Diffusion Using Power Watersheds," in *International Conference on Image Processing (ICIP'10)*, Sept. 2010, pp. 4153–4156.

[71] T. Goldstein, X. Bresson, and S. Osher, "Geometric applications of the Split Bregman Method: Segmentation and surface reconstruction," UCLA, Computational and Applied Mathematics Reports, Tech. Rep. 09-06, 2009.

[72] J. Ye, X. Bresson, T. Goldstein, and S. Osher, "A fast variational method for surface reconstruction from sets of scattered points," UCLA, Computational and Applied Mathematics Reports, Tech. Rep. 10-01, 2010.

[73] V. Lempitsky and Y. Boykov, "Global Optimization for Shape Fitting," in *Proc. IEEE Conference on Computer Vision and Pattern Recognition (CVPR)*, Minneapolis, USA, 2007.

[74] R. Levillain, T. Géraud, and L. Najman, "Why and How to Design a Generic and Efficient Image Processing Framework: The Case of the Milena Library," in *17th International Conference on Image Processing*, 2010, pp. 1941–1944.

7

Partial difference Equations on Graphs for Local and Nonlocal Image Processing

Abderrahim Elmoataz

Université de Caen Basse-Normandie
GREYC UMR CNRS 6072
6 Bvd. Maréchal Juin
F-14050 CAEN, FRANCE
Email: abderrahim.elmoataz-billah@unicaen.fr

Olivier Lézoray

Université de Caen Basse-Normandie
GREYC UMR CNRS 6072
6 Bvd. Maréchal Juin
F-14050 CAEN, FRANCE
Email: olivier.lezoray@unicaen.fr

Vinh-Thong Ta

LaBRI (Université de Bordeaux – IPB – CNRS)
351, cours de la Libération
F-33405 Talence Cedex, FRANCE
Email: vinh-thong.ta@labri.fr

Sébastien Bougleux

Université de Caen Basse-Normandie
GREYC UMR CNRS 6072
6 Bvd. Maréchal Juin
F-14050 CAEN, FRANCE
Email: sebastien.bougleux@unicaen.fr

CONTENTS

7.1   Introduction 176
7.2   Difference Operators on Weighted Graphs 177
     7.2.1   Preliminary Notations and Definitions 177
     7.2.2   Difference Operators 178
     7.2.3   Gradients and Norms 179
     7.2.4   Graph Boundary 179
     7.2.5 p-Laplace Operators 180
           7.2.5.1 p-Laplace Isotropic Operator 180

         7.2.5.2   *p*-Laplace Anisotropic Operator 181
    7.2.6   *p*-Laplacian Matrix 182
7.3   Construction of Weighted Graphs 182
    7.3.1   Graph Topology 182
         7.3.1.1   Unorganized Data 183
         7.3.1.2   Organized Data 183
    7.3.2   Graph Weights 184
7.4   *p*-Laplacian Regularization on Graphs 185
    7.4.1   Isotropic Diffusion Processes 186
    7.4.2   Anisotropic Diffusion Processes 187
    7.4.3   Related Works 188
         7.4.3.1   Related Works in Image Processing 188
         7.4.3.2   Links with Spectral Methods 189
    7.4.4   Interpolation of Missing Data on Graphs 189
         7.4.4.1   Semi-Supervised Image Clustering 190
         7.4.4.2   Geometric and Texture Image Inpainting 191
         7.4.4.3   Image Colorization 191
7.5   Examples 192
    7.5.1   Regularization-Based Simplification 192
    7.5.2   Regularization-Based Interpolation 196
7.6   Concluding Remarks 203
    Bibliography 203

7.1   Introduction

With the advent of our digital world, much different kind of data are now available. Contrary to classical images and videos, these data do not necessarily lie on a Cartesian grid and can be irregularly distributed. To represent a large number of data domains (images, meshes, social networks, etc.), the most natural and flexible representation consists in using weighted graphs by modeling neighborhood relationships. Usual ways to process such data consists in using tools from combinatorial and graph theory. However, many mathematical tools developed on usual Euclidean domains were proven to be very efficient for the processing of images and signals. As a consequence, there is much interest in the transposition of signal and image processing tools for the processing of functions on graphs. Examples of such effort include, e.g., spectral graph wavelets [1] or partial differential equations (PDEs) on graphs [2, 3]. Chapter 8 presents a framework for constructing multiscale wavelet transforms on graphs. In this chapter we are interested in a framework for PDEs on graphs.

In image processing and computer vision, techniques based on energy minimization and PDEs have shown their efficiency in solving many important problems, such as smoothing, denoising, interpolation, and segmentation. Solutions of such problems can be obtained by considering the input discrete data as continuous functions defined on a continuous domain, and by designing continuous PDEs whose solutions are discretized in order to fit with the natural discrete domain. Such PDEs-based methods have the advantages of better mathematical modeling, connections with physics, and better geometrical approximations. A complete overview of these methods can be found in [4, 5, 6] and references therein. An alternative methodology to continuous PDEs-based regularization is to formalize the problem directly in a discrete setting [7] that is not necessarily a grid. However, PDE-based methods are difficult to adapt for data that live on non-Euclidean domains since the discretization of the underlying differential operators is difficult for high dimensional data.

Therefore, we are interested in developing variational PDEs on graphs [8]. As previously mentioned, partial differential equations appear naturally when describing some important physical processes (e.g., Laplace, wave, heat, and Navier-Stokes equations). As first introduced in the seminal paper of Courant, Friederichs, and Lewy [9], such problems involving the classical partial differential equations of mathematical physics can be reduced to algebraic ones of a very much simpler structure by replacing the differentials by difference equations on graphs. As a consequence, it is possible to provide methods that mimic on graphs well-known PDE variational formulations under a **functional analysis** point of view (which is close but different to the combinatorial approach exposed in [10]). One way to tackle this is to use *partial difference equations (PdE)* over graphs [11]. Conceptually, PdEs mimic PDEs in domains having a graph structure.

Since graphs can be used to model *any* discrete data set, the development of PdEs on graphs has received a lot of attention [2, 3, 12, 13, 14]. In previous works, we have introduced nonlocal difference operators on graphs and used the framework of PdEs to transcribe PDEs on graphs. This has enabled us to unify local and nonlocal processing. In particular, in [3], we have introduced a nonlocal discrete regularization on graphs of arbitrary topologies as a framework for data simplification and interpolation (label diffusion, inpainting, colorisation). Based on these works, we have proposed the basis of a new formulation [15, 16] of mathematical morphology that considers a discrete version of PDEs-based approaches over weighted graphs. Finally, with the same ideas, we have recently adapted the Eikonal equation [17] for data clustering and image segmentation. In this chapter, we provide a self-contained review of our works on p-Laplace regularization on graphs with PdEs.

7.2 Difference Operators on Weighted Graphs

In this section, we recall some basic definitions on graphs and define difference, divergence, gradient, and p-Laplacian operators.

7.2.1 Preliminary Notations and Definitions

A weighted graph $\mathcal{G} = (\mathcal{V}, \mathcal{E}, w)$ consists in a finite set $\mathcal{V} = \{v_1, \ldots, v_N\}$ of N vertices and a finite set $\mathcal{E} = \{e_1, \ldots, e_{N'}\} \subset \mathcal{V} \times \mathcal{V}$ of N' weighted edges. We assume

$\mathcal{G}$ to be undirected, with no self-loops and no multiple edges. Let $e_{ij} = (v_i, v_j)$ be the edge of $\mathcal{E}$ that connects vertices v_i and v_j of $\mathcal{V}$. Its weight, denoted by $w_{ij} = w(v_i, v_j)$, represents the similarity between its vertices. Similarities are usually computed by using a positive symmetric function $w : \mathcal{V} \times \mathcal{V} \to \mathbb{R}^+$ satisfying $w(v_i, v_j) = 0$ if $(v_i, v_j) \notin \mathcal{E}$. The notation $v_i \sim v_j$ is also used to denote two adjacent vertices. We say that $\mathcal{G}$ is connected whenever, for any pair of vertices (v_k, v_l), there is a finite sequence $v_k = v_0, v_1, \ldots, v_n = v_l$ such that v_{i-1} is a neighbor of v_i for every $i \in \{1, \ldots, n\}$. The degree function of a vertex $deg : \mathcal{V} \to \mathbb{R}$ is defined as $deg(v_i) = \sum_{v_j \sim v_i} w(v_i, v_j)$. Let $\mathcal{H}(\mathcal{V})$ be the Hilbert space of real-valued functions defined on the vertices of a graph. A function $f : \mathcal{V} \to \mathbb{R}$ of $\mathcal{H}(\mathcal{V})$ assigns a real value $x_i = f(v_i)$ to each vertex $v_i \in \mathcal{V}$. Clearly, a function $f \in \mathcal{H}(\mathcal{V})$ can be represented by a column vector in $\mathbb{R}^{|\mathcal{V}|}$: $[x_1, \ldots, x_N]^T = [f(v_1), \ldots, f(v_N)]^T$. By analogy with functional analysis on continuous spaces, the integral of a function $f \in \mathcal{H}(\mathcal{V})$, over the set of vertices $\mathcal{V}$, is defined as $\int_{\mathcal{V}} f = \sum_{\mathcal{V}} f$. The space $\mathcal{H}(\mathcal{V})$ is endowed with the usual inner product $\langle f, h \rangle_{\mathcal{H}(\mathcal{V})} = \sum_{v_i \in \mathcal{V}} f(v_i) h(v_i)$, where $f, h : \mathcal{V} \to \mathbb{R}$. Similarly, let $\mathcal{H}(\mathcal{E})$ be the space of real-valued functions defined on the edges of $\mathcal{G}$. It is endowed with the inner product[1] $\langle F, H \rangle_{\mathcal{H}(\mathcal{E})} = \sum_{e_i \in \mathcal{E}} F(e_i) H(e_i) = \sum_{v_i \in \mathcal{V}} \sum_{v_j \sim v_i} F(e_{ij}) H(e_{ij}) = \sum_{v_i \in \mathcal{V}} \sum_{v_j \sim v_i} F(v_i, v_j) H(v_i, v_j)$, where $F, H : \mathcal{E} \to \mathbb{R}$ are two functions of $\mathcal{H}(\mathcal{E})$.

7.2.2   Difference Operators

Let $\mathcal{G} = (\mathcal{V}, \mathcal{E}, w)$ be a weighted graph, and let $f : \mathcal{V} \to \mathbb{R}$ be a function of $\mathcal{H}(\mathcal{V})$. The *difference operator* [18, 19, 20, 3] of f, noted $d_w : \mathcal{H}(\mathcal{V}) \to \mathcal{H}(\mathcal{E})$, is defined on an edge $e_{ij} = (v_i, v_j) \in \mathcal{E}$ by:

$$(d_w f)(e_{ij}) = (d_w f)(v_i, v_j) = w(v_i, v_j)^{1/2}(f(v_j) - f(v_i)) \ . \qquad (7.1)$$

The *directional derivative* (or *edge derivative*) of f, at a vertex $v_i \in \mathcal{V}$, along an edge $e_{ij} = (v_i, v_j)$, is defined as $\frac{\partial f}{\partial e_{ij}}\big|_{v_i} = \partial_{v_j} f(v_i) = (d_w f)(v_i, v_j)$. This definition is consistent with the continuous definition of the derivative of a function: $\partial_{v_j} f(v_i) = -\partial_{v_i} f(v_j)$, $\partial_{v_i} f(v_i) = 0$, and if $f(v_j) = f(v_i)$ then $\partial_{v_j} f(v_i) = 0$. The *adjoint* of the difference operator, noted $d_w^* : \mathcal{H}(\mathcal{E}) \to \mathcal{H}(\mathcal{V})$, is a linear operator defined by

$$\langle d_w f, H \rangle_{\mathcal{H}(\mathcal{E})} = \langle f, d_w^* H \rangle_{\mathcal{H}(\mathcal{V})} \ , \qquad (7.2)$$

for all $f \in \mathcal{H}(\mathcal{V})$ and all $H \in \mathcal{H}(\mathcal{E})$. The adjoint operator d_w^*, of a function $H \in \mathcal{H}(\mathcal{E})$, can by expressed at a vertex $v_i \in \mathcal{V}$ by the following expression:

$$(d_w^* H)(v_i) = \sum_{v_j \sim v_i} w(v_i, v_j)^{1/2}(H(v_j, v_i) - H(v_i, v_j)) \ . \qquad (7.3)$$

---

[1] Since graphs are considered undirected, if $e_{ij} \in \mathcal{E}$, then $e_{ji} \in \mathcal{E}$.

Proof By using the definition of the inner product in $\mathcal{H}(\mathcal{E})$, we have:

$$\langle d_w f, H \rangle_{\mathcal{H}(\mathcal{E})} = \sum_{e_{ij} \in \mathcal{E}} (d_w f)(e_{ij}) H(e_{ij}) \overset{(7.1)}{=} \sum_{e_{ij} \in \mathcal{E}} \sqrt{w_{ij}} (f(v_j) - f(v_i)) H(e_{ij}) =$$

$$\sum_{v_i \in \mathcal{V}} \sum_{v_j \sim v_i} \sqrt{w_{ij}} f(v_i) H(e_{ji}) - \sum_{v_i \in \mathcal{V}} \sum_{v_j \sim v_i} \sqrt{w_{ij}} f(v_i) H(e_{ij}) =$$

$$\sum_{v_i \in \mathcal{V}} f(v_i) \sum_{v_j \sim v_i} \sqrt{w_{ij}} (H(e_{ji}) - H(e_{ij})) = \sum_{v_i \in \mathcal{V}} f(v_i)(d_w^* H)(v_i) \overset{(7.2)}{=} \langle f, d_w^* H \rangle.$$

$$(7.4)$$

∎

The *divergence operator*, defined by $-d_w^*$, measures the net outflow of a function of $\mathcal{H}(\mathcal{E})$ at each vertex of the graph. Each function $H \in \mathcal{H}(\mathcal{E})$ has a null divergence over the entire set of vertices: $\sum_{v_i \in \mathcal{V}} (d_w^* H)(v_i) = 0$. Another general definition of the difference operator has been proposed by Zhou [21] as $(d_w f)(v_i, v_j) = \sqrt{w_{ij}} \left(\frac{f(v_j)}{\deg(v_j)} - \frac{f(v_i)}{\deg(v_i)} \right)$. However, the latter operator is not null when the function f is locally constant and its adjoint is not null over the entire set of vertices [22, 23, 24].

7.2.3 Gradients and Norms

The *weighted gradient operator* of a function $f \in \mathcal{H}(\mathcal{V})$, at a vertex $v_i \in \mathcal{V}$, is the vector operator defined by

$$(\nabla_w f)(v_i) = [\partial_{v_j} f(v_i) : v_j \sim v_i]^T = [\partial_{v_1} f(v_i), \ldots, \partial_{v_k} f(v_i)]^T, \ \forall (v_i, v_j) \in \mathcal{E}.$$

$$(7.5)$$

The $\mathcal{L}_p$ norm of this vector represents the *local variation* of the function f at a vertex of the graph. It is defined by [18, 19, 20, 3]:

$$\|(\nabla_w f)(v_i)\|_p = \left[\sum_{v_j \sim v_i} w_{ij}^{p/2} |f(v_j) - f(v_i)|^p \right]^{1/p}. \tag{7.6}$$

For $p \geq 1$, the local variation is a semi-norm that measures the regularity of a function around a vertex of the graph:

1. $\|(\nabla_w \alpha f)(v_i)\|_p = |\alpha| \|(\nabla_w f)(v_i)\|_p, \forall \alpha \in \mathbb{R}, \forall f \in \mathcal{H}(\mathcal{V}),$

2. $\|(\nabla_w (f + h))(v_i)\|_p \leq \|(\nabla_w f)(v_i)\|_p + \|(\nabla_w h)(v_i)\|_p, \forall f \in \mathcal{H}(\mathcal{V}),$

3. $\|(\nabla_w f)(v_i)\|_p = 0 \Leftrightarrow (f(v_j) = f(v_i))$ or $w(v_i, v_j) = 0, \ \forall v_j \sim v_i.$

For $p < 1$, (7.6) is not a norm since the triangular inequality is not satisfied (property 2).

7.2.4   Graph Boundary

Let $\mathcal{A}$ be a set of connected vertices with $\mathcal{A} \subset \mathcal{V}$ such that for all $v_i \in \mathcal{A}$, there exists a vertex $v_j \in \mathcal{A}$ with $e_{ij} = (v_i, v_j) \in \mathcal{E}$. We denote by $\partial^+ \mathcal{A}$ and $\partial^- \mathcal{A}$, the *external* and *internal* boundary sets of $\mathcal{A}$, respectively. $\mathcal{A}^c = \mathcal{V} \setminus \mathcal{A}$ is the complement of $\mathcal{A}$. For a given vertex $v_i \in \mathcal{V}$, $\partial^+ \mathcal{A} = \{v_i \in \mathcal{A}^c : \exists v_j \in \mathcal{A} \text{ with } e_{ij} = (v_i, v_j) \in \mathcal{E}\}$ and $\partial^- \mathcal{A} = \{v_i \in \mathcal{A} : \exists v_j \in \mathcal{A}^c \text{ with } e_{ij} = (v_i, v_j) \in \mathcal{E}\}$. Figure 7.1 illustrates these notions on two different graph structures: a 4-adjacency grid graph and an arbitrary graph. The notion of graph boundary will be used for interpolation problems on graphs.

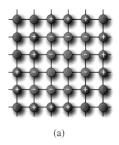 

(a)   (b)

FIGURE 7.1
Graph boundary sets on two different graphs. (a) a 4-adjacency grid graph. (b) an arbitrary undirected graph. Blue vertices correspond to $\mathcal{A}$. Vertices marked by the "−" sign correspond to the internal boundary $\partial^- \mathcal{A}$ and those marked by the "+" sign correspond to the external boundary $\partial^+ \mathcal{A}$.

7.2.5   *p*-Laplace Operators

Let $p \in (0, +\infty)$ be a real number. We introduce the isotropic and anisotropic p-Laplace operators as well as the associated p-Laplacian matrix.

7.2.5.1   *p*-Laplace Isotropic Operator

The *weighted p-Laplace isotropic operator* of a function $f \in \mathcal{H}(\mathcal{V})$, noted $\Delta^i_{w,p}$: $\mathcal{H}(\mathcal{V}) \to \mathcal{H}(\mathcal{V})$, is defined by:

$$(\Delta^i_{w,p}f)(v_i) = \tfrac{1}{2}d_w^*(\|(\nabla_\mathbf{w}\mathbf{f})(\mathbf{v_i})\|_2^{p-2}(d_w f)(v_i, v_j)) \ . \tag{7.7}$$

The isotropic p-Laplace operator of $f \in \mathcal{H}(\mathcal{V})$, at a vertex $v_i \in \mathcal{V}$, can be computed by [18, 19, 20, 3]:

$$(\Delta^i_{w,p}f)(v_i) = \tfrac{1}{2} \sum_{v_j \sim v_i} (\gamma^i_{w,p}f)(v_i, v_j)(f(v_i) - f(v_j)) \ , \tag{7.8}$$

with

$$(\gamma_{w,p}^{i}f)(v_i, v_j) = w_{ij}\left(\|(\nabla_{\mathbf{w}}\mathbf{f})(\mathbf{v_j})\|_2^{p-2} + \|(\nabla_{\mathbf{w}}\mathbf{f})(\mathbf{v_i})\|_2^{p-2}\right) . \qquad (7.9)$$

Proof From (7.7), (7.3) and (7.1), we have:

$$\begin{aligned}(\Delta_{w,p}^{i}f)(v_i) &= \tfrac{1}{2}\sum_{v_j \sim v_i}\sqrt{w_{ij}}\left(\frac{(d_w f)(v_j, v_i)}{\|(\nabla_{\mathbf{w}}\mathbf{f})(\mathbf{v_j})\|^{2-p}} - \frac{(d_w f)(v_i, v_j)}{\|(\nabla_{\mathbf{w}}\mathbf{f})(\mathbf{v_i})\|^{2-p}}\right)\\ &= \tfrac{1}{2}\sum_{v_j \sim v_i}w_{ij}\left(\frac{f(v_i)-f(v_j)}{\|(\nabla_{\mathbf{w}}\mathbf{f})(\mathbf{v_j})\|^{2-p}} - \frac{f(v_j)-f(v_i)}{\|(\nabla_{\mathbf{w}}\mathbf{f})(\mathbf{v_i})\|^{2-p}}\right) .\end{aligned} \qquad (7.10)$$

This last term is exactly (7.8). ∎

The p-Laplace isotropic operator is nonlinear, with the exception of $p = 2$ (see [18, 19, 20, 3]). In this latter case, it corresponds to the *combinatorial graph Laplacian*, which is one of the classical second order operators defined in the context of spectral graph theory [25]. Another particular case of the p-Laplace isotropic operator is obtained with $p = 1$. In this case, it is the *weighted curvature* of the function f on the graph. To avoid zero denominator in (7.8) when $p \le 1$, $\|(\nabla_{\mathbf{w}}\mathbf{f})(\mathbf{v_i})\|_2$ is replaced by $\|(\nabla_{\mathbf{w}}\mathbf{f})(\mathbf{v_i})\|_{2,\epsilon} = \sqrt{\sum_{v_j \sim v_i} w_{ij}(f(v_j)-f(v_i))^2 + \epsilon^2}$, where $\epsilon \to 0$ is a small fixed constant.

7.2.5.2 *p*-Laplace Anisotropic Operator

The *weighted p-Laplace anisotropic operator* of a function $f \in \mathcal{H}(\mathcal{V})$, noted $\Delta_{w,p}^{a} : \mathcal{H}(\mathcal{V}) \to \mathcal{H}(\mathcal{V})$, is defined by:

$$(\Delta_{w,p}^{a}f)(v_i) = \tfrac{1}{2}d_w^*(|(d_w f)(v_i, v_j)|^{p-2}(d_w f)(v_i, v_j)) . \qquad (7.11)$$

The anisotropic p-Laplace operator of $f \in \mathcal{H}(\mathcal{V})$, at a vertex $v_i \in \mathcal{V}$, can be computed by [26]:

$$(\Delta_{w,p}^{a}f)(v_i) = \sum_{v_j \sim v_i}(\gamma_{w,p}^{a}f)(v_i, v_j)(f(v_i) - f(v_j)) . \qquad (7.12)$$

with

$$(\gamma_{w,p}^{a}f)(v_i, v_j) = w_{ij}^{p/2}|f(v_i) - f(v_j)|^{p-2} . \qquad (7.13)$$

Proof From (7.11), (7.3) and (7.1), we have

$$(\Delta_{w,p}^{a}f)(v_i) = \tfrac{1}{2}\sum_{v_j \sim v_i}\sqrt{w_{ij}}\left(\frac{(d_w f)(v_j, v_i)}{|(d_w f)(v_i, v_j)|^{2-p}} - \frac{(d_w f)(v_i, v_j)}{|(d_w f)(v_i, v_j)|^{2-p}}\right) =$$

$$\sum_{v_j \sim v_i}\sqrt{w_{ij}}\left(\frac{(d_w f)(v_j, v_i)}{|(d_w f)(v_i, v_j)|^{2-p}}\right) = \sum_{v_j \sim v_i}w_{ij}^{\frac{p}{2}}|f(v_i) - f(v_j)|^{p-2}(f(v_i) - f(v_j)) .$$

$$(7.14)$$

∎

This operator is nonlinear if $p \neq 2$. In this latter case, it corresponds to the *combinatorial graph Laplacian*, as the isotropic 2-Laplacian. To avoid zero denominator in (7.12) when $p \leq 1$, $|f(v_i) - f(v_j)|$ is replaced by $|f(v_i) - f(v_j)|_\epsilon = |f(v_i) - f(v_j)| + \epsilon$, where $\epsilon \to 0$ is a small fixed constant.

7.2.6 p-**Laplacian Matrix**

Finally, the (isotropic or anisotropic) p-Laplacian matrix $\mathbf{L}_p^*$ satisfies the following properties.

(1) its expression is:

$$\mathbf{L}_p^*(v_i, v_j) = \begin{cases} \frac{\alpha}{2} \displaystyle\sum_{v_k \sim v_i} (\gamma_{w,p}^* f)(v_i, v_k) \text{ if } v_i = v_j \\ -\frac{\alpha}{2}(\gamma_{w,p}^* f)(v_i, v_j) \text{ if } v_i \neq v_j \text{ and } (v_i, v_j) \in \mathcal{E} \\ 0 \text{ if } (v_i, v_j) \notin \mathcal{E} \end{cases} \quad (7.15)$$

with $\gamma_{w,p}^* = \gamma_{w,p}^i$ and $\alpha = 1$ for the isotropic case or $\gamma_{w,p}^* = \gamma_{w,p}^a$ and $\alpha = 2$ for the anisotropic case.

(2) For every vector $\mathbf{f(u)}$, $\mathbf{f(u)}^T \mathbf{L}_p^* \mathbf{f(u)} = (\Delta_{w,p}^* f)(u)$ with $\Delta_{w,p}^* = \Delta_{w,p}^i$ or $\Delta_{w,p}^* = \Delta_{w,p}^a$.

(3) $\mathbf{L}_p^*$ is symmetric and positive semi-definite (see section 7.4 for a justification).

(4) $\mathbf{L}_p^*$ has non-negative, real-valued eigenvalues: $0 = s_1 \leq s_2 \leq \cdots \leq s_N$.

Therefore, the (isotropic or anisotropic) p-Laplacian matrix can be used for manifold learning and generalizes approaches based on the 2-Laplacian [27]. New dimensionality reduction methods can be obtained with an eigen-decomposition of the matrix $\mathbf{L}_p^*$. This has been recently explored in [28] with the isotropic p-Laplace operator and in [29] with the anisotropic p-Laplace operator for spectral clustering where a variational algorithm is proposed for generalized spectral clustering.

━━━━━━━━━━━━

7.3   Construction of Weighted Graphs

Any discrete domain can be modeled by a weighted graph where each data point is represented by a vertex $v_i \in \mathcal{V}$. This domain can represent unorganized or organized data where functions defined on $\mathcal{V}$ correspond to the data to process. We provide details on defining the graph topology and the graph weights.

7.3.1 Graph Topology

7.3.1.1 Unorganized Data

In this general case, an unorganized set of points $\mathcal{V} \subset \mathbb{R}^n$ can be seen as a function $f^0 : \mathcal{V} \to \mathbb{R}^m$. Then, defining the set of edges consists in modeling the neighborhood of each vertex based on similarity relationships between feature vectors of the data set. This similarity depends on a pairwise distance measure $\mu : \mathcal{V} \times \mathcal{V} \to \mathbb{R}^+$. A typical choice of μ for unorganized data is the Euclidean distance. Graph construction is application dependent, and no general rules can be given. However, there exists several methods to construct a neighborhood graph. The interested reader can refer to [30] for a survey on proximity and neighborhood graphs. We focus on two classes of graphs: a modified version of k-nearest neighbors graphs and τ-neighborhood graphs.

The τ-neighborhood graph, noted $\mathcal{G}_\tau$ is a weighted graph where the τ-neighborhood $\mathcal{N}_\tau$ for a vertex v_i is defined as $\mathcal{N}_\tau(v_i) = \{v_j \in \mathcal{V} \setminus \{v_i\} : \mu(v_i, v_j) \leq \tau\}$ where $\tau > 0$ is a threshold parameter.

The k-nearest neighbors graph, noted k-NNG$_\tau$ is a weighted graph where each vertex v_i is connected to its k nearest neighbors that have the smallest distance measure towards v_i according to function μ in $\mathcal{N}_\tau$. Since this graph is directed, a modified version is used to make it undirected i.e., $\mathcal{E} = \{(v_i, v_j) : v_i \in \mathcal{N}_\tau(v_j) \text{ or } v_j \in \mathcal{N}_\tau(v_i)\}$ for $v_i, v_j \in \mathcal{V}$. When $\tau = \infty$, one obtains the k-NNG$_\infty$ where the k nearest neighbors are computed for all the set $\mathcal{V} \setminus \{v_i\}$. For the sake of clarity, the k-NNG$_\infty$ will be noted k-NNG.

7.3.1.2 Organized Data

Typical cases of organized data are signals, gray-scale, or color images (eventually in 2D or 3D). Such data can be seen as functions $f^0 : \mathcal{V} \subset \mathbb{Z}^n \to \mathbb{R}^m$. Then, the distance μ used to construct the graph corresponds to a distance between spatial coordinates associated to vertices. Several distances can be considered. Among those, for a 2D image, we can quote the city-block distance: $\mu(v_i, v_j) = |x_i^1 - x_j^1| + |x_i^2 - x_j^2|$ or the Chebyshev distance: $\mu(v_i, v_j) = \max(|x_i^1 - x_j^1|, |x_i^2 - x_j^2|)$ where a vertex v_i is associated with its $(x_i^1, x_i^2)^T$ spatial coordinates. With these distances and a τ-neighborhood graph, the usual adjacency graphs used in 2D image processing are obtained where each vertex corresponds to an image pixel. Then, a 4-adjacency and an 8-adjacency grid graph (denoted $\mathcal{G}_0$ and $\mathcal{G}_1$, respectively) can be obtained with the city-block and the Chebyshev distance, respectively (with $\tau \leq 1$). More generally, $((2s+1)^2 - 1)$-adjacency graphs are obtained with a Chebyshev distance with $\tau \leq s$ and $s \geq 1$. This corresponds to adding edges between the central pixel and the other pixels within a square window of size $(2s + 1)^2$. Similar remarks apply for the construction of graphs on 3D images where vertices are associated to voxels.

The region adjacency graphs (RAGs) can also be considered for image processing. Each vertex of this graph corresponds to one region. The set of edges is obtained with an adjacency distance: $\mu(v_i, v_j) = 1$ if v_i and v_j are adjacent and $\mu(v_i, v_j) = \infty$, otherwise, together with a τ-neighborhood graph ($\tau = 1$): this corresponds to the Delaunay graph of an image partition.

Finally, we can mention the cases of polygonal curves or surfaces that have natural graph representations where vertices correspond to mesh vertices and edges are mesh edges.

7.3.2   Graph Weights

For an initial function $f^0 : \mathcal{V} \to \mathbb{R}^m$, a similarity relationship between data can be incorporated within edge weights according to a measure of similarity $g : \mathcal{E} \to [0, 1]$ with $w(e_{ij}) = g(e_{ij}), \forall e_{ij} \in \mathcal{E}$. Computing distances between vertices consists in comparing their features, generally depending on f^0. To this end, each vertex v_i is associated with a feature vector $\mathbf{F}^{f^0}_\tau : \mathcal{V} \to \mathbb{R}^{m \times q}$ where q corresponds to this vector size:

$$\mathbf{F}^{f^0}_\tau(v_i) = \left(f^0(v_j) : v_j \in \mathcal{N}_\tau(v_i) \cup \{v_i\} \right)^T . \tag{7.16}$$

Then, the following weight functions can be considered. For an edge e_{ij} and a distance measure $\rho : \mathbb{R}^{m \times q} \times \mathbb{R}^{m \times q} \to \mathbb{R}$ associated to $\mathbf{F}^{f^0}_\tau$, we can have:

$$g_1(e_{ij}) = 1 \text{ (unweighted case) },$$
$$g_2(e_{ij}) = \exp\left(-\rho\left(\mathbf{F}^{f^0}_\tau(v_i), \mathbf{F}^{f^0}_\tau(v_j) \right)^2 / \sigma^2 \right) \text{ with } \sigma > 0 , \tag{7.17}$$
$$g_3(e_{ij}) = 1 / \left(1 + \rho\left(\mathbf{F}^{f^0}_\tau(v_i), \mathbf{F}^{f^0}_\tau(v_j) \right) \right)$$

Usually, ρ is the Euclidean distance function. Several choices can be considered for the expression of $\mathbf{F}^{f^0}_\tau$ depending on the features to preserve. The simplest one is $\mathbf{F}^{f^0}_0 = f^0$.

In image processing, an important feature vector $\mathbf{F}^{f^0}_\tau$ is provided by image patches. For a gray-scale image $f^0 : \mathcal{V} \to \mathbb{R}$, the vector $\mathbf{F}^{f^0}_\tau$ defined by an image patch corresponds to the values of f^0 in a square window of size $(2\tau + 1)^2$ centered at a vertex (a pixel). Figure 7.2(a) shows examples of image patches. A given vertex (central pixel represented in red) is not only characterized by its gray-scale value but also by the gray-scale values contained in a square window (a patch) centered on it. Figure 7.2(b) shows the difference between local and nonlocal interaction in a 8-adjacency grid graph by adding nonlocal edges between vertices not spatially connected but close in a 5×5 neighborhood. This feature vector has been proposed for texture synthesis [31], and recently used for image restoration and filtering [32, 33]. In image processing, such configurations are called "nonlocal", and these latter works have shown the efficiency of nonlocal patch-based methods as compared to local ones, in particular, by better capturing complex structures such as objects boundaries or fine and repetitive patterns. The notion of "nonlocality" (as defined by [32]) includes two notions: (1) the search window of the most similar neighbors for a given pixel and (2)

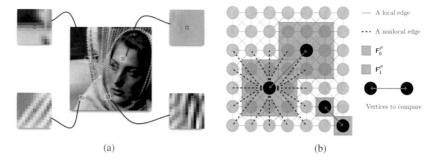

(a)   (b)

FIGURE 7.2
(a) Four examples of image patches. Central pixels (in red) are associated with vectors of gray-scale values within a patch of size 13×13 ($\mathbf{F_6^{f^0}}$). (b) Graph topology and feature vector illustration: a 8-adjacency grid graphs is constructed. The vertex surrounded by a red box shows the use of the simplest feature vector, i.e., the initial function. The vertex surrounded by a green box shows the use of a feature vector that exploits patches .

the feature vector to compare these neighbors. In [32], *nonlocal* processing consists in comparing, for a given pixel, all the patches contained in an image. In practice, to avoid this high computational cost, one can use a search window of fixed size. Local and nonlocal methods are naturally included in weighted graphs. Indeed, nonlocal patch-based configurations are simply expressed by the graph topology (the vertex neighborhood) and the edge weights (distances between vertex features). Nonlocal processing of images becomes local processing on similarity graphs. Our operators on graphs (weighted differences and gradients) naturally enable local and nonlocal configurations (with the weight function and the graph topology) and introduce new adaptive tools for image processing.

7.4   *p*-Laplacian Regularization on Graphs

Let $f^0 : \mathcal{V} \to \mathbb{R}$ be a given function defined on the vertices of a weighted graph $\mathcal{G} = (\mathcal{V}, \mathcal{E}, w)$. In a given context, the function f^0 represents an observation of a clean function $g : \mathcal{V} \to \mathbb{R}$ corrupted by a given noise n such that $f^0 = g + n$. Such noise is assumed to have zero mean and variance σ^2, which usually corresponds to observation errors. To recover the uncorrupted function g, a commonly used method is to seek for a function $f : \mathcal{V} \to \mathbb{R}$ which is regular enough on $\mathcal{G}$, and also close enough to f^0. This inverse problem can be formalized by the minimization of an energy functional, which typically involves a regularization term plus an approximation term (also called *fitting term*). We consider the following variational problem [18, 19, 20, 3]:

$$g \approx \min_{f:\mathcal{V} \to \mathbb{R}} \left\{ \mathcal{E}^*_{w,p}(f, f^0, \lambda) = R^*_{w,p}(f) + \tfrac{\lambda}{2} \| f - f^0 \|_2^2 \right\}, \tag{7.18}$$

where the regularization functional $R^*_{w,p} : \mathcal{H}(\mathcal{V}) \to \mathbb{R}$ can correspond to an isotropic $R^i_{w,p}$ or an anisotropic $R^a_{w,p}$ functionnal. The isotropic regularization functionnal $R^i_{w,p}$ is defined by the $\mathcal{L}_2$ norm of the gradient and is the discrete p-Dirichlet form of the function $f \in \mathcal{H}(\mathcal{V})$:

$$R^i_{w,p}(f) = \frac{1}{p} \sum_{v_i \in \mathcal{V}} \| (\nabla_{\mathbf{w}} \mathbf{f})(\mathbf{v_i}) \|_2^p = \tfrac{1}{p} \langle f, \Delta^i_{w,p} f \rangle_{\mathcal{H}(\mathcal{V})}$$

$$= \frac{1}{p} \sum_{v_i \in \mathcal{V}} \left[\sum_{v_j \sim v_i} w_{ij} (f(v_j) - f(v_i))^2 \right]^{\frac{p}{2}}. \tag{7.19}$$

The anisotropic regularization functionnal $R^a_{w,p}$ is defined by the $\mathcal{L}_p$ norm of the gradient:

$$R^a_{w,p}(f) = \frac{1}{p} \sum_{v_i \in \mathcal{V}} \| (\nabla_{\mathbf{w}} \mathbf{f})(\mathbf{v_i}) \|_p^p = \tfrac{1}{p} \langle f, \Delta^a_{w,p} f \rangle_{\mathcal{H}(\mathcal{V})}$$

$$= \frac{1}{p} \sum_{v_i \in \mathcal{V}} \sum_{v_j \sim v_i} w_{ij}^{p/2} |f(v_j) - f(v_i)|^p. \tag{7.20}$$

Both the previous regularization functionals show the fact that $\Delta^i_{w,p}$ and $\Delta^a_{w,p}$ are positive semi-definite (since $R^i_{w,p} \geq 0$ and $R^a_{w,p} \geq 0$). The isotropic and anisotropic versions of $\mathcal{E}^*_{w,p}$ are, respectively, denoted by $\mathcal{E}^i_{w,p}$ and $\mathcal{E}^a_{w,p}$. The trade-off between the two competing terms in the functional $\mathcal{E}^*_{w,p}$ is specified by the fidelity parameter $\lambda \geq 0$. By varying the value of λ, the variational problem (7.18) allows to describe the function f^0 at different scales, each scale corresponding to a value of λ. The degree of regularity, which has to be preserved, is controlled by the value of $p > 0$.

7.4.1 Isotropic Diffusion Processes

First, we consider the case of an isotropic regularization functional, i.e., $R^*_{w,p} = R^i_{w,p}$. When $p \geq 1$, the energy $\mathcal{E}^i_{w,p}$ is a convex functional of functions of $\mathcal{H}(\mathcal{V})$. To get the solution of the minimizer (7.18), we consider the following system of equations

$$\frac{\partial \mathcal{E}^i_{w,p}(f, f^0, \lambda)}{\partial f(v_i)} = 0, \forall v_i \in \mathcal{V}, \tag{7.21}$$

which is rewritten as:

$$\frac{\partial R^i_{w,p}(f)}{\partial f(v_i)} + \lambda(f(v_i) - f^0(v_i)) = 0, \quad \forall v_i \in \mathcal{V}. \tag{7.22}$$

The solution of the latter system is computed by using the following property. Let f be a function in $\mathcal{H}(\mathcal{V})$, one can prove from (7.19) that [3],

$$\frac{\partial R^i_{w,p}(f)}{\partial f(v_i)} = 2(\Delta^i_{w,p}f)(v_i) \ . \tag{7.23}$$

Proof Let v_1 be a vertex of $\mathcal{V}$. From (7.19), the v_1-th term of the derivative of $R^i_{w,p}(f)$ is given by:

$$\frac{\partial R^i_{w,p}(f)}{\partial f(v_1)} = \frac{1}{p}\frac{\partial}{\partial f(v_1)}\left(\sum_{v_i\in\mathcal{V}}\left(\sum_{v_j\sim v_i} w_{ij}(f(v_j)-f(v_i))^2\right)^{\frac{p}{2}}\right). \tag{7.24}$$

It only depends on the edges incident to v_1 . Let $v_l,\ldots,v_k$ be the vertices of $\mathcal{V}$ connected to v_1 by an edge of $\mathcal{E}$. Then, by using the chain rule, we have:

$$\begin{aligned}
\frac{\partial R^i_{w,p}(f)}{\partial f(v_1)} &= -\sum_{v_j\sim v_1} w_{1j}(f(v_j)-f(v_1))\|(\nabla_{\mathbf{w}}\mathbf{f})(\mathbf{v_1})\|_2^{p-2}\\
&+ w_{l1}(f(v_1)-f(v_l))\|(\nabla_{\mathbf{w}}\mathbf{f})(\mathbf{v_l})\|_2^{p-2}\\
&+\ldots+ w_{k1}(f(v_1)-f(v_k))\|(\nabla_{\mathbf{w}}\mathbf{f})(\mathbf{v_k})\|_2^{p-2},
\end{aligned} \tag{7.25}$$

which is equal to: $\sum_{v_j\sim v_1} w_{1j}(f(v_1)-f(v_j))(\|(\nabla_{\mathbf{w}}\mathbf{f})(\mathbf{v_1})\|_2^{p-2}+\|(\nabla_{\mathbf{w}}\mathbf{f})(\mathbf{v_j})\|_2^{p-2})$.
From (7.8), this latter expression is exactly $2\Delta^i_{w,p}f(v_1)$. ∎

The system of equations is then rewritten as

$$2(\Delta^i_{w,p}f)(v_i) + \lambda(f(v_i) - f^0(v_i)) = 0, \quad \forall v_i \in \mathcal{V}, \tag{7.26}$$

which is equivalent to the following system of equations:

$$\left(\lambda + \sum_{v_j\sim v_i} (\gamma^i_{w,p}f)(v_i, v_j)\right) f(v_i) - \sum_{v_j\sim v_i} (\gamma^i_{w,p}f)(v_i, v_j)f(v_j) = \lambda f^0(v_i). \tag{7.27}$$

We use the linearized Gauss-Jacobi iterative method to solve the previous system. Let n be an iteration step, and let $f^{(n)}$ be the solution at the step n. Then, the method is given by the following algorithm:

$$\begin{cases}
f^{(0)} = f^0 \\
f^{(n+1)}(v_i) = \dfrac{\lambda f^0(v_i) + \sum_{v_j\sim v_i} (\gamma^i_{w,p}f^{(n)})(v_i, v_j)f^{(n)}(v_j)}{\lambda + \sum_{v_j\sim v_i} (\gamma^i_{w,p}f^{(n)})(v_i, v_j)}, \ \forall v_i \in \mathcal{V}.
\end{cases} \tag{7.28}$$

with $\gamma^i_{w,p}$ as defined in (7.9). We present in this chapter only this simple method of resolution, but more efficient methods should be preferred [34, 35].

7.4.2   Anisotropic Diffusion Processes

Second, we consider the case of an anisotropic regularization functional, i.e., $R^*_{w,p} = R^a_{w,p}$. When $p \geq 1$, the energy $\mathcal{E}^a_{w,p}$ is a convex functional of functions of $\mathcal{H}(\mathcal{V})$. We can follow the same resolution as for the isotropic case. Indeed, to get the solution of the minimizer (7.18), we consider the following system of equations:

$$\frac{\partial R^a_{w,p}(f)}{\partial f(v_i)} + \lambda(f(v_i) - f^0(v_i)) = 0, \quad \forall v_i \in \mathcal{V}. \tag{7.29}$$

One can also prove from (7.20), using the chain rule similarly than for the isotropic case, that [26]

$$\frac{\partial R^a_{w,p}(f)}{\partial f(v_i)} = (\Delta^a_{w,p} f)(v_i) \ . \tag{7.30}$$

The previous system of equations is rewritten as

$$(\Delta^a_{w,p} f)(v_i) + \lambda(f(v_i) - f^0(v_i)) = 0, \quad \forall v_i \in \mathcal{V}. \tag{7.31}$$

Similarly as for the isotropic case, we use the linearized Gauss-Jacobi iterative method to solve the previous system. Let n be an iteration step, and let $f^{(n)}$ be the solution at the step n. Then, the method is given by the following algorithm:

$$\begin{cases} f^{(0)} = f^0 \\ f^{(n+1)}(v_i) = \dfrac{\lambda f^0(v_i) + \sum_{v_j \sim v_i} (\gamma^a_{w,p} f^{(n)})(v_i, v_j) f^{(n)}(v_j)}{\lambda + \sum_{v_j \sim v_i} (\gamma^a_{w,p} f^{(n)})(v_i, v_j)}, \forall v_i \in \mathcal{V}. \end{cases} \tag{7.32}$$

$\gamma^a_{w,p}$ as defined in (7.13).

7.4.3   Related Works

The two regularization methods describe a family of discrete isotropic or anisotropic diffusion processes [3, 26], which are parameterized by the structure of the graph (topology and weight function), the parameter p, and the parameter λ. At each iteration of the algorithm (7.28), the new value $f^{(n+1)}(v_i)$ depends on two quantities: the original value $f^0(v_i)$, and a weighted average of the filtered values of $f^{(n)}$ in a neighborhood of v_i. The familiy of induced filters exhibits strong links with numerous existing filters for image filtering and extends them to the processing of data of arbitrary dimensions on graphs (see [3, 26]). Links with spectral methods can also be exhibited.

7.4.3.1   Related Works in Image Processing

When $p = 2$, both the isotropic and anisotropic p-Laplacian are the classical combinatorial Laplacian. With $\lambda = 0$, the solution of the diffusion processes is the solution of the heat equation $\Delta_w f(v_i) = 0$, for all $v_i \in \mathcal{V}$. Moreover, since it is also the solution of the minimization of the Dirichlet energy $R^*_{w,2}(f)$, the diffusion performs also Laplacian smoothing. Many other specific filters are related to the diffusion processes we propose (see [3, 26] for details). In particular, with specific values of p, λ, graph topologies, and weights, it is easy to recover Gaussian filtering, bilateral filtering [36], the TV digital filter [37], median and minimax filtering [10] but also the NLMeans [32]. The p-Laplacian regularization we propose can as well be considered as the discrete analogue of the recently proposed continuous nonlocal anisotropic functionals [33, 38, 39]. Finally, our framework based on the isotropic or anisotropic p-Laplacian generalizes all these approaches and extends them to weighted graphs of the arbitrary topologies.

7.4.3.2 Links with Spectral Methods

When $\lambda = 0$ and $p = 2$, an iteration of the diffusion process given by $f^{(n+1)}(v_i) = \sum_{v_j \sim v_i} \Phi_{ij}\left(f^{(n)}\right) f^{(n)}(v_j)$, $\forall v_i \in \mathcal{V}$ with

$$\Phi_{ij}(f) = \frac{(\gamma^*_{w,p} f)(v_i, v_j) f(v_j)}{\sum_{v_j \sim v_i} (\gamma^*_{w,p} f)(v_i, v_j)}, \forall(v_i, v_j) \in \mathcal{E} \ . \tag{7.33}$$

For the isotropic case, $\gamma^*_{w,p} f = \gamma^i_{w,p} f$ and for the anisotropic case, $\gamma^*_{w,p} f = \gamma^a_{w,p} f$. Let $\mathbf{Q}$ be the Markov matrix defined by

$$\begin{cases} \mathbf{Q}(v_i, v_j) = & \Phi_{ij} \text{ if } (v_i, v_j) \in \mathcal{E} \\ \mathbf{Q}(v_i, v_j) = & 0 \text{ if } (v_i, v_j) \notin \mathcal{E}. \end{cases} \tag{7.34}$$

Then, an iteration of the diffusion process (7.28) with $\lambda = 0$ can be written in matrix form as $f^{(n+1)} = \mathbf{Q} f^{(n)} = \mathbf{Q^{(n)}} f^0$, where $\mathbf{Q}$ is a stochastic matrix (nonnegative, symmetric, unit row sum). An equivalent way to look at the power of $\mathbf{Q}$ in the diffusion process is to decompose each value of f on the first eigenvectors of $\mathbf{Q}$, therefore the diffusion process can be interpreted as a filtering in the spectral domain [40].

Moreover, it is also shown in [29] that one can obtain the eigenvalues and eigenvectors of the p-Laplacian by computing the global minima and minimizers of the functional $F^*_{w,p}(f) = \frac{R^*_{w,p}(f)}{\|f\|^p_p}$ according to the generalized Rayleigh-Ritz principle. This states that if $F^*_{w,p}$ has critical point at $\varphi \in \mathbb{R}^{|\mathcal{V}|}$, then φ is a p-eigenvector of $\Delta^*_{w,p}$ and the eigenvalue s is then $s = F^*_{w,p}(\varphi)$. This provides a variational interpretation of the p-Laplacian spectrum and shows that the eigenvectors of the p-Laplacian can be obtained directly in the spatial domain. Therefore, when $\lambda = 0$, the proposed diffusion process is linked to a spectral analysis of f.

7.4.4 Interpolation of Missing Data on Graphs

Many tasks in image processing and computer vision can be formulated as interpolation problems. Interpolating data consists in constructing new values for missing data in coherence with a set of known data. We consider that data are defined on a general domain represented on a graph $\mathcal{G} = (\mathcal{V}, \mathcal{E}, w)$. Let $f^0 : \mathcal{V}_0 \rightarrow \mathbb{R}$ be a function with $\mathcal{V}_0 \subset \mathcal{V}$ be the subset of vertices from the whole graph with known values. Consequently, $\mathcal{V} \setminus \mathcal{V}_0$ corresponds to the vertices with unknown values. The interpolation consists in predicting a function $f : \mathcal{V} \rightarrow \mathbb{R}$ according to f^0. This comes to recover values of f for the vertices of $\mathcal{V} \setminus \mathcal{V}_0$ given values for vertices of $\mathcal{V}_0$. Such a problem can be formulated by considering the following regularization problem on the graph:

$$\min_{f:\mathcal{V} \rightarrow \mathbb{R}} R^*_{w,p}(f) + \lambda(v_i)\|f(v_i) - f^0(v_i)\|_2^2 . \tag{7.35}$$

Since $f^0(v_i)$ is known only for vertices of $\mathcal{V}_0$, the Lagrange parameter is defined as $\lambda : \mathcal{V} \rightarrow \mathbb{R}$:

$$\lambda(v_i) = \begin{cases} \lambda & \text{if } v_i \in \mathcal{V}_0 \\ 0 & \text{otherwise.} \end{cases} \tag{7.36}$$

Then, the isotropic (7.28) and anisotropic (7.32) diffusion processes can be directly used to perform the interpolation. In the sequel, we show how we can use our formulation for image clustering, inpainting and colorization.

7.4.4.1 Semi-Supervised Image Clustering

Numerous automatic segmentation schemes have been proposed in the literature, and they have shown their efficiency. Meanwhile, recent interactive image segmentation approaches have been proposed. They reformulate image segmentation into semi-supervised approaches by label propagation strategies [41, 42, 43]. The recent power watershed formulation [44] is one of the most recent formulation, that provides a common framework for seeded image segmentation. Other applications of these label diffusion methods can be found in [21, 45]. The interpolation regularization-based functional (7.35) can be naturally adapted to address this learning problem for semi-supervised segmentation. We rewrite the problem as [46]: let $\mathcal{V} = \{v_1, \ldots, v_N\}$ be a finite set of data, where each data v_i is a vector of $\mathbb{R}^m$, and let $\mathcal{G} = (\mathcal{V}, \mathcal{E}, w)$ be a weighted graph such that data are connected by an edge of $\mathcal{E}$. The semi-supervised clustering of $\mathcal{V}$ consists in grouping the set $\mathcal{V}$ into k classes with k, the number of classes (known a priori). The set $\mathcal{V}$ is composed of labeled and unlabeled data. The aim is then to estimate the unlabeled data from labeled ones.

Let $\mathcal{C}_l$ be a set of labeled vertices, these latter belonging to the lth class. Let $\mathcal{V}_0 = \bigcup\{\mathcal{C}_l\}_{l=1,\ldots,k}$ be the set of initial *labeled* vertices, and let $\mathcal{V} \setminus \mathcal{V}_0$ be the initial *unlabeled* vertices. Each vertex of $v_i \in \mathcal{V}$ is then described by a vector of labels $\mathbf{f^0(v_i)} = (f_l^0(v_i))_{l=1,\ldots,k}^T$ with

$$f_l^0(v_i) = \begin{cases} +1 & \text{if } v_i \in \mathcal{C}_l \\ 0 & \text{otherwise.} \\ 0 & \forall v_i \in \mathcal{V} \setminus \mathcal{V}_0 \end{cases} \tag{7.37}$$

The semi-supervised clustering problem then reduces to interpolate the labels of the unlabeled vertices $(\mathcal{V}\backslash\mathcal{V}_0)$ from the labeled ones $(\mathcal{V}_0)$. To solve this interpolation problem, we consider the variational problem (7.35). We set $\lambda(v_i) = +\infty$ to avoid the modification of the initial labels.

At the end of the label propagation processes, the class membership probabilities can be estimated, and the final classification can be obtained for a given vertex $v_i \in \mathcal{V}$ by the following formulation:

$$\arg\max_{l \in 1,\ldots,k} \left\{ \frac{f_l^{(t)}(v_i)}{\sum_l f_l^{(t)}(v_i)} \right\} \tag{7.38}$$

7.4.4.2 Geometric and Texture Image Inpainting

The inpainting process consists in filling in the missing parts of an image with the most appropriate data in order to obtain harmonious and hardly detectable reconstructed zones. Recent works on image and video inpainting may fall under two main categories, namely, the geometric algorithms and the exemplar-based ones. The first category employs partial differential equations [47]. The second group of inpainting algorithms is based on the texture synthesis [31]. We can restore the missing data using the interpolation regularization-based functional (7.35) that also enables us to take into account either local and nonlocal information. The main advantage is the unification of the geometric and texture-based techniques [48]. The geometric aspect is taken into account though the graph topology, and the texture aspect is expressed by the graph weights. Therefore, by varying both these geometric and texture aspects, we can recover many approaches proposed so far in literature and extend them to weighted graphs of arbitrary topologies.

Let $\mathcal{V}_0 \subset \mathcal{V}$ be the subset of the vertices corresponding to the original known parts of the image. The inpainting purpose is to interpolate the known values of f^0 from $\mathcal{V}_0$ to $\mathcal{V}\backslash\mathcal{V}_0$. Our method consists in filling the inpainting mask from its outer line $\partial^-\mathcal{V}_0$ to its center recursively in a series of nested outlines. Our proposed regularization process (7.35) is applied iteratively on each node $v_i \in \partial^-\mathcal{V}_0$. Since at one iteration, the inpainting is performed only on vertices of $\partial^-\mathcal{V}_0$, the data-term is not used in (7.35). Moreover, we enforce the computed value of a vertex of $\partial^-\mathcal{V}_0$ to not use the estimated values of other vertices of $\partial^-\mathcal{V}_0$ even if they are neighbors. This is needed to reduce the risk of error propagation. Once convergence is reached for the entire outer line $\partial^-\mathcal{V}_0$, this subset of vertices is added to $\mathcal{V}\backslash\mathcal{V}_0$ and removed from $\mathcal{V}_0$. The weights of the graph are also updated accordingly (see [49] for an explicit modeling of the weight update step). The process is iterated until the set $\mathcal{V}_0$ is empty.

7.4.4.3 Image Colorization

Colorization is the process of adding color to monochrome images. It is usually made by hand by color experts but this process is tedious and very time-consuming. In recent years, several methods have been proposed for colorization [50, 51] that require less intensive manual efforts. These techniques colorize the image based on the user's input color scribbles and are mainly based on a diffusion process. However, most of these diffusion processes only use local pixel interactions that cannot properly describe complex structures expressed by nonlocal interactions. As we already discussed, with our formulations, the integration of nonlocal interactions is easy to obtain. We now explain how to perform image colorization with the proposed framework [52]. From a gray level image $f^l : \mathcal{V} \to \mathbb{R}$, a user provides an image of color scribbles $\mathbf{f^s} : \mathcal{V}_s \subset \mathcal{V} \to \mathbb{R}^3$ that defines a mapping from vertices to a vector of RGB color channels: $\mathbf{f^s}(\mathbf{v_i}) = [f_1^s(v_i), f_2^s(v_i), f_3^s(v_i)]^T$ where $f_j^s : \mathcal{V} \to \mathbb{R}$ is the j-th component of $\mathbf{f^s}(\mathbf{v_i})$. From these functions, one computes $\mathbf{f^c} : \mathcal{V} \to \mathbb{R}^3$ that defines a mapping from the vertices to a vector of chrominances:

$$\begin{cases} \mathbf{f^c}(\mathbf{v_i}) = \left[\dfrac{f_1^s(v_i)}{f^l(v_i)}, \dfrac{f_2^s(v_i)}{f^l(v_i)}, \dfrac{f_3^s(v_i)}{f^l(v_i)} \right]^T, \forall v_i \in \mathcal{V}_s. \\ \mathbf{f^c}(\mathbf{v_i}) = [0,0,0]^T, \forall v_i \notin \mathcal{V}_s. \end{cases} \quad (7.39)$$

We then consider the regularization of $\mathbf{f^c}(\mathbf{v_i})$ by using the interpolation regularization-based functional (7.35) with algorithms (7.28) or (7.32). Since the weights of the graph cannot be computed from f^c since it is unknown, $(\gamma_{w,p}^* f^c)(u,v)$ is replaced by $(\gamma_{w,p}^* f^l)(u,v)$ and weights are computed on the gray level image f^l. Moreover, the data-term is taken into account only on the initial known vertices of $\mathcal{V}_s$ with a small value of $\lambda(v_i)$ to enable the modification of the original color scribble to avoid annoying visual effects. At convergence, final colors are obtained by

$$f^l(v_i) \times \left[f_1^{c^{(t)}}(v_i), f_2^{c^{(t)}}(v_i), f_3^{c^{(t)}}(v_i) \right]^T, \forall v_i \in \mathcal{V}, \text{ where } t \text{ is the iteration number.}$$

7.5 Examples

In this section, we illustrate the abilities of p-Laplace (isotropic and anisotropic) regularization on graphs for simplification and interpolation of any function defined on a finite set of discrete data living in $\mathbb{R}^m$. This is achieved by constructing a weighted graph $\mathcal{G} = (\mathcal{V}, \mathcal{E}, w)$ and by considering the function to be simplified as a function $f^0 : \mathcal{V} \to \mathbb{R}^m$, defined on the vertices of $\mathcal{G}$. The simplification processes operate on vector-valued function with one process per vector component. Component-wise processing of vector-valued data can have serious drawbacks and it is preferable to drive the simplification processes by equivalent geometric attributes, taking the coupling between vector components into account. As in [3], this is done by using the same gradient for all components (a vectorial gradient).

7.5.1 Regularization-Based Simplification

We illustrate the abilities of our approach for the simplification of functions representing images or images' manifold. Several graph topologies are also considered.

Let f^0 be a color image of N pixels, with $f^0 : \mathcal{V} \subset \mathbb{Z}^2 \to \mathbb{R}^3$, which defines a mapping from the vertices to color vectors. Figure 7.3 shows sample results obtained with different parameters of the proposed isotropic and anisotropic regularization: different values of p, different values of λ, different weight functions, and different graph topologies. The PSNR value between the filtering result and the original uncorrupted image are also provided. The first row of Figure 7.3 presents the initial image and the original image corrupted by Gaussian noise of variance $\sigma = 15$. Below the first row, the first column presents results obtained on a 8-adjacency grid graph ($\mathcal{G}_1$) for isotropic and anistropic filtering with $p \in \{2,1\}$ and $\lambda = 0.005$. These results correspond to the classical results that can be obtained with the digital TV and $\mathcal{L}_2$ filters of Chan and Osher [7, 37]. When the weights are not constant, our approach extends these previous works and enables the production of better visual results by using computed edge weights as depicted in the second column of Figure 7.3. Moreover, the use of the curvature with $p = 1$ enables a better preservation of edges and fine structures with a small advantage for the isotropic regularization. Finally, all these advantages of our approach are further stressed once a nonlocal configuration is used. This enhances once again the preservation of edges and the production of flat zones. These results are presented in the last column of Figure 7.3 on a 80-adjancency graph ($\mathcal{G}_4$) with 5×5 patches as feature vectors. This shows how our proposed framework generalizes the NLMeans filter [32].

To further illustrate the excellent denoising capabilities of our approach for repetitive fine structures and texture with the integration of nonlocal interactions within the graph topology and graph weights, we provide results on another image (Figure 7.4) that exhibits high-texture information. Again, this image has been corrupted by Gaussian noise of variance $\sigma = 15$. Results show that local configurations (first column of Figure 7.4) cannot correctly restore values in highly textured areas and over-smooths homogeneous areas. By expanding the adjacency of each vertex and considering patches as feature vectors, a much better preservation of edges and fine details is obtained. The isotropic model being again for this image slightly better than the anisotropic.

If classical image simplification considers grid graphs, superpixels [53] are becoming increasingly popular for use in computer vision applications. Superpixels provide a convenient primitive from which to compute local image features. They capture redundancy in the image and greatly reduce the complexity of subsequent image processing tasks. It is therefore important to have algorithms that can directly process an image represented as a set of super-pixels instead of a set of pixels. Our approach naturally and directly enables such a processing. Given a super-pixel representation of an image, obtained by any over-segmentation method, the super-pixel image is associated with a region adjacency graph. Each region is then represented by the median color value of all the pixel belonging to this region. Let $\mathcal{G} = (\mathcal{V}, \mathcal{E}, w)$ be

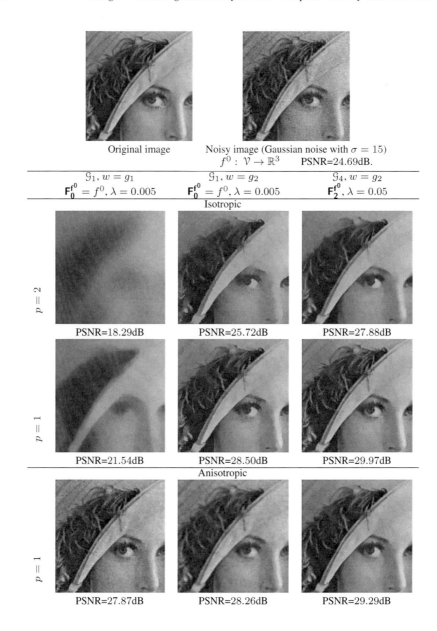

FIGURE 7.3

Isotropic and anisotropic image simplification on a color image with different parameters values of p, λ and different weight functions in local or nonlocal configurations. First row presents the original image. Next rows present simplification results where each row corresponds to a value of p.

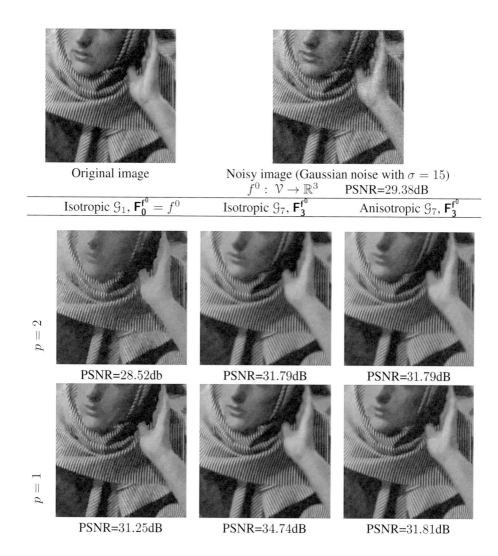

FIGURE 7.4

Isotropic and anisotropic image restoration of a textured image in local and nonlocal configurations with $w = g_2$, $\lambda = 0.005$ and different values of p.

a RAG associated to an image segmentation. Let $f^0 : \mathcal{V} \subset \mathbb{Z}^2 \to \mathbb{R}^3$ be a mapping from the vertices of $\mathcal{G}$ to the median color value of their regions. Then, a simplification of f^0 can be easily achieved by regularizing the function f^0 on $\mathcal{G}$. The advantage over performing the simplification on the grid graph comes with the reduction of the number of vertices: the RAG has much less vertices than the grid graph and the processing is very fast. Figure 7.5 presents such results. The regularization process naturally creates flat zones on the RAG by simplifying adjacent regions and therefore, after the processing, we merge regions that have exactly the same color vector. This enables to prune the RAG (this is assessed by the final number of vertices $|\mathcal{V}|$ in Figure 7.5). As for images, the obtained simplification is visually better with $p = 1$.

By nature, surface meshes have a graph structure, and we can treat them with our approach. Let $\mathcal{V}$ be a set of mesh vertices, and $\mathcal{E}$ be a set of mesh edges. If the input mesh is noisy, we can regularize vertex coordinates or any other function $f^0 : \mathcal{V} \subset \mathbb{R}^3 \to \mathbb{R}^3$ defined on the graph $\mathcal{G} = (\mathcal{V}, \mathcal{E}, w)$. Figure 7.6 presents results for a noisy triangular mesh. The simplification enables us to group similar vertices around high curvature regions while preserving sharp angles with $p = 1$.

To conclude these results for simplification, we consider the USPS data set for which a subset of 100 images from the digits 0 and 1 have been randomly chosen. The selected images are then corrupted independently by Gaussian noise ($\sigma = 40$). Figures 7.7(a) and 7.7(b) present the original and corrupted data sets. To associate a graph to the corresponding data set, a 10 Nearest Neighbor Graph (10-NNG) is constructed from the corrupted data set. This graph is depicted in Figure 7.7(c). Each image represents the whole feature vector we want to regularize, i.e., $f^0 : \mathcal{V} \to \mathbb{R}^{16 \times 16}$. Figure 7.7 shows the filtering results of p-Laplace isotropic and anisotropic regularization for different values of p ($\{2, 1\}$) and λ ($\{0, 1\}$). Each result is composed of, with $n \to \infty$: the filtering result ($f^{(n)}$), the difference between the uncorrupted images and the filtered image ($|f^{(0)} - f^{(n)}|$), the PSNR value between the filtering result and the original uncorrupted image. For $p = 2$ (with any value of λ) the filtering results are approximatively the same. The processing tends to a uniform data set of two mean digits. This type of processing can be interesting for clustering purposes on the filtered data set where the two classes of digits are more easily distinguishable. From the results, the best PSNR values are obtained with $p = 1$ and $\lambda = 1$, both for the isotropic and the anisotropic regularizations. With these parameters, the filtering better tends to better preserve the initial structure of the manifold while suppressing noise. This effect is accentuated with the use of the anisotropic p-Laplace operator. Finally, the filtering process can be considered as an interesting alternative to methods that consider the preimage problem [54]. The latter methods usually consider the manifold after a projection in a spectral domain by manifold learning [40] to solve the preimage problem [55]. With our approach, the initial manifold is recovered without the use of any projection.

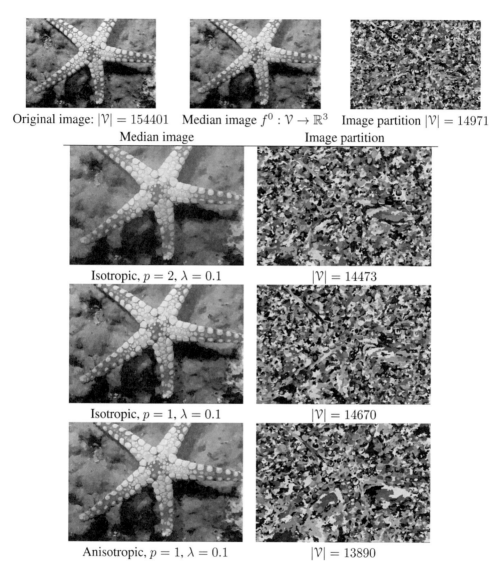

Original image: $|\mathcal{V}| = 154401$ Median image $f^0 : \mathcal{V} \to \mathbb{R}^3$ Image partition $|\mathcal{V}| = 14971$

Median image   Image partition

Isotropic, $p = 2, \lambda = 0.1$ $|\mathcal{V}| = 14473$

Isotropic, $p = 1, \lambda = 0.1$ $|\mathcal{V}| = 14670$

Anisotropic, $p = 1, \lambda = 0.1$ $|\mathcal{V}| = 13890$

FIGURE 7.5

Image simplification illustration. From an original image, one computes a pre-segmentation and the associated color median image (images in first row). Regularization and decimation are applied on the RAG of the presegmentation and simplified region maps, and color median images are obtained for different values of λ and p with $w = g_3$.

Original Mesh $f^0 : \mathcal{V} \to \mathbb{R}^3$ Isotropic, $p = 2$ Isotropic, $p = 1$,   Anisotropic, $p = 1$

FIGURE 7.6
Triangular mesh simplification illustration. The graph is obtained directly from the mesh and one has $\mathbf{F}f_0^{f^0} = f^0$, $\lambda = 0.1$, $w = g_3$.

7.5.2   Regularization-Based Interpolation

We illustrate now the abilities of our approach for the interpolation of missing values. We first consider the case of image inpainting. Figure 7.8 presents such inpainting results. The two first rows of Figure 7.8 present results for local inpainting on a 8-adjacency weighted graph (i.e., the diffusion approach) and for nonlocal inpainting on a highly connected graph ($\mathcal{G}_{15}$) with 13×13 patches (i.e., the texture replication approach). While the diffusion approach cannot correctly restore the image and produces blurred regions, the texture approach enables an almost perfect interpolation. This simple example shows the advantage of our formulation that combines both geometric and texture inpainting within a single variational formulation. The last row of Figure 7.8 shows that the approach works also for very large areas when the texture is repetitive.

Second, we consider the case of image colorization. Figure 7.9 presents such a result. An initial image is considered with input color scribbles provided by a user. Both local and nonlocal colorization results are satisfactory with an advantage towards nonlocal. Indeed, on the level of the edges, a color blending effect is observed with the local colorization whereas this effect is strongly reduced with nonlocal colorization. This can be seen on the boy's mouth, hairs and arm. Again, the advantage of our approach resides in the fact that the colorization algorithm is the same for local and nonlocal configurations. Only the graph topology and graph weights are different.

Finally, we consider the case of semi-supervised image segmentation. Figure 7.10 presents results in various configurations. The first column shows the original image, the second column shows the user input labels, and the last column presents the segmentation result. The first row presents a simple example of a color image corrupted

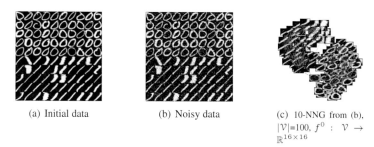

(a) Initial data	(b) Noisy data	(c) 10-NNG from (b), $\|\mathcal{V}\|$=100, $f^0 : \mathcal{V} \to \mathbb{R}^{16 \times 16}$

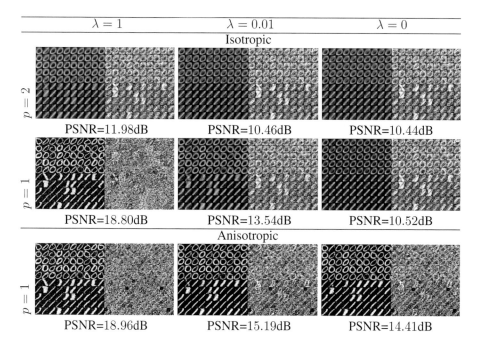

FIGURE 7.7

Isotropic (two first rows) and anisotropic (two last rows) filtering on an image manifold (USPS). Results are obtained with different values of parameters p and λ. Each result shows the filtering result (final f), method noise image ($|f^{(0)} - f^{(n)}|$) and a PSNR value. (a): original data. (b): noisy data (Gaussian noise with $\sigma = 40$), (c): 10-NNG constructed from (b).

by Gaussian noise and composed of two areas. Only two labels are provided by the user, and with a pure local approach on a 4-adjacency grid graph, a perfect result is obtained. Local configurations will provide good results when the objects to be extracted are relatively homogeneous. This is assessed by the results presented in the second row of Figure 7.10. Once the image to be segmented is much more textured, a nonlocal configuration is needed to obtain an accurate segmentation. This is shown

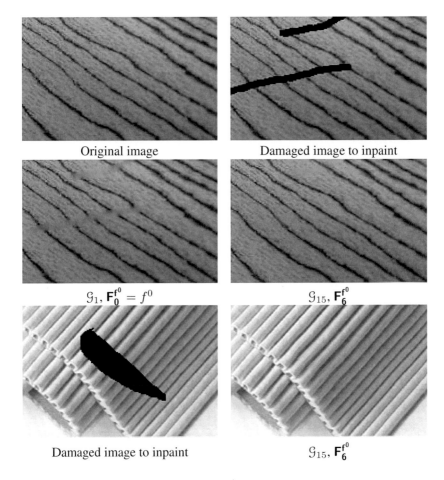

FIGURE 7.8

Isotropic image inpainting with $p = 2$, $w = g_2$ in local and nonlocal configurations.

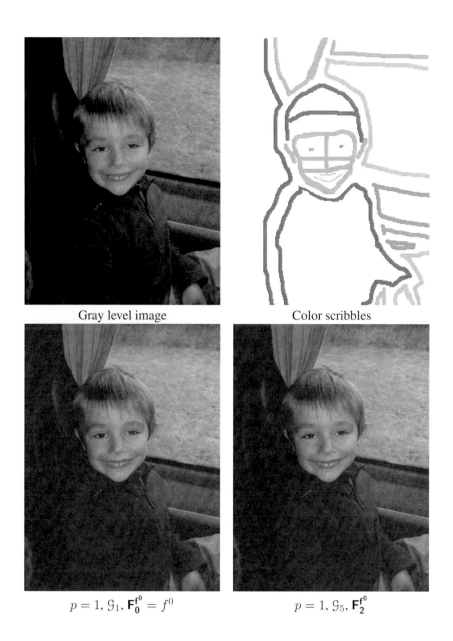

Gray level image   Color scribbles

$p = 1, \mathcal{G}_1, \mathbf{F}_0^{f^0} = f^0$ $p = 1, \mathcal{G}_5, \mathbf{F}_2^{f^0}$

FIGURE 7.9

Isotropic colorization of a grayscale image with $w = g_2$ and $\lambda = 0.01$ in local and nonlocal configurations.

Original Image   User label input   Segmentation result

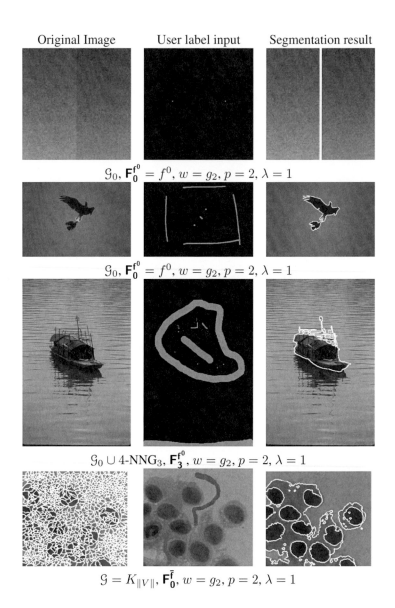

$$\mathcal{G}_0, \mathbf{F}_0^{f^0} = f^0, w = g_2, p = 2, \lambda = 1$$

$$\mathcal{G}_0, \mathbf{F}_0^{f^0} = f^0, w = g_2, p = 2, \lambda = 1$$

$$\mathcal{G}_0 \cup 4\text{-NNG}_3, \mathbf{F}_3^{f^0}, w = g_2, p = 2, \lambda = 1$$

$$\mathcal{G} = K_{\|V\|}, \mathbf{F}_0^{\bar{f}}, w = g_2, p = 2, \lambda = 1$$

FIGURE 7.10
Semi-supervised segmentation.

in the third row of Figure 7.10 where the 8-adjacency grid graph is augmented with edges from a 4-NNG$_3$ to capture nonlocal interactions with 7×7 patches as feature vectors. The last row of Figure 7.10 shows an example of label diffusion on a graph obtained from a super-pixel segmentation of an image. In this case the graph is the complete graph $K_{|\mathcal{V}|}$ of the vertices of the RAG corresponding to the super-pixel segmentation. Using the complete graph is not feasible on the whole set of pixels but is computationally efficient on a super-pixel segmentation since the number of vertices is drastically reduced (for this case the reduction factor is of 97%). Using the complete graph for label diffusion has also two strong advantages: the number of input labels is reduced (only one label per type of object to be extracted) and nonadjacent areas can be segmented with the same label even if only one label is provided (this is due to the long range connections induced by the complete graph).

7.6 Concluding Remarks

In this chapter we have shown the interest of partial difference equations on graphs for image processing and analysis. As explained in the introduction, PdEs mimic PDEs in domains having a graph structure and enable us to formulate variational problems and associated solutions on graphs of arbitrary topologies. One strong advantage towards using PdE on graphs is the ability to process any discrete data set with the advantage of naturally incorporating nonlocal interactions. We have proposed to use PdE on graphs to define a framework for p-Laplacian regularization on graphs. This framework can be considered as a discrete analogue of continuous regularization models involving PDEs while generalizing and extending them. With various experiments, we have shown the very general situations where this framework can be used for denoising, simplification, segmentation, and interpolation. In future works, other PDEs will be considered, as we have already done by adapting continuous mathematical morphology on graphs with PdEs [15].

Bibliography

[1] D. Hammond, P. Vandergheynst, and R. Gribonval, "Wavelets on graphs via spectral graph theory," *Appl. Comp. Harmonic Analysis*, vol. 30, no. 2, pp. 129 – 150, 2011.

[2] L. Grady and C. Alvino, "The piecewise smooth Mumford-Shah functional on an arbitrary graph," *IEEE Trans. on Image Processing*, vol. 18, no. 11, pp. 2547–2561, Nov. 2009.

[3] A. Elmoataz, O. Lezoray, and S. Bougleux, "Nonlocal discrete regularization on weighted graphs: a framework for image and manifold processing," *IEEE Transactions on Image Processing*, vol. 17, no. 7, pp. 1047–1060, 2008.

[4] L. Alvarez, F. Guichard, P.-L. Lions, and J.-M. Morel, "Axioms and fundamental equations of image processing," *Archive for Rational Mechanics and Analysis*, vol. 123, no. 3, pp. 199–257, 1993.

[5] T. Chan and J. Shen, *Image Processing and Analysis - Variational, PDE, Wavelets, and Stochastic Methods.* SIAM, 2005.

[6] Y.-H. R. Tsai and S. Osher, "Total variation and level set methods in image science," *Acta Numerica*, vol. 14, pp. 509–573, mai 2005.

[7] S. Osher and J. Shen, "Digitized PDE method for data restoration," in *In Analytical-Computational methods in Applied Mathematics*, E. G. A. Anastassiou, Ed. Chapman&Hall/CRC, 2000, pp. 751–771.

[8] J. Neuberger, "Nonlinear elliptic partial difference equations on graphs," *Experiment. Math.*, vol. 15, no. 1, pp. 91–107, 2006.

[9] R. Courant, K. Friedrichs, and H. Lewy, "On the partial difference equations of mathematical physics," *Math. Ann.*, vol. 100, pp. 32–74, 1928.

[10] L. Grady and J. R. Polimeni, *Discrete Calculus - Applied Analysis on Graphs for Computational Science.* Springer, 2010.

[11] A. Bensoussan and J.-L. Menaldi, "Difference equations on weighted graphs," *Journal of Convex Analysis*, vol. 12, no. 1, pp. 13–44, 2003.

[12] J. Friedman and J. Tillich, "Wave equations on graphs and the edge-based Laplacian," *Pacific Journal of Mathematics*, vol. 216, no. 2, pp. 229–266, 2004.

[13] F. Chung and S. T. Yau, "Discrete green's functions," *Journal of Combinatorial Theory, Series A*, vol. 91, no. 1-2, pp. 191–214, 2000.

[14] R. Hidalgo and M. Godoy Molina, "Navier–stokes equations on weighted graphs," *Complex Analysis and Operator Theory*, vol. 4, pp. 525–540, 2010.

[15] V. Ta, A. Elmoataz, and O. Lezoray, "Partial difference equations on graphs for mathematical morphology operators over images and manifolds," in *IEEE International Conference on Image Processing*, 2008, pp. 801–804.

[16] V. Ta, A. Elmoataz, and O. Lézoray, "Nonlocal pdes-based morphology on weighted graphs for image and data processing," *IEEE Transactions on Image Processing*, vol. 20, no. 6, pp. 1504–1516, June 2011.

[17] V. Ta, A. Elmoataz, and O. Lezoray, "Adaptation of eikonal equation over weighted graphs," in *International Conference on Scale Space Methods and Variational Methods in Computer Vision (SSVM)*, vol. LNCS 5567, 2009, pp. 187–199.

[18] A. Elmoataz, O. Lezoray, S. Bougleux, and V. Ta, "Unifying local and nonlocal processing with partial difference operators on weighted graphs," in *International Workshop on Local and Non-Local Approximation in Image Processing (LNLA)*, 2008, pp. 11–26.

[19] S. Bougleux, A. Elmoataz, and M. Melkemi, "Discrete regularization on weighted graphs for image and mesh filtering," in *Scale Space and Variational Methods in Computer Vision*, ser. LNCS, vol. 4485, 2007, pp. 128–139.

[20] ——, "Local and nonlocal discrete regularization on weighted graphs for image and mesh processing," *International Journal of Computer Vision*, vol. 84, no. 2, pp. 220–236, 2009.

[21] D. Zhou and B. Schölkopf, "Regularization on discrete spaces," in *DAGM Symposium*, ser. LNCS, vol. 3663. Springer-Verlag, 2005, pp. 361–368.

[22] M. Hein, J.-Y. Audibert, and U. von Luxburg, "Graph Laplacians and their convergence on random neighborhood graphs," *Journal of Machine Learning Research*, vol. 8, pp. 1325–1368, 2007.

[23] M. Hein and M. Maier, "Manifold denoising," in *NIPS*, 2006, pp. 561–568.

[24] M. Hein, J.-Y. Audibert, and U. Von Luxburg, "From graphs to manifolds - weak and strong pointwise consistency of graph Laplacians," in *COLT*, 2005, pp. 470–485.

[25] F. R. Chung, "Spectral graph theory," *CBMS Regional Conference Series in Mathematics*, vol. 92, pp. 1–212, 1997.

[26] V.-T. Ta, S. Bougleux, A. Elmoataz, and O. Lezoray, "Nonlocal anisotropic discrete regularization for image, data filtering and clustering," GREYC CNRS UMR 6072 – Université de Caen Basse-Normandie - ENSICAEN," HAL Technical Report, 2007.

[27] M. Belkin and P. Niyogi, "Laplacian eigenmaps for dimensionality reduction and data representation." *Neural Computation*, vol. 15, no. 6, pp. 1373–1396, 2003.

[28] O. Lezoray, V. Ta, and A. Elmoataz, "Manifold and data filtering on graphs," in *International Symposium on Methodologies for Intelligent Systems, International Workshop on Topological Learning*, 2009, pp. 19–28.

[29] T. Bühler and M. Hein, "Spectral clustering based on the graph p-laplacian," in *International Conference on Machine Learning*, 2009, pp. 81–88.

[30] J. O'Rourke and G. Toussaint, "Pattern recognition," in *Handbook of discrete and computational geometry*, J. Goodman and J. O'Rourke, Eds. Chapman & Hall/CRC, New York, 2004, ch. 51, pp. 1135–1162.

[31] A. A. Efros and T. K. Leung, "Texture synthesis by non-parametric sampling," in *International Conference on Computer Vision*, vol. 2, 1999, pp. 1033–1038.

[32] A. Buades, B. Coll, , and J.-M. Morel, "Non-local image and movie denoising," *International Journal of Computer Vision*, vol. 76, no. 2, pp. 123–139, 2008.

[33] G. Gilboa and S. Osher, "Non-local linear image regularization and supervised segmentation," *Multiscale Modeling & Simulation*, vol. 6, no. 2, pp. 595–630, 2007.

[34] A. Chambolle, "An algorithm for total variation minimization and applications," *Journal of Mathematical Imaging and Vision*, vol. 20, no. 1-2, pp. 89–97, 2004.

[35] A. Chambolle and T. Pock, "A first-order primal-dual algorithm for convex problems with applications to imaging," *Journal of Mathematical Imaging and Vision*, vol. 40, no. 1, pp. 120–145, 2011.

[36] C. Tomasi and R. Manduchi, "Bilateral filtering for gray and color images," in *International Conference on Computer Vision*. IEEE Computer Society, 1998, pp. 839–846.

[37] T. Chan, S. Osher, and J. Shen, "The digital TV filter and nonlinear denoising," *IEEE Transactions on Image Processing*, vol. 10, no. 2, pp. 231–241, 2001.

[38] G. Gilboa and S. Osher, "Nonlocal operators with applications to image processing," UCLA, Tech. Rep. CAM Report 07–23, July 2007.

[39] ——, "Nonlocal operators with applications to image processing," *Multiscale Modeling & Simulation*, vol. 7, no. 3, pp. 1005–1028, 2008.

[40] R. Coifman, S. Lafon, A. Lee, M. Maggioni, B. Nadler, F. Warner, and S. Zucker, "Geometric diffusions as a tool for harmonic analysis and structure definition of data," *Proc. of the National Academy of Sciences*, vol. 102, no. 21, pp. 7426–7431, 2005.

[41] F. Wang, J. Wang, C. Zhang, and H. C. Shen, "Semi-Supervised Classification Using Linear Neighborhood Propagation," *IEEE Computer Society Conference on Computer Vision and Pattern Recognition - Volume 1 (CVPR'06)*, vol. 1, pp. 160–167, 2006.

[42] L. Grady, "Random walks for image segmentation," *IEEE Transactions on Pattern Analysis and Machine Intelligencen*, vol. 28, no. 11, pp. 1768–1783, 2006.

[43] A. K. Sinop and L. Grady, "A Seeded Image Segmentation Framework Unifying Graph Cuts And Random Walker Which Yields A New Algorithm," in *International Conference on Computer Vision*, 2007, pp. 1–8.

[44] C. Couprie, L. Grady, L. Najman, and H. Talbot, "Power watersheds: A new image segmentation framework extending graph cuts, random walker and optimal spanning forest," in *International Conference on Computer Vision*, Sept. 2009, pp. 731–738.

[45] M. Belkin, P. Niyogi, V. Sindhwani, and P. Bartlett, "Manifold Regularization: A Geometric Framework for Learning from Labeled and Unlabeled Examples," *Journal of Machine Learning Research*, vol. 7, pp. 2399–2434, 2006.

[46] V.-T. Ta, O. Lezoray, A. Elmoataz, and S. Schüpp, "Graph-based tools for microscopic cellular image segmentation," *Pattern Recognition*, vol. 42, no. 6, pp. 1113–1125, 2009.

[47] M. Bertalmío, G. Sapiro, V. Caselles, and C. Ballester, "Image inpainting," in *SIGGRAPH*, 2000, pp. 417–424.

[48] M. Ghoniem, Y. Chahir, and A. Elmoataz, "Geometric and texture inpainting based on discrete regularization on graphs," in *ICIP*, 2009, pp. 1349–1352.

[49] G. Facciolo, P. Arias, V. Caselles, and G. Sapiro, "Exemplar-based interpolation of sparsely sampled images," in *EMMCVPR*, 2009, pp. 331–344.

[50] A. Levin, D. Lischinski, and Y. Weiss, "Colorization using optimization," *ACM Transactions on Graphics*, vol. 23, no. 3, pp. 689–694, 2004.

[51] L. Yatziv and G. Sapiro, "Fast image and video colorization using chrominance blending," *IEEE Transactions on Image Processing*, vol. 15, no. 5, pp. 1120–1129, 2006.

[52] O. Lezoray, A. Elmoataz, and V. Ta, "Nonlocal graph regularization for image colorization," in *International Conference on Pattern Recognition (ICPR)*, 2008.

[53] A. Levinshtein, A. Stere, K. N. Kutulakos, D. J. Fleet, S. J. Dickinson, and K. Siddiqi, "Turbopixels: Fast superpixels using geometric flows," *IEEE Transactions on Pattern Analysis and Machine Intelligence*, vol. 31, pp. 2290–2297, 2009.

[54] J. T. Kwok and I. W. Tsang, "The pre-image problem in kernel methods," in *International Conference on Machine Learning*, 2003, pp. 408–415.

[55] N. Thorstensen, F. Segonne, and R. Keriven, "Preimage as karcher mean using diffusion maps : Application to shape and image denoising," in *Proceedings of International Conference on Scale Space and Variational Methods in Computer Vision*, ser. LNCS, Springer, Ed., vol. 5567, 2009, pp. 721–732.

8

Image Denoising with Nonlocal Spectral Graph Wavelets

David K. Hammond

NeuroInformatics Center
University of Oregon
Eugene, OR, USA
Email: hammond@uoregon.edu

Laurent Jacques

ICTEAM Institute, ELEN Department
Université catholique de Louvain
Louvain-la-Neuve, Belgium
Email: laurent.jacques@uclouvain.be

Pierre Vandergheynst

Institute of Electrical Engineering
Ecole Polytechnique Féderale de Lausanne
Lausanne, Switzerland
Email: pierre.vandergheynst@epfl.ch

CONTENTS

8.1	Introduction	208
	8.1.1 Related Work	209
8.2	Spectral Graph Wavelet Transform	210
	8.2.1 Fast SGWT via Chebyshev Polynomial Approximation	212
	8.2.2 SGWT Inverse	215
	8.2.3 SGWT Design Details	215
8.3	Nonlocal Image Graph	216
	8.3.1 Acceleration of Computation with Weeds	217
8.4	Hybrid Local/Nonlocal Image Graph	220
	8.4.1 Oriented Local Connectivity	221
	8.4.2 Unions of Hybrid Local/Non-Local SGWT Frames	223
	8.4.3 Image Graph Wavelets	224
8.5	Scaled Laplacian Model	225
8.6	Applications to Image Denoising	226
	8.6.1 Construction of Nonlocal Image Graph from Noisy Image	227
	8.6.2 Scaled Laplacian Thresholding	228
	8.6.3 Weighted ℓ_1 Minimization	229
	8.6.4 Denoising results	230

8.7   Conclusions 233
8.8   Acknowledgments 234
      Bibliography 234

8.1   Introduction

Effectively modeling the structure of image content is a major theme in image processing. A long and successful tradition of research in image modeling involves applying some transform to the image, and then describing the behavior of the resulting coefficients. In particular, the use of localized, oriented multiscale wavelet transforms and a whole host of variations has been very widespread. A key component of the success of wavelet methods is that images typically contain highly localized features (such as edges) interspersed with relatively smooth regions, resulting in sparse behavior of wavelet coefficients. This type of local regularity in image content is well described by localized wavelets.

Another important type of regularity present in images is self-similarity. Many natural images have similar localized patterns present in spatially distant regions of the image. Self-similarity also occurs in images of man-made structures with repeating features, such as an image containing a brick wall or a building facade with repeated similar windows.

For the problem of restoring images corrupted by noise, it is reasonable that knowledge of which image regions are similar should aid image recovery. Intuitively, averaging over image regions which have similar underlying structure should effect reduction in noise while preserving desired image content. In this work, we describe a novel algorithm for image denoising which seeks to capture nonlocal image regularity while leveraging established transform-based techniques. This is achieved through assuming sparsity of image coefficients under a transform that is constructed to respect the nonlocal structure of the image.

Our approach is based on explicitly describing the image self-similarity by constructing a nonlocal image graph. This is a weighted graph where the vertices are the original image pixels, and the edge weights represent the strength of similarity between image patches. We use a newly described method, the Spectral Graph Wavelet Transform (SGWT), for constructing wavelet transforms on arbitrary weighted graphs [1]. Using the SGWT construction with the nonlocal image graph gives the nonlocal graph wavelet transform. We also explore a class of hybrid weighted graphs which smoothly mix the nonlocal image graph with local connectivity structure. With the SGWT, this produces a hybrid local/nonlocal wavelet transform.

We examine two methods for image denoising, the scaled Laplacian method based on a simple thresholding rule, and a method based on ℓ_1 minimization. The former method is based on modeling the coefficients with a Laplacian distribution, after a simple rescaling needed to account for inhomogeneity in the norms of the nonlocal graph wavelets. With the scaled Laplacian approach, the coefficients are thresholded and the transform is then inverted to give the denoised image. Even though this method operates only a simple operation on each coefficient independently, it gives denoising performance comparable to wavelet methods using much more complicated modeling of the dependencies between coefficients.

The closely related ℓ_1 minimization based algorithm is based on the same underlying nonlocal graph wavelet transform. Rather than treating each coefficient separately, it proceeds by minimizing a single convex functional containing a quadratic data fidelity term and a weighted ℓ_1 prior penalty for the coefficients. We compute the minimum of this functional using an iterative forward-backward splitting procedure. We find the ℓ_1 procedure to give improvements over the scaled Laplacian method in perceived image quality, at the expense of additional computational complexity.

8.1.1 Related Work

The idea of exploiting redundancy present in images due to self-similarity has a long history. A large part of the literature on statistical image modeling concerns describing the strongly related scale-invariance properties of natural images [2, 3, 4]. Quite recently, a number of image processing and denoising methods have sought to exploit image self-similarity through patch-based methods. Of particular relevance for this work is the nonlocal means algorithm for image denoising originally introduced by Buades et al. [5].

Many extensions and variations of the original nonlocal means have been been studied [6, 7, 8]. The basic nonlocal means algorithm proceeds by first measuring the similarity between pairs of noisy image patches by computing the ℓ_2 norms of differences of patches. These are then used to compute weights, such that the weight is high (close to unity) for two similar patches and close to zero for two dissimilar patches. Finally, each noisy pixel is replaced by a weighted average of all other pixels. If a particular patch is highly similar to many other patches in the image, then it will be replaced by a structure that is averaged over a large number of regions, resulting in reduction of noise. The weights used in the nonlocal means algorithm are essentially the same as the edge weights for the nonlocal image graph used in the current work.

Another approach to using patch similarity for denoising is the BM3D collaborative filtering algorithm introduced by Dabov et al. [9]. In this approach, similar patches of the noisy image are stacked vertically to form a 3-D data volume. A separable 3-D transform is applied, then the stacked volumes are denoised by thresholding. The power of this method comes from the interaction of wavelet thresholding with the redundancy presented across the stacks of similar image patches, resulting in an implicit averaging of similar image structure across the patches.

The spectral graph wavelet transform used in this work relies on tools from spectral graph theory, specifically the use of eigenvectors of the graph Laplacian operator. Several authors have employed the graph Laplacian for image processing before. Zhang and Hancock studied image smoothing using the heat kernel corresponding to the graph Laplacian, using a graph whose edge weights depended on the difference of local neighboring windows [10]. Szlam et al. investigated similar heat kernel based smoothing in a more general context, including image denoising examples using nonlocal image graphs more similar to the current work [11]. Peyré used nonlocal graph heat diffusion for denoising, as well as studying thresholding in the basis of eigenvectors of the nonlocal graph Laplacian [12]. Elmoataz et al. [13] studied regularization on arbitrary graph domains, including image processing applications employing nonlocal graphs, with a variational framework employing the p-Laplacian, a nonlinear (for general p) operator reducing to the graph Laplacian for $p = 2$.

8.2 Spectral Graph Wavelet Transform

The spectral graph wavelet transform (SGWT) is a framework for constructing multi-scale wavelet transforms defined on the vertices of an arbitrary finite weighted graph. The SGWT is described in detail in [1]; a brief description of it will be given here.

Weighted graphs provide an extremely flexible way of describing a large number of data domains. For example, vertices in a graph may correspond to individual people in a social network or cities connected by a transportation network of roads. Wavelet transforms on such weighted graph structures thus have the potential for broad applicability to many problems beyond the image denoising application considered in this work. Other authors have introduced methods for wavelet-like structures on graphs. These include vertex-domain methods based on n-hop distance [14], lifting schemes [15], and methods restricted to trees [16]. Gavish et al. have recently described an orthogonal wavelet construction on hierarchical trees, and demonstrated its use for estimating functions on arbitrary graphs by first applying hierarchical clustering of nodes [17]. The above approaches differ from the SGWT primarily in that they are based in the vertex domain, rather than constructed using spectral graph theory. Other constructions using spectral graph theory have been developed, notably the "diffusion wavelets" of Maggioni and Coifman [18]; the primary difference of the SGWT from their approach is in the simplicity of the SGWT construction, and that the SGWT produces an overcomplete wavelet frame rather than an orthogonal basis.

We consider an undirected weighted graph $\mathcal{G}$ with N vertices. Any scalar valued function $\mathbf{f}$ defined on the vertices of the graph can be associated with a vector in $\mathbb{R}^N$, where the coordinate f_i is the value of $\mathbf{f}$ on the ith vertex. For any real scale parameter $t > 0$, the SGWT will define the wavelets $\psi_{t,n} \in \mathbb{R}^N$ which are centered on each of the vertices n. Key properties of these graph wavelets are that they are zero mean, they are localized around the central vertex n, and they have decreasing support as t decreases.

The SGWT construction is based upon the graph Laplacian operator $\mathbf{L}$, defined as follows. Let $\mathbf{W} \in \mathbb{R}^{N \times N}$ be the symmetric adjacency matrix for $\mathcal{G}$, so that $w_{i,j} \geq 0$ is the weight of the edge between vertices i and j. The degree of vertex i is defined as the sum of the weights of all edges incident to i, i.e., $d_i = \sum_j w_{i,j}$. Set the diagonal degree matrix $\mathbf{D}$ to have $\mathbf{D}_{i,i} = d_i$. The graph Laplacian operator is then $\mathbf{L} = \mathbf{D} - \mathbf{W}$.

The operator $\mathbf{L}$ should be viewed as the graph analogue of the standard Laplacian operator $-\Delta$ for flat Euclidean domains. In particular, the eigenvectors of $\mathbf{L}$ are analogous to the Fourier basis elements $e^{i\mathbf{k}\cdot\mathbf{x}}$, and may be used to define the graph Fourier transform. As $\mathbf{L}$ is a real symmetric matrix, it has a complete set of orthonormal eigenvectors $\chi_\ell \in \mathbb{R}^N$ for $\ell = 0, \cdots, N-1$ with associated real eigenvalues λ_ℓ. We order these in nondecreasing order, so that $\lambda_0 \leq \lambda_1 \cdots \leq \lambda_{N-1}$. For the graph Laplacian it can be shown the eigenvalues are nonnegative, and that $\lambda_0 = 0$. Now for any function $\mathbf{f} \in \mathbb{R}^N$, the graph Fourier transform is defined by

$$\hat{f}(\ell) = \langle \chi_\ell, \mathbf{f} \rangle = \sum_n \chi_\ell^*(n)\mathbf{f}(n) . \tag{8.1}$$

We interpret $\hat{f}(\ell)$ as the ℓth graph Fourier coefficient of $\mathbf{f}$. As the χ_ℓ's are orthonormal, it is straightforward to see that $\mathbf{f}$ may be recovered from its Fourier transform by $\mathbf{f} = \sum_\ell \hat{f}(\ell)\chi_\ell$.

The use of the graph Fourier transform for defining the SGWT may be motivated by examining the classical continuous wavelet transform on the real line in the Fourier domain. These wavelets are generated by taking a "mother" wavelet $\psi(x)$, and then applying scaling and translation to obtain the wavelet $\psi_{t,a} = \frac{1}{t}\psi((x-a)/t)$ [19]. For a given function f, the wavelet coefficients at scale t and location a are then given by inner products, i.e., $W_f(t,a) = \int \frac{1}{t}\psi^*((x-a)/t)f(x)dx$. For fixed scale t, this integral may be rewritten as a convolution evaluated at the point a. Letting $\bar{\psi}_t(x) = \frac{1}{t}\psi^*(-x/t)$, we see $W_f(t,a) = (\bar{\psi}_t \star f)(a)$. Now consider taking the Fourier transform over the a variable. On the real line, we may employ the convolution theorem, and this becomes

$$\widehat{W_f}(t,\omega) = \hat{\bar{\psi}}_t(\omega)\hat{f}(\omega) = \hat{\psi}^*(t\omega)\hat{f}(\omega) . \tag{8.2}$$

Thus, the operation of mapping f to its wavelet coefficients $W_f(t,a)$ consists of taking the Fourier transform of f, multiplying by $\hat{\psi}^*(t\omega)$, and applying the inverse Fourier transform. The key point here is that the operation of scaling by t has been transferred from the original domain to a dilation of the function $\hat{\psi}^*(t\omega)$ in the Fourier domain. This is significant, as a major problem in adapting the classical wavelet transform to graph domains is the inherent difficulty of defining scaling of a function on an irregular graph.

The SGWT operator for fixed scale t is defined in analogy with the previous discussion. Its specific form is fixed by the choice of a nonnegative wavelet kernel

$g(x)$, analogous to the Fourier transformed wavelet $\hat{\psi}^*$. This kernel g should behave as a band-pass function, specifically we require $g(0) = 0$ and $\lim_{x \to \infty} g(x) = 0$. For any scale $t > 0$, the SGWT operator $T_g^t : \mathbb{R}^N \to \mathbb{R}^N$ is defined by $T_g^t(\mathbf{f}) = g(t\mathbf{L})\mathbf{f}$. The result of applying the real-valued function $g(t\cdot)$ to the operator $\mathbf{L}$ yields a linear operator, which for symmetric $\mathbf{L}$ may be defined by its action on the eigenvectors. Specifically, we have $g(t\mathbf{L})(\chi_\ell) \equiv g(t\lambda_\ell)\chi_\ell$. By linearity, $g(t\mathbf{L})$ applied to arbitrary $\mathbf{f} \in \mathbb{R}^N$ can be expressed in terms of the graph Fourier coefficients, as

$$T_g^t \mathbf{f} \equiv g(t\mathbf{L})\mathbf{f} = \sum_\ell g(t\lambda_\ell)\hat{f}(\ell)\chi_\ell \,. \tag{8.3}$$

Examining this expression, we see that applying T_g^t to $\mathbf{f}$ is equivalent to taking the graph Fourier transform of $\mathbf{f}$, multiplying by the function $g(t\lambda)$, and then inverting the transform. This is in exact analogy to the Fourier domain description of the classical wavelet transform implied by equation (8.2).

Equation (8.3) defines the N spectral graph wavelet coefficients of $\mathbf{f}$ at scale t. We denote these coefficients by $W_{\mathbf{f}}(t, n)$, i.e., so that $W_{\mathbf{f}}(t, n) = (T_g^t \mathbf{f})_n$. The wavelets may be recovered by applying this operator to a delta impulse. If we set $\boldsymbol{\delta}_n \in \mathbb{R}^N$ to have value 1 at vertex n and zero's elsewhere, then the wavelet centered at vertex n is given by $\psi_{t,n} = T_g^t \delta_n$. It is straightforward to verify that this is consistent both with the previous definition of the coefficients and the desire for the coefficients to be generated by inner products, i.e., that $W_{\mathbf{f}}(t, n) = \langle \psi_{t,n}, \mathbf{f} \rangle$. In Figure 8.1 we show an example graph with 300 vertices randomly sampled from a smooth manifold, and edges given by connecting all vertices (with unit edge weight) whose distance is below a chosen threshold. We show a single scaling function, and wavelets at two different scales.

The overall stability of the SGWT is improved by including scaling functions to represent the low frequency content of the signal. This is done by introducing a scaling function kernel h, a low-pass function satisfying $h(0) > 0$, and $h(x) \to 0$ as $x \to \infty$. As with the wavelets, we define the scaling function operator $T_h = h(\mathbf{L})$, scaling functions $\phi_n = T_h \delta_n$, and the scaling function coefficients of f to be given by $T_h \mathbf{f}$.

The above theory describes the SGWT for continuous scales t. In practice, we will discretize t to a finite number of scales $t_1 > \cdots > t_J > 0$, as detailed in Section 8.2.3. Once these are fixed, we shall often abuse notation by referring to the wavelet $\psi_{j,n}$ (or coefficient $W_{\mathbf{f}}(j, n)$) as shorthand for $\psi_{t_j,n}$ (or $W_{\mathbf{f}}(t_j, n)$).

We may then consider the overall transform operator $\mathbf{T} : \mathbb{R}^N \to \mathbb{R}^{(J+1)N}$ formed by concatenating the scaling function coefficients and each of the J sets of wavelet coefficients. In ([1], Theorem 5.6) this operator was shown to be a frame, with frame bounds A and B that can be estimated as follows. First define $G(\lambda) = h^2(\lambda) + \sum_{j=1}^J g^2(t_j\lambda)$. Then for any $\mathbf{f} \in \mathbb{R}^N$, the inequalities $A\|\mathbf{f}\|^2 \leq \|\mathbf{T}\mathbf{f}\|^2 \leq B\|\mathbf{f}\|^2$ hold, where $A = \min_{\lambda \in [0, \lambda_{N-1}]} G(\lambda)$, and $B = \max_{\lambda \in [0, \lambda_{N-1}]} G(\lambda)$.

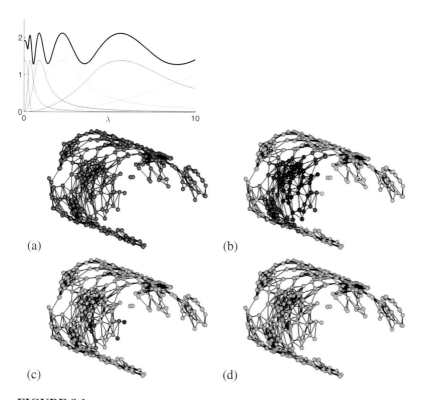

FIGURE 8.1
Left: Scaling function $h(\lambda)$ (blue curve), wavelet generating kernels $g(t_j\lambda)$, and sum of squares G (black curve), for $J = 5$ scales, $\lambda_{\max} = 10$, $K_{\mathrm{lp}} = 20$. Details in Section 8.2.3. Right: Example graph showing specified center vertex (a), scaling function (b) and wavelets at two scales (c,d). Figures adapted from [1].

8.2.1   Fast SGWT via Chebyshev Polynomial Approximation

Direct computation of the SGWT by applying Equation (8.3) requires explicit computation of the entire set of eigenvectors and eigenvalues of $\mathbf{L}$. This approach scales extremely poorly for large graphs, requiring $O(N^2)$ memory and $O(N^3)$ computational complexity. Direct computation of the SGWT through diagonalizing $\mathbf{L}$ is feasible only for graphs with fewer than a few thousand vertices. This limitation would completely destroy the practical applicability of the SGWT for image processing problems, which routinely involve data with hundreds of thousands of dimensions (i.e., number of pixels).

The SGWT coefficients at each scale are given by $g(t_j\mathbf{L})\mathbf{f}$. The fast algorithm avoids the need for diagonalizing $\mathbf{L}$ by approximating $g(t_jx)$ by a degree m polynomial $p_j(x)$, where the approximation holds on an interval $[0, \lambda_{\max}]$ containing the spectrum of $\mathbf{L}$. Once this approximate polynomial is computed, the approximate SGWT coefficients at scale j are given by $p_j(\mathbf{L})\mathbf{f}$.

We compute $p_j(x)$ from Chebyshev polynomial expansion (see, e.g., [20] for overview). The Chebyshev polynomials $T_k(x)$ may be generated from the stable recurrence relation $T_k(x) = 2xT_{k-1}(x) - T_{k-2}(x)$ with $T_0 = 1$ and $T_1 = x$. For $x \in [-1, 1]$, they satisfy $-1 \le T_k(x) \le 1$, and are a natural basis for approximating functions on the domain $[-1, 1]$. As we require an approximation valid on $[0, \lambda_{\max}]$, we employ the shifted Chebyshev polynomials $\overline{T}_k(x) = T_k(\frac{x-a}{a})$ with $a = \lambda_{\max}/2$. For sufficiently regular $g(t_jx)$, the expansion $g(t_jx) = \frac{1}{2}c_{j,0} + \sum_{k=1}^{\infty} c_{j,k}\overline{T}_k(x)$ holds for $x \in [0, \lambda_{max}]$, with $c_{j,k} = \frac{2}{\pi}\int_0^\pi \cos(k\theta)g(t_j(a(\cos(\theta)+1)))d\theta$. We compute the $c_{j,k}$ by numerical integration, and then obtain $p_j(x)$ by truncating the above series to m terms. The wavelet coefficients from the fast SGWT approximation are then given by

$$\tilde{W}_{\mathbf{f}}(t_j, n) = \left(\frac{1}{2}c_{j,0}\mathbf{f} + \sum_{k=1}^{m} c_{j,k}\overline{T}_k(\mathbf{L})\mathbf{f}\right)_n . \tag{8.4}$$

We calculate scaling function coefficients similarly with the analogous polynomial approximation for $h(x)$.

Crucially, we may use the Chebyshev recurrence relation to compute the terms $\overline{T}_k(\mathbf{L})\mathbf{f}$ appearing above, accessing $\mathbf{L}$ only through matrix-vector multiplication. The shifted Chebyshev polynomials satisfy $\overline{T}_k(x) = \frac{2}{a}(x-1)\overline{T}_{k-1}(x) - \overline{T}_{k-2}(x)$; it follows that $\overline{T}_k(\mathbf{L})\mathbf{f} = \frac{2}{a}(\mathbf{L}-\mathbf{I})\left(\overline{T}_{k-1}(\mathbf{L})\mathbf{f}\right) - \overline{T}_{k-2}(\mathbf{L})\mathbf{f}$. Viewing each vector $\overline{T}_k(\mathbf{L})\mathbf{f}$ as a single symbol, this implies that $\overline{T}_k(\mathbf{L})\mathbf{f}$ can be computed from the $\overline{T}_{k-1}(\mathbf{L})\mathbf{f}$ and $\overline{T}_{k-2}(\mathbf{L})\mathbf{f}$ with computational cost dominated by a single matrix-vector multiplication by $\mathbf{L}$. Using sparse matrix representation, the computational cost of applying $\mathbf{L}$ is proportional to the number of nonzero edges, making the polynomial approximation algorithm for the SGWT efficient in the important case when the $\mathbf{L}$ is sparse.

Further details of the algorithm are given in [1]. In addition, it is shown there that the computational complexity for the fast SGWT is $O(m|E| + Nm(J+1))$, where J is the number of wavelet scales, m is the order of the polynomial approximation, and $|E|$ is the number of nonzero edges in the underlying graph $\mathcal{G}$. In particular, for classes of graphs where $|E|$ scales linearly with N, such as graphs of bounded maximal degree, the fast SGWT has computational complexity $O(N)$.

We will require an upper bound $\lambda_{\max}$ of the largest eigenvalue of $\mathbf{L}$, which may be computed at far lower cost than computing the entire spectrum. We compute a rough estimate of λ_{N-1} using Arnoldi iteration, which accesses $\mathbf{L}$ only through matrix vector multiplication, then augment this estimate by 1% to get $\lambda_{\max}$.

8.2.2 SGWT Inverse

A wide class of signal processing applications, including denoising methods described later in this work, involve manipulating the coefficients of a signal in a certain transform, and later inverting the transform. For the SGWT to be useful for more than simply signal analysis, it is important to be able to recover a signal corresponding to a given set of coefficients.

The SGWT is an overcomplete transform, mapping an input vector $\mathbf{f}$ of size N to the $N(J+1)$ coefficients $\mathbf{c} = \mathbf{T}\mathbf{f}$. As is well known, this means that $\mathbf{T}$ will have an infinite number of left-inverses $\mathbf{B}$ s.t. $\mathbf{B}\mathbf{T}\mathbf{f} = \mathbf{f}$. A natural choice is to use the pseudoinverse $\mathbf{T}^{+} = (\mathbf{T}^{T}\mathbf{T})^{-1}\mathbf{T}^{T}$, which satisfies the minimum-norm property

$$\mathbf{T}^{+}\mathbf{c} = \arg\min_{\mathbf{f}\in\mathbb{R}^{N}} \|\mathbf{c} - \mathbf{T}\mathbf{f}\|_{2}.$$

For applications which involve manipulation of the wavelet coefficients, it is very likely to need to apply the inverse to a set of coefficients which no longer lie directly in the image of $\mathbf{T}$. In this case, the pseudoinverse corresponds to orthogonal projection onto the image of $\mathbf{T}$, followed by inversion on the image of $\mathbf{T}$.

Given a set of coefficients $\mathbf{c}$, the pseudoinverse will be given by solving the square matrix equation $(\mathbf{T}^{T}\mathbf{T})\mathbf{f} = \mathbf{T}^{T}\mathbf{c}$. This system is too large to invert directly, but may be solved iteratively using conjugate gradients. The computational cost of conjugate gradients is dominated by matrix-vector multiplication by $\mathbf{T}^{T}\mathbf{T}$ at each step. As described further in [1], the fast Chebyshev approximation scheme described in Section 8.2.1 may be adapted to compute efficiently the application of either $\mathbf{T}^{T}$ or $\mathbf{T}^{T}\mathbf{T}$. We use conjugate gradients, together with the fast Chebyshev approximation scheme, to compute the SGWT inverse.

The convergence speed of the nonpreconditioned conjugate gradient algorithm is determined by the spectral condition number κ of $T^{T}T$, in particular, the error after n steps is bounded by a constant times $(\frac{\sqrt{\kappa}-1}{\sqrt{\kappa}+1})^{n}$ [21]. We note that $\kappa \leq \frac{A}{B}$, where A and B are frame bounds for T; the design of the SGWT wavelet and scaling function kernels is influenced by the desire to keep A/B small.

8.2.3 SGWT Design Details

The SGWT framework places very few restrictions on the choice of the wavelet and scaling function kernels. Motivated by simplicity, we choose g to behave as a monic power near the origin, and to have power law decay for large x. In between, we set g to be a cubic spline such that g and g' are continuous. Specifically, in this work we used

$$g(x) = x^2 \, \zeta_{[0,1)}(x) + [(x-1)(x-2)(x-3)+1] \, \zeta_{[1,2)}(x) + 4x^{-2} \, \zeta_{[2,+\infty)}(x), \quad (8.5)$$

where $\zeta_A(x)$ is the indicator of $A \subset \mathbb{R}$; equal to 1 if $x \in A$ and 0 otherwise.

The wavelet scales t_j (with $t_j > t_{j+1}$) are selected to be logarithmically equis-paced between the minimum and maximum scales t_J and t_1. These are themselves adapted to the upper bound $\lambda_{\max}$ of the spectrum of $\mathbf{L}$, described in Section 8.2.1. The placement of the maximum scale t_1 as well as the scaling function kernel h will be determined by the selection of $\lambda_{\min} = \lambda_{\max}/K_{\mathrm{lp}}$, where K_{lp} is a design parameter of the transform. We then set t_1 so that $g(t_1 x)$ has power-law decay for $x > \lambda_{\min}$, and set t_J so that $g(t_J x)$ has monic polynomial behavior for $x < \lambda_{\max}$. This is achieved for $t_1 = 2/\lambda_{\min}$ and $t_J = 2/\lambda_{\max}$. We take $h(x) = \gamma \, \exp(-(\frac{x}{0.6\lambda_{\min}})^4)$, where γ is set such that $h(0)$ has the same value as the maximum value of g. This set of scaling function and wavelet generating kernels, for parameters $\lambda_{\max} = 10$, $K_{\mathrm{lp}} = 20$, and $J = 5$, are shown in Figure 8.1.

8.3 Nonlocal Image Graph

The nonlocal image graph is a weighted graph with number of vertices N equal to the number of pixels in the original image. This association assumes some labeling of the two-dimensional pixels, they may for example be labeled in raster-scan order. We will define the edge weights so that the weights will lie between 0 and 1, and will provide a measure of the similarity between image patches.

The number of pixels N for any reasonable sized image will be so large that we cannot reasonably expect to explicitly diagonalize the corresponding graph Laplacian. As we wish to use the SGWT defined on the nonlocal graph, we must produce a graph which is sparse enough to be handled by the fast Chebyshev polynomial approximation scheme.

Given a patch radius K, we let $\mathbf{p}_i \in \mathbb{R}^{(2K+1)^2}$ be the $(2K+1) \times (2K+1)$ square patch of pixels centered on pixel i. For pixels within distance K of the image boundary, patches may be defined by extending the image outside of its original boundary by reflection. Set $d_{i,j} = \|\mathbf{p}_i - \mathbf{p}_j\|$ to be the norm of the differences of patches. Following the weighting used in much of the NL-means literature, we set the edge weights for the (unsparsified) nonlocal image graph to be $w_{i,j} = \exp(-d_{i,j}^2/2\sigma_p^2)$.

In order to sparsify the adjacency matrix of our nonlocal graph, we employ a procedure bounding the degree of each graph vertex, rather than thresholding the values $w_{i,j}$ to some fixed threshold. In particular, this limits the creation of disconnected vertex clusters. The adjacency matrix $\mathbf{W}^{\text{nl}}$ of this sparsified nonlocal image graph is simply obtained by keeping the edges corresponding to the M nearest neighbors (in the patch domain) to each node (excluding self connection). For each node i, we denote the index set of these M nearest nodes $\mathcal{N}(i)$. $\mathbf{W}^{\text{nl}}$ is then obtained by symmetrizing the edge connections, that is,

$$\mathbf{W}^{\text{nl}}_{i,j} = \begin{cases} w_{i,j} \text{ if } i \in \mathcal{N}(j) \text{ or } j \in \mathcal{N}(i) \\ 0 \text{ otherwise .} \end{cases}$$

After this procedure, each vertex of $\mathbf{W}^{\text{nl}}$ will have at least M nonzero edges incident to it, and the total number of nonzero edges will be upper bounded by $2NM$. Note that due to the symmetrization of the matrix, it is possible for vertices to have more than M incident nonzero edges.

For convenience in the sequel, we denote by $\mathcal{D}(i)$ the ordered set of M patch distances $\{d_{i,j} \text{ for } j \in \mathcal{N}(i)\}$. As the function $\exp(-\frac{(\cdot)^2}{2\sigma_p^2})$ determining $w_{i,j}$ from $d_{i,j}$ is decreasing, the elements of $\mathcal{D}(i)$ consist of the M smallest patch distances from the patch at vertex i. Note that the sets $\mathcal{N}(i)$ and $\mathcal{D}(i)$ do not depend on σ_p.

The parameter σ_p is fixed by specifying the desired mean of all of the nonzero elements of $\mathbf{W}^{\text{nl}}$. Given a target mean $\mu_{\text{NL}} \in (0, 1)$, σ_p is set so that

$$\frac{1}{|\{\mathbf{W}^{\text{nl}}_{i,j} : \mathbf{W}^{\text{nl}}_{i,j} \neq 0\}|} \left(\sum_{i,j} \mathbf{W}^{\text{nl}}_{i,j} \right) = \mu_{\text{NL}} .$$

We chose to determine σ_p this way, rather than simply fixing a numerical value for it, in order to allow automatic selection of an appropriate σ_p as the input image and other parameters are changed (such as the patch size K). In practice, the value $\mu_{\text{NL}} = 0.75$ gives good results.

8.3.1 Acceleration of Computation with Weeds

Naive computation of the sets $\mathcal{N}(i)$ and $\mathcal{D}(i)$ could be done simply by looping over i, computing the entire set of patch distances $d_{i,j}$, sorting them, and retaining the smallest M. This is computationally intensive. In this section we describe an accelerated algorithm for more efficiently computing these sets. We note that a number of other authors have explored various strategies for reducing the computational cost of the closely related nonlocal means filter. Representative approaches include dimensionality reduction of raw patches via PCA [22], and structured organization of patches into a cluster tree [23].

Our "seeds acceleration" algorithm proceeds by exploiting the geometry of the patch space $\mathbb{R}^{(2K+1)^2}$ to get lower bounds on distances between patches. These lower bounds will depend on the selection of a set of seed patches to which all distances will be computed. By using the lower bounds to show on the fly that certain patch distances will not appear amongst the M smallest, we will suppress the need to compute them.

Let $\mathcal{S} = \{\mathbf{s}_1, \cdots, \mathbf{s}_F\}$ be a set of F seeds, with each $\mathbf{s}_k \in \mathbb{R}^{(2K+1)^2}$. We first compute and store the NF values $\xi_{i,k} = \|\mathbf{p}_i - \mathbf{s}_k\|$. We then employ the triangle inequality, which shows $\|\mathbf{p}_i - \mathbf{s}_k\| \leq \|\mathbf{p}_i - \mathbf{p}_j\| + \|\mathbf{p}_j - \mathbf{s}_k\|$, and $\|\mathbf{p}_j - \mathbf{s}_k\| \leq \|\mathbf{p}_j - \mathbf{p}_i\| + \|\mathbf{p}_i - \mathbf{s}_k\|$. These in turn imply

$$\left| \left(\|\mathbf{p}_i - \mathbf{s}_k\| - \|\mathbf{p}_j - \mathbf{s}_k\| \right) \right| \leq \|\mathbf{p}_i - \mathbf{p}_j\| \equiv d_{i,j} . \tag{8.6}$$

Using this inequality for all $k \leq J$ shows

$$\max_k (|\xi_{i,k} - \xi_{j,k}|) \leq d_{i,j} .$$

Denote this quantity $\bar{d}_{i,j} = \max_k (|\xi_{i,k} - \xi_{j,k}|)$. Computing $\bar{d}_{i,j}$ for fixed i, j is very fast (complexity $O(F)$) as it requires no floating point operations other than absolute value and comparison. Crucially, it is significantly faster than directly computing $d_{i,j}$.

We now consider computation of all of the index and distance sets $\mathcal{N}(i)$ and $\mathcal{D}(i)$. The naive algorithm without using seeds would be Algorithm 1 in Table 8.1. This is wasteful, though, as the entire list L is sorted even though only the M smallest elements are needed. Instead, we can only keep track of the smallest M elements while looping through j. This can be done efficiently using a heap data structure. We are now in a position to use seeds to prune unnecessary computation. If the computed lower bound $\bar{d}_{i,j}$ is large enough to ensure that $d_{i,j}$ exceeds the current heap maximum, then computation of $d_{i,j}$ may be safely skipped. This yields the accelerated Algorithm 2 in Table 8.1.

We note that as the speedup of the seeds acceleration algorithm is derived from skipping computation of values of $d_{i,j}$ for j that cannot possibly be among the M smallest, it is not an approximate algorithm. In particular, it will produce exactly the same edge and distance sets as the naive algorithm (unless there are ties in the distances, which may be broken differently).

The performance of the accelerated algorithm is highly dependent on the selection of the seed points $\mathbf{s}_k$. For any patch $\mathbf{p}_i$, the lower bounds (8.6) on $d_{i,j}$ given by seed $\mathbf{s}_k$ will be more informative the closer $\mathbf{s}_k$ is to $\mathbf{p}_i$. This implies that a set of seed points will be good if the seeds are near as many patches as possible. On the other hand, if two seeds are close to each other, the information they provide in their lower bounds is redundant. This implies the contrasting criterion that the seeds should be far from each other.

Motivated by these two considerations, we introduce a maxmin heuristic for selecting seed points. This heuristic is an example of the "farthest point" strategy developed previously in the context of progressive image sampling by Eldar [24], and also

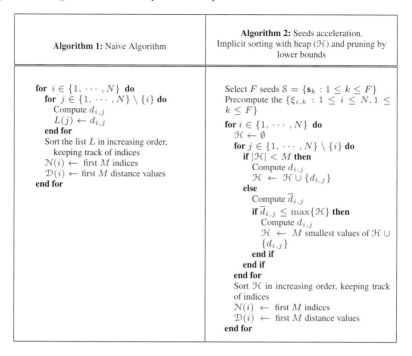

TABLE 8.1
Naive and seeds accelerated algorithms

applied by Peyré et al. in the context of efficient computation of geodesic distances [25]. We select the first seed $\mathbf{s}_1$ to be the patch corresponding to the very middle of the image. We then compute and store the N values $\xi_{i,1} = \|\mathbf{p}_i - \mathbf{s}_1\|$. The next seed $\mathbf{s}_2$ is picked as the point furthest from $\mathbf{s}_1$, i.e., $s_2 = \arg\max_i \xi_{i,1}$. At each successive step, the $k + 1^{\text{th}}$ seed is picked to be the point maximizing the distance to the set $\{\mathbf{s}_1, \mathbf{s}_2, \cdots, \mathbf{s}_k\}$, i.e.,

$$\mathbf{s}_{k+1} = \arg\max_i \left(\min_{k' \le k} \xi_{i,k'} \right).$$

This procedure is iterated until all F seeds are computed. Note that the precomputed distances $\xi_{i,k}$ are produced during the execution of this heuristic.

We have found the performance of the seed acceleration algorithm using seeds computed with the above maxmin heuristic to be slightly better than using the same number of randomly selected seeds. Even using randomly selected seeds, however, provides a marked improvement over the naive algorithm.

The performance of the seeds acceleration can be quantified by the ratio ρ of patch distances $d_{i,j}$ computed by the accelerated algorithm divided by the number of patch distances computed by the naive algorithm. Alternatively, one may directly

		Lena	Barbara	Boat	House
maxmin	ρ	5.2 %	4.9 %	12.5 %	14.3 %
	runtime	89.07 s	87.77 s	111.67 s	121.02 s
random	ρ	6.1 %	6.0 %	13.1 %	14.9 %
	runtime	92.71 s	92.31 s	112.12 s	122.44 s
naive	runtime	425.16 s	430.84 s	434.08 s	425.87 s

TABLE 8.2
Performance of seeds acceleration algorithm, averaged over 3 trials for $K = 5$ and
$M = 10$. Runtimes (in seconds) were measured on a parallel implementation using
4 Intel Xeon 2.0 GHz cpu cores.

examine the reduction in run time. The overall run time will not be reduced as much
as the number of patch distance computations, reflecting the additional overhead of
computing the lower bounds $\overline{d}_{i,j}$. Given ρ, a quick computational complexity es-
timation provides that Algorithm 2 runs in $O\left(NM(2K+1)^2 + (N-M)NF + \rho(N-M)N(2K+1)^2\right)$. Considering that the proportionality constant are the same
in Algorithms 1 and 2, the ratio of computational time is about $\rho + \frac{F}{(2K+1)^2} + \frac{M}{N}$.
Table 8.2 approximately follows this rough estimation.

In practice, on 256×256 clean test images with patch radius $K = 5$, the seeds
acceleration algorithm using 15 seeds allows reduction of run time by a factor of 3-8
compared to the naive algorithm for computing the sparsified adjacency matrix, as
shown in Table 8.2. We note that the performance improvement is very dependent
on the structure of the input image. In particular, when testing on images of random
noise, computation time is not reduced at all! This makes sense as the lower bounds
are useful only if seeds are near where "most" of the image patches are, for random
images the patches are too distributed.

8.4   Hybrid Local/Nonlocal Image Graph

As the spectral graph wavelet transform is parametrized by the underlying weighted
graph, great flexibility is available for influencing the wavelets by choosing the design of the graph. While much of the emphasis of this work is on construction of
the nonlocal image graph, one may also construct local graph wavelets simply by
using a connectivity graph that includes only local connections vertices. Given both
the nonlocal adjacency $\mathbf{W}^{nl}$ and a local adjacency $\mathbf{W}^{loc}$, we may form a hybrid lo-
cal/nonlocal graph by convex combination , e.g., $\mathbf{W}^{hyb}(\lambda) = \lambda \mathbf{W}^{loc} + (1-\lambda)\mathbf{W}^{nl}$.
Here λ smoothly parametrizes the degree of locality present in the graph, so that at
$\lambda = 0$ the graph is purely nonlocal, while at $\lambda = 1$ it is purely local. The structure
of the SGWT makes it extremely easy to explore the effects of such a hybrid connectivity. As long as both the local and nonlocal adjacencies are sufficiently sparse

to allow the fast Chebyshev transform to work efficiently, no other changes need be made to any other parts of the SGWT machinery.

In the hybrid graph, the strength of connection between two locations depends both on the similarity of the image content of the two regions and the similarity of the coordinates defining the two locations. This concept is similar in spirit to the "semi-local processing" conditions as described in [12, 23], where image patches are "lifted" to an augmented patch including both the patch center coordinates and image intensity data, before computing patch distance measures. The concept of splitting graph connections into local and nonlocal edge sets is related to the slow/fast patch graph model studied in [26]. Finally, computing filter weights incorporating both domain and range similarity underlies the older bilateral filtering approach [27].

8.4.1 Oriented Local Connectivity

For image processing applications, oriented wavelet filters are generally more effective than spatially isotropic filters, as much of the significant content in images is highly oriented, such as edges. This observation motivates the design of a local adjacency graph that would give oriented filters when used with the SGWT. The standard 4-point local connectivity, with equal edge weights in all directions, gives rise to spectral graph wavelets which are isotropic, and thus unlikely to be highly efficient as a sparsity basis.

We will generate oriented local graph wavelets by choosing a local connectivity which is itself oriented. The design of these oriented local connectivities will be determined by requiring that the associated graph Laplacian approximate a continuous *oriented* second derivative operator.

We describe the target continuous operator in terms of both a dominant orientation parameter θ, and a second parameter $\delta \in [0, 1]$ specifying the degree of anisotropy, or "orientedness". For any angle θ, let the perfectly oriented second derivative in the θ direction be D_θ^2, so that $(D_\theta^2 f)(x, y) \equiv \frac{d^2}{d\epsilon^2} f(x + \epsilon \cos \theta, y + \epsilon \sin \theta)|_{\epsilon=0}$ for any (x, y) in the plane. For any $\delta < 1$, the target operator $D(\theta, \delta)$ will include some fraction of the second derivative in the perpendicular direction. Specifically, we define

$$D(\theta, \delta) = \tfrac{1+\delta}{2} D_\theta^2 + \tfrac{1-\delta}{2} D_{\theta+\pi/2}^2 .$$

This operator can be expressed in terms of partial derivatives in the standard coordinate directions, as

$$D(\theta, \delta)f = \tfrac{1}{2}(1 + \delta \cos 2\theta) f_{xx} + (\delta \sin 2\theta) f_{xy} + \tfrac{1}{2}(1 - \delta \cos 2\theta) f_{yy} . \quad (8.7)$$

We choose the local connectivity $\mathbf{W}_{\theta,\delta}^{\text{loc}}$ to be the 8-point connectivity with weights such that the graph Laplacian approximates $D(\theta, \delta)$. The local connectivity will be

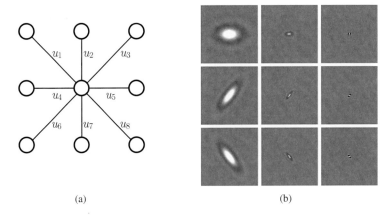

(a)   (b)

FIGURE 8.2
(a) 8-nearest-neighbor connectivity. The complete graph is produced by reproducing these edges across the entire image. (b) Spectral graph wavelets computed using local oriented adjacencies $\mathbf{W}^{\text{loc}}_{\theta,\delta}$. Shown are scaling functions and wavelets from selected scales for transform with $J = 20$ wavelet scales, $S = 3$ orientations, using $\delta = 0.44$. Top row, $\theta = 0$, middle row, $\theta = \pi/3$, bottom row, $\theta = 2\pi/3$. Left column, scaling function; middle column, $j = 10$, right column, $j = 20$.

completely specified by the values of the weights of any vertex to its 8 nearest neighbors. For convenience, these may be labeled $u_1, u_2, \cdots, u_8$, as shown in Figure 8.2 (a). As we are considering undirected graphs, we must have equality of edge weights for edges in opposite directions, e.g., $u_1 = u_8$, $u_2 = u_7$, $u_3 = u_6$, $u_4 = u_5$. We may consider the first four weights as components of the vector $\mathbf{u} \in \mathbb{R}^4$.

We match the discrete graph Laplacian implied by the choice of $\mathbf{u}$ to the continuous operator $D(\theta, \delta)$ by looking at the local Taylor expansion. We now consider f to be a continuous function sampled on the image pixel lattice. Let f_i for $0 \leq i \leq 8$ denote the value of f at the vertices shown in Figure 8.2, and let x_i and y_i be the integer offsets for vertex i, i.e., $(x_1, y_1) = (-1, 1)$, $(x_2, y_2) = (0, 1)$, etc. If Δ is the grid spacing, the second order Taylor expansion shows

$$f_i - f_0 = \Delta\left(x_i f_x + y_i f_y\right) + \tfrac{1}{2}\Delta^2\left(x_i^2 f_{xx} + 2x_i y_i f_{xy} + y_i^2 f_{yy}\right) + o(\Delta^3).$$

By above, applying the graph Laplacian to f at the center vertex gives

$$\sum_i u_i(f_i - f_0) = \Delta^2\left(f_{xx}\sum_i \tfrac{1}{2} u_i x_i^2 + f_{xy}\sum_i u_i x_i y_i + f_{yy}\sum_i \tfrac{1}{2} u_i y_i^2\right) + o(\Delta^3).$$
$$(8.8)$$

We assume unit grid spacing, i.e., $\Delta = 1$. Equating coefficients of f_{xx}, f_{xy} and f_{yy} in (8.7) and (8.8) yields

$$\begin{cases} u_1 + u_3 + u_4 = \tfrac{1}{2}(1 + \delta\cos 2\theta) \\ -u_1 + u_3 = \tfrac{1}{2}\delta\sin 2\theta \\ u_1 + u_2 + u_3 = \tfrac{1}{2}(1 - \delta\cos 2\theta) \end{cases} \quad \Leftrightarrow \quad \mathbf{M}\mathbf{u} = \mathbf{v}(\theta, \delta), \qquad (8.9)$$

where we set $\mathbf{M} = \begin{pmatrix} 1 & 0 & 1 & 1 \\ -1 & 0 & 1 & 0 \\ 1 & 1 & 0 & \end{pmatrix}$ and $\mathbf{v} = (1/2, 0, 1/2)^T + \frac{\delta}{2}(\cos\theta/2, \sin\theta/2, -\cos\theta/2)^T$.

This is an underdetermined linear system for the four free components of $\mathbf{u}$, which will have an infinite number of solutions for any $\mathbf{v}(\theta, \delta)$. A natural choice for ensuring a unique $\mathbf{u}$ is to select the solution with the smallest ℓ_2 norm. Unfortunately, this may produce a solution with negative values for the graph weights, which we require to be nonnegative. Including these two considerations yields a convex optimization program for $\mathbf{u}$:

$$\mathbf{u}^*(\theta, \delta) = \arg\min_{\mathbf{u}} \ \|\mathbf{u}\|^2 \text{ subject to } \mathbf{Mu} = \mathbf{v}(\theta, \delta), \text{ and } u_i \geq 0 . \qquad (8.10)$$

For δ close to 1, the program (8.10) fails to be feasible for angles θ away from aligned with the coordinate axes. We have observed through numerical simulation that the program is feasible for all θ whenever $\delta < \widetilde{\delta} \approx 0.44$. Note that taking δ above $\widetilde{\delta}$ may still be possible for a specific set of θ values. We solve the program (8.10) numerically using the MATLAB cvx toolbox [28]. Using these values of $\mathbf{u}(\theta, \delta)$ to construct the weighted graph for the entire image yields the local connectivity $\mathbf{W}^{\text{loc}}_{\theta, \delta}$.

For completeness, we show several images of the spectral graph wavelets computed using $\mathbf{W}^{\text{loc}}_{\theta, \delta}$ in Figure 8.2(b). The important point of these local oriented wavelets is not necessarily that we expect them to be highly effective for image representation on their own, but rather that they allow the introduction of oriented filter character in a manner compatible with the SGWT construction.

8.4.2 Unions of Hybrid Local/Non-Local SGWT Frames

The SGWT generates a frame of wavelets and scaling functions from a single adjacency matrix. Using a hybrid adjacency with an oriented local connectivity would give a single frame where all of the wavelets had the same directionality. Using such a frame for image restoration tasks is likely to introduce a bias toward orientations in this specific direction. We would prefer a wavelet transform that samples orientations uniformly on $[0, \pi]$. A straightforward way to do this is to form the union of multiple SGWT frames corresponding to hybrid adjacencies with uniformly sampled orientations. Let $\mathbf{T}^{\theta, \delta} : \mathbb{R}^N \to \mathbb{R}^{N(J+1)}$ be the SGWT operator generated using the hybrid adjacency matrix $\mathbf{W}^{\text{hyb}}_{\theta, \delta} \equiv \lambda \mathbf{W}^{\text{loc}}_{\theta, \delta} + (1 - \lambda)\mathbf{W}^{\text{nl}}$. Denote by S the number of orientation directions to sample, and set $\theta_k = \frac{(k-1)\pi}{S}$ for $1 \leq k \leq S$. We then define the overall wavelet transform operator $\mathbf{T} : \mathbb{R}^N \to \mathbb{R}^{NS(J+1)}$ by $\mathbf{Tf} = ((\mathbf{T}^{\theta_1, \delta}\mathbf{f})^T, (\mathbf{T}^{\theta_2, \delta}\mathbf{f})^T, \cdots, (\mathbf{T}^{\theta_S, \delta}\mathbf{f})^T)^T$.

While in principle separate scaling function and wavelet kernels h and g could be used for the separate component SGWT frames, we use the same kernels as well as the same sampled scales t_j for each of the S subframes. As the selection of scales is determined by the spectrum of the graph Laplacian (as described in Section 8.2), choosing the scales uniformly only makes sense if the maximal eigenvalues of the S graph Laplacians formed from each of the $\mathbf{W}^{\text{hyb}}_{\theta_k, \delta}$ are similar. In practice, we find this to be the case. The oriented local graph Laplacians are approximations of continuous second derivative operators $D(\theta, \delta)$ which are related to each other by rotation, and should thus have exactly the same spectra. Intuitively, the fact that the maximal eigenvectors of the hybrid Laplacian operators, including the nonlocal component, do not vary strongly with θ reflects a lack of orientation bias in the original image.

The union of frames transformation is itself a frame, where the redundancy is increased by a factor S. If each of the sub-frames satisfy the frame bounds A and B, then the union of frames transformation will have frame bounds SA and SB, and will be $S(J+1)$ times overcomplete. While highly redundant transforms may be not be appropriate for some signal processing applications such as compression, overcompleteness does not pose a fundamental problem for use of a transform for image denoising. Overcomplete transforms have been widely and successfully used for denoising, e.g., [29, 30, 31, 32], some level of transform redundancy is in fact critical for avoiding problems due to loss of translation invariance [33, 34].

The hybridization with λ may also be described directly with the graph Laplacian operators, i.e. we may write $\mathbf{L}^{\text{hyb}}_{\theta_k, \delta} = \lambda \mathbf{L}^{\text{loc}}_{\theta_k, \delta} + (1 - \lambda)\mathbf{L}^{\text{nl}}$. One issue with this parametrization is that, depending on the parameters of the graph construction, the operator norm of $\mathbf{L}^{\text{nl}}$ may be very different in magnitude than that of the $\mathbf{L}^{\text{loc}}_{\theta_k, \delta}$'s. In such a case, taking $\lambda = 0.5$ will not really give an equal mixing of local and nonlocal character. For convenience, we introduce the parameter λ'. We wish that the effective nonlocal contribution be proportional to λ', and the effective local contribution be proportional to $1 - \lambda'$. We enforce this by determining λ so that $(1 - \lambda)\|\mathbf{L}^{\text{nl}}\| = (1 - \lambda')C$, and $\lambda\|\mathbf{L}^{\text{loc}}\| = \lambda'C$ for some constant C. In the above, we have suppressed the dependence on the angle θ_k for $\|\mathbf{L}^{\text{loc}}\|$, as discussed earlier we find there to be little dependence on k and we may just take $\|\mathbf{L}^{\text{loc}}\| = \max_k \|\mathbf{L}^{\text{loc}}_{\theta_k, \delta}\|$. Solving the above set of equations gives

$$\lambda = \frac{\|\mathbf{L}^{\text{nl}}\|}{(\frac{1-\lambda'}{\lambda'})\|\mathbf{L}^{\text{loc}}\| + \|\mathbf{L}^{\text{nl}}\|}.$$

Finally, for convenience of notation we note the following. Coefficients of the graph wavelet transforms formed from a single SGWT frame may be indexed by scale parameter j and location index n, while coefficients of the hybrid local/nonlocal wavelet transforms require an additional orientation index k. In the sequel, we will often replace the scale and orientation indices by a single multi-index β associated to one choice of (t_j, θ_k). This allows us to refer to the wavelet $\psi_{\beta,n}$ and coefficient $x_{\beta,n}$ at (scale and/or orientation) band β, position n, independent of the specific form of the transform.

8.4.3   Image Graph Wavelets

As an illustrative example of what the resulting hybrid local/nonlocal image graph wavelets look like, we display several in Figure 8.3 computed for the well-known

original   s.f., $\theta = 0$ $j = 3, \theta = 0$ $j = 6, \theta = 0$ s.f., $\theta = \pi$ $j = 3, \theta = \pi$ $j = 6, \theta = \pi$

FIGURE 8.3
Images of the graph wavelets with hybrid local/nonlocal adjacency, for the boat image. Shown are scaling functions and wavelets from selected scales for a transform with $J = 20$ wavelet scales, $S = 2$ orientations, using $\delta = 0.5$, for two different wavelet centers. Red dots on original image indicate wavelet center locations.

"boat" test image. These wavelets were computed using the same parameters as are later used in our denoising applications. A relatively small value of $\lambda' = 0.15$ was used, so they are mostly nonlocal spiced with a small pinch of local character. Note that the wavelets really are nonlocal – this is especially evident with the scaling functions which show support on parts of the image very distant from the wavelet center. It is also evident that the support for both the scaling functions and wavelets consists of image regions that are similar in structure to the center patch. Note for the top row where the center patch is on the ground how the support of the scaling functions are mostly ground and similar blank sky, while for the bottom row the supports consist of regions with strong horizontal structure similar to the patch center. It can also be verified that for increasing j (corresponding to decreasing scale t_j), the wavelets are increasingly localized. For even smaller scales (not pictured), this localization proceeds further.

8.5   Scaled Laplacian Model

Coefficients of natural signals in common localized bases (e.g., (bi)orthogonal wavelets, wavelet frames, steerable wavelets, curvelets, ...) typically exhibit sparse behavior. This has commonly been modeled by using peaked, heavy-tailed distributions as prior models for coefficients. A canonical example of this is the Laplacian distribution $p(x) = \frac{1}{2s} \exp(\frac{-|x|}{s})$. Laplacian models have been very commonly used in the literature to describe the marginal statistics of wavelet coefficients of images, e.g., [35, 36, 37].

Typically, the parameter s is allowed to depend on the wavelet scale. Including this is necessary to allow the variance of the signal coefficients to depend on scale. This effect arises from the power spectral properties of the original image signal (typically showing larger power at low frequencies) . It may also arise from normalization of the wavelets themselves, as for many overcomplete wavelet transforms, the norms of the wavelets may depend on the wavelet scale.

Unlike classical wavelet transforms which are translation invariant at each scale (i.e., all wavelets at a particular subband are translates of a single waveform), the spectral graph wavelets centered at different vertices, for a particular wavelet scale t, do not all have the same norm. As the coefficients $c_{\beta,n} = \langle \psi_{\beta,n}, f \rangle$ are linear in the norm of each wavelet, it is problematic to model the coefficients at one scale with a single Laplacian density. Allowing the Laplacian parameter s to vary independently at each vertex location n would give a model with too many free parameters that we would not be able to fit.

These considerations motivate the scaled Laplacian model, where we model each coefficient as Laplacian with a parameter that is proportional to the wavelet norm. This proportionality constant is allowed to depend on the wavelet band β, giving one free parameter for the model at each wavelet band. The model is

$$ \mathrm{p}_{\beta,n}(x) = \frac{1}{2\alpha_j \sigma_{\beta,n}} \exp \left(\frac{-|x|}{\alpha_\beta \sigma_{\beta,n}} \right) , $$

where $\sigma_{\beta,n} = \|\psi_{\beta,n}\|$ is the norm of the wavelet at band β and vertex n.

This model implies that at each band, the marginal distribution of the coefficients rescaled by $1/\sigma_{\beta,n}$ should have a Laplacian distribution. We find these marginal distributions are qualitatively reasonably fit by a Laplacian, as shown in Figure 8.4. For the higher j scales, the rescaled wavelet marginals appear slightly more kurtotic than the Laplacian fit. These may in fact be better modeled by a generalized Gaussian density of the form $p(x) \propto e^{-|x/s|^p}$ with an exponent $p < 1$. Employing such a scaled GGD model for $p < 1$ would lead to different denoising algorithms, but we do not consider this here. We show results for the purely nonlocal graph wavelet transform, but we observe very similar results for the hybrid local/nonlocal transforms.

We note that many authors have constructed statistical image models employing spatially varying statistical parameters, including [31, 37, 38]. In these works, translation-invariant transforms were used, and the motivation for allowing spatial adaptation of the model is directly to explain the nonstationarity of image content. In contrast, in our work the adaptation of the Laplacian model is motivated by the need to account for the fact that the nonlocal graph wavelets are not translation invariant, and have inhomogeneous norms. This inhomogeneity does arise from the nonstationarity of image content, but not straightforwardly.

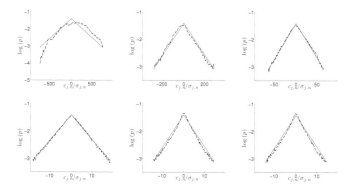

FIGURE 8.4

Log-histograms of rescaled graph wavelet coefficients, from the purely nonlocal transform with 6 wavelet scales, for the clean "boat" image. From top left to bottom right are wavelet scales $j = 1$ through $j = 6$. Solid red lines show best Laplacian fit.

8.6 Applications to Image Denoising

As an application of the nonlocal and hybrid graph wavelets, we study the problem of removing noise from natural images. In this work we consider additive Gaussian white noise. We fix our notation as follows. We let $\mathbf{x}, \mathbf{y}$, and $\mathbf{n}$ denote the (unknown) clean image, noisy image, and noise respectively. This gives the noise model $\mathbf{y} = \mathbf{x} + \mathbf{n}$, where $\mathbf{n}$ is a zero-mean white Gaussian noise with variance σ_I^2 in the pixel domain. Our denoising algorithms are described in wavelet space. Let $\mathbf{T}$ be the graph wavelet operator, which will be either the SGWT with nonlocal Laplacian, or a union of SGWT frames for the hybrid local/nonlocal adjacencies described in Section 8.4.2. We will use the notation $\mathbf{c}, \mathbf{d}$, and $\mathbf{e}$ to refer to the wavelet coefficients of the clean image, noisy image, and noise process, respectively, so that $\mathbf{c} = \mathbf{Tx}, \mathbf{d} = \mathbf{Ty}$, and $\mathbf{e} = \mathbf{Tn}$.

8.6.1 Construction of Nonlocal Image Graph from Noisy Image

For a realistic image denoising problem, we cannot measure the true nonlocal image graph as computing it would require the availability of the clean image data. The question of how to estimate the nonlocal image graph is a key concern for our method, as well as for many other works using nonlocal methods, as the effectiveness of the overall nonlocal image denoising procedure is highly dependent on the quality of the estimate underlying nonlocal graph. This presents a chicken-and-egg problem, as denoising the image demands a good quality nonlocal graph, but computing the graph depends on good estimates of the clean image patches.

We have found that directly applying the denoising procedures described in this section using the nonlocal graph computed directly from noisy image patches leads to very poor denoising performance. This is somewhat surprising considering that standard nonlocal means performs well using the same noisy graph. This necessitates the development of some method for robustly estimating the nonlocal graph in the presence of noise. While we have not satisfactorily solved this problem, we still wish to demonstrate the potential of the nonlocal graph wavelet methods for capturing image structure. In this work, we sidestep this issue by using some different image denoising method, dubbed the "predenoiser", to first estimate the image. We then compute the nonlocal image graph from this "predenoised" image. Once the nonlocal image graph is computed, the predenoised image is discarded and we then proceed purely with our graph wavelet techniques. In this work, we use the Gaussian Scale Mixture model of Portilla et al. [31] to perform the predenoising. As a direction for future research, we may consider jointly estimating the graph weights and the denoised image, through a block relaxation scheme following [39].

While incorporating predenoising yields a highly effective overall denoising algorithm, it introduces some question of how much the performance is depending on the effectiveness of the estimation of the nonlocal graph. To address this, we also examine denoising under the condition where the nonlocal graph is given by an "oracle", i.e., computed from the original clean image. While this does not result in a true denoising methodology, as it requires access to the original clean image, it does provide an upper bound on the performance of the overall method.

8.6.2 Scaled Laplacian Thresholding

Our first denoising algorithm consists of taking the forward graph wavelet transform, applying a spatially varying soft thresholding operation to the wavelet coefficients, and then inverting the transform. We derive the soft thresholding rule as a Bayesian maximum a posteriori (MAP) estimator, assuming the signal coefficients follow the scaled Laplacian model described in Section 8.5.

Given an input noisy image, we must first estimate the parameters α_β at each wavelet band β. In the wavelet domain, the noise model is $d_{\beta,n} = c_{\beta,n} + e_{\beta,n}$, where the noise coefficient $e_{\beta,n} = \langle \psi_{\beta,n}, \mathbf{n} \rangle$. Assuming signal and noise are independent, we have

$$\mathbb{E}[d_{\beta,n}^2] = 2\alpha_\beta^2 \sigma_{\beta,n}^2 + \sigma_I^2 \sigma_{\beta,n}^2 \Rightarrow \alpha_\beta^2 = \left(2\sum_n \sigma_{\beta,n}^2\right)^{-1}\left(\sum_n \mathbb{E}[d_{\beta,n}^2] - \sigma_I^2 \sum_n \sigma_{\beta,n}^2\right).$$

Replacing $\sum_n \mathbb{E}[d_{\beta,n}^2]$ by the plug-in estimate $\sum_n d_{\beta,n}^2$ gives a formula for the estimator $\tilde{\alpha}_\beta^2$. For actual data it is possible to obtain a negative value for $\tilde{\alpha}_\beta^2$, in this case we set $\tilde{\alpha}_\beta = 0$.

The scaled Laplacian thresholding rule is derived by assuming the independence of wavelet coefficients of both the signal and noise at different spatial locations.

In that case, each coefficient $d_{\beta,n}$ consists of the sum of the desired signal, a zero mean Laplacian with parameter $\alpha_\beta \sigma_{\beta,n}$, and zero mean Gaussian noise with variance $\sigma_I^2 \sigma_{\beta,n}^2$. The MAP estimator for $c_{\beta,n}$ in this case is given by soft-thresholding (see, e.g., [40]), i.e.,

$$c_{\beta,n}^{\text{MAP}}(d_{\beta,n}) = \arg\min_x (d_{\beta,n} - x)^2 + \tfrac{2\sigma_I^2}{\alpha_\beta} \sigma_{\beta,n} |x| =: S_{\tau_{\beta,n}}(d_{\beta,n}), \quad (8.11)$$

with threshold $\tau_{\beta,n} = \tfrac{\sigma_I^2}{\alpha_\beta} \sigma_{\beta,n}$, where $S_\tau(y) = (|y| - \tau)_+ \text{sign}(y)$ with $(\lambda)_+ = \max(0, \lambda)$. The denoised image is recovered by applying the inverse transform, described in Section 8.2, to the estimated coefficients $\mathbf{c}^{\text{MAP}}$.

The scaled Laplacian model requires knowledge of the norms $\sigma_{\beta,n}$ of the wavelet coefficients. This is somewhat problematic given how the SGWT is computed, as the wavelets are not explicitly formed in memory. Computing these norms by naively applying the SGWT to delta impulses at all image locations would be extremely slow. Instead, we estimate the $\sigma_{\beta,n}$ by computing the variances of the wavelet coefficients of pseudorandom noise. Let $\mathbf{e}^{(k)} = \mathbf{T}\mathbf{n}^{(k)}$, where $\mathbf{n}^{(k)}$ is a sample of zero mean unit white Gaussian noise in the image domain. We estimate $\sigma_{\beta,n}$ by $\widetilde{\sigma}_{\beta,n} = \tfrac{1}{P} \sum_{k=1}^{P} \left(e_{\beta,n}^{(k)}\right)^2$, using $P = 100$ samples.

8.6.3 Weighted ℓ_1 Minimization

While the scaled Laplacian thresholding may be cleanly motivated by statistical modeling, it is limited in that each wavelet coefficient is processed independently of all the others. As a related alternative denoising algorithm, we develop a variational weighted ℓ_1 minimization procedure which allows for coupling between different wavelet coefficients.

This weighted ℓ_1 functional is obtained from considering the minimization problems from (8.11). Formally, all of these uncoupled minimizations may be written together as

$$\arg\min_{\mathbf{c}} \sum_{\beta,n} (d_{\beta,n} - c_{\beta,n})^2 + \sum_{\beta,n} 2\tau_{\beta,n} |c_{\beta,n}|.$$

The first term above is the sum of squares of the portion of the wavelet coefficients assigned to the noise process. We arrive at our ℓ_1 program by replacing this term by the sum of squares of the estimated noise in the image domain, where the recovered image is given by $\mathbf{T}^T \mathbf{c}$. This yields

$$\mathbf{c}^* \in \arg\min_{\mathbf{c}} \|\mathbf{y} - \mathbf{T}^T \mathbf{c}\|^2 + 2\|\mathbf{c}\|_{\tau,1}, \quad (8.12)$$

where $\mathbf{T}^T \mathbf{c}$ represents the recovered image, $\|\mathbf{c}\|_{\tau,1} = \sum_{\beta,n} \tau_{\beta,n} |c_{\beta,n}|$ is the τ-weighted ℓ_1-norm of the coefficients $\mathbf{c}$, for the thresholds described previously. The denoised image is then given by $\mathbf{x}^* = \mathbf{T}^T \mathbf{c}^*$.

The objective function in (8.12) is convex in **c**. This is a BPDN/Lasso problem in Lagrangian form with a synthesis sparsity prior and thus admits a global minimum which can be found by well established minimization techniques like the iterative soft thresholding (IST) [41], the forward-backward (FB) splitting [42], or related approaches [43, 44]. We use the FB method for all of our experiments.

8.6.4 Denoising results

We have performed numerical simulations demonstrating the efficacy of the graph wavelet methods for image denoising. We also explore the change in performance of our denoising methods when the local orientation component of the hybrid local/nonlocal is removed. Our test images were all 256×256 pixels in size, with grayscale values ranging between 0 and 255. In all our experiments, reconstructed image qualities have been assessed by the peak signal-to-noise ratio, or PSNR, given by $10 \log_{10} 255^2 N / \|\mathbf{x} - \mathbf{x}^*\|^2$, where $\mathbf{x} \in \mathbb{R}^N$ and $\mathbf{x}^* \in \mathbb{R}^N$ stand for the original and the reconstructed image.

The extreme flexibility of the methods used in this work give rise to a number of parameters for detailing the specifics of both the SGWT and the graph construction. For convenience, we list them here. For all of our denoising examples, we have used the SGWT with $J = 20$ scales, with $K_{\mathrm{1p}} = 200$ and polynomial order $m = 80$ for the fast SGWT transform. The nonlocal image graph was constructed using 11×11 patches (i.e., patch radius $K = 5$), $M = 10$, and $\mu_{NL} = 0.75$. For the hybrid local/nonlocal graph, we used $S = 2$ orientation directions, with $\delta = 0.5$, and normalized hybridization parameter $\lambda' = 0.15$. Note that while $\delta = 0.5$ exceeds the upper bound $\tilde{\delta}$ ensuring feasibility of (8.10) for all θ, this is not a problem as (8.10) is feasible with $\delta = 0.5$ for the particular (horizontal and vertical) values of θ implied by $S = 2$.

Results for the Scaled Laplacian thresholding (SL) and weighted ℓ_1 minimization (ℓ_1-min) algorithms on four standard test images are shown in Table (8.3). Three different levels of noise were tested, with standard deviation $\sigma = 20$, 40, and 80. The variants using the oracle to compute the nonlocal graph, as well as variants not using predenoising (i.e., computing the nonlocal graph directly from the noisy image) are also shown. We include comparison against the 8-band GSM used for predenoising, as well as the result of nonlocal means applied using the same nonlocal graph used in the SL and ℓ_1-min cases. We also compare against the BM3D method [9], which currently defines the state-of-the art for denoising PSNR performance.

We first note that the PSNR performance of both the SL and ℓ_1-min methods are competitive with the 8-band GSM, providing higher performance in a few cases, but typically within 0.2 dB, while the denoising performance of the BM3D method is consistently higher. While the similarity in denoising performance may seem to suggest that the graph wavelet methods are simply reproducing the predenoised image, the results presented really do represent a distinct methodology. This can be seen by examining images of the results, in particular, the visual quality of the GSM

		Noisy	GSM8	NLM	BM3D	SL†	ℓ_1-min†	SL	ℓ_1-min	SL*	ℓ_1-min*
barbara	$\sigma = 20$	22.12	30.78	30.07	**31.74**	29.74	29.48	30.70	30.75	31.42	31.47
	$\sigma = 40$	16.10	27.31	25.46	28.03	26.38	26.18	27.37	27.46	28.62	**28.85**
	$\sigma = 80$	10.08	23.89	20.02	24.68	23.05	23.14	24.08	24.11	25.93	**26.36**
boat	$\sigma = 20$	22.12	30.11	29.14	**30.61**	28.82	28.86	29.54	29.68	30.27	30.36
	$\sigma = 40$	16.10	26.68	24.90	27.04	25.39	25.65	26.28	26.41	27.36	**27.82**
	$\sigma = 80$	10.08	23.58	19.42	23.81	22.63	22.91	23.36	23.24	24.98	**25.54**
lena	$\sigma = 20$	22.12	31.39	30.38	**32.13**	30.22	29.43	31.14	31.13	31.80	31.77
	$\sigma = 40$	16.10	27.69	25.49	28.27	26.38	26.20	27.69	27.71	28.74	**28.91**
	$\sigma = 80$	10.08	24.27	19.93	25.18	23.00	23.23	24.43	24.35	26.00	**26.29**
house	$\sigma = 20$	22.12	32.25	30.09	**33.84**	30.71	30.50	32.01	31.97	31.72	31.99
	$\sigma = 40$	16.10	29.08	24.97	**30.66**	27.97	27.57	28.91	28.77	29.28	29.53
	$\sigma = 80$	10.08	25.85	19.57	27.19	24.73	24.73	25.82	25.62	26.83	**27.25**

TABLE 8.3

Performance of proposed denoising algorithm and variants (reported by PSNR for 4 standard 256×256 test images). In above, GSM8: 8-band GSM method of Portilla et al., used for predenoising; NLM: nonlocal means using same weights as SL and ℓ_1-min; BM3D: Block-matching 3D filtering of Dabov et al.; SL† and ℓ_1-min†: variants of SL and ℓ_1-min where nonlocal graph computed without predenoising; SL: Scaled Laplacian thresholding; ℓ_1-min: weighted ℓ_1 minimization; SL* and ℓ_1-min*: oracle variants. In bold, the highest PSNR per row. Underlined, the best PSNR amongst the non-oracle methods.

predenoiser and the ℓ_1-min method are different. In Figure 8.5, carefully comparing images (c) and (e) shows the presence of unsightly ripple artifacts in the GSM image that are smooth in the ℓ_1-min image, such as near the table edges.

Clearly, the good performance of the graph wavelet methods does depend on calculating a good estimate of the nonlocal graph. As shown in Table 8.3, the SL and ℓ_1-min methods calculated using the nonlocal graph computed directly from the noisy image show poorer performance, typically losing over 1dB. In contrast, the oracle methods which use the perfect nonlocal graph show extremely good performance. This is especially noticeable for the high noise ($\sigma = 80$) case, exceeding the GSM method by more that 2 dB.

While the PSNR performance of the SL and ℓ_1-min methods using predenoising are very similar, their visual qualities are different. The SL method results seem oversmoothed compared to the ℓ_1-min results, which preserve sharper edges and generally appear to have more high frequency content. Interestingly, the PSNR of the ℓ_1-min method is consistently better than SL for the oracle case, indicating that the ℓ_1-min method is more sensitive to the quality of the nonlocal graph. The computational cost of the two methods is shown in Table 8.4. For the SL method, the run-time is dominated by computing the nonlocal graph and the wavelet norms, while for the ℓ_1-min method the FB iteration is the largest part of the overall run-time.

The nonlocal graph used in our methods has weights that are computed the same way as those used in the nonlocal means method for image denoising. As both the nonlocal graph wavelet methods and nonlocal means can be described using the same underlying graph structure, it is interesting to compare against the results of nonlocal

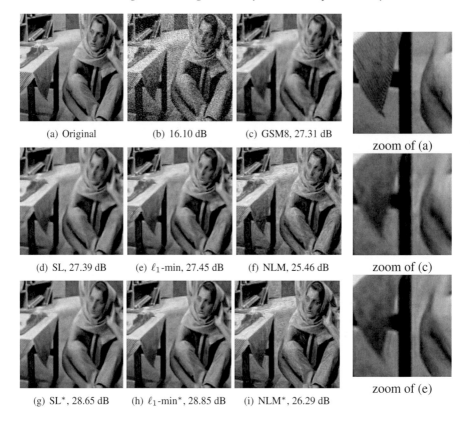

(a) Original	(b) 16.10 dB	(c) GSM8, 27.31 dB	zoom of (a)
(d) SL, 27.39 dB	(e) ℓ_1-min, 27.45 dB	(f) NLM, 25.46 dB	zoom of (c)
(g) SL*, 28.65 dB	(h) ℓ_1-min*, 28.85 dB	(i) NLM*, 26.29 dB	zoom of (e)

FIGURE 8.5
Denoising results. NonlocalmMeans* is for nonlocal means with oracle graph.

means computed with the same nonlocal graph. Given the adjacency matrix **W**, the nonlocal means proceeds by replacing each pixel of the noisy image by the weighted average of its graph neighbors. Recall that **W** is defined with zeros on the diagonal; the nonlocal means denoising result is

$$\mathbf{x}^*_{\text{NLM}}(j) = \frac{\mathbf{y}(j) + \sum_i w_{i,j}\mathbf{y}(i)}{1 + \sum_i w_{i,j}} .\tag{8.13}$$

We show the result of this nonlocal means in Figure 8.5, and note that the performance of the graph wavelet methods is much better. It should be noted that the nonlocal means results shown here are not the best results that can be obtained with the nonlocal means algorithm. The graph construction here has been optimized to give good results for the graph wavelet methods. We have observed that different choices for the graph sparsification and μ_{NL} can improve the nonlocal means performance. However, the dramatic difference in performance when using the same graph emphasizes that the graph wavelet methods are really doing something beyond simple nonlocal means.

	$\sigma = 20$	$\sigma = 40$	$\sigma = 80$
ℓ_1-min	0.133	0.117	0.116
oracle ℓ_1-min	-0.164	-0.159	-0.067

Predenoising (GSM8)	48.73 s
Nonlocal graph weights	272.15 s
Local graph weights	0.21 s
$\lambda_{\max}$, Chebyshev coefficients $c_{j,n}$	1.89 s
Wavelet norms ($\sigma_{\beta,n}$)	263.06 s

SL	64.65 s
ℓ_1-min	601.12 s

TABLE 8.4
Left: PSNR($\lambda' = 0.15$) - PSNR($\lambda' = 0$), averaged over 16 test images under different noise conditions. Right: CPU time for different stages of denoising, averaged over same four test images as in Table 8.3. Times for all steps in left column must be incurred for both SL and ℓ_1-min methods, right column gives the additional CPU time required for either SL or ℓ_1-min. Run-time measured with a MATLAB implementation running on one core of a 2.66 GHz Intel Xeon processor.

In order to assess the usefulness of introducing orientation through the local/nonlocal hybridization described in Section 8.4, we have studied the change in performance when the local oriented adjacency is removed (i.e., setting $\lambda' = 0$). Results averaged over a larger set of 16 test images are shown in Table 8.4. Including the oriented local component provides a moderate improvement on average when the nonlocal graph is computed via predenoising. Interestingly, including the local component actually degrades performance when the oracle graph is used. This suggests that the usefulness of the local component is to provide some regularization of the imperfectly estimated nonlocal graph. For the oracle case, as the nonlocal graph is perfectly estimated, the local component is not needed.

8.7 Conclusions

We have introduced a novel image model based on describing the wavelet coefficients under a new nonlocal image graph wavelet transform. The underlying nonlocal image graph has vertices associated with the image pixels, and edge weights which measure the similarity of different image patches. We constructed an overcomplete frame of wavelets on this graph using the spectral graph wavelet transform, and showed that the nonlocal graph wavelet coefficients of images are well modeled by a scaled Laplacian probability model. We also detailed a way for building local oriented wavelets with the spectral graph wavelet transform, enabling the construction of hybrid local/nonlocal graph wavelets.

As a demonstration of the power of the image graph wavelets, we have developed two related methods for image denoising based on the scaled Laplacian model. A straightforward Bayesian MAP inference in the wavelet domain leads to the scaled Laplacian thresholding method. The second weighted ℓ_1 method uses the scaled

Laplacian prior as part of a global convex objective function, leading to improved recovery of higher frequency image features.

There are many opportunities for future work using the nonlocal image graph wavelets. At present, we are actively investigating the use of nonlocal image graph wavelets for other image processing problems. Both image deconvolution and image super-resolution problems may be cast in a convex variational framework as the ℓ_1-min method. An additional important question is how to improve the estimation of the nonlocal graph, hopefully to avoid the somewhat awkward use of the predenoising method.

8.8   Acknowledgments

DH is funded by the US DoD TATRC Research Grant # W81XWH-09-2-0114.

LJ is funded by the Belgian Science Policy (Return Grant, BELSPO, Belgian Interuniversity Attraction Pole IAP-VI BCRYPT).

Bibliography

[1] D. K. Hammond, P. Vandergheynst, and R. Gribonval, "Wavelets on graphs via spectral graph theory," *Applied and Computational Harmonic Analysis*, vol. 30, pp. 129–150, 2011.

[2] D. L. Ruderman and W. Bialek, "Statistics of natural images: Scaling in the woods," *Physical Review Letters*, vol. 73, pp. 814–817, 1994.

[3] D. B. Mumford and B. Gidas, "Stochastic models for generic images," *Quarterly of Applied Mathematics*, vol. 59, pp. 85–111, 2001.

[4] A. B. Lee, D. Mumford, and J. Huang, "Occlusion models for natural images: A statistical study of a scale-invariant dead leaves model," *International Journal of Computer Vision*, vol. 41, pp. 35–59, 2001.

[5] A. Buades, B. Coll, and J. M. Morel, "A review of image denoising algorithms, with a new one," *Multiscale Modeling & Simulation*, vol. 4, pp. 490–530, 2005.

[6] G. Gilboa and S. Osher, "Nonlocal linear image regularization and supervised segmentation," *Multiscale Modeling & Simulation*, vol. 6, pp. 595–630, 2007.

[7] A. Buades, B. Coll, and J.-M. Morel, "Nonlocal image and movie denoising," *International Journal of Computer Vision*, vol. 76, pp. 123–139, 2008.

[8] L. Pizarro, P. Mrazek, S. Didas, S. Grewenig, and J. Weickert, "Generalised nonlocal image smoothing," *International Journal of Computer Vision*, vol. 90, pp. 62–87, 2010.

[9] K. Dabov, A. Foi, V. Katkovnik, and K. Egiazarian, "Image denoising by sparse 3D transform-domain collaborative filtering," *IEEE Transactions on Image Processing*, vol. 16, pp. 2080–2095, 2007.

[10] F. Zhang and E. R. Hancock, "Graph spectral image smoothing using the heat kernel," *Pattern Recogn*, vol. 41, pp. 3328–3342, 2008.

[11] A. D. Szlam, M. Maggioni, and R. R. Coifman, "Regularization on graphs with function-adapted diffusion processes," *Journal of Machine Learning Research*, vol. 9, pp. 1711–1739, 2008.

[12] G. Peyré, "Image processing with nonlocal spectral bases," *Multiscale Modeling & Simulation*, vol. 7, pp. 703–730, 2008.

[13] A. Elmoataz, O. Lezoray, and S. Bougleux, "Nonlocal discrete regularization on weighted graphs: A framework for image and manifold processing," *IEEE Transactions on Image Processing*, vol. 17, pp. 1047–1060, 2008.

[14] M. Crovella and E. Kolaczyk, "Graph wavelets for spatial traffic analysis," *INFOCOM*, pp. 1848–1857, 2003.

[15] M. Jansen, G. P. Nason, and B. W. Silverman, "Multiscale methods for data on graphs and irregular multidimensional situations," *Journal of the Royal Statistical Society: Series B*, vol. 71, pp. 97–125, 2009.

[16] A. B. Lee, B. Nadler, and L. Wasserman, "Treelets – an adaptive multi-scale basis for sparse unordered data," *Annals of Applied Statistics*, vol. 2, pp. 435–471, 2008.

[17] M. Gavish, B. Nadler, and R. Coifman, "Multiscale wavelets on trees, graphs and high dimensional data: Theory and applications to semi supervised learning," in *International Conference on Machine Learning*, 2010.

[18] R. R. Coifman and M. Maggioni, "Diffusion wavelets," *Applied and Computational Harmonic Analysis*, vol. 21, pp. 53–94, 2006.

[19] I. Daubechies, *Ten Lectures on Wavelets*. Society for Industrial and Applied Mathematics, 1992.

[20] G. M. Phillips, *Interpolation and Approximation by Polynomials*. Springer-Verlag, 2003.

[21] G. Golub and C. V. Loan, *Matrix Computations*. Johns Hopkins University Press, 1983.

[22] T. Tasdizen, "Principal neighborhood dictionaries for nonlocal means image denoising," *IEEE Transactions on Image Processing*, vol. 18, pp. 2649–2660, 2009.

[23] T. Brox, O. Kleinschmidt, and D. Cremers, "Efficient nonlocal means for denoising of textural patterns," *Image Processing, IEEE Transactions on*, vol. 17, pp. 1083–1092, 2008.

[24] Y. Eldar, M. Lindenbaum, M. Porat, and Y. Zeevi, "The farthest point strategy for progressive image sampling," *IEEE Transactions on Image Processing*, vol. 6, pp. 1305–1315, 1997.

[25] G. Peyré and L. Cohen, "Geodesic remeshing using front propagation," *International Journal of Computer Vision*, vol. 69, pp. 145–156, 2006.

[26] K. M. Taylor and F. G. Meyer, "A random walk on image patches," 2011, arXiv:1107.0414v1 [physics.data-an].

[27] C. Tomasi and R. Manduchi, "Bilateral filtering for gray and color images," *IEEE International Conference on Computer Vision*, vol. 0, p. 839, 1998.

[28] M. Grant and S. Boyd, "CVX: Matlab software for disciplined convex programming," http://cvxr.com/cvx.

[29] I. Selesnick, R. Baraniuk, and N. Kingsbury, "The dual-tree complex wavelet transform," *Signal Processing Magazine, IEEE*, vol. 22, pp. 123–151, 2005.

[30] J.-L. Starck, E. Candes, and D. Donoho, "The curvelet transform for image denoising," *IEEE Transactions on Image Processing*, vol. 11, pp. 670–684, 2002.

[31] J. Portilla, V. Strela, M. J. Wainwright, and E. P. Simoncelli, "Image denoising using scale mixtures of Gaussians in the wavelet domain," *IEEE Transactions on Image Processing*, vol. 12, pp. 1338–1351, 2003.

[32] M. Elad and M. Aharon, "Image denoising via sparse and redundant representations over learned dictionaries," *IEEE Transactions on Image Processing*, vol. 15, pp. 3736–3745, 2006.

[33] R. R. Coifman and D. L. Donoho, "Translation invariant de-noising," in *Wavelets and Statistics*, A. Antoniadis and G. Oppenheim, Eds. Springer-Verlag, 1995, pp. 125–150.

[34] M. Raphan and E. Simoncelli, "Optimal denoising in redundant representations," *IEEE Transactions on Image Processing*, vol. 17, pp. 1342–1352, 2008.

[35] S. Mallat, "A theory for multiresolution signal decomposition: the Wavelet representation," *IEEE Pattern Analysis and Machine Intelligence*, vol. 11, 1989.

[36] P. Moulin and J. Liu, "Analysis of multiresolution image denoising schemes using generalized Gaussian and complexity priors," *IEEE Transactions on Information Theory*, vol. 45, pp. 909–919, 1999.

[37] H. Rabbani, "Image denoising in steerable pyramid domain based on a local Laplace prior," *Pattern Recognition*, vol. 42, pp. 2181–2193, 2009.

[38] S. Lyu and E. Simoncelli, "Modeling multiscale subbands of photographic images with fields of Gaussian scale mixtures," *IEEE Transactions on Pattern Analysis and Machine Intelligence*, vol. 31, pp. 693 –706, 2009.

[39] G. Peyré, S. Bougleux, and L. Cohen, "Non-local regularization of inverse problems," in *European Conference on Computer Vision (ECCV)*, 2008, pp. 57–68.

[40] A. Chambolle, R. De Vore, N.-Y. Lee, and B. Lucier, "Nonlinear wavelet image processing: variational problems, compression, and noise removal through wavelet shrinkage," *IEEE Transactions on Image Processing*, vol. 7, pp. 319–335, 1998.

[41] I. Daubechies, M. Defrise, and C. D. Mol, "An iterative thresholding algorithm for linear inverse problems with a sparsity constraint," *Communications on Pure and Applied Mathematics*, vol. 57, pp. 1413–1457, 2004.

[42] P. Combettes and V. Wajs, "Signal recovery by proximal forward-backward splitting," *Multiscale Model Sim*, vol. 4, pp. 1168–1200, 2005.

[43] J. Bioucas-Dias and M. Figueiredo, "A new twist: Two-step iterative shrinkage/thresholding algorithms for image restoration," *Image Processing, IEEE Transactions on*, vol. 16, pp. 2992–3004, 2007.

[44] M. Figueiredo, R. Nowak, and S. Wright, "Gradient projection for sparse reconstruction: Application to compressed sensing and other inverse problems," *Selected Topics in Signal Processing*, vol. 1, pp. 586–597, 2007.

9

Image and Video Matting

Jue Wang

Advanced Technology Labs
Adobe Systems
801 N 34th St, Seattle, WA 98103, USA
Email: juewang@adobe.com

CONTENTS

9.1 Introduction 237
    9.1.1 User Constraint 238
    9.1.2 Earlier Approaches 240
9.2 Graph Construction for Image Matting 241
    9.2.1 Defining Edge Weight w_{ij} 242
    9.2.2 Defining Node Weight w_i 245
9.3 Solving Image Matting Graphs 250
    9.3.1 Solving MRF 250
    9.3.2 Linear Optimization 251
9.4 Data Set 252
9.5 Video Matting 253
    9.5.1 Overview 254
    9.5.2 Graph-Based Trimap Generation 254
    9.5.3 Matting with Temporal Coherence 258
9.6 Conclusion 259
    Bibliography 261

9.1 Introduction

Matting refers to the problem of accurately separating a foreground object from the background by determining both full and partial pixel coverage, in both still images and video sequences. Mathematically, the color vector $\mathbf{c}_p$ of a pixel p in the input image is modeled as a convex combination of a foreground color $\mathbf{f}_p$ and a background color $\mathbf{b}_p$:

$$\mathbf{c}_p = \alpha_p \mathbf{f}_p + (1 - \alpha_p)\mathbf{b}_p, \tag{9.1}$$

where $\alpha_p \in [0, 1]$ is the *alpha value* of the pixel. The collection of alpha values of all pixels in the image is called the *alpha matte*. This equation, first introduced by Porter

237

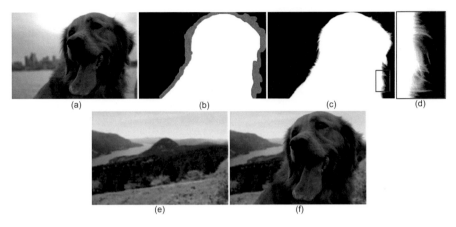

FIGURE 9.1
An image matting example: (a) original image, (b) user specified trimap, where white means foreground, black means background, and gray means unknown, (c) extracted alpha matte, white means higher alpha value, (d) a close-up view of the highlighted region on the alpha matte, (e) a new background image, (f) a new composite.

and Duff in 1984 [1], is called the *Compositing Equation*. In the matting problem, usually only the observed color $\mathbf{c}_p$ is known, and the goal is to accurately recover the alpha value α_p and the underlying foreground color $\mathbf{f}_p$ [1], so that the foreground object can be fully separated from the background. Once estimated, the alpha matte can be used as an operational mask in numerous image and video editing applications, such as applying special digital filters to the foreground object, or replacing the original background image with a new one to create a novel composite. Figure 9.1 shows an example of using matting techniques to extract the foreground object and compose it onto a new background image.

Matting can be viewed as an extension to the classic binary segmentation problem, where each pixel is assigned fully to either the foreground or the background, i.e., $\alpha_p \in \{0, 1\}$. In natural images, although the majority of pixels are usually either definite foreground or definite background, accurately estimating fractional alpha values for pixels on the foreground edge is essential for extracting fuzzy object boundaries such as hair or fur, as shown in the example in Figure 9.1.

9.1.1   User Constraint

Matting is inherently an under-constrained problem. Assuming the input image has three color channels, seven unknown variables (three for $\mathbf{f}_p$, three for $\mathbf{b}_p$, and one for

---

[1]In many applications, recovering α_p alone is sufficient. We will mainly focus on how to recover the alpha matte in this chapter.

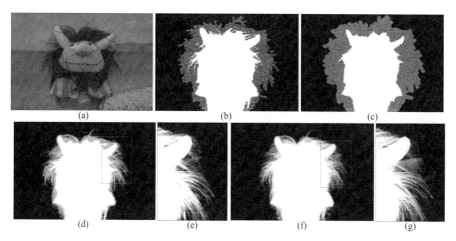

FIGURE 9.2
Image matting with different trimaps: (a) input image, (b) a tight trimap, (c) a coarse trimap, (d) matte generated using the tight trimap, (e) a close-up view of the highlighted region in (d), (f) matte generated using the coarse trimap, (g) a close-up view of the highlighted region in (f). Both mattes are generated using the Robust matting algorithm [2].

α_p) need to be estimated from three color channel values of c_p. Without any additional constraint, the problem is ill-posed and many possible solutions exist. For instance, a simple solution satisfying the Compositing Equation is to set $f_p = b_p = c_p$, and choose α_p arbitrarily. This is obviously not the correct solution for separating the foreground. In order to estimate an alpha matte that represents the correct foreground coverage, most matting approaches rely on both user guidance and prior knowledge on natural image statistics to constrain the solution space. A typical user constraint provided to a matting system is a *trimap*, where each pixel in the image is assigned to three possible values: definite foreground $\mathcal{F}$ ($\alpha_p = 1$), definite background $\mathcal{B}$ ($\alpha_p = 0$), and unknown $\mathcal{U}$ (α_p to be determined). Figure 9.1b shows an example of a user-specified trimap. Given the input trimap, the task of matting is thus reduced to estimating alpha values for unknown pixels, under the constraint of the known pixels in $\mathcal{F}$ and $\mathcal{B}$.

The accuracy of the input trimap has a direct impact on the quality of the final alpha matte. In general, an accurately specified trimap, where the unknown region covers only truly semitransparent pixels, often leads to a more accurate alpha matte than a loosely defined trimap, due to the reduced number of unknown variables. An example is shown in Figure 9.2, where the same matting algorithm is applied to the same input image with two different trimaps. The trimap in Figure 9.2b is more accurate than the one shown in Figure 9.2c. Therefore, it leads to a more accurate alpha matte, as shown in Figure 9.2d to Figure 9.2g.

Given that manually specifying an accurate trimap is still a tedious process, fast image segmentation approaches have been adopted in matting systems for efficient trimap generation. The Grabcut system [3] can generate a fairly accurate binary segmentation of the foreground object starting from a user-specified bounding box, using iterated graph cuts optimization. Under a similar optimization framework, the Lazy Snapping system [4] generates a binary segmentation from a few foreground and background scribbles. Once the binary segmentation is obtained, one can simply erode and dilate the foreground contour to create an unknown region for matting. Rhemann et al. [5] proposed a more accurate trimap segmentation tool, which explicitly segments the image into three regions $\mathcal{F}$, $\mathcal{B}$, and $\mathcal{U}$, using a parametric max-flow algorithm. With the help of these techniques, accurate trimaps can be efficiently generated with a small amount of user interaction.

In this chapter, for image matting, we assume the trimap is already given, and focus on how to estimate fractional alpha values in the $\mathcal{U}$ region. For video matting, we will describe how to generate temporally consistent trimaps for video frames, as it is the main bottleneck for applying matting techniques in video.

9.1.2 Earlier Approaches

Earlier matting approaches try to estimate alpha values for individual pixels independently without any spatial regularization. A common strategy is that for an unknown pixel p, sampling a set of known foreground and background colors nearby as priors for $\mathbf{f}_p$ and $\mathbf{b}_p$. Assuming the image colors vary smoothly, these samples can be treated as reasonably accurate estimates of $\mathbf{f}_p$ and $\mathbf{b}_p$. Once $\mathbf{f}_p$ and $\mathbf{b}_p$ are estimated, solving α_p from the compositing equation becomes trivial.

Various statistical models have been used to estimate $\mathbf{f}_p$ and $\mathbf{b}_p$ from color samples. Ruzon and Tomasi developed a parametric sampling algorithm [6], where foreground and background color samples are modeled as mixture of Gaussians, and the observed color $\mathbf{c}_p$ is modeled as coming from an intermediate distribution between the foreground and background Gaussian distributions. Mishima [7] developed a blue screen matting system which uses color samples in a nonparametric way. Detailed explanations of these methods can be found in Wang and Cohen's survey [8].

A notable method among early approaches is the Bayesian matting [9] approach, which formulates the estimation of $\mathbf{f}_p$, $\mathbf{b}_p$, and α_p in a Bayesian framework, and solves the matte using the maximum a posteriori (MAP) technique. It is the first to cast the matting problem into a well-formulated statistical inference framework. Mathematically, for an unknown pixel p, its matting solution is formulated as:

$$\arg\max_{\mathbf{f}_p,\mathbf{b}_p,\alpha_p} P(\mathbf{f}_p,\mathbf{b}_p,\alpha_p|\mathbf{c}_p) = \arg\max_{\mathbf{f}_p,\mathbf{b}_p,\alpha_p} L(\mathbf{c}_p|\mathbf{f}_p,\mathbf{b}_p,\alpha_p)+L(\mathbf{f}_p)+L(\mathbf{b}_p)+L(\alpha_p),$$

$$(9.2)$$

where $L(\cdot)$ is the log likelihood $L(\cdot) = \log P(\cdot)$. The first term on the right side is measured as:

$$L(\mathbf{c}_p|\mathbf{f}_p,\mathbf{b}_p,\alpha_p) = -\parallel \mathbf{c}_p - \alpha_p\mathbf{f}_p - (1-\alpha_p)\mathbf{b}_p \parallel^2 /\sigma_p^2, \qquad (9.3)$$

where the color variance σ_p is measured locally. This is simply the fitting error according to the compositing equation. To estimate $L(\mathbf{f}_p)$, foreground colors in the nearby region are first partitioned into groups, and in each group an oriented Gaussian is estimated by computing the mean $\bar{\mathbf{f}}$ and covariance $\mathbf{\Sigma}_F$. $L(\mathbf{f}_p)$ is then defined as:

$$L(\mathbf{f}_p) = -(\mathbf{f}_p - \bar{\mathbf{f}})^T \mathbf{\Sigma}_F^{-1}(\mathbf{f}_p - \bar{\mathbf{f}})/2. \tag{9.4}$$

$L(\mathbf{b}_p)$ is calculated in the same way using background samples. $L(\alpha_p)$ is treated as a constant. The solution of the inference problem is obtained by iteratively estimating $(\mathbf{f}_p, \mathbf{b}_p)$ and α_p.

Despite their success in simple cases, matting approaches that only depend on color sampling often fail on more complicated images where foreground and background color distributions are overlapping, and the input trimap is loosely defined. This is due to two fundamental limitations of this framework. First, estimating alpha values independently for individual pixels without any spatial regularization is prone to image noise. Second, when the trimap is loosely defined, the sampled foreground and background colors are no longer good priors for estimating $\mathbf{f}_p$ and $\mathbf{b}_p$, leading to estimation bias. Modern matting approaches try to avoid these limitations by enforcing local spatial regularization on the estimated alpha matte, which allows them to perform more robustly on difficult examples. We will focus on these methods from now on.

9.2 Graph Construction for Image Matting

Modern image matting approaches usually start by modeling the input image as an undirected and weighted graph $\mathcal{G} = (\mathcal{V}, \mathcal{E})$, where each $v_i \in \mathcal{V}$ corresponds to an unknown pixel, i.e., a pixel in the $\mathcal{U}$ region of the input trimap[2]. An edge $e_i \in \mathcal{E}$ is usually defined between a node and its four spatial neighbors, although in some work the edges are defined among pixels in larger spatial neighborhoods such as 3×3 or 5×5 [10]. Additionally, some approaches define a node weight w_i for each node v_i, which encodes some prior knowledge of the alpha value of v_i, given its observed color $\mathbf{c}_i$ and nearby foreground and background colors. The structure of such a graph is shown in Figure 9.3. Using this graph representation, matting is transferred into a graph labeling problem, where the objective is to solve for the optimal α_is that can minimize the total energy of the graph.

Although many image matting approaches share the same common graph structure, the key difference among them is the formulation of the edge weight w_{ij}, and optionally the node weight w_i. Table 9.1 summaries some representative image matting techniques that will be discussed in this section, and their choices of the edge

---

[2]For notation simplicity, we will use v_i to denote both an unknown pixel and its corresponding node in the graph.

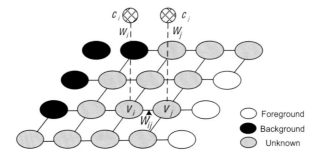

FIGURE 9.3
A typical graph setup for image matting.

Method	Reference	Edge weight	Node weight	Optimization method
Random walk matting	[11]	w_{ij}^{lpp} (Eqn.9.7)	No	Linear system
Easy matting	[12]	w_{ij}^{easy} (Eqn.9.8)	w_i^{easy} (Eqn.9.21)	Linear system
Iterative BP matting	[13]	w_{ij}^{bp} (Eqn.9.9)	w_i^{bp} (Eqn.9.22)	Belief propagation
Closed-form matting	[10]	w_{ij}^{cf} (Eqn.9.13)	No	Linear system
Robust matting	[2]	w_{ij}^{cf} (Eqn.9.13)	w_i^{rm} (Eqn.9.27)	Linear system
Learning-based matting	[14]	w_{ij}^{ln} (Eqn.9.20)	No	Linear system
Global sampling matting	[15]	w_{ij}^{cf} (Eqn.9.13)	w_i^{gs} (Eqn.9.32)	Linear system

TABLE 9.1
Summary of image matting techniques and their graph structures.

and the node weights. Specifically, different edge weights will be discussed in detail in Section 9.2.1, and various node weights will be illustrated in Section 9.2.2. Finally, how to solve the graph labeling problem using the define edge and node weights will be described in Section 9.3.

9.2.1 Defining Edge Weight w_{ij}

One straightforward idea of mapping pixel colors to the edge weight e_{ij} is to use the following classic Euclidean norm:

$$w_{ij} = \exp\left(-\frac{\|\mathbf{c}_i - \mathbf{c}_j\|^2}{\sigma_{ij}^2}\right), \qquad (9.5)$$

where $\mathbf{c}_i$ is the RGB color of v_i, σ_{ij} is a parameter which can be either manually selected by the user, or automatically computed based on local image statistics [16]. This function penalizes large alpha changes in flat image regions where the color difference between two neighboring pixels is small. This metric has been widely used in graph-based binary image segmentation systems [17]. Grady et al. [11] adopted this formulation for image matting, but instead of measuring the Euclidean norm in the RGB color space, they proposed to use the Locality Preserving Projections (LPP) technique [18] to define a conjugate norm, which is more reliable for describing perceptual object boundaries than the RGB color space. Mathematically, the projections defined by the LPP algorithm are given by solving the following generalized eigenvector problem:

$$\mathbf{Z}\mathbf{L}\mathbf{Z}^T\mathbf{x} = \lambda\mathbf{Z}\mathbf{D}\mathbf{Z}^T\mathbf{x}, \tag{9.6}$$

where $\mathbf{Z}$ is a $3 \times N$ (the number of pixels in the image) matrix with each $\mathbf{c}_i$ color vector as a column, $\mathbf{D}$ is the diagonal matrix defined as $\mathbf{D}_{ii} = d_i \doteq \sum w_{ij}$, and $\mathbf{L}$ is the sparse Laplacian matrix where $\mathbf{L}_{ii} = d_i$, and $\mathbf{L}_{ij} = -w_{ij}$ for $j \neq i$. Denote its solution by $\mathbf{Q}$ where each eigenvector is a row, the final edge weight is computed as:

$$w_{ij}^{\text{lpp}} = \exp\left(-\frac{(\mathbf{c}_i - \mathbf{c}_j)^T\mathbf{Q}^T\mathbf{Q}(\mathbf{c}_i - \mathbf{c}_j)}{\sigma_{ij}^2}\right). \tag{9.7}$$

The edge weights defined above are static given an input image, and they are not related to the values of the random variables α_i and α_j. In some other approaches, the edge weight is explicitly defined as a function of α_i and α_j to enforce the alpha matte to be locally smooth. The Easy matting system [12] defines the edge weight in a quadratic form as:

$$w_{ij}^{\text{easy}} = \lambda \cdot \frac{(\alpha_i - \alpha_j)^2}{\|\mathbf{c}_i - \mathbf{c}_j\|}, \tag{9.8}$$

where λ is a user defined constant. Minimizing the sum of edge weights over the graph will force the alpha values of adjacent pixels to be similar if local image gradient is small. Under the same motivation, Wang and Cohen [13] constructed a Markov Random Field (MRF) where the joint distribution over α_i and α_j is given by the Boltzmann distribution of a similar quadratic energy function as:

$$w_{ij}^{\text{bp}} = \exp\left(-\frac{(\alpha_i - \alpha_j)^2}{\sigma_{\text{const}}^2}\right). \tag{9.9}$$

All the edge weights defined above implicitly assume that the input image is locally smooth. In the closed-form matting algorithm [10], the local smoothness is explicitly modeled using a *color line model*. That is, in a small local window Υ (3×3 or 5×5), the foreground and background colors ($\mathbf{f}_i$s and $\mathbf{b}_i$s) are linear mixtures of two latent colors:

$$\mathbf{f}_i = \beta_i^f\mathbf{f}_{l1} + (1 - \beta_i^f)\mathbf{f}_{l2} \quad \text{and} \quad \mathbf{b}_i = \beta_i^b\mathbf{b}_{l1} + (1 - \beta_i^b)\mathbf{b}_{l2}, \quad \forall i \in \Upsilon, \tag{9.10}$$

where $\mathbf{f}_{l1}, \mathbf{f}_{l2}, \mathbf{b}_{l1}$, and $\mathbf{b}_{l2}$ are latent colors. Combining this constraint with the Compositing Equation, it is easy to show that alpha values in Υ can be expressed as:

$$\alpha_i = \sum_k a^k\mathbf{c}_i^k + b, \forall i \in \Upsilon, \tag{9.11}$$

where k refers to color channels, and a^k and b are functions of β_i^f, β_i^b, $\mathbf{f}_{l1}$, $\mathbf{f}_{l2}$, $\mathbf{b}_{l1}$ and $\mathbf{b}_{l2}$, thus they are constants in the window. Based on this constraint, Levin et al. [10] defined a quadratic matting cost function as:

$$J(\alpha, a, b) = \sum_{j \in I} \left(\sum_{i \in \Upsilon_j} \left(\alpha_i - \sum_k a_j^k \mathbf{c}_i^k - b_j \right)^2 + \epsilon \sum_k a_j^{k2} \right), \qquad (9.12)$$

where the second term is a regularization term mainly for the purpose of numerical stability. It also has a desirable side effect of biasing the solution toward smoother alpha mattes, since $a_j = 0$ means that α is constant over the jth window. Given this cost function, the edge weight can be derived as:

$$w_{ij}^{cf} = \frac{1}{|\Upsilon_k|} \sum_{k|(i,j) \in \Upsilon_k} \left(1 + (\mathbf{c}_i - \boldsymbol{\mu}_k)^T (\boldsymbol{\Sigma}_k + \frac{\varepsilon}{|\Upsilon_k|} \mathbf{I}_3)^{-1} (\mathbf{c}_j - \boldsymbol{\mu}_k) \right), \qquad (9.13)$$

where $\boldsymbol{\Sigma}_k$ is a 3×3 covariance matrix, $\boldsymbol{\mu}_k$ is a 3×1 mean vector of the colors in the window Υ_k, and $\mathbf{I}_3$ is the 3×3 identity matrix. Consequently, a graph Laplacian matrix can be computed as:

$$\mathbf{L}_{ij}^{cf} = \begin{cases} -w_{ij}^{cf} & : \quad \text{if } i \neq j, (i,j) \in \Upsilon_k, \\ \sum_{l,l \neq i} w_{il}^{cf} & : \quad \text{if } i = j, \\ 0 & : \quad \text{otherwise.} \end{cases} \qquad (9.14)$$

which is called the *matting Laplacian*. It is easy to show that the Laplacian matrix $\mathbf{L}^{cf}$ is symmetric and positive semidefinite.

There are several unique characteristics of the edge weight w_{ij}^{cf}. First, each pixel has a nonzero weight with all other pixels in a local (5×5 if Υ is 3×3) neighborhood, thus the Laplacian matrix is a much denser one than that of a typical four-neighbor image graph. In each row of the Laplacian matrix $\mathbf{L}^{cf}$, there are 25 nonzero elements, compared with 5 in a four-neighbor graph Laplacian. Second, the edge weight w_{ij}^{cf} can be negative, which is different from the strictly nonnegative edge weights defined in Equation 9.5 to 9.9. It is thus worth emphasizing that some of the commonly used nonnegative graph analysis methods do not apply to $\mathbf{L}^{cf}$. For instance, in a nonnegative graph, the degree of a vertex v_i is $d_i = \sum w_{ij}$ for all edges e_{ij} incident on v_i, and is often used for normalizing all nonzero values in the ith row of the graph Laplacian [17]. For the matting Laplacian $\mathbf{L}^{cf}$, computing the degree of a vertex and using it for normalization is no longer suitable, as positive and negative values cancel out each other. Another side effect of using negative edge weights is that the computed alpha values are no longer guaranteed to be within $[0, 1]$, as discussed in [19]. Using this method it is quite often to get alpha values that are out of bounds, thus in practice one has to clip the alpha values at 0 and 1 to generate visually correct alpha mattes.

Since the matting Laplacian $\mathbf{L}^{cf}$ is strictly derived from the color line model, it often leads to accurate alpha mattes when the underlying color model is satisfied. In practice, if the input image is composed of smooth regions without strong textures, the color line model often holds true. Given its generality and accuracy, it has been widely used in recent matting approaches where it is combined with other techniques to achieve high-quality results. It has also been applied in many other applications such as image dehazing [20] and light source separation [21] as a general edge-aware interpolation tool.

The limitations of the matting Laplacian $\mathbf{L}^{cf}$ have also been extensively studied. Singaraju et al. [22] demonstrated that the color line model provides an overfit and leads to ambiguity when the intensity variations of the foreground and background layers are much simpler than the color line. Specifically, they studied two compact color models, point-point color model and line-point color model. The former applies when both the foreground and background intensities are locally constant (the point constraint), and the latter applies when one layer satisfies the point constraint while the other satisfies the color line constraint. These two color models lead to modified edge weights which work better than the original matting Laplacian in these special cases.

To deal with more complicated cases where a linear color model is incapable of accurately describing the color variations, Zheng et al. [14] proposed a semi-supervised learning approach, where the well-known kernel trick [23] is used to deal with nonlinear local color distributions. This approach assumes that the alpha value of an unknown pixel v_i is a linear combination of the alpha values of its neighboring pixels, e.g., a 7×7 window centered at the pixel:

$$\alpha_i = \boldsymbol{\xi}_i^T \alpha, \tag{9.15}$$

where α is the alpha value vector of all pixels in the image, and $\boldsymbol{\xi}_i$ is a coefficient vector where the majority values are 0 except for pixels which are in the neighborhood of v_i. By introducing a new matrix $\mathbf{G}$ through stacking $\boldsymbol{\xi}_i$s: $\mathbf{G} = [\boldsymbol{\xi}_1, ..., \boldsymbol{\xi}_n]$, the above equation can be rewritten as:

$$\alpha = \mathbf{G}^T \alpha. \tag{9.16}$$

This leads to a solution of α by minimizing the following quadratic cost function:

$$\arg \min_{\alpha} \|\alpha - \mathbf{G}^T \alpha\|^2, \tag{9.17}$$

which can be reformulated as:

$$E^{\ln}(\alpha) = \alpha^T (\mathbf{I}_n - \mathbf{G})(\mathbf{I}_n - \mathbf{G})^T \alpha. \tag{9.18}$$

The Laplacian matrix of this approach is then defined as:

$$\mathbf{L}^{\ln} = (\mathbf{I}_n - \mathbf{G})(\mathbf{I}_n - \mathbf{G})^T, \tag{9.19}$$

and the corresponding edge weight is

$$w_{ij}^{\ln} = \mathbf{L}^{\ln}(i, j). \tag{9.20}$$

The matrix $\mathbf{G}$ is obtained by training a non-linear alpha-color model from known pixels in the trimap, either locally or globally, as detailed in the original paper [14].

9.2.2 Defining Node Weight w_i

For each v_i, the node weight w_i measures how compatible an estimated α_i is with its observed color $\mathbf{c}_i$, and nearby foreground and background colors. Not all matting approaches define and use this term, but it has been shown [2] that the node weight, if defined properly, can lead to more accurate alpha estimation.

A common approach for defining w_i is to first sample a set of nearby known foreground and background colors, denoted as $\mathbf{c}_k^f$ and $\mathbf{c}_l^b$, $0 < k, l < N$, as shown in Figure 9.4. Assuming the foreground and background colors vary smoothly, these color samples assemble reasonable probabilistic distributions of $\mathbf{f}_i$ and $\mathbf{b}_i$. Combined with $\mathbf{c}_i$, these samples can be used to test the feasibility of an estimated α_i using the compositing equation. Using this idea, Guan et al. [12] defined the node weight in the Easy Matting system as:

$$w_i^{\text{easy}} = \frac{1}{N^2} \sum_{k=1}^{N} \sum_{l=1}^{N} \|\mathbf{c}_i - \alpha_i \mathbf{c}_k^f - (1 - \alpha_i)\mathbf{c}_l^b\|^2 / \sigma_i^2, \tag{9.21}$$

where σ_i is the distance variance among $\mathbf{c}_i$ and $\alpha_i \mathbf{c}_k^f + (1 - \alpha_i)\mathbf{c}_l^b$. This node weight favors an α_i that can best explain the observed color $\mathbf{c}_i$ as a linear combination of the sampled colors $\mathbf{c}_k^f$ and $\mathbf{c}_l^b$. Similarly, Wang and Cohen [13] defined a node weight using exponential functions as:

$$w_i^{\text{bp}} = \frac{1}{N^2} \sum_{k=1}^{N} \sum_{l=1}^{N} \mu_k^f \mu_l^b \cdot \exp\left(-\|\mathbf{c}_i - \alpha_i \mathbf{c}_k^f - (1 - \alpha_i)\mathbf{c}_l^b\|^2 / \sigma_i^2\right), \tag{9.22}$$

where μ_k^f and μ_l^b are additional weights for color samples based on their spatial distances to v_i.

The node weights defined in Equation 9.21 and 9.22 treat every color sample equally. In practice, however, when the size of the sample set is large and the foreground and background color distributions are complex, the samples set may present a large color variance. It is often the case that only a small number of samples are good for estimating α_i. To decide which samples are good for defining w_i, Wang and Cohen [2] proposed a color sample selection method. In this approach, "good" sample pairs are defined as those that can explain $\mathbf{c}_i$ as a convex combination of themselves. Mathematically, as illustrated in Figure 9.4, for a pair of foreground and background colors $(\mathbf{c}_k^f, \mathbf{c}_l^b)$, a distance ratio is defined as:

$$R_d(\mathbf{c}_k^f, \mathbf{c}_l^b) = \frac{\|\mathbf{c}_i - \hat{\alpha}_i \mathbf{c}_k^f - (1 - \hat{\alpha}_i)\mathbf{c}_l^b\|}{\|\mathbf{c}_k^f - \mathbf{c}_l^b\|}, \tag{9.23}$$

where $\hat{\alpha}_i$ is the alpha value estimated from this sample pair as:

$$\hat{\alpha}_i = \frac{(\mathbf{c}_i - \mathbf{c}_l^b)(\mathbf{c}_k^f - \mathbf{c}_l^b)}{\|\mathbf{c}_k^f - \mathbf{c}_l^b\|^2}. \tag{9.24}$$

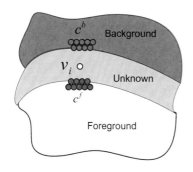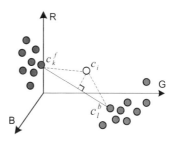

FIGURE 9.4
Left: to define the node weight for an unknown pixel v_i, a set of spatially nearby foreground and background colors $\mathbf{c}^f$ and $\mathbf{c}^b$ are collected. Right: a sample pair $\mathbf{c}_k^f$ and $\mathbf{c}_l^b$ is considered to be a good estimation of the true foreground and background colors of v_i if $\mathbf{c}_k^f$, $\mathbf{c}_l^b$, and $\mathbf{c}_i$ together satisfy the linear constraint in the RGB color space.

$R_d(\mathbf{c}_k^f, \mathbf{c}_l^b)$ essentially measures the linearity of $\mathbf{c}_k^f$, $\mathbf{c}_l^b$, and I_i in the color space. It is easy to show that $R_d(\mathbf{c}_k^f, \mathbf{c}_l^b)$ is small if the three colors approximately lie on a color line, and vise versa. Based on the distance ratio, a confidence value for the sample pair is defined as:

$$f(\mathbf{c}_k^f, \mathbf{c}_l^b) = \exp\left(-\frac{R_d(\mathbf{c}_k^f, \mathbf{c}_l^b)}{\sigma_c^2}\right) \cdot \gamma(\mathbf{c}_k^f) \cdot \gamma(\mathbf{c}_l^b), \qquad (9.25)$$

where σ_c is a constant set at 0.1. $\gamma(\mathbf{c}_k^f)$ is the weight for the color $\mathbf{c}_k^f$:

$$\gamma(\mathbf{c}_k^f) = \exp\left(-\frac{\|\mathbf{c}_k^f - \mathbf{c}_i\|^2}{\min_k(\|\mathbf{c}_k^f - \mathbf{c}_i\|^2)}\right), \qquad (9.26)$$

which favors foreground samples that are close to the target pixel. $\gamma(\mathbf{c}_l^b)$ is defined in a similar way.

Given a set of foreground and background samples, this approach exhaustively examines every possible foreground background sample combination, and finally chooses a few sample pairs (typically 3) with the highest confidence values. Denote the average confidence value of these pairs as $\overline{f}_i$, and the average alpha value estimated from these pairs using Equation 9.24 as $\overline{\alpha}_i$, the final node weight is computed as

$$w_i^{\mathrm{rm}} = \overline{f}_i \cdot (\alpha_i - \overline{\alpha}_i)^2 + (1 - \overline{f}_i)(\alpha_i - H(\overline{\alpha}_i - 0.5))^2, \qquad (9.27)$$

where $H(x)$ is the Heaviside step function which outputs 1 when $x > 0$ and 0 otherwise. This node weight encourages the final alpha value α_i to be close to $\overline{\alpha}_i$ when the sampling confidence $\overline{f}_i$ is high, and α_i to be either 0 or 1 when the confidence

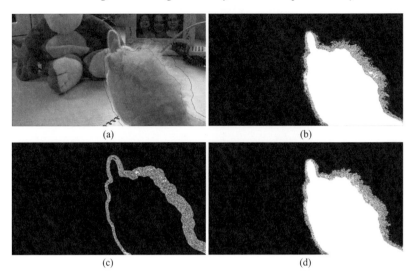

(a) (b)

(c) (d)

FIGURE 9.5
An example of the color sampling method used in the Robust Matting system [2]. (a) Input image with overlayed trimap, (b) initial alpha matte computed by Equation 9.24 using best sample pairs, (c) the confidence map computed by Equation 9.25, white means higher confidence values, (d) final alpha matte after minimizing the energy in Equation 9.40.

is low. When the confidence is low, it suggests that c_i cannot be well approximated as a linear interpolation of known foreground and background colors, and it is thus more likely to be a new foreground or background color.

Figure 9.5 shows an example of using this sampling method for alpha matting. Given the input image and the specified trimap as shown in Figure 9.5a, the alpha matte $\overline{\alpha}_i$ computed using Equation 9.24, and the confidence map $\overline{f}_i$ computed using Equation 9.25 are visualized in Figure 9.5b and 9.5c. The final alpha matte of the Robust Matting algorithm after minimizing the energy function in Equation 9.40 is shown in Figure 9.5d.

Based on the color sampling method described above, Rhemann et al. [24] proposed an improved sampling procedure for defining the node weight. In Wang and Cohen's work, foreground and background samples are selected solely based on their spatial distances to v_i, without considering the underlying image structures. To improve this, Rhemann et al. proposed to select foreground samples based on their geodesic distances [25] to v_i in the image space. This distance measure encourages the foreground samples to not only be close to v_i spatially, but also belong to the same connected image component as v_i. In this approach, a confidence function that is slightly different from Equation 9.25 is defined as:

$$f_2(\mathbf{c}_k^f, \mathbf{c}_l^b) = \exp\left(-\frac{R_d(\mathbf{c}_k^f, \mathbf{c}_l^b) \cdot \gamma'(\mathbf{c}_k^f) \cdot \gamma'(\mathbf{c}_l^b)}{\sigma_c^2}\right), \tag{9.28}$$

where the sample weights are defined as

$$\begin{cases} \gamma'(\mathbf{c}_k^f) = \exp\left(-\frac{\max_k(\|\mathbf{c}_k^f - \mathbf{c}_i\|^2)}{\|\mathbf{c}_k^f - \mathbf{c}_i\|^2}\right), \\ \gamma'(\mathbf{c}_l^b) = \exp\left(-\frac{\max_k(\|\mathbf{c}_l^b - \mathbf{c}_i\|^2)}{\|\mathbf{c}_l^b - \mathbf{c}_i\|^2}\right). \end{cases} \tag{9.29}$$

Both sampling procedures described above exhaustively examine every possible combination of foreground and background samples, thus their computational cost is high. For instance, if N ($N = 20$ in [2]) samples are collected for both the foreground and the background, N^2 pair evaluations have to be performed for every unknown pixel. To reduce the computational cost while maintaining the sampling accuracy, Gastal and Oliveira proposed a Shared Sampling method [26], which is motivated by the fact that pixels in a small neighborhood usually share the same attributes. This algorithm first selects at most k_g (a small number) foreground and background samples for each pixel, resulting in at most k_g^2 test pairs, from which the best pair is selected. The algorithm also makes sure that sample sets of neighboring pixels are disjoint. Then, in a small neighborhood of k_r pixels, each pixel analyzes the best choices of its k_r spatial neighbors, and chooses the best sample pair as its final decision. Thus, while in practice $k_g^2 + k_r$ pair evaluations are performed for each pixel, due to the affinity among neighboring pixels, this is roughly equivalent to performing $k_g^2 \times k_r$ pair evaluations. In their system, k_g and k_r are set to be 4 and 200, respectively, thus $4 \times 4 + 200 = 216$ pair evaluations can achieve the effect of evaluating $16 \times 200 = 3200$ pairs. Using this efficient sampling method plus some local smoothing postprocessing steps, Gastal and Oliveira developed a real-time matting system which achieves high accuracy on the publicly available online matting benchmark [27].

The sampling procedures described so far are all local sampling methods, i.e., for an unknown pixel, only a limited number of nearby foreground and background colors are collected for alpha estimation. He et al. [15] pointed out that due to the limited size of the sample set and sometimes complex foreground structures, local sampling may not always cover the true foreground and background colors of unknown pixels. They further proposed a global sampling approach, where for evaluating the alpha value of an unknown pixel, all known foreground and background colors in the image are used as samples. Specifically, from all possible sample pairs, the best one is chosen which minimizes the cost function:

$$E^{gs}(\mathbf{c}_k^f, \mathbf{c}_l^b) = \kappa\|\mathbf{c}_i - \hat{\alpha}_i \mathbf{c}_k^f - (1 - \hat{\alpha}_i)\mathbf{c}_l^b\| + \eta(\mathbf{x}_k^f) + \eta(\mathbf{x}_l^b), \tag{9.30}$$

where $\hat{\alpha}_i$ is the alpha value estimated from $\mathbf{c}_k^f$ and $\mathbf{c}_l^b$ using Equation 9.24, κ is a balancing weight and $\eta(\mathbf{x}_k^f)$ is the spatial energy computed from the spatial location of $\mathbf{c}_k^f$ and c_i as:

$$\eta(\mathbf{x}_k^f) = \frac{\|\mathbf{x}_k^f - \mathbf{x}_i\|}{\min_k \|\mathbf{x}_k^f - \mathbf{x}_i\|}. \tag{9.31}$$

$\eta(\mathbf{x}_l^b)$ is the spatial energy computed for $\mathbf{c}_l^b$ in the same way. To handle the computational complexity introduced by the large number of samples, their system poses the sampling task as a correspondence problem in a special "FB search space", and uses a generalized fast patch matching algorithm [28] to efficiently find the best sample pair in that space, denoted as $\hat{\mathbf{c}}_k^f$ and $\hat{\mathbf{c}}_l^b$. After sample search, the node weight is finally defined as:

$$w_i^{\text{gs}} = \exp\left(-|\mathbf{c}_i - \hat{\alpha}_i \mathbf{c}_k^f - (1 - \hat{\alpha}_i)\mathbf{c}_l^b\|\right)(\alpha_i - \hat{\alpha}_i)^2. \tag{9.32}$$

Note that the global sampling method intrinsically assumes that the foreground and background color distributions are spatially invariant, so that the color samples that are spatially far away from the target pixel may still be valid estimations of the true foreground and background colors of the pixel. If the foreground or background colors are spatially varying, using remote color samples may introduce additional color ambiguity which will lead to less accurate alpha estimation.

9.3   Solving Image Matting Graphs

Once the edge and node weights are properly defined for the matting graph, solving the alpha matte becomes a graph labeling problem. Depending on exactly how the edge and node weights are formulated, various optimization techniques can be applied to obtain the solution. Here we review a few representative techniques that have been widely used in existing matting systems.

9.3.1   Solving MRF

Wang and Cohen proposed an iterative optimization approach [13] to compute the alpha matte from sparsely specified user scribbles. In this method, the matting graph is formulated as a Markov Random Field (MRF), and the total energy to be minimized is defined as:

$$E^{\text{bp}}(\alpha) = \sum_{v_i \in \mathcal{V}} w_i^{\text{bp}} + \lambda \cdot \sum_{v_i, v_j \in \mathcal{V}} w_{ij}^{\text{bp}}, \tag{9.33}$$

where w_i^{bp} is the node weight defined in Equation 9.22, and w_{ij}^{bp} is the edge weight in Equation 9.9. This approach also quantizes the continuous alpha value in $[0, 1]$ into multiple discrete levels so that discrete optimization is applicable. With the MRF defined in this way, finding alpha labels corresponds to the MAP estimation problem, which can be practically solved using the loopy Belief Propagation (BP) algorithm [29].

To generate accurate alpha mattes from sparsely defined user trimaps, in this approach the energy minimization method described above is applied iteratively in an active region. The active region is created and updated iteratively by expanding the user scribbles toward the rest of the image until all unknown pixels have been covered. This ensures that pixels nearby the user scribbles are computed first, and they will in turn affect other pixels that are far away from the scribbles. In each iteration of the algorithm, the data weight w_i^{bp} and edge weight w_{ij}^{bp} are updated based on the alpha matte computed in the previous iteration, and a new energy $E^{\mathrm{bp}}(\alpha)$ is minimized using the BP algorithm.

9.3.2   Linear Optimization

One of the major limitations of the MRF formulation is its high computational complexity. Furthermore, iteratively applying the BP optimization may lead the algorithm to converge to a local minimum. To avoid these limitations, some approaches carefully define their energy functions in such a way that they can be efficiently solved by closed-form optimization techniques.

Grady et al. [11] defined a matting graph where the edge weight w_{ij}^{lpp} is formulated as Equation 9.7, and solved the graph labeling problem using Random Walks. In this algorithm, α_i is modeled as the probability that a random walker starting from v_i will reach a pixel in the foreground before striking a pixel in the background, when biased to avoid crossing the foreground boundary. Denote the degree of v_i as:

$$d_i^{\mathrm{rm}} = \sum w_{ij}^{\mathrm{lpp}}, \tag{9.34}$$

for all edges e_{ij} incident on v_i. The probability that a random walker at v_i transitions to v_j is computed as $p_{ij} = w_{ij}^{\mathrm{lpp}}/d_i$. Theoretical studies [30] show that the solution to the random walker problem is exactly the solution to the inhomogeneous Dirichlet problem from potential theory, given Dirichlet boundary conditions that $\alpha_j = 1$ if v_j is in the foreground region of the trimap, and $\alpha_j = 0$ if v_j is in the background region. Specifically, these probabilities are an exact, steady-state, global minimum to the Dirichlet energy functional:

$$E^{\mathrm{rm}}(\boldsymbol{\alpha}) = \boldsymbol{\alpha}^T \mathbf{L}^{\mathrm{rm}} \boldsymbol{\alpha}, \tag{9.35}$$

subject to the boundary conditions. $\mathbf{L}^{\mathrm{rm}}$ is the graph Laplacian given by:

$$\mathbf{L}_{ij}^{\mathrm{rm}} = \begin{cases} d_i^{\mathrm{rm}} & : \quad \text{if} \quad i = j, \\ -w_{ij}^{\mathrm{lpp}} & : \quad \text{if i and j are neighbors,} \\ 0 & : \quad \text{otherwise.} \end{cases} \tag{9.36}$$

The energy functional in Equation 9.35 can be efficiently minimized by solving a sparse, symmetric, positive-definite linear system, and has been further implemented on GPU for real-time interaction [11].

Similar to Wang and Cohen's approach [13], the Easy Matting system [12] also employs an iterative optimization framework. In the kth iteration, the total energy to be minimized is defined as

$$E^{\mathrm{easy}}(\alpha, k) = \sum_{v_i \in \mathcal{V}} w_i^{\mathrm{easy}} + \lambda_k \cdot \sum_{v_i, v_j \in \mathcal{V}} w_{ij}^{\mathrm{easy}}, \tag{9.37}$$

where w_i^{easy} is the node weight defined in Equation 9.21, and w_{ij}^{easy} is the edge weight defined in Equation 9.8. Note that since both terms are carefully defined in quadratic forms, this energy function can be efficiently minimized by solving a large linear system. Also, unlike $E^{\mathrm{bp}}(\alpha)$ defined in Equation 9.33 which uses a constant weight λ for balancing the two energy terms, the weight λ_k in Equation 9.37 is dynamically adjusted as

$$\lambda_k = e^{-(k-\beta)^3}, \tag{9.38}$$

where k is the number of iterations, and β is a predefined constant which is set to be 3.4 in the system. In this setting, λ_k becomes smaller when the iteration number k increases. This allows the foreground and background regions to grow faster from sparsely specified user scribbles in early iterations, by using large λ_k values early on. In later iterations when the foreground and background regions encounter, a smaller λ_k will allow the node weight w_i^{easy} to play a bigger role on determining alpha values for pixels on the foreground edge.

Based on the edge weight defined in Equation 9.13, Levin et al. proposed a matting energy as:

$$E^{\mathrm{cf}}(\alpha) = \alpha^T \mathbf{L}^{\mathrm{cf}} \alpha, \tag{9.39}$$

where $\mathbf{L}^{\mathrm{cf}}$ is the matting Laplacian defined in Equation 9.14. This is again a problem of minimizing a quadratic error score, which can be solved as a linear system. Built upon this, the Robust Matting algorithm [2] defines its energy function as:

$$E^{\mathrm{rm}}(\alpha) = \sum_{v_i \in \mathcal{V}} w_i^{\mathrm{rm}} + \lambda \cdot \alpha^T \mathbf{L}^{\mathrm{cf}} \alpha, \tag{9.40}$$

where w_i^{rm} is the node weight defined in Equation 9.27. The shared matting approach [26] and the global sampling approach [15] minimize a similar energy function, except that the node weight w_i is determined using the shared sampling approach or the global sampling approach described in Section 9.2.2.

| Highly transparent | Medium transparent |
| Strongly transparent | Little transparent |

FIGURE 9.6
The height test images with different properties in the online image matting benchmark [27].

9.4   Data Set

To quantitatively compare various image matting approaches, Rhemann et al. [27] proposed the first online benchmark[3] for image matting. This benchmark provides a high resolution ground truth data set, which contains 8 test and 27 training images, along with predefined trimaps. These images were shot in a controlled environment, and the ground truth mattes were extracted using the triangulation method [31], by shooting the same foreground object against multiple single-colored backgrounds. The test images have different properties such as hard and soft boundaries, translucency, or different boundary topologies, as shown in Figure 9.6.

The online system also provides all scripts and data necessary to allow people to submit new results on the test images, and compare them with results of other approaches in the system. Four different error metrics have been implemented when comparing results to the ground truth: absolute differences (SAD), mean squared error (MSE), and two other perceptually motivated measures named as connectivity and gradient errors. Since introduced, this benchmark has been used by many recent image matting systems for objective and quantitative evaluation.

---

[3]The web URL for the benchmark is www.alphamatting.com.

9.5   Video Matting

9.5.1   Overview

Video matting refers to the problem of estimating alpha mattes of dynamic foreground objects from a video sequence. Compared with image matting, it is a considerably harder problem due to two new challenges: interaction efficiency and temporal coherence.

In image matting systems, the user is usually required to manually specify an accurate trimap in order to achieve accurate mattes. However, this quickly becomes tedious for video, as a short video sequence may contain hundreds or thousands of frames, and manually specifying a trimap for each frame requires simply too much work. To minimize the required user interaction, video matting approaches typically ask the user to provide trimaps only on sparsely distributed keyframes, and then use automatic methods to propagate trimaps to other frames. Automatic trimap generation thus becomes the key component of a video matting system.

In the video matting task, in addition to be accurate on each frame, the alpha mattes computed in consecutive frames are required to be temporally coherent. In fact, temporal coherence is often more important than single frame accuracy, as the human visualization system (HVS) is very sensitive to temporal inconsistency presented in a video sequence [32]. Simply applying image matting techniques frame-by-frame without considering temporal coherence often results in temporally jittering alpha mattes. Maintaining temporal coherence of the resulting mattes thus becomes another fundamental requirement for video matting.

9.5.2   Graph-Based Trimap Generation

Video matting approaches usually apply binary foreground object segmentation for trimap generation. Once a binary segmentation is obtained on each frame, the unknown region of the trimap can be easily generated by dilating and eroding the foreground region. An example is shown in Figure 9.9, where for the input image Figure 9.9b, a binary segmentation in Figure 9.9b is first created, which leads to the trimap shown in Figure 9.9c.

In the video object cut-and-paste system [33], each video frame is at first automatically segmented into atomic regions using the watershed image segmentation algorithm [34], mainly for improving the computational efficiency. These atomic regions are treated as basic elements for graph construction. The user is then required to provide accurate foreground object segmentation on a few keyframes as initial guidance, by using existing interactive image segmentation tools. Between each pair of successive keyframes, a 3D graph $\mathcal{G} = (\mathcal{V}, \mathcal{E})$ is then built on atomic regions, as shown in Figure 9.7.

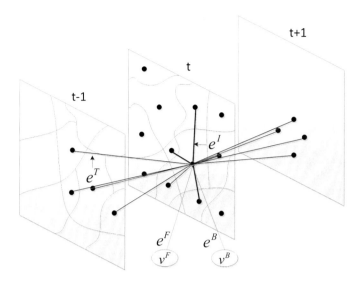

FIGURE 9.7
Graph construction for trimap generation in the video object cut-and-paste system [33].

In graph $\mathcal{G}$, the node set $\mathcal{V}$ includes all atomic regions on all frames between the two keyframes. There are also two virtual nodes v_F and v_B, which correspond to definite foreground $\mathcal{F}$ and definite background $\mathcal{B}$ as hard constraints. There are three types of edges in the graph: intraframe edges e^I, interframe edges e^T, and edges between atomic regions and virtual nodes e^F and e^B. An intraframe edge e^I_{ij} connects two adjacent atomic regions v^t_i and v^t_j on frame t, while an interframe edge e^T_{ij} connects v^t_i and v^{t+1}_j, if the two regions have both similar colors and overlapping spatial locations. Every atomic region connects to the virtual nodes v^F and v^B.

For edges e^F_i and e^B_i, the weights w^F_i and w^B_i encode the user-provided hard constraints, as well as the color similarity between v_i and user-marked regions. Mathematically, they are defined as:

$$w^F_i = \begin{cases} 0 & : \text{if } v_i \in \mathcal{F}, \\ \infty & : \text{if } v_i \in \mathcal{B}, \\ d^F(v_i) & : \text{otherwise;} \end{cases} \tag{9.41}$$

and

$$w^B_i = \begin{cases} \infty & : \text{if } v_i \in \mathcal{F}, \\ 0 & : \text{if } v_i \in \mathcal{B}, \\ d^B(v_i) & : \text{otherwise.} \end{cases} \tag{9.42}$$

If v_i is marked as either $\mathcal{F}$ ($\alpha_i = 1$) or $\mathcal{B}$ ($\alpha_i = 0$) on the keyframes, then the edge weight is 0 to the corresponding node and ∞ to the other. Otherwise, w^F_i and w^B_i

are determined by its color distances to the known foreground and background colors, denoted as $d^F(v_i)$ and $d^B(v_i)$. To compute the color distance, Gaussian Mixture Models (GMMs) [35] are used to describe the foreground and background color distributions, collected from the ground-truth segmentations on keyframes. Denote the kth component of the foreground GMM as $(w_k^f, \mu_k^f, \Sigma_k^f)$, representing the weight, the mean color, and the color covariance matrix, respectively, for a given color $\mathbf{c}$, $d^F(\mathbf{c})$ is computed as:

$$d^F(\mathbf{c}) = \min_k \left[\hat{D}(w_k^f, \Sigma_k^f) + \overline{D}(\mathbf{c}, \mu_k^f, \Sigma_k^f) \right], \tag{9.43}$$

where

$$\hat{D}(w, \Sigma) = -\log w + \frac{1}{2} \log det\Sigma, \tag{9.44}$$

and

$$\overline{D}(\mathbf{c}, \mu, \Sigma) = \frac{1}{2}(\mathbf{c} - \mu)^T \Sigma^{-1}(\mathbf{c} - \mu). \tag{9.45}$$

For the node v_i (note that v_i is an atomic region containing many pixels), its distance to the foreground GMMs $d^F(v_i)$ is defined as the average distance of $d^F(\mathbf{c}_j \in v_i)$, where $\mathbf{c}_j$ is the color of a pixel inside the atomic region. The background distance $d^F(v_i)$ is defined in the same way using the background GMMs.

For the intraframe and interframe edges e^I and e^T, the edge weights are defined as

$$w_{ij} = |\alpha_i - \alpha_j| \cdot \exp\left(-\beta \|\bar{\mathbf{c}}_i - \bar{\mathbf{c}}_j\|^2\right), \tag{9.46}$$

where α_i and α_j are labels for v_i and v_j, which are constrained to be either 0 or 1 for the purpose of binary segmentation. β is a parameter that weights the color contrast, which can be set as a constant or computed adaptively using the robust method proposed by Blake et al. [36]. $\bar{\mathbf{c}}_i$ and $\bar{\mathbf{c}}_j$ are the average colors of pixels inside v_i and v_j.

The total energy to be minimized for the graph labeling problem is

$$E(\boldsymbol{\alpha}) = \sum_{v_i \in \mathcal{V}} \left(\alpha_i w_i^F + (1 - \alpha_i)w_i^B\right) + \lambda_1 \sum_{e_{ij} \in e^I} w_{ij} + \lambda_2 \sum_{e_{ij} \in e^T} w_{ij}, \tag{9.47}$$

where α_i can only be 0 or 1 in this binary labeling problem, and λ_1 and λ_2 are used to adjust the weights of the energy terms. For instance, a higher λ_2 will enforce the solution to be more temporally coherent. This energy function can be minimized using the graph cuts algorithm [16].

After the graph labeling problem is solved, a local tracking and refinement step is further employed to improve the segmentation accuracy. Finally, a trimap is created for each frame by dilating and eroding the segmented foreground region, and a coherent matting approach [37], which is a variant of the Bayesian matting algorithm [9], is used to generate the final alpha mattes. Wang et al. [38] introduced another 3D graph labeling approach for trimap generation in video. Their system first applies the

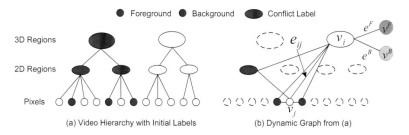

FIGURE 9.8

Video hierarchy and dynamic graph construction in the 3D video cutout system [38].
(a). Video hierarchy created by two-step mean shift segmentation. User-specified
labels are then propagated upward from the bottom level. (b). A graph is dynamically
constructed based on the current configuration of the labels. Only highest level nodes
that have no conflict are selected. Each node in the hierarchy also connects to two
virtual nodes v^F and v^B.

2D mean shift image segmentation algorithm [39] on each frame to generate over-
segmented regions, similar to the atomic regions in the video object cut-and-paste
system. The system then treats the 2D mean shift regions as super-pixels by representing each region using the mean position and color of all pixels inside it, and
applies the mean shift algorithm again to group 2D regions into 3D spatiotemporal
regions. This two-step clustering method results in a strict hierarchical structure of
the input video, shown in Figure 9.8a, where each pixel belongs to a 2D region, and
each 2D region belongs to a 3D spatiotemporal region.

A unique characteristic of Wang et al.'s system is the dynamic graph construction.
In this approach, an optimization graph is dynamically constructed in real time given
the user input and the precomputed video hierarchy. Suppose the user has marked
some pixels as foreground and some as background, as shown in Figure 9.8a. The
labels are then automatically propagated upward in the hierarchy to all higher level
nodes. Conflict with be introduced in this propagation process since upper level nodes
may contain both foreground and background pixels. To construct the optimization
graph, the system only picks out highest level nodes which have no conflict, and connects neighboring nodes using edges, as shown in Figure 9.8b. An edge is established
between any two nodes which are adjacent to each other in the 3D video cube. The
graph labeling problem is then solved by an optimization process, which assigns a
label to every node in the graph. The computed labels are then propagated downward
to every singe pixel to create the final segmentation. If the segmentation contains
errors, the user can then mark more pixels as hard constraints, which will result in a
new graph for optimization. This dynamic graph construction allows the system to
always use the smallest number of nodes for optimization, and at the same time to
satisfy all the user-provided constraints. The segmentation efficiency is thus greatly
improved. For instance, the system can reportedly segment a 200-frame 720×480
video sequence in less than 10 seconds [8].

Similar to the graph constructed in Figure 9.7, each node in the dynamic graph (Figure 9.8b) also connects to two virtual nodes v^F and v^B. If v_i is marked as foreground by the user, then $w_i^F = 0$ and $w_i^B = \infty$. Similarly, $w_i^F = \infty$ and $w_i^B = 0$ if v_i is specified as background. Otherwise they are computed as

$$
\begin{cases}
w_i^F &= \frac{D_i^B}{D_i^F + D_i^B}, \\
w_i^B &= \frac{D_i^F}{D_i^F + D_i^B},
\end{cases}
\tag{9.48}
$$

where D_i^F and D_i^B are color distances measured between v_i and known foreground and background colors. The system first trains GMMs on known foreground and background colors, and then computes the color distances by fitting the average pixel color in v_i (v_i contains multiple pixels if it is a higher level node), denoted as $\bar{\mathbf{c}}_i$, to the GMMs as

$$
D_i^F = 1 - \sum_k w_k^f \exp\left((\bar{\mathbf{c}}_i - \mu_k^f)^T \mathbf{\Sigma}_k^{-1} (\bar{\mathbf{c}}_i - \mu_k^f)/2 \right),
\tag{9.49}
$$

where $(w_k^f, \mu_k^f, \mathbf{\Sigma}_k^f)$ represent the weight, the mean color, and the color covariance matrix of the kth component of the foreground GMM. D_i^B is computed in a similar way using the background GMM.

For the edge weight w_{ij}, the system uses the classic exponential term as defined in Equation 9.5. Additionally, if the input video is known to have a static background, a local background color model, as well as a local background link model, are defined and incorporated into the global node and edge weights, which can greatly help the system better recognize background nodes. The detailed definitions of the local node and edge weights can be found in the original paper [38].

The dynamically constructed graph is finally solved by the graph cuts algorithm, which assigns every node in the graph a foreground or background label. The labels are then propagated to the bottom level pixels to create a complete segmentation. The whole process iterates as the user provides more scribbles to correct segmentation errors, until satisfactory results are achieved.

9.5.3   Matting with Temporal Coherence

Once trimaps are created for all video frames, image matting techniques can be applied frame-by-frame to generate the final alpha mattes. However, as discussed earlier, this naive approach does not guarantee temporal coherence and can easily introduce temporal jitter to the mattes. Extra temporal coherence constraints thus have to be employed for video matting.

As described in Section 9.1.2, for an unknown pixel, Bayesian matting algorithm samples nearby foreground and background colors, and applies them in a Bayesian framework for estimating its alpha value. Chuang et al. [40] extended this approach for video matting, by constraining the color samples to be temporally coherent.

Specifically, once the foreground object is masked out on each frame, the remaining background fragments are registered and assembled to form a composite mosaic [41], which can then be reprojected into each original frame to form a dynamic clean plate. The dynamic clean plate essentially gives every unknown pixel an accurate background estimation, resulting in improved accuracy of the foreground matte. Since the reconstructed clean plates are temporally coherent, the temporal coherence of the final alpha mattes is also greatly improved.

Xue et al. proposed a more explicit temporal coherence term in their matting approach in the Video SnapCut system [42]. The key idea of this algorithm is to warp the alpha matte computed on frame $t - 1$ to frame t using estimated motion between the two frames, and treat it as a prior for computing the matte on frame t. For the matting graph defined in this approach, the edge weight is defined using the closed-form solution in Equation 9.13. The node weight for pixel v_i on frame t contains two components: a color prior α_i^C computed from locally sampled nearby foreground and background colors, and a temporal prior $\alpha_{s(i)}^{t-1}$ which is the alpha value at pixel $s(i)$ on frame $t - 1$. The pixel $s(i)$ is the corresponding pixel of v_i on frame $t - 1$, according to the computed optical flow between the two frames. The alpha matte is solved by minimizing the following energy function:

$$E(\alpha^t) = \sum_i \left[\lambda_i^T (\alpha_i - \alpha_{s(i)}^{t-1})^2 + \lambda_i^C (\alpha_i - \alpha_i^C)^2 \right] + \alpha^{tT} \mathbf{L}^{cf} \alpha^t, \qquad (9.50)$$

where λ_i^T and λ_i^C are locally adaptive weights. Specifically, λ_i^C measures the confidence of the sampled foreground and background colors on frame t and the alpha value estimated from them, which is computed as in Equation 9.25. λ_i^T measures the confidence of the temporal prior $\alpha_{s(i)}^{t-1}$, based on how similar the foreground shapes are in local windows around v_i on frame t, and $s(i)$ on frame $t - 1$. The detailed formulations can be found in the original paper [42]. $\mathbf{L}^{cf}$ is the matting Laplacian defined in Equation 9.14.

Figure 9.9 shows an example of how the explicit temporal coherence term (the first term in Equation 9.50) can help improve the temporal coherence of the alpha mattes. Given the input frames t and $t + 1$ shown in Figure 9.9b and 9.9e, binary segmentation results (Figure 9.9c and 9.9f) for these two frames are first created, and two trimaps are then generated from the binary segmentation, as shown in Figure 9.9d and 9.9g. Suppose α^t, the alpha matte on frame t is already computed. If α^{t+1} is computed without the temporal coherence term, the resulting matte is erroneous and not consistent with α^t, as shown in Figure 9.9i. On the contrary, with the temporal coherence term, α^{t+1} has less errors and is more consistent with α^t, as shown in Figure 9.9j. This example shows that with the explicit temporal coherence term, the alpha estimation is more robust against dynamic backgrounds that have complex colors and textures.

9.6   Conclusion

Image and video matting is an active research topic which not only is theoretically interesting, but also has huge potentials in numerous real-world applications, ranging from image editing to film production. The state-of-the-art in matting research has been significantly advanced in recent years, by formulating matting as a graph labeling problem and solving it using graph optimization methods. In this chapter, we show that many image matting approaches share a rather common graph structure, and the merit of each method lies in its unique way to define the edge and node weights in the graph. We further show how to extend graph analysis to video for generating accurate trimaps and alpha mattes in a temporally coherent way.

Despite the significant progress that has been made, image and video matting still remains to be an unsolved problem in difficult cases. From the analysis presented in this chapter, it is clear that most matting approaches are built upon some smoothness image priors, either implicitly or explicitly. In difficult cases where the foreground and background regions contain high contrast textures, most existing matting approaches may not work well, since their underlying color smoothness assumptions are violated. Furthermore, compared with image matting, video matting poses additional challenges. A good video matting system has to provide accurate mattes on each frame. More importantly, the alpha mattes on adjacent frames have to be temporally coherent. Existing video matting solutions still cannot deal with dynamic objects with large semitransparent regions, such as long hair blowing in the wind against a moving background. We expect novel graphical models and new graph analysis methods to be be developed in the future to address these limitations.

Bibliography

[1] T. Porter and T. Duff, "Compositing digital images," in *Proc. ACM SIGGRAPH*, vol. 18, July 1984, pp. 253–259.

[2] J. Wang and M. Cohen, "Optimized color sampling for robust matting," in *Proc. IEEE Conf. Computer Vision and Pattern Recognition*, 2007, pp. 1–8.

[3] C. Rother, V. Kolmogorov, and A. Blake, ""grabcut": Interactive foreground extraction using iterated graph cuts," *ACM Trans. Graphics*, vol. 23, no. 3, pp. 309–314, 2004.

[4] Y. Li, J. Sun, C.-K. Tang, and H.-Y. Shum, "Lazy snapping," *ACM Trans. Graphics*, vol. 23, no. 3, pp. 303–308, 2004.

[5] C. Rhemann, C. Rother, A. Rav-Acha, and T. Sharp, "High resolution matting via interactive trimap segmentation," in *Proc. IEEE Conf. Computer Vision and Pattern Recognition*, 2008, pp. 1–8.

[6] M. Ruzon and C. Tomasi, "Alpha estimation in natural images," in *Proc. IEEE Conf. Computer Vision and Pattern Recognition*, 2000, pp. 18–25.

[7] Y. Mishima, "Soft edge chroma-key generation based upon hexoctahedral color space," in *U.S. Patent 5,355,174*, 1993.

[8] J. Wang and M. Cohen, "Image and video matting: A survey," *Foundations and Trends in Computer Graphics and Vision*, vol. 3, no. 2, pp. 97–175, 2007.

[9] Y.-Y. Chuang, B. Curless, D. H. Salesin, and R. Szeliski, "A Bayesian approach to digital matting," in *Proc. IEEE Conf. Computer Vision and Pattern Recognition*, 2001, pp. 264–271.

[10] A. Levin, D. Lischinski, and Y. Weiss, "A closed-form solution to natural image matting," *IEEE Trans. Pattern Analysis and Machine Intelligence*, vol. 30, no. 2, pp. 228–242, 2008.

[11] L. Grady, T. Schiwietz, S. Aharon, and R. Westermann, "Random walks for interactive alpha-matting," in *Proc. Visualization, Imaging, and Image Processing*, 2005, pp. 423–429.

[12] Y. Guan, W. Chen, X. Liang, Z. Ding, and Q. Peng, "Easy matting: A stroke based approach for continuous image matting," in *Computer Graphics Forum*, vol. 25, no. 3, 2006, pp. 567–576.

[13] J. Wang and M. Cohen, "An iterative optimization approach for unified image segmentation and matting," in *Proc. International Conference on Computer Vision*, 2005, pp. 936–943.

[14] Y. Zheng and C. Kambhamettu, "Learning based digital matting," in *Proc. International Conference on Computer Vision*, 2009.

[15] K. He, C. Rhemann, C. Rother, X. Tang, and J. Sun, "A global sampling method for alpha matting," in *Proc. IEEE Conf. Computer Vision and Pattern Recognition*, June 2011, pp. 1–8.

[16] Y. Boykov, O. Veksler, and R. Zabih, "Fast approximate energy minimization via graph cuts," *IEEE Trans. Pattern Analysis and Machine Intelligence*, vol. 23, no. 11, pp. 1222–1239, 2001.

[17] J. Shi and J. Malik, "Normalized cuts and image segmentation," *IEEE Trans. Pattern Analysis and Machine Intelligence*, pp. 888–905, 2000.

[18] X. He and P. Niyogi, "Locality preserving projections," in *Proc. Advances in Neural Information Processing Systems*, 2003.

[19] D. Singaraju and R. Vidal, "Interactive image matting for multiple layers," in *Proc. IEEE Conf. Computer Vision and Pattern Recognition*, 2008, pp. 1–7.

[20] K. He, J. Sun, and X. Tang, "Single image haze removal using dark channel prior," in *Proc. IEEE Conf. Computer Vision and Pattern Recognition*, 2009, pp. 1956–1963.

[21] E. Hsu, T. Mertens, S. Paris, S. Avidan, and F. Durand, "Light mixture estimation for spatially varying white balance," *ACM Trans. Graphics*, vol. 27, pp. 1–7, 2008.

[22] D. Singaraju, C. Rother, and C. Rhemann, "New appearance models for natural image matting," in *Proc. IEEE Conf. Computer Vision and Pattern Recognition*, 2009, pp. 659–666.

[23] B. Scholkopf and A. J. Smola, *Learning with Kernels: Support Vector Machines, Regularization, Optimization, and Beyond.* Cambridge, MA, USA: MIT Press, 2001.

[24] C. Rhemann, C. Rother, and M. Gelautz, "Improving color modeling for alpha matting," in *Proc. British Machine Vision Conference*, 2008, pp. 1155–1164.

[25] X. Bai and G. Sapiro, "Geodesic matting: A framework for fast interactive image and video segmentation and matting," *International Journal on Computer Vision*, vol. 82, no. 2, pp. 113–132, 2008.

[26] E. S. L. Gastal and M. M. Oliveira, "Shared sampling for real-time alpha matting," *Computer Graphics Forum*, vol. 29, no. 2, pp. 575–584, May 2010.

[27] C. Rhemann, C. Rother, J. Wang, M. Gelautz, P. Kohli, and P. Rott, "A perceptually motivated online benchmark for image matting," in *Proc. IEEE Conf. Computer Vision and Pattern Recognition*, 2009, pp. 1826–1833.

[28] C. Barnes, E. Shechtman, A. Finkelstein, and D. B. Goldman, "Patchmatch: A randomized correspondence algorithm for structural image editing," *ACM Trans. Graphics*, vol. 28, no. 3, pp. 1–11, July 2009.

[29] Y. Weiss and W. Freeman, "On the optimality of solutions of the max-product belief propagation algorithm in arebitrary graphs," *IEEE Trans. Information Theory*, vol. 47, no. 2, pp. 303–308, 2001.

[30] S. Kakutani, "Markov processes and the Dirichlet problem," in *Proc. Japanese Academy*, vol. 21, 1945, pp. 227–233.

[31] A. R. Smith and J. F. Blinn, "Blue screen matting," in *Proc. ACM SIGGRAPH*, 1996, pp. 259–268.

[32] P. Villegas and X. Marichal, "Perceptually-weighted evaluation criteria for segmentation masks in video sequences," *IEEE Trans. Image Processing*, vol. 13, no. 8, pp. 1092–1103, 2004.

[33] J. S. Y. Li and H. Shum, "Video object cut and paste," *ACM Trans. Graphics*, vol. 24, pp. 595–600, 2005.

[34] L. Vincent and P. Soille, "Watersheds in digital spaces: an efficient algorithm based on immersion simulations," *IEEE Trans. Pattern Analysis and Machine Intelligence*, vol. 13, no. 6, pp. 583–598, 1991.

[35] D. Titterington, A. Smith, and U. Makov, *Statistical Analysis of Finite Mixture Distributions.* John Wiley & Sons, 1985.

[36] A. Blake, C. Rother, M. Brown, P. Perez, and P. Torr, "Interactive image segmentation using an adaptive GMMRF model," in *Proc. European Conference on Computer Vision*, 2004, pp. 428–441.

[37] H. Shum, J. Sun, S. Yamazaki, Y. Li, and C. Tang, "Pop-up light field: An interactive image-based modeling and rendering system," *ACM Trans. Graphics*, vol. 23, no. 2, pp. 143–162, 2004.

[38] J. Wang, P. Bhat, R. A. Colburn, M. Agrawala, and M. F. Cohen, "Interactive video cutout," *ACM Trans. Graphics*, vol. 24, pp. 585–594, 2005.

[39] D. Comaniciu and P. Meer, "Mean shift: A robust approach toward feature space analysis," *IEEE Trans. Pattern Analysis and Machine Intelligence*, vol. 24, no. 5, pp. 603–619, 2002.

[40] Y.-Y. Chuang, A. Agarwala, B. Curless, D. Salesin, and R. Szeliski, "Video matting of complex scenes," *ACM Trans. Graphics*, vol. 21, pp. 243–248, 2002.

[41] R. Szeliski and H. Shum, "Creating full view panoramic mosaics and environment maps," in *Proc. ACM SIGGRAPH*, 1997, pp. 251–258.

[42] X. Bai, J. Wang, D. Simons, and G. Sapiro, "Video snapcut: Robust video object cutout using localized classifiers," *ACM Trans. Graphics*, vol. 28, no. 3, pp. 1–11, July 2009.

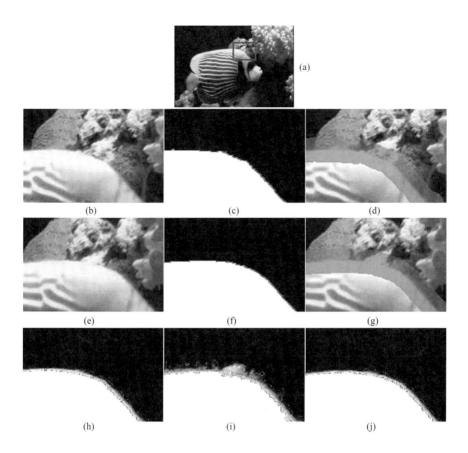

FIGURE 9.9
An example of coherent matting in video. (a) The input sequence, the red box shows
the highlighted region in the rest of the figure; (b) the highlighted region on frame
t; (c) binary segmentation on frame t; (d) trimap generated on frame t; (e) the highlighted region on frame $t + 1$; (f) binary segmentation on frame $t + 1$; (g) trimap
generated on frame $t + 1$; (h) computed α^t, (i) computed α^{t+1} without the temporal
coherence term in Equation 9.50, (j) computed α^{t+1} with the temporal coherence
term.

10

Optimal Simultaneous Multisurface and Multiobject Image Segmentation

Xiaodong Wu

Departments of Electrical & Computer Engineering, and
Radiation Oncology
The University of Iowa
Iowa City, IA, USA
Email: xiaodong-wu@uiowa.edu

Mona K. Garvin

Department of Electrical & Computer Engineering
The University of Iowa
Iowa City, IA, USA
Email: mona-garvin@uiowa.edu

Milan Sonka

Departments of Electrical & Computer Engineering,
Radiation Oncology, and Ophthalmology & Visual Sciences
The University of Iowa
Iowa City, IA, USA
Email: milan-sonka@uiowa.edu

CONTENTS

10.1 Introduction 266
10.2 Motivation and Problem Description 267
10.3 Methods for Graph-Based Image Segmentation 268
    10.3.1 Optimal Surface Detection Problems 268
    10.3.2 Optimal Single Surface Detection 270
        10.3.2.1 The Intralayer Self-Closure Property of the OSSD Problem 271
        10.3.2.2 The Graph Transformation Scheme for the OSSD Problem 271
        10.3.2.3 Optimal Single Surface Detection by Computing a Minimum-
Cost Closed Set 272
    10.3.3 Optimal Multiple Surface Detection 275
        10.3.3.1 Overview of the OMSD Algorithm 276
        10.3.3.2 The self-closure structure of the OMSD Problem 276
        10.3.3.3 The Graph Transformation Scheme for the OMSD Problem 277
        10.3.3.4 Computing Optimal Multiple Surfaces for the OMSD Prob-
lem 279
    10.3.4 Optimal Surface Detection with Convex Priors 281

10.3.4.1 The Graph Transformation Scheme for the Convex EOSD
Problem 283
10.3.4.2 Computing Optimal Multiple Surfaces for the Convex EOSD
Problem 284
10.3.5 Layered Optimal Graph Image Segmentation for Multiple Objects and
Surfaces — LOGISMOS 285
10.3.5.1 Object Pre-segmentation 286
10.3.5.2 Construction of Object-Specific Graphs 286
10.3.5.3 Multiobject Interactions 286
10.3.5.4 Electric Lines of Force 287
10.3.5.5 ELF-Based Cross-Object Surface Mapping 288
10.3.5.6 Cost Function and Graph Optimization 290
10.4 Case Studies 290
10.4.1 Segmentation of Retinal Layers in Optical Coherence Tomography
Volumes 290
10.4.1.1 Graph Structure 290
10.4.1.2 Cost Function Design 291
10.4.2 Simultaneous Segmentation of Prostate and Bladder in Computed Tomography Volumes 291
10.4.3 Cartilage and Bone Segmentation in the Knee Joint 293
10.4.3.1 Bone Presegmentation 295
10.4.3.2 Multisurface Interaction Constraints 295
10.4.3.3 Multiobject Interaction Constraints 297
10.4.3.4 Knee Joint Bone–Cartilage Segmentation 297
10.4.3.5 Knee Joint Bone/Cartilage Segmentation 297
10.5 Conclusion 299
10.6 Acknowledgments 299
Bibliography 299

10.1 Introduction

Images are increasingly occurring as higher-dimensional data. While the early effort was devoted to developing methods for image processing and analysis in 2D, the focus has later shifted to stereo images, 2D + time (motion video) image data, and to multispectral (or multiband) 3D images. Substantial effort has been devoted to developing automated or semi-automated image segmentation techniques over the past decades [1]. With the increased availability of X-ray computed tomography (CT), magnetic resonance (MR), positron-emission tomography (PET), single-photon emission computed tomography (SPECT), ultrasound, optical coherence tomography (OCT), and other medical imaging modalities, higher-dimensional data are omnipresent in routine clinical medicine. Clearly, 3D, 4D, and 5D volumetric images are becoming common. In this context, 3D images represent volumetric data typically organized as a x–y–z matrices. When adding temporal image acquisition aspects, the image data become four-dimensional (x–y–z–t matrices) – MR, CT, or ultrasound images of cardiac ventricles obtained over cardiac cycle may serve as an example of such 4D datasets. When such 4D image data are acquired at several time instances over time, for example, weekly over a period of several weeks, 5D image data result (x–y–z–t_1–t_2 matrices) — pulmonary CT images acquired over one breathing cycle and longitudinally over several imaging sessions are an example of such 5D image data. It is easily possible to imagine how higher-dimensional image data can be formed. Same as conventional 2D image data, high-dimensional image data need to be processed, segmented, analyzed, and used for intelligent decision making. However, the n-D character of the underlying image data requires development of fundamentally different approaches, in which contextual information across all dimensions of such images is considered and inherently incorporated. In no other image analysis step is this more apparent than in image segmentation.

Incorporating image-based contextual information and performing image segmentation in an optimal fashion are two powerful concepts that form the basis of the n-dimensional graph-based image segmentation approaches described below [2, 3, 4]. After introducing the basic algorithm on a simple 2D example, a general methodology for optimal segmentation of single surfaces is introduced that is applicable to n-D image data. Extensions facilitating optimal segmentation of multiple interacting surfaces for a single or multiple interacting objects are reported. The generality of the approach is demonstrated on closed-surface and multiobject segmentation cases and applicability to n-D image data is shown. Since the optimality of the solution is guaranteed with respect to the cost function, special emphasis is given to cost function design, including edge-based, region-based, and edge-region-combined cost functions. Methods for incorporating shape priors are discussed leading to arc-based graph representation and associated arc-based graph segmentation solutions.

10.2   Motivation and Problem Description

As mentioned, n-dimensional images are routinely occurring in medical imaging applications worldwide. With the ever-increasing numbers of such acquired image data, with ever-increasing numbers of image slices forming such images, and with increased in-plane image resolution, the image data amounts that require expert analysis by radiologists, orthopedists, internists, and other medical professionals are growing rapidly and exceed the capability of physicians to analyze them visually. Performing organ/object segmentations in 3D, 4D, or higher-D is infeasible for a human observer in a clinical setting due to the time constraints and the overall tedious character of manual tracing. The functionality and performance of the reported automated methods is demonstrated in a variety of medical image analysis tasks providing a convincing evidence of the broad applicability of the introduced algorithmic concepts.

The segmentation problems we are addressing in this chapter can be divided in the following categories, all of them applicable to n-D image data:

- Single-surface terrain-like segmentation, for example diaphragm segmentation in 3D or 4D CT images.

- Multisurface terrain-like segmentation, e.g., segmentation of multiple retinal layers from 3D OCT images.

- Single-/multisurface segmentation of tubular objects, like detection of vascular inner and outer wall in 3D/4D intravascular ultrasound.

- Single-/multisurface segmentation of closed surfaces, e.g., identification of a liver surface from 3D CT or detection of endo- and epi-cardial surfaces of cardiac left ventricle from 3D/4D/5D ultrasound.

- Single-/multisurface segmentation of bifurcating tubular structures, e.g., inner and outer walls of intrathoracic airway trees across bifurcations from 3D CT images.

- Single-/multisurface segmentation of several mutually interacting objects, like simultaneous detection of prostate and bladder from CT or simultaneous segmentation of knee bones and cartilages from MR images.

In all of these and many similar cases, the underlying image data are used to construct a graph in which optimal surfaces are identified according to a node- or arc-based cost function. Importantly, all cases for which the reported approach is applicable require that the graph columns intercept the sought surface (or surfaces) exactly once. Having this constraint is advantageous in applications for which it is imperative to preserve object topology, which is different from the label-based approaches of [5, 6]. To satisfy this requirement, the graph must be formed correspondingly, either by using a priori information about the approximate shape and/or topology of the resulting object(s), or by performing topologically correct approximate presegmentation that is used for graph construction. Details of these processes are given below.

10.3 Methods for Graph-Based Image Segmentation

10.3.1 Optimal Surface Detection Problems

Let $\mathcal{I}$ be a given 3-D volumetric image with size of $n = X \times Y \times Z$. Figure 10.1(a) shows a 3-D retinal OCT volume. For each (x, y) pair, $0 \le x < X$ and $0 \le y < Y$, the voxel subset $\{\mathcal{I}(x, y, z) \mid 0 \le z < Z\}$ forms a *column* parallel to the z-axis, denoted by $\text{Col}(x, y)$. Two columns are *adjacent* if their (x, y) coordinates satisfy some neighborhood conditions. For instance, under the 4-neighbor setting, the column $\text{Col}(x, y)$ is adjacent to $\text{Col}(x', y')$ if $|x-x'| + |y-y'| = 1$. Henceforth, we use a model of the 4-neighbor adjacency; this simple model can be easily extended to

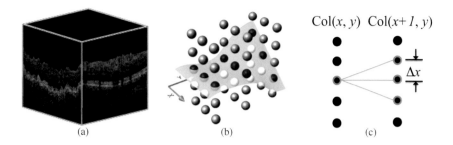

FIGURE 10.1
The optimal surface detection problem. (a) A 3-D retinal OCT volume. (b) The surface orientation. (c) Two adjacent columns with the smoothness parameters $\Delta x = 1$.

other adjacency settings. Note that each of the sought **z**-monotone surfaces contains exactly one voxel in each column of $\mathcal{I}$. Denote by $S(x, y)$ the z-coordinate of the voxel $\mathcal{I}(x, y, z)$ on a surface S. Figure 10.1(b) shows the surface orientation. The surface *smoothness constraint* is specified by two *smoothness parameters*, Δx and Δy, which define the maximum allowed change in the z-coordinate of a surface along each unit distance change in the **x** and **y** dimensions, respectively. If $\mathcal{I}(x, y, z')$ and $\mathcal{I}(x{+}1, y, z'')$ (respectively, $\mathcal{I}(x, y{+}1, z'')$) are two (neighboring) voxels on a feasible surface, then $|z'{-}z''| \leq \Delta x$ (respectively, $|z'{-}z''| \leq \Delta y$) (Figure 10.1(c)). More precisely, a *surface* in I is defined as a function $S : [0..X) \times [0..Y) \rightarrow [0..Z)$, such that for any pair (x, y), $|S(x, y) - S(x + 1, y)| \leq \Delta x$ and $|S(x, y) - S(x, y + 1)| \leq \Delta y$, where $[a..b)$ denotes the set of integers from a to $b - 1$. In multiple surface detection, for each pair of the sought surfaces S and S', we use two parameters, $\delta^l \geq 0$ and $\delta^u \geq 0$, to represent the surface *separation constraint*, which defines the relative positioning and the distance range of the two surfaces. That is, if $\mathcal{I}(x, y, z) \in S$ and $\mathcal{I}(x, y, z') \in S'$, we have $\delta^l \leq z' - z \leq \delta^u$ for every (x, y) pair (Figure 10.2). A set of surfaces are considered *feasible* if each individual surface in the set satisfies the given smoothness constraints and if each pair of surfaces satisfies the surface separation constraints.

Designing appropriate cost functions is of paramount importance for any optimization-based segmentation method. The cost function for each voxel in the input image usually reflects either edge-based and/or region-based properties of the target surface, depending on the imaging setting.

Edges defined by image gradients are commonly used for image segmentation. A typical *edge-based cost function* aims to accurately position the boundary surface in the volumetric image. Incorporating regional information, which alleviates the sensitivity of the initial model and improves the robustness, is becoming increasingly important in image segmentation [7, 8, 9, 10, 11, 12, 13]. We thus assign costs to every voxel of $\mathcal{I}$ to integrate the edge- and region-based cost functions, as follows. Each voxel $\mathcal{I}(x, y, z)$ has an *edge-based cost* $b_i(x, y, z)$, which is an arbitrary real value,

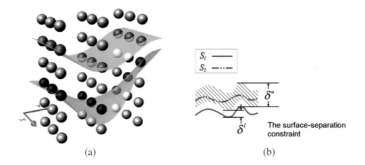

FIGURE 10.2
(a) Two sought surfaces divide the image $\mathcal{I}$ into three regions. (b) The surface inter-relation modeling with the minimum (δ^l) and maximum (δ^u) surface distances.

for the detection of the i-th surface. Note that λ surfaces $\mathcal{S} = \{S_1, S_2, \ldots, S_\lambda\}$ in $\mathcal{I}$ induce $\lambda + 1$ regions $\{R_0, R_1, \ldots, R_\lambda\}$ (Figure 10.2(a)). For each region R_i ($i = 0, 1, \ldots, \lambda$), every voxel $\mathcal{I}(x, y, z)$ is assigned a real-valued *region-based cost* $c_i(x, y, z)$. The edge-based cost of each voxel in $\mathcal{I}$ is inversely related to the likelihood that it may appear on a desired surface, while the region-based costs $c_i(\cdot)$ ($i = 0, 1, \ldots, \lambda$) measure the inverse likelihood of a given voxel preserving the expected regional properties (e.g., homogeneity) of the partition $\{R_0, R_1, \ldots, R_\lambda\}$. Both the edge-based and region-based costs can be determined by using simple low-level image features [1, 8, 9, 12]. Then, the total cost $\alpha(\mathcal{S})$ induced by the λ surfaces in $\mathcal{S}$ is defined as

$$\alpha(\mathcal{S}) = \sum_{i=1}^{\lambda} b_i(S_i) + \sum_{i=0}^{\lambda} c_i(R_i) = \sum_{i=1}^{\lambda} \sum_{\mathcal{I}(x,y,z) \in S_i} b_i(x, y, z) + \sum_{i=0}^{\lambda} \sum_{\mathcal{I}(x,y,z) \in R_i} c_i(x, y, z).$$
$$(10.1)$$

10.3.2   Optimal Single Surface Detection

This section presents the polynomial time algorithm for solving the *optimal single surface detection (OSSD)* problem, in which $\lambda = 1$ and no region-based costs are considered, that is, the energy function is $\alpha(S) = \sum_{\mathcal{I}(x,y,z) \in S} b(x, y, z)$. We first exploit the intralayer self-closure structure of the OSSD problem, and then formulate it as a minimum-cost closed set problem based on a nontrivial graph transformation scheme.

A *closed set* $\mathcal{C}$ in a directed graph with arbitrary vertex costs $w(\cdot)$ is a subset of vertices such that all successors of any vertex in $\mathcal{C}$ are also contained in $\mathcal{C}$[14, 15].

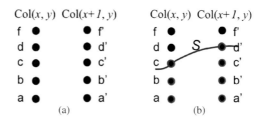

FIGURE 10.3

Illustrating the intralayer self-closure structure of the OSSD problem. The surface smoothness parameter $\Delta x = 1$. (a) The lowest neighbors. If voxel $c \in \text{Col}(x, y)$ is on a feasible surface, then only one of the voxels $\{b', c', d'\}$ on $\text{Col}(x + 1, y)$ can be on the same surface. Thus, b' is the lowest neighbor of voxel c on $\text{Col}(x + 1, y)$. The lowest neighbor of voxel $a' \in \text{Col}(x + 1, y)$ on $\text{Col}(x, y)$ is voxel a. (b) The intralayer self-closure structure. S is a feasible surface. The red voxels form the set $\text{Lw}(S)$. The lowest neighbor of each voxel in $\text{Lw}(S)$ is also in $\text{Lw}(S)$.

The *cost* of a closed set $\mathcal{C}$, denoted by $w(\mathcal{C})$, is the total cost of all vertices in $\mathcal{C}$. Note that a closed set can be empty (with a cost zero). The minimum-cost closed set problem seeks a closed set in the graph whose cost is minimized.

10.3.2.1 The Intralayer Self-Closure Property of the OSSD Problem

The algorithm for the OSSD problem hinges on the following observation about the self-closure structure of any feasible OSSD solution. For a voxel $\mathcal{I}(x, y, z)$ and each adjacent column $\text{Col}(x', y')$ of $\text{Col}(x, y)$, the *lowest neighbor* of $\mathcal{I}(x, y, z)$ on $\text{Col}(x', y')$ is the voxel on $\text{Col}(x', y')$ with the smallest z-coordinate that can appear together with $\mathcal{I}(x, y, z)$ on the same feasible surface in $\mathcal{I}$. More precisely, the lowest neighbor of $\mathcal{I}(x, y, z)$ on its adjacent column $\text{Col}(x \pm 1, y)$ (respectively, $\text{Col}(x, y \pm 1)$) is $\mathcal{I}(x \pm 1, y, \max\{0, z - \Delta x\})$ (respectively, $\mathcal{I}(x, y \pm 1, \max\{0, z - \Delta y\})$). In Figure 10.3(a), the lowest neighbor of voxel c on $\text{Col}(x + 1, y)$ is b' and the lowest neighbor of voxel a' on $\text{Col}(x, y)$ is voxel a, where $\Delta x = 1$.

We say that a voxel $\mathcal{I}(x, y, z)$ is *below* a surface S if $S(x, y) > z$, and denote by $\text{Lw}(S)$ all the voxels of $\mathcal{I}$ that are on or below S. A key observation is that *for any feasible surface S in $\mathcal{I}$, the lowest neighbors of every voxel in $\text{Lw}(S)$ are also contained in $\text{Lw}(S)$*. Figure 10.3(b) shows the intralayer self-closure structure for two adjacent columns. For instance, the lowest neighbor b' of voxel $c \in \text{Lw}(S)$ on $\text{Col}(x+1, y)$ is in $\text{Lw}(S)$ and the lowest-neighbor c of voxel $d' \in \text{Lw}(S)$ on $\text{Col}(x, y)$ is also in $\text{Lw}(S)$. This intralayer self-closure property is crucial to our algorithm and suggests to relate our target problem to the minimum closed set problem. In our approach, instead of finding an optimal surface S^* directly, we seek in $\mathcal{I}$ a voxel set $\text{Lw}(S^*)$, which uniquely defines the surface S^*.

10.3.2.2　The Graph Transformation Scheme for the OSSD Problem

This section presents the construction of the node-weighted directed graph $\mathcal{G} = (\mathcal{V}, \mathcal{E})$ from the input image $\mathcal{I}$, which enables to detect an optimal single surface by computing a minimum-cost closed set. This construction crucially relies on the intralayer self-closure structure.

The directed graph $\mathcal{G}$ is constructed, as follows. Every node $v(x, y, z) \in \mathcal{V}$ represents exactly one voxel $\mathcal{I}(x, y, z) \in \mathcal{I}$. The graph $\mathcal{G}$ can be viewed as a geometric graph defined on a 3-D grid. To abuse the notation, we also use $\text{Col}(x, y)$ to denote the nodes in $\mathcal{G}$ corresponding to the voxels of $\text{Col}(x, y)$ in $\mathcal{I}$. The arcs of $\mathcal{G}$ (an arc is a directed edge in a graph) are then introduced to guarantee that all nodes corresponding the voxel set $\text{Lw}(S)$ of a feasible surface S form a closed set in $\mathcal{G}$. First, we need to make sure that if voxel $\mathcal{I}(x, y, z_s)$ is on S, then all the nodes $v(x, y, z)$ of $\mathcal{G}$ below $v(x, y, z_s)$ (with $z \leq z_s$) on the same column must be in a closed set $\mathcal{C}$. Thus, for each column $\text{Col}(x, y)$, every node $v(x, y, z)$ ($z > 0$) has a directed arc to the node $v(x, y, z - 1)$ immediately below it. These arcs are called the *intracolumn arcs*, which, in fact, enforce the monotonicity of the sought surface (the sought surface intersects each column exactly once). Next, we need to incorporate into $\mathcal{G}$ the smoothness constraints along the **x**- and **y**-dimensions. Notice that if voxel $\mathcal{I}(x, y, z_s)$ is on surface S, then its neighboring voxels on S along the **x**-dimension must be no "lower" than voxel $\mathcal{I}(x, y, z_s - \Delta x)$. Hence, node $v(x, y, z_s) \in \mathcal{C}$ indicates that nodes $v(x + 1, y, z_s - \Delta x)$ and $v(x - 1, y, z_s - \Delta x)$ must be in $\mathcal{C}$. Thus, for each node $v(x, y, z)$, we put arcs from $v(x, y, z)$ to $v(x + 1, y, z - \Delta x)$ and to $v(x - 1, y, z - \Delta x)$, respectively. Note that node $v(x + 1, y, z - \Delta x)$ (respectively, $v(x - 1, y, z - \Delta x)$) corresponds to the lowest neighbor of voxel $\mathcal{I}(x, y, z)$ on its adjacent column $\text{Col}(x + 1, y)$ (respectively, $\text{Col}(x - 1, y)$). Those arcs are called the *intercolumn arcs*. Here, to avoid cluttering the exposition of our key ideas, we do not consider the boundary conditions, which can be handled as well. The same construction is done for the **y**-dimension.

To incorporate the edge-based costs, a node cost $w(x, y, z)$ is assigned to each node $v(x, y, z)$ in $\mathcal{G}$, as follows.

$$w(x, y, z) = \begin{cases} b(x, y, z) & \text{if } z = 0 \\ b(x, y, z) - b(x, y, z - 1) & \text{for } z = 1, 2, \ldots, Z - 1 \end{cases} \quad (10.2)$$

where $b(x, y, z)$ is the cost of voxel $\mathcal{I}(x, y, z)$. This completes the construction of $\mathcal{G}$, which has n nodes and $O(n)$ arcs. Note that the size of $\mathcal{G}$ is independent of the smoothness parameters Δx and Δy. Figure 10.4 exemplifies the construction on two adjacent columns of voxels.

10.3.2.3　Optimal Single Surface Detection by Computing a Minimum-Cost Closed Set

The graph $\mathcal{G}$ thus constructed allows us to detect an optimal single surface in $\mathcal{I}$, by computing a minimum-cost nonempty closed set in $\mathcal{G}$. The analysis of the optimal

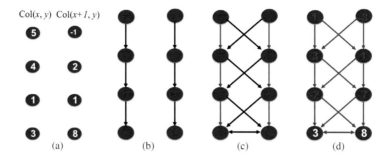

FIGURE 10.4
Illustrating the graph construction for the OSSD problem. The construction is exemplified on two adjacent columns. The surface smoothness parameter $\Delta x = 1$. (a) Two adjacent columns of voxels. Each number is the edge-based cost of the corresponding voxel. (b) The intracolumn arcs (black) pointing downward enforce the monotonicity of the sought surface. (c) The intercolumn arcs (black) enforce the surface smoothness constraints. (d) Node costs assignment scheme.

surface detection problem in a more general setting in [2, 16] reveals the following facts: (1) Any closed set $\mathcal{C} \neq \varnothing$ in $\mathcal{G}$ defines a feasible surface in $\mathcal{I}$ whose total cost equals that of $\mathcal{C}$; (2) any feasible surface S in $\mathcal{I}$ corresponds to a closed set $\mathcal{C} \neq \varnothing$ in $\mathcal{G}$ whose cost equals that of S. Consequently, a nonempty closed set in $\mathcal{G}$ with the minimum cost can specify an optimal single surface in $\mathcal{I}$.

Given any closed set $\mathcal{C} \neq \varnothing$ in $\mathcal{G}$, we define a feasible surface S in $\mathcal{I}$, as follows. For each (x, y)-pair, let $\mathcal{C}(x, y) \subset \mathcal{C}$ be the set of nodes on the column $\mathrm{Col}(x, y)$ of $\mathcal{G}$. Based on the construction of $\mathcal{G}$, it is not hard to see that $\mathcal{C}(x, y) \neq \varnothing$. Let $z_h(x, y)$ be the largest z-coordinate of the nodes in $\mathcal{C}(x, y)$. Define the surface S as $S(x, y) = z_h(x, y)$ for every (x, y)-pair. Figure 10.5 shows an example of recovering a feasible surface from a closed set.

Hence, we can compute a minimum-cost closed set $\mathcal{C}^* \neq \varnothing$ in $\mathcal{G}$, which specifies an optimal single surface S^* in $\mathcal{I}$. However, the minimum closed set $\mathcal{C}^*$ in G' can be empty (with a cost zero), and when this is the case, $\mathcal{C}^* = \varnothing$ gives little useful information on recovering the surface. Fortunately, our careful construction of $\mathcal{G}$ still enables us to overcome this difficulty. If the minimum closed set in $\mathcal{G}$ is empty, then it implies that the cost of every nonempty closed set in $\mathcal{G}$ is nonnegative. We want to do a transformation to reduce the cost of every closed set in $\mathcal{G}$ by a constant to make sure that at least one is of negative cost. To do so, we investigate a special closed set $\mathcal{C}_0$ consisting of the "bottom" nodes of all columns in $\mathcal{G}$, that is, $\mathcal{C}_0 = \{v(x, y, 0) \mid 0 \leq x < X, 0 \leq y < Y\}$. The construction of the graph $\mathcal{G}$ indicates that $\mathcal{C}_0$ is a closed set and is a subset of any nonempty closed set in $\mathcal{G}$. Hence, to obtain a minimum *nonempty* closed set in $\mathcal{G}$, we do the following: Let M be the total cost of nodes in $\mathcal{C}_0$; if $M > 0$, pick an arbitrary node $v(x_0, y_0, 0) \in \mathcal{C}_0$ and assign

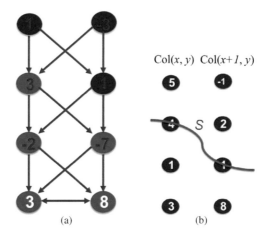

Col(x, y)   Col(x+1, y)

(a)   (b)

FIGURE 10.5

Illustrating the recovery of a feasible surface from a closed set. (a) shows a closed set in the constructed graph in Figure 10.4(d), which consists of all green nodes. The surface specified by the closed set in (a) is shown in (b).

a new weight $w(x_0, y_0, 0) - M - 1$ to $v(x_0, y_0, 0)$ (Figure 10.6(a)). We call this a *translation operation* on $\mathcal{G}$. Thus, the total cost of the closed set C_0 (after a translation operation on $\mathcal{G}$) is negative. Since C_0 is a subset of any non-empty closed set in $\mathcal{G}$, the cost of any nonempty closed set is reduced by $(M + 1)$. Hence, the minimum-cost nonempty closed set in $\mathcal{G}$ is not changed after the translation operation, and it cannot be empty. Then, we can simply find a minimum closed set C^* in $\mathcal{G}$ after performing a translation operation on $\mathcal{G}$. C^* is a minimum-cost nonempty closed set in $\mathcal{G}$ before the translation.

As in [14, 15, 2, 16], we find a minimum-cost closed set $C^* \neq \varnothing$ in $\mathcal{G}$ by formulating it as computing a minimum s-t cut in a weighted directed graph $\mathcal{G}'$ transformed from $\mathcal{G}$. Let $\mathcal{V}^+$ and $\mathcal{V}^-$ denote the set of nodes in $\mathcal{G}$ with nonnegative and negative costs, respectively. Define a new directed graph $\mathcal{G}_{st} = (\mathcal{V} \cup \{s, t\}, \mathcal{E} \cup \mathcal{E}_{st})$. The node set of $\mathcal{G}_{st}$ is the node set $\mathcal{V}$ of $\mathcal{G}$ plus a source s and a sink t. The arc set of $\mathcal{G}_{st}$ is the arc set $\mathcal{E}$ of $\mathcal{G}$ plus a new arc set $\mathcal{E}_{st}$. $\mathcal{E}_{st}$ consists of the following arcs: The source s is connected to each node $v \in \mathcal{V}^-$ by an arc of cost $-w(v)$; every node $v \in \mathcal{V}^+$ is connected to the sink t by an arc of cost $w(v)$ (Figure 10.6(a)). Let $(A, \bar{A})$ denote a finite-cost s-t cut in $\mathcal{G}_{st}$ with $s \in A$ and $t \in \bar{A}$, and $C(A, \bar{A})$ denote the total cost of the cut. Note that the arcs in the cut $(A, \bar{A})$ are either in $(A \cap \mathcal{V}^+, t)$ or in $(s, \bar{A} \cap \mathcal{V}^-)$. Let $w(\mathcal{V}')$ denote the total cost of nodes in a subset $\mathcal{V}' \subseteq \mathcal{V}$.

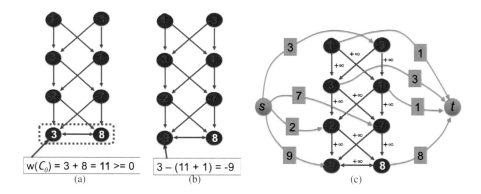

FIGURE 10.6
Illustrating the computation of a minimum-cost nonempty closed set. (a) The "bottom" nodes of all columns forms a minimal closed set which is a subset of any nonempty closed set of the graph in Figure 10.4(d). (b) Performing a translation operation on the graph in (a). (b) The constructed s-t graph for computing a minimum-cost nonempty closed set.

$$
\begin{aligned}
C(A, \bar{A}) &= \sum_{v \in \bar{A} \cap V^-} (-w(v)) + \sum_{v \in A \cap V^+} w(v) \\
&= \sum_{v \in V^-} (-w(v)) - \sum_{v \in A \cap V^-} (-w(v)) + \sum_{v \in A \cap V^+} w(v) \quad (10.3) \\
&= -w(V^-) + \sum_{v \in A} w(v)
\end{aligned}
$$

Note that the term $-w(V^-)$ is fixed and is the sum over all nodes with negative costs in $\mathcal{G}$. The term $\sum_{v \in S} w(v)$ is the total cost of all nodes in the source set A of the cut $(A, \bar{A})$. But, $A - \{s\}$ is a closed set in $\mathcal{G}$ [14, 15]. Thus, the cost of a cut $(A, \bar{A})$ in G_{st} and the cost of the corresponding closed set in $\mathcal{G}$ differ by a constant, and the source set of a minimum cut in G_{st} corresponds to a minimum-cost closed set in $\mathcal{G}$.

Therefore, the minimum s-t cut in $\mathcal{G}_{st}$ defines a minimum-cost closed set C^* in $\mathcal{G}$, which can be used to specify an optimal single surface S^* in $\mathcal{I}$.

10.3.3   Optimal Multiple Surface Detection

This section presents the algorithm for solving the *optimal multiple surface detection (OMSD)* problem, i.e., simultaneous detection of $\lambda > 1$ interrelated surfaces in a 3-D image $\mathcal{I}$ such that the total cost of the λ surfaces defined in Equation 10.1 is minimized [3, 16]. In addition to the intralayer self-closure structure for each single surface, we further explore the *interlayer self-closure structure* of the pairwise inter-acting surfaces, which again enables us to model the OMSD problem as a minimum-cost closed set problem.

10.3.3.1 Overview of the OMSD Algorithm

The OMSD algorithm is based on a sophisticated graph transformation scheme, which enables us to simultaneously identify $\lambda > 1$ optimal interrelated net surfaces as a whole by computing a minimum-cost closed set in a weighted directed graph $\mathcal{G}$ that we transform from $\mathcal{I}$. The algorithm uses the following three main steps.

Step 1: Graph Construction. Build a node-weighted directed graph $\mathcal{G} = (\mathcal{V}, \mathcal{E})$, which contains λ node-disjoint subgraphs $\mathcal{G}_i = (\mathcal{V}_i, \mathcal{E}_i)$. Each subgraph $\mathcal{G}_i$ is used to search for the i-th surface in $\mathcal{I}$. The separation constraints of the surfaces are enforced by introducing a subset of directed arcs between any two adjacent subgraphs, $\mathcal{G}_i$ and $\mathcal{G}_{i+1}$ ($i = 1, 2, \ldots, \lambda - 1$). The construction of the graph $\mathcal{G}$ (see Section 10.3.3.3) hinges on the self-closure structure exploited in Section 10.3.3.2.

Step 2: Computing a Minimum-Cost Closed Set. Compute a minimum-cost nonempty closed set C^* in $\mathcal{G}$, which can be done by formulating it as computing a minimum s-t cut in an edge-weighted directed graph transformed from $\mathcal{G}$.

Step 3: Surfaces Reconstruction. The optimal set of λ surfaces is reconstructed from the minimum-cost closed set C^* with each surface being specified by $C^* \cap \mathcal{V}_i$.

10.3.3.2 The self-closure structure of the OMSD Problem

Assume that $\mathcal{S} = \{S_1, S_2, \ldots, S_\lambda\}$ is a feasible set of λ surfaces in $\mathcal{I}$ with S_{i+1} being "on top" of S_i. For each pair of the sought surfaces, S_i and S_{i+1}, two parameters $\delta_i^l \geq 0$ and $\delta_i^u \geq 0$ [1] are used to specify the surface separation constraint (note that we may define the separation constraint between any pair of the surfaces in $\mathcal{S}$). First, consider each individual surface $S_i \in \mathcal{S}$. Recall that $\mathrm{Lw}(S_i)$ denotes the subset of all voxels of $\mathcal{I}$ that are on or below S_i. As in Section 10.3.2.1, we observe that each $\mathrm{Lw}(S_i)$ has the intralayer self-closure structure.

However, the task here is more involved since the λ surfaces in $\mathcal{S}$ are interre-lated. The following observation reveals the common essential structure between the smoothness and the separation constraints, leading us to further examine the closure structure *between* the $\mathrm{Lw}(S_i)$'s. We may view the 3-D image $\mathcal{I}$ as a set of X 2-D slices embedded in the **yz**-plane. Thus, a feasible surface S of $\mathcal{I}$ is decomposed into X **z**-monotone curves with each in a 2-D slice. We observe that each feasible **z**-monotone curve is subjective to the smoothness constraint in the corresponding slice,

---

[1] If desired, δ_i^l and δ_i^u may be defined to be location specific (i.e., δ_i^l and δ_i^u may vary from column to column).

and any pair of adjacent **z**-monotone curves expresses to meet the analogical separation constraints. This observation suggests that the surface separation constraint in
a d-D image may be viewed as the surface smoothness constraint in the $(d+1)$-D
image consisting of the stack of a sequence of λ d-D images. Hence, we intend to
map the detection of λ optimal surfaces in d-D to the problem of finding a single
optimal surface in $(d+1)$-D.

To distinguish the self-closure structures, we define below the *upstream* and
downstream voxels of any voxel $\mathcal{I}(x, y, z)$ in $\mathcal{I}$ for the given surface separation
parameters δ_i^l and δ_i^u: if $\mathcal{I}(x, y, z) \in S_i$, then the i-th upstream (respectively,
downstream) voxel of $\mathcal{I}(x, y, z)$ is the voxel on column $\mathrm{Col}(x, y)$ with the smallest z-coordinate that can be on S_{i+1} (respectively, S_{i-1}). More precisely, for every voxel $\mathcal{I}(x, y, z)$ and $1 \leq i < \lambda$ (respectively, $1 < i \leq \lambda$), the *i-th upstream* (respectively, *downstream*) voxel of $\mathcal{I}(x, y, z)$ is $\mathcal{I}(x, y, z + \delta_i^l)$ (respectively, $\mathcal{I}(x, y, \max\{0, z - \delta_{i-1}^u\})$) if $z + \delta_i^l < Z$ (respectively, $z - \delta_{i-1}^l \geq 0$).
Consider any voxel $\mathcal{I}(x, y, z) \in Lw(S_i)$. Since $z \leq S_i(x, y)$, the i-th upstream
voxel of $\mathcal{I}(x, y, S_i(x, y))$ is "above" the i-th upstream voxel of $\mathcal{I}(x, y, z)$ (i.e.,
$S_i(x, y) + \delta_i^l \geq z + \delta_i^l$). In the meanwhile, $S_{i+1}(x, y) \geq S_i(x, y) + \delta_i^l$ (by the
definition of the i-th upstream voxel). Hence, the i-th upstream voxel of $\mathcal{I}(x, y, z)$,
$\mathcal{I}(x, y, z + \delta_i^l)$, is in $Lw(S_{i+1})$. Using a similar argument, the i-th downstream voxel
of $\mathcal{I}(x, y, z)$, $\mathcal{I}(x, y, \max\{0, z - \delta_{i-1}^u\})$, is in $Lw(S_{i-1})$. Thus, we have the following
interlayer self-closure structure of $Lw(S_i)$'s: *Given any set S of λ feasible surfaces
in $\mathcal{I}$, the i-th upstream (respectively, downstream) voxel of each voxel in $Lw(S_i)$ is in
$Lw(S_{i+1})$ (respectively, $Lw(S_{i-1})$), for every $1 \leq i < \lambda$ (respectively, $1 < i \leq \lambda$).*

Both the intralayer and interlayer self-closure structures together bridge the
OMSD problem and the minimum closed set problem. Instead of directly search-
ing for an optimal set of λ surfaces, $S^* = \{S_1^*, S_2^*, \ldots, S_\lambda^*\}$, we intend to find λ
optimal subsets of voxels in $\mathcal{I}$, $Lw(S_1^*) \subset Lw(S_2^*) \subset \ldots \subset Lw(S_\lambda^*)$, such that each
$Lw(S_i^*)$ uniquely defines a surface $S_i^* \in S^*$.

10.3.3.3 The Graph Transformation Scheme for the OMSD Problem

This section presents the construction of the node-weighted directed graph $\mathcal{G} = (\mathcal{V}, \mathcal{E})$, enabling us to simultaneously identify an optimal set of $\lambda > 1$ interrelated
surfaces as a whole by computing a minimum-cost closed set. An example of the
construction is shown in Figure 10.7.

The graph $\mathcal{G}$ contains λ node-disjoint subgraphs $\{\mathcal{G}_i = (\mathcal{V}_i, \mathcal{E}_i) \mid i = 1, 2, \ldots, \lambda\}$; each $\mathcal{G}_i$ is for the search of the i-th surface S_i. $\mathcal{V} = \bigcup_{i=1}^{\lambda} \mathcal{V}_i$ and
$\mathcal{E} = \bigcup_{i=1}^{\lambda} \mathcal{E}_i \cup \mathcal{E}_s$. The surface separation constraints between any two consecu-
tive surfaces S_i and S_{i+1} are enforced in $\mathcal{G}$ by a subset of edges in $\mathcal{E}_s$, which connect
the corresponding subgraphs $\mathcal{G}_i$ and $\mathcal{G}_{i+1}$. Each subgraph $\mathcal{G}_i$ is constructed in the
same way as in Section 10.3.2.2 by observing the intralayer self-closure structure of
the problem (Figure 10.7(b)).

We then put directed arcs into $\mathcal{E}_s$ between $\mathcal{G}_i$ and $\mathcal{G}_{i+1}$, to enforce the surface separation constraints. From the interlayer self-closure property, if voxel $\mathcal{I}(x, y, z) \in S_i$, then its i-th upstream voxel $\mathcal{I}(x, y, z + \delta_i^l)$ must be on or below the surface S_{i+1} (i.e., $\mathcal{I}(x, y, z + \delta_i^l) \in \mathrm{Lw}(S_{i+1})$). Thus, for each node $v_i(x, y, z)$ with $z < Z - \delta_i^l$ on the column $\mathrm{Col}_i(x, y)$ of $\mathcal{G}_i$, a directed arc is put in $\mathcal{E}_s$ from $v_i(x, y, z)$ to $v_{i+1}(x, y, z + \delta_i^l)$ on $\mathrm{Col}_{i+1}(x, y)$ of $\mathcal{G}_{i+1}$. Intuitively, these arcs ensure that the surface S_{i+1} must be at a distance of *at least* δ_i^l "above" S_i (i.e., for each (x, y)-pair, $S_{i+1}(x, y) - S_i(x, y) \geq \delta_i^l$). On the other hand, each node $v_{i+1}(x, y, z)$ with $z \geq \delta_i^l$ on $\mathrm{Col}_{i+1}(x, y)$ has a directed arc in $\mathcal{E}_s$ to $v_i(x, y, z')$ on $\mathrm{Col}_i(x, y)$ with $z' = \max\{0, z - \delta_i^u\}$ (note that $\mathcal{I}(x, y, z')$ is the $(i + 1)$-st downstream voxel of $\mathcal{I}(x, y, z)$), making sure that S_{i+1} must be at a distance of *no larger* than δ_i^u "above" S_i (i.e., for each (x, y)-pair, $S_{i+1}(x, y) - S_i(x, y) \leq \delta_i^u$). This construction is applied to every pair of the corresponding columns of any two subgraphs $\mathcal{G}_i$ and $\mathcal{G}_{i+1}$, for $i = 1, 2, \ldots, \lambda - 1$. Figure 10.7(c) shows an example of introducing those intergraph arcs.

Recall that we aim to compute a minimum-cost non-empty closed set in $\mathcal{G}$, which can specify λ optimal surfaces in I. However, the graph $\mathcal{G}$ constructed up to this point does not yet work for this purpose. In the graph construction above, one may notice that any node $v_i(x, y, z)$ with $z \geq Z - \delta_i^l$ has no arc to any node on $\mathrm{Col}_{i+1}(x, y)$, and any node $v_{i+1}(x, y, z)$ with $z < \delta_i^l$ has no arc with any node on $\mathrm{Col}_i(x, y)$. Each of those nodes of $\mathcal{G}$ is called a *deficient node*. The voxels corresponding to the deficient nodes cannot be on any feasible surface. Thus, those deficient nodes along with their incident arcs are safe to be removed (Figure 10.7(d)). The removal of the deficient nodes may also cause other nodes to be deficient (i.e., their corresponding voxels cannot be on any feasible surface). We need to continue eliminate those new deficient nodes. We simply denote also by $\mathcal{G}$ the graph after the deficient node pruning process. Note that in the resulting graph $\mathcal{G}$ if any column $\mathrm{Col}_i(x, y) = \varnothing$, then there is no feasible solution to the OMSD problem. In the rest of this section, we assume that the OMSD problem has feasible solutions. Then, for every (x, y)-pair and $i = 1, 2, \ldots, \lambda$, let $z_i^{\mathrm{bot}}(x, y)$ and $z_i^{\mathrm{top}}(x, y)$ be the smallest and largest z-coordinates of the nodes on the column $\mathrm{Col}_i(x, y)$ of G_i, respectively. The deficient node pruning process may remove some useful arcs whose ending nodes need to be changed. For any node $v_i(x, y, z)$ with $z < z_i^{\mathrm{bot}}(x, y)$ in $\mathcal{G}$ before the pruning process, if it has an incoming arc, the ending node of the arc needs to be changed to the node $v_i(x, y, z_i^{\mathrm{bot}}(x, y))$ in $\mathcal{G}$ after the pruning process (Figure 10.7(d)).

We next further assign a cost $w(\cdot)$ to each node in $\mathcal{G}$, as follows. For every (x, y)-pair, the cost of the node $v_i(x, y, z)$ is assigned as $w_i(x, y, z)$, with

$$
w_i(x, y, z) = \begin{cases} b_i(x, y, z) + \sum_{j=0}^{z} [c_{i-1}(x, y, j) - c_i(x, y, j)] \\ \quad \text{if } z = z_i^{\mathrm{bot}}(x, y), \\ [b_i(x, y, z) - b_i(x, y, z - 1)] + [c_{i-1}(x, y, z) - c_i(x, y, z)] \\ \quad \text{for } z = z_i^{\mathrm{bot}}(x, y) + 1, \ldots, z_i^{\mathrm{top}}(x, y). \end{cases}
$$

$$(10.4)$$

This completes the construction of $\mathcal{G}$. It takes $O(\lambda n)$ time to construct the graph, which has $O(\lambda n)$ nodes and $O(\lambda n)$ arcs.

10.3.3.4 Computing Optimal Multiple Surfaces for the OMSD Problem

The graph $\mathcal{G}$ thus constructed allows us to find an optimal set of λ surfaces in $\mathcal{I}$, by computing a nonempty minimum-cost closed set in $\mathcal{G}$. In order to do that, as in Ref. [16, 3], we can prove the following facts: (1) Any closed set $\mathcal{C} \neq \varnothing$ in $\mathcal{G}$ defines a set $\mathcal{S}$ of λ feasible surfaces $\{S_1, S_2, \dots, S_\lambda\}$ in $\mathcal{I}$; (2) any set $\mathcal{S} = \{S_1, S_2, \dots, S_\lambda\}$ of λ feasible surfaces in $\mathcal{I}$ corresponds to a closed set $\mathcal{C} \neq \varnothing$ in $\mathcal{G}$. Furthermore, we can show that the cost $\alpha(\mathcal{S})$ of $\mathcal{S}$ differs by a fixed value from the total cost $w(\mathcal{C})$ of $\mathcal{C}$. Let $\mathcal{C}_i = \mathcal{C} \cap V_i$ and denote by $\mathcal{C}_i(x, y)$ the set of nodes of $\mathcal{C}_i$ on the column $\mathrm{Col}_i(x, y)$ of G_i. Note that if a node $v_i(x, y, z_c) \in \mathcal{C}_i(v)$, then all nodes in $\{v_i(x, y, z) \mid v_i(x, y, z) \in \mathrm{Col}_i(x, y), z \leq z_c\}$ are also in $\mathcal{C}_i(x, y)$. Hence, the total node cost of $\mathcal{C}_i(x, y)$ is $w(\mathcal{C}_i(x, y)) = \sum_{z=z_i^{\mathrm{bot}}(x,y)}^{S_i(x,y)} w_i(x, y, z)$. Thus, we have

$$
\alpha(\mathcal{S}) = \sum_{i=1}^{\lambda} b_i(S_i) + \sum_{i=0}^{\lambda} c_i(R_i) = \sum_{i=1}^{\lambda} \sum_{\mathcal{I}(x,y,z) \in S_i} b_i(x, y, z) + \sum_{i=0}^{\lambda} \sum_{\mathcal{I}(x,y,z) \in R_i} c_i(x, y, z)
$$

$$
= \sum_{i=1}^{\lambda} \sum_{(x,y)} \left\{ b_i(x, y, 0) + \sum_{z=1}^{S_i(x,y)} [b_i(x, y, z) - b_i(x, y, z-1)] \right\}
$$

$$
+ \left(\sum_{i=1}^{\lambda} \sum_{(x,y)} \sum_{z=0}^{S_i(x,y)} [c_{i-1}(x, y, z) - c_i(x, y, z)] + \sum_{(x,y)} \sum_{z=0}^{Z-1} c_\lambda(x, y, z) \right) =
$$

$$
\sum_{i=1}^{\lambda} \sum_{(x,y)} \left\{ \underbrace{\left(b_i(x, y, 0) + \sum_{z=1}^{z_i^{\mathrm{bot}}(x,y)} [b_i(x, y, z) - b_i(x, y, z-1)] + \sum_{z=0}^{z_i^{\mathrm{bot}}(x,y)} [c_{i-1}(x, y, z) - c_i(x, y, z)] \right)}_{\text{the cost of the "bottom-most" node } v_i(x, y, z_b) \, (z_b = z_i^{\mathrm{bot}}(x,y)) \text{ of } \mathrm{Col}_i(x,y) \text{ in } G_i} \right.
$$

$$
\left. + \sum_{z=z_i^{\mathrm{bot}}(x,y)+1}^{S_i(x,y)} \underbrace{([b_i(x, y, z) - b_i(x, y, z-1)] + [c_{i-1}(x, y, z) - c_i(x, y, z)])}_{\text{the cost of node } v_i(x, y, z) \in \mathrm{Col}_i(x,y)} \right\}
$$

$$
+ \sum_{(x,y)} \sum_{z=0}^{Z-1} c_\lambda(x, y, z)
$$

$$
= \sum_{i=1}^{\lambda} \sum_{(x,y)} \sum_{z=z_i^{\mathrm{bot}}(x,y)}^{S_i(x,y)} w_i(x, y, z) + \sum_{(x,y)} \sum_{z=0}^{Z-1} c_\lambda(x, y, z)
$$

$$
= \sum_{i=1}^{\lambda} \sum_{(x,y)} w(\mathcal{C}_i(x, y)) + \sum_{(x,y)} \sum_{z=0}^{Z-1} c_\lambda(x, y, z)
$$

$$
= w(\mathcal{C}) + \sum_{(x,y)} \sum_{z=0}^{Z-1} c_\lambda(x, y, z) \tag{10.5}
$$

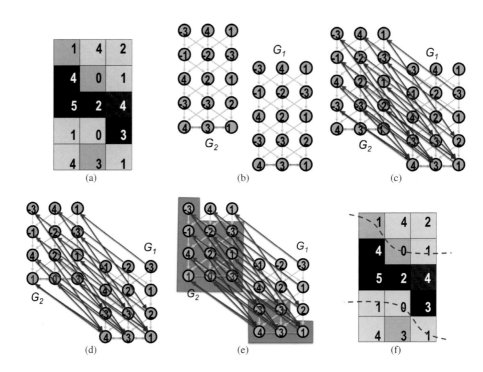

FIGURE 10.7

Illustrating the optimal multiple surface detection algorithm. For the purpose of visualization, a 2-D image (a) is used. Only edge-based costs are used and no region-based costs are considered in this example. The number in each pixel is its edge-based cost. Two interacting surfaces are desired to be detected. The surface smoothness parameter is $\Delta x = 1$, the minimum surface distance $\delta^l = 1$, and the maximum surface distance $\delta^u = 3$. (b) shows two subgraphs constructed from the image in (a), each is used to search one surface. The introduced arcs (red and blue) shown in (c) to enforce the minimum and maximum distance between two sought surfaces. The topmost row of nodes in G_1 and the bottom-most row of nodes in G_2 may not present on any feasible surface and are removed as in (d). (e) shows a minimum-cost nonempty closed set in $\mathcal{G}$, which consists of all nodes in the blue boxes. The recovered two surfaces are shown in (f).

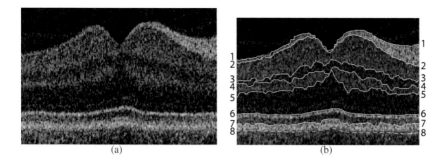

(a)   (b)

FIGURE 10.8
Segmentation of multiple (8) surfaces on one slice from spectral-domain 3-D OCT volume. Note the fovea location "dip" in the middle of the image slice.

Note that the term $\sum_{(x,y)} \sum_{z=0}^{Z-1} c_\lambda(x, y, z)$ is fixed and is the total sum of the λ-th in-region costs of all voxels in $\mathcal{I}$. Consequently, a minimum-cost nonempty closed set $\mathcal{C}^*$ in $\mathcal{G}$ (Figure 10.7(e)), which can be computed as in Section 10.3.2.3, can specify an optimal set $\mathcal{S}^*$ of λ surfaces $\{S_1^*, S_2^*, \dots, S_\lambda^*\}$ in $\mathcal{I}$. Each S_i^* is specified by $\mathcal{C}^* \cap V_i$ as in Section 10.3.2.3 (Figure 10.7(f)).

10.3.4   Optimal Surface Detection with Convex Priors

Up to this point, in our optimal layered graph search model, the node weights in a graph represent the desired segmentation properties such as edge- and region-based image costs, and the desired surface smoothness is hardwired as connectedness of neighboring columns. This representation is limiting the ability to incorporate a broader variety of a priori knowledge in the segmentation process. The connectedness of one voxel to the voxels of its neighboring columns is basically of equal importance in our current model, which prevents us from fully utilizing image edge information as well as from fully taking advantage of shape priors. In some applications, one may prefer to detect surfaces with certain configurations or shapes. For example, the fovea in macular OCT images (Figure 10.8) has a specific shape, the preference of which may be incorporated in the optimization using a weighted agreement between the potential surface solution and the expected shape. In addition, the hardwired smoothness constraints may oversmooth the target surfaces, which makes it difficult to capture their abrupt changes. One way to alleviate this drawback is to use varying smoothness parameters obtained from a training set. It worked quite well in a retinal OCT segmentation as reported in [17]. However, that approach applies only if the dual shape–location dependency is consistent, which is fortunately satisfied in retinal OCT.

To obtain a generally applicable approach, we utilize the weights of both graph nodes and arcs to represent the desired segmentation properties for optimal single- and multiple-surface segmentation, which can incorporate a wide spectrum of constraints into the problem formulation. Let $\Gamma = [0..X-1] \times [0..Y-1]$ denote the grid domain of image $\mathcal{I}$. For optimal surface detection, in addition to requiring each surface S to satisfy the hard smoothness constraints as in Section 10.3.1, we introduce into the objective function a *soft smoothness a priori shape compliance* energy term $\mathcal{E}_{\text{smooth}}(S)$ on the grid domain Γ for each surface S, with $\mathcal{E}_{\text{smooth}}(S) = \int_\Gamma \phi(\nabla S)$, where ϕ is a smoothness penalty function, for instance, penalizing the first derivatives of the grid Γ deformation. We consider a discrete approximation of $\mathcal{E}_{\text{smooth}}(S)$ extended by the piecewise property. Assume that $\mathcal{N}$ is a given neighborhood system on Γ. For any $p(x, y) \in \Gamma$, let $S(p)$ denote the z-coordinate of the voxel $\mathcal{I}(x, y, z)$ on the surface S. Then, the discrete a priori shape compliance smoothness energy $\mathcal{E}_{\text{smooth}}(S)$ can be expressed as $\sum_{(p,q)\in\mathcal{N}} f_{p,q}(|S(p) - S(q)|)$, where $f_{p,q}$ is a nondecreasing function associated with two neighboring columns of p and q that penalizes the shape changes of S on p and q. The node-weights are assigned as in our current model to reflect the desired segmentation properties. The *enhanced optimal surface detection* (EOSD) problem seeks to find a feasible set $\mathcal{S}$ of λ surfaces in $\mathcal{I}$ such that (1) each individual surface satisfies the hard smoothness constraints; (2) each pair of the surfaces satisfies the surface separation constraints; and (3) the cost $\alpha(\mathcal{S})$ induced by $\mathcal{S}$, with

$$
\alpha(\mathcal{S}) = \underbrace{\sum_{i=1}^{\lambda} \sum_{\mathcal{I}(x,y,z)\in S_i} b_i(x, y, z)}_{\text{edge term}} + \underbrace{\sum_{i=0}^{\lambda} \sum_{\mathcal{I}(x,y,z)\in R_i} c_i(x, y, z)}_{\text{region term}}
$$

$$
+ \quad \underbrace{\sum_{i=1}^{\lambda} \sum_{(p,q)\in\mathcal{N}} f_{p,q}^{(i)}(|S_i(p) - S_i(q)|)}_{\text{smoothness shape compliance term}}, \tag{10.6}
$$

is minimized. The objective function $\alpha(\mathcal{S})$ actually is that for the OSD problem (Equation (10.1) in Section 10.3.1) plus the total smoothness a priori shape compliance energy of λ surfaces in $\mathcal{S}$ (i.e., $\sum_{i=1}^{\lambda} \mathcal{E}_{\text{smooth}}(S_i)$).

The most related problem is the metric labeling problem in computer science or called the Markov Random Field (MRF) optimization problem in computer vision. Our EOSD problem is a substantial generalization of the metric labeling problem, in which $\lambda = 1$ and no region term is involved. The metric labeling problem captures a broad range of classification problems where the quality of a labeling depends on the pairwise relations between the underlying set of objects, such as image restoration [18, 19], image segmentation [20, 21, 22, 23], visual correspondence [24, 25], and deformable registration [26]. After being introduced by Kleinberg and Tardos [27], it has been studied extensively in theoretical computer science [28, 29, 30, 31]. The best-known approximation algorithm for the problem is an $O(\log L)$ (L is the number of labels, or in our case, $L = Z$) [30, 27] and has no $\Omega(\sqrt{\log L})$ approximation unless

NP has quasi-polynomial time algorithms [29]. Due to the application nature of the problem, researchers in image processing and computer vision have also developed a variety of good heuristics that use classical combinatorial optimization techniques, such as network flow and local search (e.g., [18, 32, 21, 20, 24, 33]), for solving some special cases of the metric labeling problem.

Due to the NP-hardness of the metric labeling problem, the EOSD problem is NP-hard for nonconvex smoothness penalty functions. In this section, we focus on convex smoothness penalty functions (i.e., $f_{p,q}^{(i)}(\cdot)$ is a convex and nondecreasing function) that are widely used in medical image processing and in Markov Random Fields. The convex metric labeling problem is known to be polynomially solvable (e.g., [2, 33, 34]). Wu and Chen studied a more general case in [2] than both Ahuja et al. [34] and Ishikawa [33] in the sense that they considered nonuniform connectivity between columns. To solve the EOSD problem with convex smoothness penalty functions (*convex EOSD*, for short), we extend the technique developed for solving the convex metric labeling problem (in which smoothness penalty functions are convex) while incorporating the OMSD algorithm presented in Section 10.3.3 [35]. We reduce the problem to computing a minimum-cost s-excess set in a directed graph. Instead of forcing no arc (directed graph edge) leaving the sought node set, the minimum s-excess problem [2, 34] charges a penalty onto each arc leaving the set (i.e., the tail of the arc is in the set, while the head is not), which can be solved by using a minimum s-t cut algorithm.

10.3.4.1   The Graph Transformation Scheme for the Convex EOSD Problem

The graph $\mathcal{G} = (\mathcal{V}, \mathcal{E})$ consists of λ node-disjoint subgraphs $\{\mathcal{G}_i = (\mathcal{V}_i, \mathcal{E}_i) \mid i = 1, 2, \ldots, \lambda\}$; each $\mathcal{G}_i$ is for the search of the i-th surface S_i, as constructed in Section 10.3.2. We thus enforce in $\mathcal{G}$ both surface smoothness and surface separation constraints and incorporate the edge term and the region term of Equation 10.6.

The remaining problem is how to incorporate the soft smoothness a priori shape compliance term in Equation 10.6. Note that the minimum s-excess problem can be solved by computing a minimum s-t cut. Essentially, we need to "distribute" the cost $f_{p,q}(|S(p) - S(q)|)$ to the corresponding cut between the columns in $\mathcal{G}$ corresponding to the columns p and q in $\mathcal{I}$. Two intertwined questions need to be answered: how to put arcs between two adjacent columns, and how to assign a nonnegative cost to each arc (negative arc costs make the computation of a minimum s-excess computationally intractable). Fortunately, the (discrete equivalent of) second derivative of $f_{p,q}^{(i)}(\cdot)$ defined as in Equation 10.7, has the desired features that $f''_{p,q}^{(i)}(h) \geq 0$ for $h = 0, 1, 2, \ldots, \Delta_x - 1(\Delta_y - 1)$.

$$
\begin{aligned}
f''_{p,q}^{(i)}(0) &= f_{p,q}^{(i)}(1) - f_{p,q}^{(i)}(0) \\
f''_{p,q}^{(i)}(h) &= [f_{p,q}^{(i)}(h+1) - f_{p,q}^{(i)}(h)] - [f_{p,q}^{(i)}(h) - f_{p,q}^{(i)}(h-1)], \\
&\qquad h = 1, 2, \ldots, \Delta_x - 1(\Delta_y - 1).
\end{aligned}
\tag{10.7}
$$

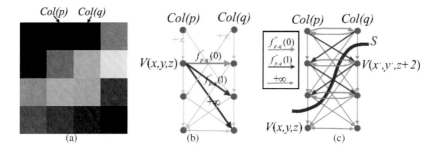

FIGURE 10.9
Graph construction for the convex smoothness penalty function. (a) Example of two
adjacent columns. The smoothness parameter $\Delta_x = 2$. (b) Weighted arcs introduced for each node $V(x, y, z)$. Gray arcs of weight $+\infty$ reflect the hard smoothness constraints. (c) Weighted arcs are built between nodes corresponding to the
two adjacent columns. The feasible surface S cuts the arcs with a total weight of
$f_{p,q}(2) = f''_{p,q}(0) + [f''_{p,q}(0) + f''_{p,q}(1)]$, which determines smoothness shape compliance term of $\alpha(S)$ for the two adjacent columns.

Based on $f''^{(i)}_{p,q}(\cdot)$, a cost distribution scheme is developed [2, 33, 34] to incorporate
the soft smoothness a priori shape compliance term (Figure 10.9). Assume that the
surface smoothness parameters are Δx and Δy. For each subgraph $\mathcal{G}_i$, we introduce
additional intercolumn arcs: for each $h = 0, 1, \ldots, \Delta x - 1$ $(\Delta y - 1)$, $v_i(x, y, z)$ has
an arc to $v_i(x', y', z - h)$ with an arc-weight of $f''^{(i)}_{p,q}(h)$ (Figure 10.9(b)), where
$p(x, y)$ and $q(x', y')$ are two adjacent columns with $|x - x'| + |y - y'| = 1$ (note
that we consider 4-neighborhood setting). Thus, the size of the graph $\mathcal{G}$ is of $O(\lambda n)$
nodes and $O((\Delta x + \Delta y)\lambda n)$ arcs.

10.3.4.2   Computing Optimal Multiple Surfaces for the Convex EOSD Problem

In the following, we show that using this construction, the total cost of the arcs that
are cut by a feasible surface S_i between two adjacent columns $Col(p)$ and $Col(q)$
equals the shape prior penalty $f^{(i)}_{p,q}(|S_i(p) - S_i(q)|)$. Without loss of generality, as-
sume that $f^{(i)}_{p,q}(0) = 0$ (otherwise, we can subtract $f^{(i)}_{p,q}(0)$ from the cost of each arc
from $Col(p)$ to $Col(q)$ without affecting the optimal solution). Let $z_p = S_i(p)$ and
$z_q = S_i(q)$. We next only consider the case that $z_p \geq z_q$ (the case $z_p \leq z_q$ can be
symmetrically the same). If $z_p = z_q$, S_i does not cut any inter-column arc between
$Col(p)$ to $Col(q)$. Thus, the induced penalty is zero, which is the same as the cost
$f^{(i)}_{p,q}(0)$. We now consider the case with $z_p > z_q$. For each z with $z_q < z \leq z_p$,
$v_i(x, y, z)$ has a directed arc in $\mathcal{G}_i$ to each node $v_i(x', y', \varpi)$ for $z_q < \varpi \leq z$, with a
cost of $f''^{(i)}_{p,q}(z - \varpi)$. Below, we list all directed arcs cut by the surface S_i in such a
way that each row is for the arcs originating from the same node in $Col(p)$ and each

column is for the arcs pointing to the same node in $\mathrm{Col}(q)$, as follows:

$$
\begin{pmatrix} v_i(x,y,z_p), \\ v_i(x',y',z_p) \end{pmatrix}
\quad
\begin{pmatrix} v_i(x,y,z_p), \\ v_i(x',y',z_p-1), \\ v_i(x,y,z_p-1), \\ v_i(x',y',z_p-1) \end{pmatrix}
\quad \cdots \cdots \quad
\begin{pmatrix} v_i(x,y,z_p), \\ v_i(x',y',z_q+1) \\ v_i(x,y,z_p-1), \\ v_i(x',y',z_q+1) \end{pmatrix}
$$

$$
\vdots \qquad\qquad \vdots
$$

$$
\begin{pmatrix} v_i(x,y,z_q+1), \\ v_i(x',y',z_q+1) \end{pmatrix}
$$

The total cost of those cut arcs between $\mathrm{Col}(p)$ and $\mathrm{Col}(q)$ equals

$$
\begin{aligned}
& f''^{(i)}_{p,q}(0) + [f''^{(i)}_{p,q}(0) + f''^{(i)}_{p,q}(1)] + \cdots + \\
& [f''^{(i)}_{p,q}(0) + \cdots + f''^{(i)}_{p,q}(z_p - z_q - 2) + f''^{(i)}_{p,q}(z_p - z_q - 1)] \\
= \; & [f^{(i)}_{p,q}(1) - f^{(i)}_{p,q}(0)] + [f^{(i)}_{p,q}(2) - f^{(i)}_{p,q}(1)] + \cdots + \\
& [f^{(i)}_{p,q}(z_p - z_q) - f^{(i)}_{p,q}(z_p - z_q - 1)] \\
= \; & f^{(i)}_{p,q}(z_p - z_q) \qquad \text{(note that } f^{(i)}_{p,q}(0) = 0) \qquad\qquad (10.8)
\end{aligned}
$$

Figure 10.9(c) shows the directed arcs cut by the surface S for the case $z_p \le z_q$.

Hence, we prove that the total cost of the arcs that are cut by a feasible surface S_i between two adjacent columns $\mathrm{Col}(p)$ and $\mathrm{Col}(q)$ equals the shape prior penalty $f^{(i)}_{p,q}(|S_i(p) - S_i(q)|)$.

Then, together with a similar argument in Section 10.3.3, we are able to show the following facts: (1) Any nonempty s-excess set $\mathcal{X}$ with a finite cost in $\mathcal{G}$ defines λ feasible surfaces in $\mathcal{I}$ whose total cost differs from that of $\mathcal{X}$ by a fixed value; (2) any set S of λ feasible surfaces in $\mathcal{I}$ corresponds to a nonempty s-excess set $\mathcal{X}$ in $\mathcal{G}$ whose cost differs from that of S by a fixed value. Consequently, a nonempty s-excess set in $\mathcal{G}$ with the minimum cost can specify an optimal set of λ surfaces in $\mathcal{I}$ while minimizing the energy function (10.6).

A minimum s-excess set in $\mathcal{G}$ can be computed by using a minimum s-t cut algorithm, as in [22]. If the minimum s-excess set in $\mathcal{G}$ thus obtained is empty, we can first perform on $\mathcal{G}$ a translation operation similar to that in Section 10.3.2, and then apply the s-t cut based algorithm to obtain a minimum *nonempty* s-excess set in $\mathcal{G}$. As in Section 10.3.2, we define a directed graph $\mathcal{G}_{st}$ from $\mathcal{G}$ and then compute a minimum s-t cut $(A^*, \bar{A}^*)$ in $\mathcal{G}_{st}$. Then, $\mathcal{X}^* = A^* - \{s\}$ is the minimum-cost s-excess set in $\mathcal{G}$, which can be used to specify an optimal set of λ surfaces as in Section 10.3.3.

10.3.5 Layered Optimal Graph Image Segmentation for Multiple Objects and Surfaces — LOGISMOS

The optimal graph-based segmentation approach can offer many advantages when employed for multiobject multisurface segmentation. Such a method allows optimally segmenting multiple surfaces that mutually interact within individual objects and/or between objects. Similar to the multisurface case presented above, intrasurface, intersurface, and interobject relationships are represented by context-specific graph arcs.

When segmenting complex shapes, the LOGISMOS approach [4] typically starts with an object presegmentation step the purpose of which is to identify the topology of the desired segmentaation surfaces. Using the presegmentation information, a single graph holding all relationships and surface cost elements is constructed, in which the segmentation of all desired surfaces is performed simultaneously in a single optimization process. While the description given below specifically refers to 3D image segmentation, the LOGISMOS method is fundamentally n-dimensional.

10.3.5.1 Object Pre-segmentation

The LOGISMOS method begins with a coarse presegmentation of the image data, but there is no prescribed method that must be used. The only requirement is that presegmentation yields robust approximate surfaces of the individual objects, having the same (correct) topology as the underlying objects and being sufficiently close to the true surfaces. The definition of "sufficiently close" is problem-specific and needs to be considered in relationship with how the layered graph is constructed from the approximate surfaces. Note that it is frequently sufficient to generate a single presegmented surface per object, even if the object itself exhibits more than one mutually interacting surface of interest. Depending upon the application, level sets, deformable models, active shape/appearance models, or other segmentation techniques can be used to yield object presegmentations.

10.3.5.2 Construction of Object-Specific Graphs

If the object-specific graph is constructed from a result of object presegmentation, the approximate presegmented surface may be meshed and the graph columns constructed as normals to individual mesh faces. The lengths of the columns are then derived from the expected maximum distances between the presegmented approximate surface and the true surface, so that the correct solution can be found within the constructed graph. Maintaining the same graph structure for individual objects, the base graph is formed using the presegmented surface mesh $\mathcal{M}$. V_B is the vertex set on $\mathcal{M}$ and E_B is the edge set. A graph column is formed by equally sampling several nodes along the normal direction of a vertex in V_B. The base graph is formed by connecting the bottom nodes by the connection relationship of E_B. In the multiple closed surface detection case, a duplication of the base graph is constructed each time when searching for an additional surface. The duplicated base graphs are connected by undirected arcs to form a new base graph, which ensures that the interacting surfaces can be detected simultaneously. Additional directed *intracolumn arcs*, *intercolumn arcs*, and *intersurface arcs* incorporate surface smoothness Δ and surface separation δ constraints into the graph.

10.3.5.3   Multiobject Interactions

When multiple objects with multiple surfaces of interest are in close apposition, a multiobject graph construction is adopted. This begins with considering pairwise interacting objects, with the connection of the base graphs of these two objects to form a new base graph. Object interaction is frequently local, limited to only some portions of the two objects' surfaces. Here we will assume that the region of pairwise mutual object interaction is known. A usual requirement may be that surfaces of closely located adjacent objects do not cross each other, that they are at a specific maximum/minimum distance, or similar. Object-interacting surface separation constraints are implemented by adding *interobject arcs* at the interacting areas. Interobject surface separation constraints are also added to the interacting areas to define the separation requirements that shall be in place between two adjacent objects. The interobject arcs are constructed in the same way as the intersurface arcs. The challenge in this task is that no one-to-one correspondence exists between the base graphs (meshes) of the interacting object pairs. To address this challenge, corresponding columns i and j need to be defined between the interacting objects. The corresponding columns should have the same directions. Considering signed distance offset d between the vertex sets $\mathcal{V}_i$ and $\mathcal{V}_j$ of the two objects, interobject arcs $\mathcal{E}^o$ between two corresponding columns can be defined as:

$$
\begin{aligned}
\mathcal{E}^o = \{ &\langle \mathcal{V}_i(k), \mathcal{V}_j(k - d + \delta^l) \rangle | \forall k : \\
&\max (d - \delta^l, 0) \le k \le \min (K - 1 + d - \delta^l, K - 1) \} \\
\cup \{ &\langle \mathcal{V}_j(k), \mathcal{V}_i(k + d - \delta^u) \rangle | \forall k : \\
&\max (\delta^u - d, 0) \le k \le \min (K - 1 + d - \delta^u, K - 1) \} \quad (10.9)
\end{aligned}
$$

where k is the vertex index number; δ^l and δ^u are interobject separation constraints with $\delta^l \le \delta^u$.

Nevertheless, it may be difficult to find corresponding columns between two regions of different topology. The approach presented below offers one possible solution. Since more than two objects may be mutually interacting, more than one set of pairwise interactions may coexist in the constructed graph.

10.3.5.4   Electric Lines of Force

Starting with the initial presegmented shapes, a cross-object search direction must be defined for each location along the surface. The presently adopted method for defining corresponding columns via cross-object surface mapping relies upon electric field theory for a robust and general definition of such search direction lines to guarantee nonintersecting character of the search-direction lines [36]. Recall Coulomb's law from basic physics

$$
E_i = \frac{1}{4\pi\varepsilon_0} \frac{Q}{r^2} \hat{\mathbf{r}} \,, \quad (10.10)
$$

where E_i is the electric field at point i. Q is the charge of point i; r is the distance from point i to the evaluation point; $\hat{r}$ is the unit vector pointing from the point i to the evaluation point. ε_0 is the vacuum permittivity. Since the total electric field E is the sum of E_i's

$$E = \sum_i E_i ,\tag{10.11}$$

the electric field has the same direction as the electric line(s) of force (ELF). When an electric field is generated from multiple source points, the electric lines of force exhibit a nonintersection property, which is of major interest in the context of finding corresponding columns.

When computing ELF for a computer generated 3D triangulated surface, the surface is composed of a limited number of vertices that are usually not uniformly distributed. These two observations greatly reduce the effect of charges located in close proximity. To cope with this undesirable effect, a positive charge Q_i is assigned to each vertex v_i. The value of Q_i is determined by the area sum of triangles t_j where $v_i \in t_j$. When changing r^2 to r^m ($m > 2$), the nonintersection property still holds. The difference is that more distant vertices will be penalized in ELF computing. Therefore, a slightly larger m will increase the robustness of local ELF computation. Discarding the constant term, the electric field is defined as

$$\hat{E} = \sum_i \frac{\sum_j \text{AREA}(t_j)}{r_i{}^m} \hat{r}_i ,\tag{10.12}$$

where $v_i \in t_j$ and $m > 2$.

Assuming there is a closed surface in an n-D space, the point having a zero electric field is the solution of equation $\hat{E} = 0$. In an extreme case, the closed surface will converge to the solution points when searching along the ELF. Except for these points, the nonintersecting ELF will fill the entire space. Since ELF are nonintersecting, it is easy to interpolate ELF at nonvertex locations on a surface. The interpolation can greatly reduce total ELF computation load when upsampling a surface. In 2D, linear interpolation from two neighboring vertices and their corresponding ELF can be implemented. In 3D, use of barycentric coordinates is preferred to interpolate points within triangles.

When a closed surface is used for ELF computation, isoelectric potential surfaces can be found. Except for the solution points $\hat{E} = 0$, all other points belong to an isoelectric potential surface and the ELF passing through such a point can be easily interpolated. The interpolated ELF intersects the initial closed surface. Consequently, this technique can be used to create connections between a point in space and a closed surface, yielding cross-surface mapping.

10.3.5.5 ELF-Based Cross-Object Surface Mapping

The nonintersection property of ELF is useful to find one-to-one mapping between adjacent objects with different surface topologies. Interrelated ELF of two coupled surfaces can be determined in two steps:

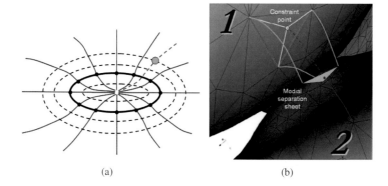

(a)   (b)

FIGURE 10.10

Cross-object surface mapping by ELF. (a) The ELF (blue lines) are pushed forward from a surface composed of black vertices. The dashed black surfaces indicate the location of isoelectric potential contours. The red-dashed ELF is the traced-back line from a green point to the solid black surface. The traced-back line is computed by interpolating two neighboring pushed-forward ELF. (b) Constraint-point mapping of coupled 3D surfaces is performed in the following 5 steps: (i) Green and red ELF are pushed forward from surface 1 and 2, respectively. (ii) The intersections between the ELF and medial separating sheet form a blue triangle and a red point. (iii) The red point is traced back along dotted red line to surface 1. (iv) When the dotted red line intersects surface 1, it forms a light-blue constraint point on surface 1. (v) The constraint point is connected at surface 1 by yellow edges. *Reprinted, with IEEE permission, from [4].*

- Push forward—regular ELF path computation using Equation (10.12).

- Trace back—interpolation process to form an ELF path from a point in space to a closed surface as outlined above.

The general idea of mapping two coupled surfaces using ELF includes defining an ELF path using the push forward process, identifying the medial-sheet intersection point on this path, generating a constraint point on the opposite surface, and connecting the constraint point with already existing close vertices on this surface – so called constraint-point mapping [36]. Figure 10.10a shows ELF pushed forward from a surface and traced back from a point to that surface. Figure 10.10b shows an example of mapping two coupled surfaces in 3D.

Note that each vertex in the object-interaction area can therefore be used to create a constraint point affecting the coupled surface. Importantly, the corresponding pairs of vertices (the original vertex and its constraint point) from two interacting objects in the contact area identified using the ELF are guaranteed to be in a one-to-one relationship and all-to-all mapping, irrespective of surface vertex density. As a result, the desirable property of maintaining the previous surface geometry (e.g., the orange triangle in Figure 10.10b) is preserved. Therefore, the mapping procedure avoids surface regeneration and merging [37] (which is usually difficult) and enhances robustness with respect to local roughness of the surface, when compared with our previously introduced nearest point based mapping techniques [38, 39].

10.3.5.6   Cost Function and Graph Optimization

The resulting segmentation is driven by the cost functions associated with the graph vertices. Design of vertex-associated costs is problem specific, and costs may reflect edge-based, region-based, or combined edge–region image properties as already described above. Same as in the earlier cases, the optimization problem can be converted to finding the minimum nonempty closed set in the modified graph. As a result of such single optimization process, a globally optimal solution provides all segmentation surfaces for all involved interacting objects while satisfying all surface and object interaction constraints.

10.4   Case Studies

10.4.1   Segmentation of Retinal Layers in Optical Coherence Tomography Volumes

As a first case study, we consider the task of automatically segmenting multiple retinal layers in spectral-domain optical coherence tomography volumes, as illustrated in Figure 10.8. Segmenting these layers is necessary to quantify changes that occur in these layers as a result of ocular diseases causing blindness, such as glaucoma, diabetic retinopathy, and age-related macular degeneration [40]. Such a task is well suited for utilizing the layer-based graph approach discussed in this chapter as the ultimate goal is to simultaneously define approximately seven-to-eleven layered 3D surfaces within these volumes. Since many of the separating surfaces may seem similar to one another at a local level (e.g., surfaces that indicate a transition from a relatively dark region to a relatively bright region), the ability to simultaneously segment these surfaces in 3D is important, as is possible using the layered graph approach described in this chapter.

10.4.1.1   Graph Structure

As an example of a layered graph segmentation problem, we can assume that in an OCT volume $I(x, y, z)$, we desire to find a set of surfaces, where each surface i can be represented by a function $f_i(x, y)$. Because for every (x, y) location, there is only one z-value on the surface, in constructing the subgraph associated with each surface, we can simply define one column $Col(x, y)$ of nodes for each (x, y) location. This thus results in the set of nodes of the surface subgraph directly corresponding to the voxel locations in the original volume, and we can use a similar notation to index the nodes of the subgraph using $N_i(x, y, z)$ (where x, y, and z correspond to the voxel locations and i corresponds to the surface) as we would index the image volume using $I(x, y, z)$. In finding a simultaneous set of n surfaces, we thus have n subgraphs with nodes $N_1(x, y, z), \ldots, N_n(x, y, z)$. Edges are added between the columns as discussed in general in Section 10.3.2 and 10.3.3 to enforce feasibility constraints (i.e., smoothness and surface separation constraints). Better results can be obtained if we allow this constraints to vary as a function of (x, y) location [41, 42]. Such locally varying constraints can be learned from a training set as described in [41, 42].

While the layered graph approach enables the simultaneous segmentation of *all* surfaces, further efficiency can be obtained if the surfaces are segmented in groups in a multiresolution fashion [42, 43]. For instance, one might first simultaneously segment a smaller set of the most visible surfaces in a low-resolution representation of the volume, while refining the segmentation (and segmenting more surfaces) more locally in higher resolutions of the volume.

10.4.1.2 Cost Function Design

As reported in [44, 42], segmenting the layers in OCT volumes can benefit from incorporating both edge and regional information into the cost function design (i.e., having both on-surface and in-region costs). In fact, the segmentation of OCT layers was the first application to utilize both of these terms using the graph approach of this chapter. In [45], on-surface costs were defined by computing signed edge terms to favor either a bright-to-dark transition or dark-to-bright-transition, depending on the surface. In-region cost terms were based on fuzzy membership functions in dark, medium, or bright intensity classes so that voxels that most matched the expected intensity range of the region would have the lowest costs. The relative weights of each of these terms was defined based on a training set.

10.4.2 Simultaneous Segmentation of Prostate and Bladder in Computed Tomography Volumes

In the United States, prostate cancer is the most common cancer in men, accounting for about 28% of all newly diagnosed cases [46]. Precise target delineation is critical for a successful 3-D radiotherapy treatment planning for prostate cancer treatment. Automatic segmentation of pelvic structure is of particular difficulty. It involves soft tissues that present a large variability in shape and size. Those soft tissues also have similar intensity and have mutual influence in position and shape. To overcome all these difficulties, both shape prior information and surface context information are incorporated using an arc-weighted graph representation to simultaneously segment prostate and bladder in 3D.

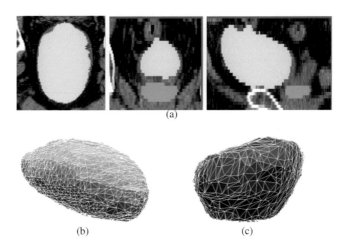

(a)

(b)   (c)

FIGURE 10.11
(a) Presegmentation for bladder (yellow) and prostate (blue) in the transverse (left), coronal (middle) and sagittal (right) views. (b) Triangulated mesh for the bladder. (c) Triangulated mesh for the prostate.

Our method consists of the following steps [47, 35, 48]. First, we obtain a presegmentation of bladder and prostate as the initial shape model, which provides useful information about the topological structures of the target objects (Figure 10.11(a)). A 3-D geodesic active contour method is employed [49] for presegmenting the bladder. The mean shape of the prostate from the training datasets is simply fitted to the never-before seen CT images using a rigid transform. Two triangulated meshes $M_1(V_1, E_1)$ and $M_2(V_2, E_2)$ are built for the bladder and prostate based on the shape model, respectively (Figure 10.11(b),(c)). Based on these two triangulated meshes, the arc-weighted graph is constructed using the method described in Section 10.3.4. Specifically, two weighted subgraphs $\mathcal{G}_i$ are built from the mesh M_i as follows. For each vertex $v \in V_i$, a column of K nodes is created in $\mathcal{G}_i$. The positions of nodes reflect the positions of corresponding voxels in the image domain. The length of the column is set according to the required search range. The number of nodes K on each column is determined by the required resolution. The direction of the column is set as the triangle normal. The feasible surface S_i in the graph $\mathcal{G}_i$ is defined as the surface containing exactly one node in each column. To avoid the overlap of two target surfaces, a "partially interacting area" is defined according to the distance between two meshes, which indicates that the two target surfaces may mutually interact each other at that area. To model the interaction relation, the two graphs $\mathcal{G}_1$ and $\mathcal{G}_2$ "share" some common node columns in that partially interacting area, and the target surfaces S_1 and S_2 both cut those columns, as shown in Figure 10.12.

Cost function design plays an important role for successful surface detection. In

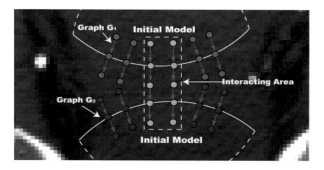

FIGURE 10.12
Graph construction for mutually interacting objects. An example 2-D slice is presented. Note that in the interacting region, for each column with green nodes, there actually exists two columns with the same position, one for Graph $\mathcal{G}_1$, one for Graph $\mathcal{G}_2$.

our segmentation, the gradient-based cost function was combined with class uncertainty information for edge-based costs [48]. Given intensity information, the posterior probability that the voxel belongs to the target object is learned from the training set, which is used as the region-based costs. The quadratic shape prior penalty function is obtained from the experiments on the training datasets [48].

The experiments were conducted for simultaneous segmentation of the bladder and the prostate. 3-D CT images from different patients with prostate cancer were used. The image sizes ranged from $80 \times 120 \times 30$ to $190 \times 180 \times 80$ voxels. The image spacing resolution ranged from $0.98 \times 0.98 \times 3.00$ mm^3 to $1.60 \times 1.60 \times 3.00$ mm^3. Out of 21 volumes, 8 were randomly selected as the training data and our segmentation was performed on the remaining 13 datasets.

For quantitative validation, the result was compared with the expert-defined manual contours. For volumetric error measurement, the Dice similarity coefficient (DSC) was computed using $D = 2|V_m \cap V_c|/(|V_m| + |V_c|)$, where V_m denotes the manual volumetric result and V_c denotes the computed result. For surface distance error, both mean and the maximum unsigned surface distance error were computed for the bladder and the prostate surfaces between the computed result and the manual delineation. The result is shown in Table 10.1.

For visual performance assessment, the illustrative result with both computed contours and manual contours is displayed in Figure 10.13(a). The 3-D representation was shown in Figure 10.13(d).

$Surface$	DSC	$Mean$ (mm)	$Maximum$ (mm)
Prostate	0.797	1.01 ± 0.94	5.46 ± 0.96
Bladder	0.900	0.99 ± 0.77	5.88 ± 1.29

TABLE 10.1

Overall quantitative results. Mean $\pm$ SD in mm for the unsigned surface distance error.

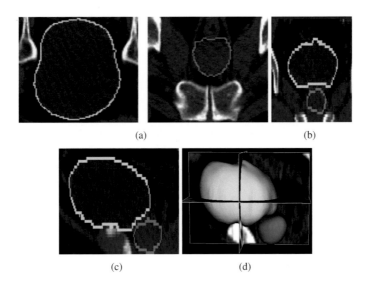

(a)   (b)

(c)   (d)

FIGURE 10.13

The bladder (yellow) and the prostate (blue) segmentation results. (a) Transverse view. (b) Coronal view. (c) Sagittal view. (d) 3-D representation of the bladder (yellow) and the prostate (blue).

10.4.3   Cartilage and Bone Segmentation in the Knee Joint

Section 10.3.5 introduced our LOGISMOS method for multiobject, multisurface segmentation. Here we demonstrate its functionality on a knee bone/cartilage segmentation example. In this orthopedic application, a typical scenario includes the need to segment surfaces of the periosteal and subchondral bone and of the overlying articular cartilage from MRI scans with high accuracy and in a globally consistent manner (Figure 10.14).

Three bones articulate in the knee joint: the femur, the tibia, and the patella. Each of these bones is partly covered by cartilage in regions where individual bone pairs slide over each other during joint movements. For assessment of the knee joint cartilage health, it is necessary to identify six surfaces: femoral bone, femoral cartilage, tibial bone, tibial cartilage, patellar bone, and patellar cartilage. In addition to each

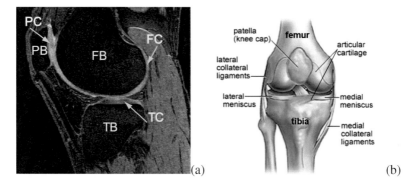

(a)   (b)

FIGURE 10.14
Human knee. (a) Example MR image of a knee joint — femur, patella, and tibia bones with associated cartilage surfaces are clearly visible. FB = femoral bone, TB = tibial bone, PB = patellar bone, FC = femoral cartilage, TC = tibial cartilage, PC = patellar cartilage. (b) Schematic view of knee anatomy; adapted from [50]. *Reprinted, with IEEE permission, from [4].*

connected bone and cartilage surface mutually interacting on a given bone, the bones interact in a pairwise manner – cartilage surfaces of the tibia and femur and of the femur and patella are in close proximity (or in frank contact) for any given knee joint position. Clearly, the problem of simultaneous segmentation of the six surfaces belonging to three interacting objects is well suited for application of the LOGISMOS method.

10.4.3.1   Bone Presegmentation

Figure 10.15 shows the flowchart of our approach [4]. As the first step, the volume of interest (VOI) of each bone, together with its associated cartilage, is identified using an AdaBoost classification approach in 3D MR images. Three VOIs per knee joint image result within which the individual bones (femur, tibia, patella) are located.

After localizing the object VOIs, approximate surfaces of the individual bones must be obtained by first roughly fitting the mean bone shape models directly to the automatically identified VOIs (upper panels of Figure 10.15b). A single surface detection graph was constructed based on the fitted mean shapes – graph columns were built along nonintersecting electric lines of force to increase the robustness of the graph construction (Section 10.3.5.4). The surface costs were associated with each graph node based on the inverted surface likelihood probabilities provided by the random forest classifiers. After repeating this step iteratively until convergence (usually 3–5 iterations were needed), the approximate surfaces of each bone were automatically identified, without considering any bone-to-bone context, see lower panel of Figure 10.15b.

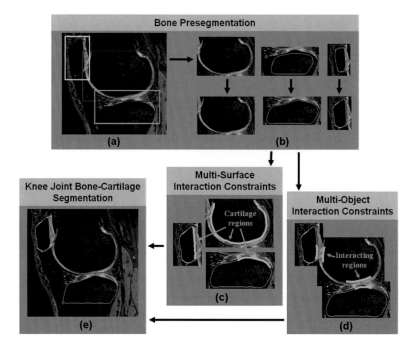

FIGURE 10.15
The flowchart of LOGISMOS-based segmentation of articular cartilage for all bones in the knee joint. (a) Detection of bone volumes of interest using AdaBoost approach. (b) Approximate bone segmentation using single-surface graph search. (c) Generation of multisurface interaction constraints. (d) Construction of multiobject interaction constraints. (e) LOGISMOS-based simultaneous segmentation of 6 bone & cartilage surfaces in 3D. *Reprinted, with IEEE permission, from [4].*

10.4.3.2   Multisurface Interaction Constraints

Image locations adjacent to and outside of the bone may belong to cartilage, meniscus, synovial fluid, or other tissue, and they thus exhibit different image appearance (Figure 10.15c). Since the cartilage generally covers only those parts of the respective bones which may articulate with another bone, two surfaces (cartilage and bone) are defined only at those locations, while single (bone) surfaces are to be detected in noncartilage regions. To facilitate a topologically robust problem definition across a variety of joint shapes and cartilage disease stages, two surfaces are detected for each bone, and the single–double surface topology differentiation reduces into differentiation of zero and nonzero distances between the two surfaces. In this respect, the noncartilage regions along the external bone surface were identified as regions in which zero distance between the two surfaces was enforced so that the two surfaces collapsed onto each other, effectively forming a single bone surface. In the cartilage regions, the zero-distance rule was not enforced, providing for both a subchondral bone and articular cartilage surface segmentation.

10.4.3.3   Multiobject Interaction Constraints

In addition to dual-surface segmentation that must be performed for each individual bone, the bones of the joint interact in the sense that cartilage surfaces from adjacent bones cannot intersect each other, cartilage and bone surfaces must coincide at the articular margin, the maximum anatomically feasible cartilage thickness shall be observed, etc. The regions in which adjacent cartilage surfaces come into contact are considered the interacting regions (Figure 10.15d). In the knee, such interacting regions exist between the tibia and the femur (tibiofemoral joint) and between the patella and the femur (patellofemoral joint). To automatically find these interacting regions, an isodistance medial separation sheet is identified in the global coordinate system midway between adjacent presegmented bone surfaces. If a vertex is located on an initial surface while having a search direction intersecting the sheet, the vertex is identified as belonging to the region of surface interaction. The separation sheet can be identified using signed distance maps even if the initial surfaces intersect. Following the ELF approach described above, one-to-one and all-to-all corresponding pairs are generated between femur–tibia contact area as well as femur–patella contact area by constraint-point mapping technique. The corresponding pairs and their ELF connections are used for interobject graph link construction [38, 39].

10.4.3.4   Knee Joint Bone–Cartilage Segmentation

After completion of the above steps, the segmentation of multiple surfaces of multiple mutually interacting objects is performed simultaneously and globally optimally subject to the interaction constraints (Figure 10.15e). Specifically, double surface segmentation graphs were constructed individually for each bone using that bone's initial surface. The three double surface graphs were further connected by interobject graph arcs between the corresponding columns identified during the previous step as belonging to the region of close-contact object interaction. The minimum distance between the interacting cartilage surfaces from adjacent bones was set to zero to avoid cartilage overlap.

10.4.3.5   Knee Joint Bone/Cartilage Segmentation

Our LOGISMOS method was applied to bone/cartilage segmentation of 60 randomly selected knee MR images from the publicly available Osteoarthritis Initiative (OAI) database, which is available for public access at http://www.oai.ucsf.edu/. The MR images used were acquired with a 3T scanner following a standardized procedure. A sagittal 3D dual-echo steady state (DESS) sequence with water excitation and the following imaging parameters: image stack of $384 \times 384 \times 160$ voxels, with voxel size of $0.365 \times 0.365 \times 0.70$ mm.

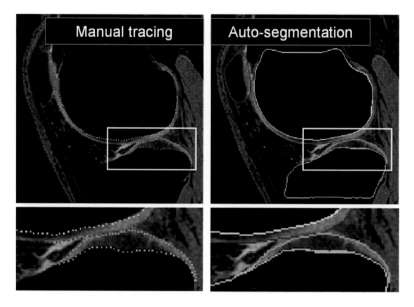

FIGURE 10.16

MR image segmentation of a knee joint – a single contact-area slice from a 3D MR dataset is shown. Segmentation of all six surfaces was performed simultaneously in 3D. (left) Original image data with expert-tracing overlaid. (right) Computer segmentation result. Note that the double-line boundary of tibial bone is caused by intersecting the segmented 3D surface with the image plane. *Reprinted, with IEEE permission, from [4].*

Figure 10.16 shows an example of a knee joint contact area slice from the 3D MR dataset. Note the contact between the femoral and tibial cartilage surfaces, as well as the contact between the femoral and patellar cartilage surfaces. Furthermore, there is an area of high-intensity synovial fluid adjacent to the femoral cartilage that is not part of the cartilage tissue and should not be segmented as such. The right panel of Figure 10.16 shows the resulting segmentation demonstrating very good delineation of all six segmented surfaces and correct exclusion of the synovial fluid from the cartilage surface segmentations. Since the segmentations are performed simultaneously for all six bone and cartilage surfaces in the 3D space, computer segmentation directly yields the 3D cartilage thickness for all bone surface locations. Typical segmentation results are given in Figure 10.17.

The signed and unsigned surface positioning errors of the obtained cartilage segmentations were quantitatively measured over the cartilage regions. The average signed surface positioning errors for the six detected surfaces ranged from 0.04 to 0.16 mm, while the average unsigned surface positioning error ranged from 0.22 to 0.53 mm. The close-to-zero signed positioning errors attest to a small bias of surface detection. The unsigned positioning errors show that the local fluctuations around the correct location are much smaller than the longest face of MR image voxels (0.70 mm). Our results therefore achieved virtually no surface positioning bias and subvoxel local accuracy for each of the six detected surfaces.

When assessing the performance using Dice coefficients (DSC), the obtained DSC values were 0.84, 0.80 and 0.80, for the femoral, tibial, and patellar cartilage surfaces, respectively.

10.5   Conclusion

The presented framework for simultaneous optimal segmentation of single and multiple surfaces possibly belonging to multiple objects is very powerful. The method can be directly extended to n-D, and the intersurface or interobject relationships can include higher-dimensional interactions, e.g., mutual object motion, interactive shape changes over time, and similar. Overall, the described method is general and useful for a broad range of applications.

10.6   Acknowledgments

This work was supported, in part, by NIH grants R01–EB004640, K25–CA123112, R44–AR052983, P50 AR055533 and by NSF grants CCF-0830402, CCF-0844765.

The Osteoarthritis Initiative (OAI) is a public–private partnership comprised of five contracts (N01-AR-2-2258; N01-AR-2-2259; N01-AR-2-2260; N01-AR-2-2261; N01-AR-2-2262) funded by the NIH and conducted by the OAI Study Investigators. Private funding partners include Merck Research Laboratories; Novartis Pharmaceuticals Corporation, GlaxoSmithKline; and Pfizer, Inc.

Bibliography

[1] M. Sonka, V. Hlavac, and R. Boyle, *Image Processing, Analysis, and Machine Vision (3rd ed.)*. Thomson Engineering, 2008.

[2] X. Wu and D. Chen, "Optimal net surface problems with applications," in *Proc. 29th International Colloquium on Automata, Languages and Programming (ICALP)*, ser. Lecture Notes in Computer Science, vol. 2380.   Springer, July 2002, pp. 1029–1042.

[3] K. Li, X. Wu, D. Chen, and M. Sonka, "Optimal surface segmentation in volumetric images–a graph-theoretic approach," *IEEE Trans. on Pattern Analysis and Machine Intelligence*, vol. 28, no. 1, pp. 119–134, 2006.

[4] Y. Yin, X. Zhang, R. Williams, X. Wu, D. Anderson, and M. Sonka, "LOGISMOS - Layered Optimal Graph Image Segmentation of Multiple Objects and Surfaces: Cartilage segmentation in the knee joints," *IEEE Trans. Medical Imaging*, vol. 29, no. 12, pp. 2023 – 2037, 2010.

[5] Y. Boykov and G. Funka-Lea, "Graph cuts and efficient n-d image segmentation," *International Journal of Computer Vision*, vol. 70, no. 2, pp. 109–131, 2006.

[6] A. Delong and Y. Boykov, "Globally optimal segmentation of multi-region objects," in *International Conference on Computer Vision (ICCV), Kyoto, Japan*, 2009, pp. 285–292.

[7] A. Chakraborty, H. Staib, and J. Duncan, "Deformable boundary finding in medical images by integrating gradient and region information," *IEEE Trans. on Medical Imaging*, vol. 15, no. 6, pp. 859–870, 1996.

[8] S. Zhu and A. Yuille, "Region competition: Unifying snakes, region growing, and bayes/mdl for multiband image segmentation," *IEEE Trans. on Pattern Analysis and Machine Intelligence*, vol. 18, pp. 884–900, 1996.

[9] A. Yezzi, A. Tsai, and A. Willsky, "A statistical approach to snakes for bimodal and trimodal imagery," in *Proc. of Int. Conf. on Computer Vision (ICCV)*, Corfu, Greece, 1999, pp. 898–903.

[10] N. Paragios and R. Deriche, "Coupled geodesic active regions for image segmentation: A level set approach," in *Proc. of the European Conference in Computer Vision (ECCV)*, vol. II, 2001, pp. 224–240.

[11] T. F. Chan and L. A. Vese, "Active contour without edges," *IEEE Trans. Image Processing*, vol. 10, pp. 266–277, 2001.

[12] N. Paragios, "A variational approach for the segmentation of the left ventricle in cardiac image analysis," *Int. J. of Computer Vision*, vol. 46, no. 3, pp. 223–247, 2002.

[13] Y. Boykov and V. Kolmogorov, "Computing geodesics and minimal surfaces via graph cuts," in *Proc. of Int. Conf. on Computer Vision (ICCV)*, Nice, France, October 2003, pp. 26–33.

[14] J. Picard, "Maximal closure of a graph and applications to combinatorial problems," *Management Science*, vol. 22, pp. 1268–1272, 1976.

[15] D. Hochbaum, "A new-old algorithm for minimum-cut and maximum-flow in closure graphs," *Networks*, vol. 37, no. 4, pp. 171–193, 2001.

[16] X. Wu, D. Chen, K. Li, and M. Sonka, "The layered net surface problems in discrete geometry and medical image segmentation," *Int. J. Comput. Geometry Appl.*, vol. 17, no. 3, pp. 261–296, 2007.

[17] M. K. Haeker, M. D. Abràmoff, X. Wu, R. Kardon, and M. Sonka, "Use of varying constraints in optimal 3-D graph search for segmentation of macular optical coherence tomography images," in *Medical Image Computing and Computer-Assisted Intervention (MICCAI 2007)*, ser. Lecture Notes in Computer Science, N. Ayache, S. Ourselin, and A. Maeder, Eds., vol. 4791. Berlin/New York: Springer, 2007, pp. 244–251.

[18] Y. Boykov, O. Veksler, and R. Zabih, "Markov Random Fields with efficient approximations," in *Proc. of the IEEE Conf. on Computer Vision and Pattern Recognition*, 1998, pp. 648–655.

[19] ——, "Fast approximate energy minimization via graph cuts," *IEEE Trans. on Pattern Analysis and Machine Intelligence*, vol. 23, no. 11, pp. 1222–1239, 2001.

[20] H. Ishikawa and D. Geiger, "Segmentation by grouping junctions," in *Proc. of the IEEE Conf. on Computer Vision and Pattern Recognition*, Santa Barbara, CA, 1998, pp. 125–131.

[21] D. Greig, B. Porteous, and A. Seheult, "Exact maximum a posteriori estimation for binary image," *J. Roy. Statist. Soc. Ser. B*, vol. 51, pp. 271–279, 1989.

[22] D. Hochbaum, "An efficient algorithm for image segmentation, markov randomfields and related problems," *J. of the ACM*, vol. 48, pp. 686–701, 2001.

[23] B. Glocker, N. Komodakis, N. Paragios, C. Glaser, G. Tziritas, and N. Navab, "Primal/dual linear programming and statistical atlases for cartilage segmentation," in *Medical Image Computing and Computer-Assisted Intervention (MICCAI 2007)*, ser. Lecture Notes in Computer Science, N. Ayache, S. Ourselin, and A. Maeder, Eds., vol. 4792. Springer, 2007, pp. 536–543.

[24] S. Roy and I. Cox, "A maximum-flow formulation of the n-camera stereo correspondence problem," in *Proc. of Int. Conf. on Computer Vision (ICCV)*, 1998, pp. 492–499.

[25] V. Kolmogorov and R. Zabih, "Computing visual correspondence with occlusions using graph cuts," in *Proc. of Int. Conf. on Computer Vision (ICCV)*, Vancouver, Canada, July 2001, pp. 508–515.

[26] B. Glocker, N. Komodakis, N. Paragios, G. Tziritas, and N. Navab, "Inter and intra-modal deformable registration: Continuous deformations meet efficient optimal linear programming," in *Proc. of the 20th Int. Conf. on Information Processing in Medical Imaging (IPMI)*, vol. LNCS 4584. Springer, 2006, pp. 408–420.

[27] J. Kleinberg and E. Tardos, "Approximation algorithms for classification problems with pairwise relationships: Metric labeling and Markov random fields," in *Proc. of the 40th IEEE Symp. on Foundations of Computer Science*, 1999, pp. 14–23.

[28] A. Archer, J. Fakcharoenphol, C. Harrelson, R. Krauthgamer, K. Talvar, and E. Tardos, "Approximate classification via earthmover metrics," in *Proc. of the 15th Annual ACM-SIAM Symposium on Discrete Algorithms*, 2004, pp. 1079–1089.

[29] J. Chuzhoy and S. Naor, "The hardness of metric labeling," *SIAM Journal on Computing*, vol. 36, no. 5, pp. 1376–1386, 2007.

[30] C. Chekuri, A. Khanna, J. Naor, and L. Zosin, "A linear programming formulation and approximation algorithms for the metric labeling problem," *SIAM Journal of Discrete Mathematics*, vol. 18, no. 3, pp. 608–625, 2005.

[31] A. Gupta and E. Tardos, "Constant factor approximation algorithms for a class of classification problem," in *Proc. of the 32nd Annual ACM Symp. on Theory of Computing (STOC)*, 2000, pp. 652–658.

[32] R. Dubes and A. Jain, "Random field models in image analysis," *J. Appl. Stat.*, vol. 16, pp. 131–164, 1989.

[33] H. Ishikawa, "Exact optimization for Markov random fields with convex priors," *IEEE Trans. on Pattern Analysis and Machine Intelligence*, vol. 25, no. 10, pp. 1333–1336, 2003.

[34] R. Ahuja, D. Hochbaum, and J. Orlin, "A cut based algorithm for the convex dual of the minimum cost network flow problem," *Algorithmica*, vol. 39, pp. 189–208, 2004.

[35] Q. Song, X. Wu, Y. Liu, M. Sonka, and M. Garvin, "Simultaneous searching of globally optimal interacting surfaces with shape priors," in *Proc. of the 20rd IEEE Conference on Computer Vision and Pattern Recognition (CVPR)*, vol. 4584, San Francisco, CA, USA, June 2010, pp. 2879–2886.

[36] Y. Yin, Q. Song, and M. Sonka, "Electric field theory motivated graph construction for optimal medical image segmentation," in *7th IAPR-TC-15 Workshop on Graph-based Representations in Pattern Recognition*, 2009, pp. 334–342.

[37] D. Kainmueller, H. Lamecker, S. Zachow, M. Heller, and H.-C. Hege, "Multi-object segmentation with coupled deformable models," in *Proc. of Medical Image Understanding and Analysis (MIAU)*, 2008, pp. 34–38.

[38] Y. Yin, X. Zhang, and M. Sonka, "Fully three-dimensional segmentation of articular cartilage performed simultaneously in all bones of the joint," in *Osteoarthritis and Cartilage*, vol. 15(3), 2007, p. C177.

[39] Y. Yin, X. Zhang, D. D. Anderson, T. D. Brown, C. V. Hofwegen, and M. Sonka, "Simultaneous segmentation of the bone and cartilage surfaces of a knee joint in 3D," in *SPIE Symposium on Medical Imaging*, vol. 7258, 2009, p. 72591O.

[40] M. D. Abràmoff, M. K. Garvin, and M. Sonka, "Retinal imaging and image analysis," *IEEE Reviews in Biomedical Engineering*, vol. 3, pp. 169–208, 2010.

[41] M. Haeker (Garvin), M. D. Abràmoff, X. Wu, R. Kardon, and M. Sonka, "Use of varying constraints in optimal 3-D graph search for segmentation of macular optical coherence tomography images," in *Proceedings of the 10th International Conference on Medical Image Computing and Computer-Assisted Intervention (MICCAI 2007)*, ser. Lecture Notes in Computer Science, vol. 4791. Springer-Verlag, 2007, pp. 244–251.

[42] M. K. Garvin, M. D. Abràmoff, X. Wu, S. R. Russell, T. L. Burns, and M. Sonka, "Automated 3-D intraretinal layer segmentation of macular spectral-domain optical coherence tomography images," *IEEE Trans. Med. Imag.*, vol. 28, no. 9, pp. 1436–1447, Sept. 2009.

[43] K. Lee, M. Niemeijer, M. K. Garvin, Y. H. Kwon, M. Sonka, and M. D. Abràmoff, "Segmentation of the optic disc in 3D-OCT scans of the optic nerve head," *IEEE Trans. Med. Imag.*, vol. 29, no. 1, pp. 159–168, Jan. 2010.

[44] M. Haeker (Garvin), X. Wu, M. D. Abràmoff, R. Kardon, and M. Sonka, "Incorporation of regional information in optimal 3-D graph search with application for intraretinal layer segmentation of optical coherence tomography images," in *Information Processing in Medical Imaging (IPMI)*, ser. Lecture Notes in Computer Science, vol. 4584. Springer, 2007, pp. 607–618.

[45] M. K. Garvin, M. D. Abramoff, X. Wu, S. R. Russell, T. L. Burns, and M. Sonka, "Automated 3-D intraretinal layer segmentation of macular spectral-domain optical coherence tomography images," *IEEE Trans. Med. Imaging*, vol. 28, no. 9, pp. 1436–1447, 2009.

[46] A. Jemal, R. Siegel, J. Xu, and E. Ward, "Cancer statistics, 2010," *CA Cancer Journal for Clinicians*, vol. 60, no. 5, pp. 277–300, 2010.

[47] Q. Song, X. Wu, Y. Liu, M. Smith, J. Buatti, and M. Sonka, "Optimal graph search segmentation using arc-weighted graph for simultaneous surface detection of bladder and prostate," in *Proc. International Conference on Medical Image Computing and Computer-Assisted Intervention*, 2009, pp. 827–835.

[48] Q. Song, Y. Liu, Y. Liu, P. Saha, M. Sonka, and X. Wu, "Graph search with appearance and shape information for 3-D prostate and bladder segmentation," in *Proc. International Conference on Medical Image Computing and Computer-Assisted Intervention*, 2010.

[49] V. Caselles, R. Kimmel, and G. Sapiro, "Geodesic active contours," *International Journal of Computer Vision*, vol. 22, pp. 61–97, 1997.

[50] http://www.ACLSolutions.com.

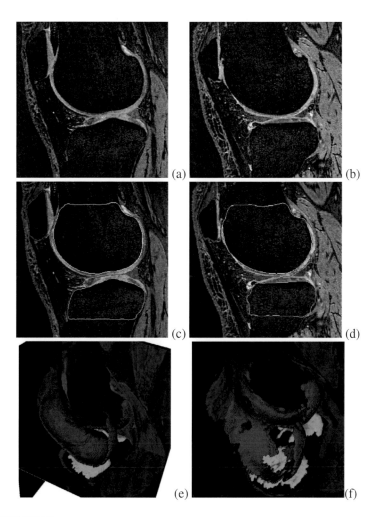

FIGURE 10.17

3D segmentation of knee cartilages. Images from a knee minimally affected by osteoarthritis shown on the left. Severe cartilage degeneration shown on the right. (a,b) Original images. (c,d) The same slice with bone/cartilage segmentation. (e,f) Cartilage segmentation shown in 3D, note the cartilage thinning and "holes" in panel (f). *Reprinted, with IEEE permission, from [4].*

11

Hierarchical Graph Encodings

Luc Brun

GREYC UMR CNRS 6072
ENSICAEN
Caen, France
Email: luc.brun@ensicaen.fr

Walter Kropatsch

Pattern Recognition and Image Processing Group
Vienna University of Technology
Vienna, Austria
Email: krw@prip.tuwien.ac.at

CONTENTS

11.1 Introduction 306
11.2 Regular Pyramids 306
11.3 Irregular Pyramids Parallel construction schemes 310
    11.3.1 Maximal Independent Set 312
    11.3.2 Maximal Independent Vertex Set 313
        11.3.2.1 Definition of Reduction Window 315
    11.3.3 Data-Driven Decimation Process 316
    11.3.4 Maximal Independent Edge Set 317
    11.3.5 Maximal Independent Directed Edge Set 320
    11.3.6 Comparison of MIVS, MIES, and MIDES 322
11.4 Irregular Pyramids and Image properties 323
    11.4.1 Relating Regions 324
    11.4.2 Graph models 325
        11.4.2.1 Simple Graph 326
        11.4.2.2 Dual Graph Model 327
        11.4.2.3 Combinatorial Maps 330
    11.4.3 Simple Graph Pyramids 332
    11.4.4 Dual Graph Pyramids 334
        11.4.4.1 Contraction Kernel 334
        11.4.4.2 Dual Graph pyramids and Topological relationships 336
    11.4.5 Combinatorial Pyramids 337
        11.4.5.1 Relationships between regions 338
        11.4.5.2 Implicit Encoding of a Combinatorial Pyramid 339
        11.4.5.3 Dart's Embedding and Segments 341
11.5 Conclusion 344
    Bibliography 347

11.1   Introduction

Regular image pyramids have been introduced in 1981/82 [1] as a stack of images with decreasing resolutions. Such pyramids present interesting properties for image processing and analysis such as [2] the reduction of noise, the processing of local and global features within the same frame, the use of local processes to detect global features at low resolution, and the efficiency of many computations on this structure. Regular pyramids are usually made of a very limited set of levels (typically $\log(n)$ where n is the diameter of the input image). Therefore, if each level is constructed in parallel from the level below, the whole pyramid may be built in $\mathcal{O}(\log(n))$ parallel steps. The systematic construction establishes a strong relation between the levels enabling a quick top-down access to every pixel of the original image. However, the regular structure of such pyramids induce several drawbacks such as the limited number of regions which may be encoded at a given level or the fact that a small shift of an initial image may induce important modifications of its associated regular pyramid.

Irregular pyramids overcome these negative properties while keeping the main advantages of their regular ancestors. Such pyramids are defined as a stack of successively reduced graphs. Each vertex of such a pyramid being defined from a connected set of vertices in the graph below. Typically, such graphs encode the pixel's neighborhood, the region adjacency, or a particular semantic of image objects. Since 2D images correspond to a sampling of the plane, such graphs are embedded in the plane by definition.

The construction of such pyramids raises two questions: (1) How to build efficiently each level of the pyramid from the one below? (2) Which properties of the plane are captured at each level? This chapter provides answers to both questions together with an overview of hierarchical models. We first present in Section 11.2 the regular pyramid framework. We then present the main parallel construction schemes of irregular pyramids in Section 11.3. These construction schemes determine an important "vertical" property of an irregular pyramid: the speed at which the size of graphs decreases between the different levels of the pyramid. The main graph encodings used within the irregular pyramid framework together with image properties captured by such models are presented in Section 11.4. Image properties captured by a given graph model may be considered as horizontal's properties of a pyramid. Each combination of a parallel construction scheme (Section 11.3) with a graph encoding (Section 11.4) defines an irregular pyramid model with specific vertical and horizontal properties.

11.2 Regular Pyramids

A *regular pyramid* is defined as a sequence of images with an exponentially decreasing resolution. Each image of such a sequence is called a level of the pyramid. The first level, called the base of the pyramid, corresponds to the original image, while the top of the pyramid corresponds usually to a single pixel whose value is a weighted mean of pixels belonging to the base. Using image neighborhood relationships, a *reduction window* connects any pixel of the pyramid to a set of pixels defined at the level below. Any pixel of a pyramid is considered as the parent of the pixels belonging to its reduction window (filled circles in Figure 11.1). Conversely, all pixels within a reduction window are considered as the **children** of the pixel associated to this reduction window in the above level. The parent/child or hierarchical relationship may be extended between any two level using the transitive closure of the relationship induced by the reduction window. The set of pixels in the base level image associated to a given pixel of the pyramid is called its *receptive field*. Let us, for example, consider the top-left pixel in Figure 11.1(c). Its reduction window and receptive field are the top left 2×2 and 4×4 windows of the image in Figure 11.1(b) and (a) respectively.

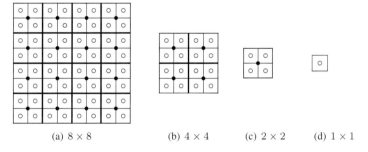

(a) 8×8 (b) 4×4 (c) 2×2 (d) 1×1

FIGURE 11.1
A $2 \times 2/4$ regular pyramid.

One additional important parameter of a regular pyramid is the *decimation ratio* (also called **reduction factor**) which encodes the ratio between the size of two successive images. This ratio remains the same between any two successive levels of the pyramid. A *reduction function* determines the value of any pixel of the pyramid from the value of its children defining its reduction window (Figure 11.2). A regular pyramid may thus be formally defined as the ratio $N \times N/q$, where $N \times N$ represent the size of the reduction window, while q represents the decimation ratio. Different types of pyramid may be distinguished according to the value of $N \times N/q$:

- If $N \times N/q < 1$, the pyramid is named a *nonoverlapping holed pyramid*. Within such a pyramid, some pixels have no parent [3] (e.g., the center pixel in Figure 11.3(a));

- If $N \times N/q = 1$, the pyramid is called a *nonoverlapping pyramid without hole* (see, e.g., in Figure 11.3(b)). Within such pyramids, each pixel in the reduction window has exactly one parent [3];

- If $N \times N/q > 1$, the pyramid is named an *overlapping pyramid* (see, e.g., in Figure 11.3(c)). Each pixel of such a pyramid has on average N^2/q *parents* [1]. If each child selects one parent, the set of children in the reduction window of each parent is rearranged. Consequently, the receptive field may take any form inside the original receptive field.

Regular pyramids have several interesting properties enumerated by Bister [2]:

1. Reduction functions are usually defined as low-pass filters, hence inducing a robustness against noise in high levels of the pyramid;

2. Algorithms insensitive to image resolution may be readily designed using the pyramid framework;

3. Global properties in the base level image become local at higher levels in the hierarchy and may thus be detected using local filters;

4. Top-down analysis of the content of a pyramid may be achieved efficiently using a divide and conquer strategy;

5. Objects to be detected may be retrieved using low resolution images with a simplified image content.

Despite these interesting properties, regular pyramids have several limitations. Indeed, the bounded size and fixed shape of the reduction window induce rigid constraints on the data structure which may not allow it to readily handle the variability of image content. Regular pyramids are, for example, quite sensitive to small image

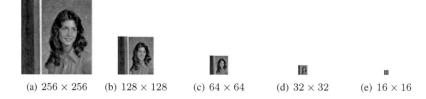

   (a) 256×256 (b) 128×128 (c) 64×64 (d) 32×32 (e) 16×16

FIGURE 11.2
A $2 \times 2/4$ pyramid whose reduction function is defined as a Gaussian centered on the center of the reduction window. The size of each image is indicated below.

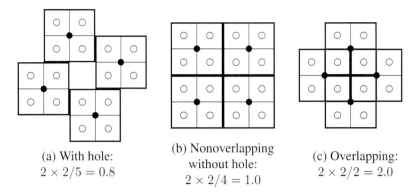

(a) With hole:
$2 \times 2/5 = 0.8$

(b) Nonoverlapping
without hole:
$2 \times 2/4 = 1.0$

(c) Overlapping:
$2 \times 2/2 = 2.0$

FIGURE 11.3
Three different types of regular pyramids.

(a) Number of regions

(b) Elongated objects

FIGURE 11.4
Two $1 \times 4/2$ image pyramids based on a 1×4 base level image. These two figures illustrate the limited number of objects which may be defined at a given level of a pyramid (a) and the difficulties encountered by such pyramids to encode elongated objects(b).

shifts due to the fixed locations of reduction window (this is called the *shift depen-dence problem*). Moreover, the fixed size of reduction windows bounds artificially the number of regions that may be encoded at a given level and does not allow us to readily encode elongated objects.

This last drawback is illustrated in Figure 11.4(a). The base of the pyramid is a 1×4 image encoding a single line composed of 4 pixels with different colors. These pixels correspond to different regions that should be keep by a segmentation process. How-ever, using an image pyramid with a reduction factor equal to 2, only 2 pixels can survive at level 1. Two regions defined at the base level should thus be removed at level one, this result being independent of the gray level differences between pixels. Figure 11.4(b) illustrates the difficulty encountered by regular pyramids to encode elongated objects. The 1×4 base level image is composed of pixels with close values which should be grouped into a single entity by a segmentation process. Us-ing a $1 \times 4/2$ image pyramid, these 4 pixels are merged into two pixels at level 1. This merging operation artificially increases the gray level difference between pixels, hence producing two distinct regions at level 1.

 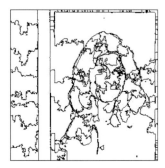

(a) Original Image   (b) Segmented Image

FIGURE 11.5
Image segmentation of Girl test image using the pyramid linking algorithm [2] with an overlapping $5 \times 5/4$ pyramid.

This phenomenon is also illustrated in Figure 11.5 on a real image using Bister algorithm [2] on a $5 \times 5/4$ pyramid. As shown in Figure 11.4(b), several elongated regions located on the top and the left part of the image together with the hair of girl are artificially split into several small regions.

Let us finally temper the above remarks by noting that Bister [2] always managed to adapt the regular pyramid to the class of images to process. However, such an adaptation of the reduction window, decimation ratio, reduction function... should be performed for each image class on an heuristic basis. We can additionally note that some drawbacks of regular pyramids may become advantages on some specific applications. For example, the fact that regular pyramids do not readily allow us to encode elongated objects may become an advantage in fields where objects to be detected are known to be compact. Such an example of application is provided by Rosenfeld [4] who uses a regular pyramid in order to detect vehicles in infrared images. Within this particular context, vehicles are known to be compact and may be detected as single pixels at some levels of the pyramid.

11.3 Irregular Pyramids Parallel construction schemes

Irregular pyramids have been first proposed by Meer and Montanvert et al. [5] in order to overcome negative properties of their regular ancestors while keeping their main advantages. More particularly, two important advantages of regular pyramids over nonhierarchical image processing algorithms should be preserved:

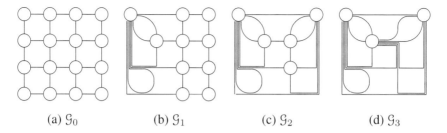

(a) $\mathcal{G}_0$ (b) $\mathcal{G}_1$ (c) $\mathcal{G}_2$ (d) $\mathcal{G}_3$

FIGURE 11.6
First levels of an irregular pyramid using the 4-adjacency of pixels.

- Property I. The bottom-up computation of any pyramid may be performed in parallel, each pixel computing its value from its child independently;

- Property II. Due to the fixed size of the decimation ratio, the height of a pyramid is equal to a log of the image size. Therefore, using a parallel computation scheme any pyramid may be computed in $\mathcal{O}(\log(|I|))$ parallel steps, where $|I|$ denotes the size of the base level image.

On the other hand, the main drawbacks of regular pyramids are the fixed shape of reduction windows and the fixed neighborhood of each pixel which do not allow us to readily adapt the pyramid to the data and to preserve adjacency relationships along the pyramid. In order to overcome these drawbacks, Meer and Montanvert et al. [5] defined irregular pyramids as stacks of successively reduced graphs ($\mathcal{G}_0, \mathcal{G}_1, \mathcal{G}_2, \mathcal{G}_3$ in Figure 11.6). Such pyramids may adapt the shape of the reduction window to the data and preserve adjacency relationships between regions through graph encoding since a vertex may have an arbitrary number of neighbors. In reference to their regular ancestors, the *decimation ratio* of irregular pyramids is usually defined as the mean value of the ratio between the number of vertices of two successive graphs $\left(\frac{|\mathcal{V}_l|}{|\mathcal{V}_{l+1}|}\right)$ when this last term is approximately constant along the pyramid. The construction of a graph $\mathcal{G}_{l+1} = (\mathcal{V}_{l+1}, \mathcal{E}_{l+1})$ from $\mathcal{G}_l = (\mathcal{V}_l, \mathcal{E}_l)$, is performed through the following steps:

- Step I. Define a partition of $\mathcal{V}_l$ into a set of connected components. A vertex of $\mathcal{G}_{l+1}$ is created and attached to each connected component of this partition. The connected component attached to each vertex of $\mathcal{G}_{l+1}$ is called its *reduction window* in reference to regular pyramids;

- Step II. Define adjacency relationships between vertices of $\mathcal{G}_{l+1}$ from adjacency relationships between the connected components of $\mathcal{V}_l$'s partition.

This section is devoted to parallel reduction schemes which are uniquely concerned by the first step of the above construction scheme, the second step being detailed in

Section 11.4. Sequential reduction schemes such as [6] violate Property I (Section 11.3) and are thus not considered in this chapter. This section describes global reduction schemes of a graph. Such global reduction schemes may, however, be easily restricted according to some segmentation criterion (e.g., [7]) by removing from the initial set of edges $\mathcal{E}_0$ all edges encoding adjacency relationships between vertices which should not merge according to a segmentation criterion.

Note that the number of connected components defining a partition of $\mathcal{G}_l$ defines the cardinality of $\mathcal{V}_{l+1}$ and hence the decimation ratio (also called *reduction factor*) of irregular pyramids. Several parallel methods have been proposed to ensure a constant value to the decimation ratio $\left(\frac{|\mathcal{V}_l|}{|\mathcal{V}_{l+1}|} \right)$, all these methods being based on the notion of **maximal independent set**.

11.3.1 Maximal Independent Set

Let us consider a finite or enumerable set $\mathcal{X}$ provided with a symmetric neighborhood function $\mathcal{N} : \mathcal{X} \to \mathcal{P}(X)$, where $\mathcal{P}(\mathcal{X})$ denotes the powerset of $\mathcal{X}$. An *independent set* of $\mathcal{X}$ is a subset $\mathcal{I}$ of $\mathcal{X}$ such that no elements of $\mathcal{I}$ are related by the neighborhood function $\mathcal{N}(.)$:

$$\forall (x, y) \in \mathcal{I}^2, \ x \notin \mathcal{N}(y) \tag{11.1}$$

An independent set is said to be maximal if it is not the subset of any independent set. For example, the set of even numbers, is a maximal independent set over the set of natural integers provided with the neighborhood relationship : $\mathcal{N}(n) = \{n - 1, n + 1\}$. Note that, both even and odd integers define a maximal independent set using this neighborhood relationship. A maximal independent set is thus not unique. Moreover, given a maximal independent set $\mathcal{I}$, any element x in $\mathcal{X} - \mathcal{I}$ must be adjacent to some element of $\mathcal{I}$, otherwise x could be added to $\mathcal{I}$ hence violating the maximal property. Conversely, if $\mathcal{I}$ is a nonmaximal independent set, at least one element x may be added to $\mathcal{I}$. This element is not adjacent to any element of $\mathcal{I}$ due to condition 11.1. The maximality of an independent set may thus be characterized by:

$$\forall x \in \mathcal{X} - \mathcal{I}, \ \exists y \in \mathcal{I} \, | \, x \in \mathcal{N}(y) \tag{11.2}$$

Within the irregular pyramid framework elements of a maximal independent set are usually called surviving elements. Condition 11.1, states that two adjacent elements cannot both survive. On the other hand, condition 11.2 states that any nonsurviving element should be adjacent to a surviving one. A maximal independent set may thus be interpreted as a subsampling of the initial set $\mathcal{X}$ according to the neighborhood relationship $\mathcal{N}$. Let us, for example, consider the case where $\mathcal{X}$ corresponds to the set of pixels of an image, while the neighborhood relationship encodes the 8 connectivity of pixels. A maximal independent set may then be achieved by selecting one pixel every two lines and two columns. One may easily check in Figure 11.7 that surviving pixels (■) are not adjacent (condition 11.1) and that any nonsurviving pixel (□) is adjacent to at least one surviving pixel (condition 11.2).

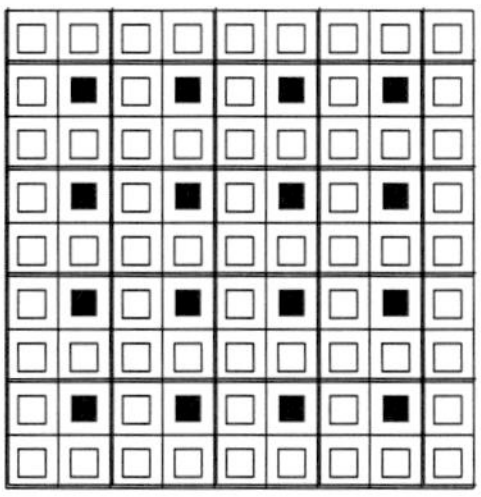

FIGURE 11.7
A maximal independent set (■) over a planar sampling grid with the 8 neighborhood.

The remainder of this chapter will be devoted to finite sets. Given such a finite set X, an independent set will be called maximum if its cardinal is maximum over all possible maximal independent sets. Moreover, in such a case, neighborhood relationships may be encoded by a graph $\mathcal{G} = (X, \mathcal{E})$, where $(u, v) \in X^2$ belongs to $\mathcal{E}$ if and only if $u \in \mathcal{N}(v)$.

11.3.2 Maximal Independent Vertex Set

A *maximal independent vertex set* (MIVS) of a graph $\mathcal{G}_l = (\mathcal{V}_l, \mathcal{E}_l)$ is defined as a maximal independent set over the set $\mathcal{V}_l$ using the neighborhood relationship induced by $\mathcal{E}_l$. If we denote the maximal independent set by $\mathcal{V}_{l+1}$, conditions 11.1 and 11.2 may be written as follows:

$$\forall (v, v') \in \mathcal{E}_l : \ (v, v') \notin \mathcal{E}_l \tag{11.3}$$

$$\forall v \in \mathcal{V}_l - \mathcal{V}_{l+1} \ \exists v' \in \mathcal{V}_{l+1} : \ (v, v') \in \mathcal{E}_l \tag{11.4}$$

The construction scheme of such a maximal independent set was introduced by Meer [8] as an iterative stochastic process based on the outcome of a random variable attached to each vertex and uniformly distributed over $[0, 1]$. Any vertex corresponding to a local maximum of the random variable is then considered as a surviving vertex. One iteration of this process allows us to satisfy condition (11.3) and hence to obtain an independent set, since two adjacent vertices cannot both correspond to local maxima. However, one vertex not corresponding to a local maximum may have no surviving vertices in its neighborhood. In other terms, such a process may lead to a nonmaximal independent set. Such a configuration is illustrated on Figure 11.8 where values inside each vertex represent the outcome of the random variable. Vertices with value 9 represent local maxima of the random variable and are thus selected as survivors. Vertices 7 and 8 adjacent to a surviving vertex cannot be selected as survivors due to condition (11.3). Vertex 6 does not correspond to a maximum of the random variable. This vertex must nevertheless be selected as a survivor in order to fulfill condition (11.4) and obtain a maximal independent set.

We should thus iterate the selection of local maxima until both conditions (11.3)

FIGURE 11.8
Construction of a maximal independent vertex set on a $1D$ graph. Surviving vertices are shown by double circles.

and (11.4) are satisfied. Such an iterative process requires us to attach three variables x_i, p_i, q_i to each vertex $v_i \in \mathcal{V}_l$. Variable x_i encodes the outcome of the random variable attached to v_i, while p_i and q_i correspond to Boolean variables whose values encode the following states:

- If p_i is true, v_i is considered as a **surviving vertex**;

- If q_i is true, v_i is a candidate which may become a surviving vertex at some further iteration. Conversely, if q_i is false, v_i is considered as a nonsurviving vertex.

Let $(p_i^{(k)})_{k \in \{1,\ldots,n\}}$ and $(q_i^{(k)})_{k \in \{1,\ldots,n\}}$ denote values taken by variables p_i and q_i along the iteration of our iterative algorithm. Variable p_i and q_i are initialized as follow

$$
\begin{aligned}
p_i^{(1)} &= x_i = \max_{v_j \in \mathcal{N}(v_i)}\{x_j\} \\
q_i^{(1)} &= \bigwedge_{v_j \in \mathcal{N}(v_i)} \overline{p_j}^{(1)}
\end{aligned}
\tag{11.5}
$$

where $\mathcal{N}(v_i)$ denotes the neighborhood of v_i and $\bigwedge$ corresponds to the logical "and" operator. We suppose by convention that $v_i \in \mathcal{N}(v_i)$.

Equation (11.5) states that a vertex survives ($p_i^{(1)} =$ true) if it corresponds to a maxima of the random variable. One vertex is a candidate ($q_i^{(1)} =$ true) if it is not already a survivor and if none of its neighbors survive.

The iterative update of the predicate p_i and q_i is performed using the following rules:

$$
\begin{aligned}
p_i^{(k+1)} &= p_i^{(k)} \vee (q_i^{(k)} \wedge x_i = \max\{x_j \mid v_j \in \mathcal{N}(v_i) \wedge q_j^{(k)}\}) \\
q_i^{(k+1)} &= \bigwedge_{v_j \in \mathcal{N}(v_i)} \overline{p_j}^{(k+1)}
\end{aligned}
\tag{11.6}
$$

where $\vee$ denotes the logical "or" operator.

One vertex surviving at iteration k thus remain a survivor at further iterations ($p_i^{(k+1)} = p_i^{(k)} \vee \ldots$). Moreover, a candidate ($q_i^{(k)} =$ true) may become a survivor if it corresponds to a local maxima of all the candidates in its neighborhood ($x_i = \max\{x_j \mid v_j \in \mathcal{N}(v_i) \wedge q_j^{(k)}\}$). Finally, a candidate remains in this state ($q_i^{(k+1)} =$

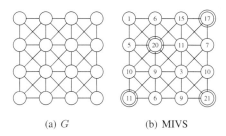
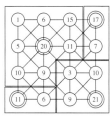

    (a) *G*   (b) MIVS   (c) Reduction windows

FIGURE 11.9
Construction of a maximal independent vertex set using a random variable on a graph
$\mathcal{G}$ encoding a 4×4 8-connected sampling grid (a). The outcome of the random variable is displayed inside each vertex. Surviving vertices are surrounded by nested circles (b). Reduction windows associated to surviving vertices (c).

true) if it does not become a survivor and if none of its neighbors are selected as a survivor.

Note that the set of vertices such that $p_i^{(k)}$ = true defines an independent set at any step k of this iterative algorithm. Each step decreases the number of candidates and increases the number of survivors until no more survivors may be added. The set of survivors then define a maximal independent set. Let us additionally note that this process is purely local. Indeed, for each iteration, each vertex updates the values of p_i and q_i according to the one of its neighbors. Such a process may thus be easily encoded on a parallel machine. Figure 11.9 (b) shows the outcomes of the random variable on a graph defined from the 4×4 8-connected sampling grid (Figure 11.9 (a)). Surviving vertices are surrounded by an extra circle.

Jolion [9] improves the adaptability of the above decimation process by defining surviving vertices as local maxima of an interest operator. For example, within the segmentation framework, Jolion defines the operator of interest as a decreasing function of the gray level variance computed in the neighborhood of each vertex. This operator provides a location of surviving vertices in homogeneous regions.

11.3.2.1 Definition of Reduction Window

Given the set of surviving vertices, constraint (11.3) of a MIVS insures that each nonsurviving vertex is adjacent to at least one survivor. The selection of a surviving parent by each nonsurviving vertex may be performed using various heuristics. Meer [8] and Montanvert et al. [5] connect each nonsurviving vertex to its surviving neighbor with the greatest value of the random variable. Jolion [9] uses a contrast measure such as the gray level difference, to link each nonsurviving vertex to its least contrasted surviving neighbor. Each surviving vertex is thus the parent of all nonsurviving vertices attached to it. Figure 11.9(c) shows the reduction window defined from Figure 11.9(b) by connecting each nonsurviving vertex to its surviving neighbor of greatest value.

Given the partition of $\mathcal{V}_l$ into a set of reduction window with their associated survivors $\mathcal{V}_{l+1}$, the reduced graph $\mathcal{G}_{l+1} = (\mathcal{V}_{l+1}, \mathcal{E}_{l+1})$ is defined using heuristics defined in Section 11.4.

11.3.3   Data-Driven Decimation Process

As mentioned in Section 11.3.2, a vertex selected as a survivor remains in this state until the end of the iterative construction scheme of a MIVS (equation 11.6). Moreover, the neighbors of such a survivor are classified as noncandidates and remain also in this state until the end of the iterative process. Therefore, one surviving vertex together with its nonsurviving neighbors could be reduced into a single vertex of the reduced graph as soon as it is classified as a survivor at some iteration step. However, using the construction scheme defined in Section 11.3.2, the definition of a reduced graph using a MIVS and requires that all the vertices of a graph are marked as survivors or noncandidate. For important graphs, this latency which is proportional to the number of iterations may be important.

The *data-driven decimation process* (**D3P**) proposed by Jolion [10] is based on a single iteration of the iterative decimation process (Equation 11.6) at each level of the pyramid. The two variables p_i and q_i defined in Section 11.3.2 keep the same meaning but are now global to the whole pyramid. These variables are initialized to true at the base level of the pyramid:

$$p_i^1 = q_i^1 = \text{true}, \forall v_i \in \mathcal{V}_0 \tag{11.7}$$

where $\mathcal{G}_0 = (\mathcal{V}_0, \mathcal{E}_0)$ encodes the base of the pyramid.

The update of variables p_i and q_i between each level is performed using the following equations:

$$
\begin{aligned}
p_i^{(k+1)} &= \left((p_i^{(k)} \vee q_i^{(k)}) \wedge x_i = \max\{x_j \mid q_j^{(k)} \wedge v_j \in \mathcal{N}_k(v_i)\} \right) \\
q_i^{(k+1)} &= \wedge_{v_j \in \mathcal{N}_k(v_i)} \overline{p_j}^{(k+1)} \wedge \mathcal{N}_k(v_i) \neq \{v_i\}.
\end{aligned}
\tag{11.8}
$$

In other terms, a vertex v_i is selected as a survivor ($p_i^{(k+1)} = \text{true}$), if it was a survivor or a candidate in graph $\mathcal{G}_k$ ($p_i^{(k)} \vee q_i^{(k)}$) and if it corresponds to a maximum of the random variable among the candidates in $\mathcal{N}_k(v_i)$ ($x_i = \max\{x_j \mid q_j^{(k)} \wedge v_j \in \mathcal{N}_k(v_i)\}$). A vertex of $\mathcal{G}_k$ becomes a candidate if it is not adjacent to any surviving vertex and if it is not isolated ($\mathcal{N}_k(v_i) \neq \{v_i\}$).

The set of surviving vertices determined by Equation (11.8) defines an independent set over $\mathcal{G}_k$. This set is, however, usually nonmaximal. More precisely, we may decompose the set $\mathcal{V}_k$ into three subsets based on variables $p_i^{(k+1)}$ and $q_i^{(k+1)}$: Survivors ($p_i^{(k+1)} = \text{true}$), candidates ($q_i^{(k+1)} = \text{true}$), and noncandidates

$(q_i^{(k+1)} =$ false). Noncandidate vertices are adjacent to at least one survivor (Equation 11.8). These vertices together with surviving vertices are grouped into reduction windows reduced into a single vertex in graph $\mathcal{G}_{k+1}$. Candidate vertices represent the set of vertices which would have been classified as survivors or nonsurvivors at further iteration steps using the iterative construction scheme of a MIVS (equation 11.6). Using the D3P construction scheme, the classification of such vertices is delayed to higher levels of the pyramid. The set of vertices defined at level $k + 1$ is thus equal to the set of survivors together with the set of candidates:

$$\mathcal{V}_{k+1} = \{v_i \in \mathcal{V}_k \mid p_i^{(k+1)} \vee q_i^{(k+1)}\}. \tag{11.9}$$

The set $\mathcal{E}_{k+1}$ is defined according to various heuristics defined in Sections 11.4.3 and 11.4.4. Equation (11.8) is then applied on the reduced graph $\mathcal{G}_{k+1} = (\mathcal{V}_{k+1}, \mathcal{E}_{k+1})$ until the top of the pyramid.

Let us recall (Section 11.3.2) that Jolion [9] defines surviving vertices as local maxima of an interest operator. One main advantage of the D3P is that vertices corresponding to strong local maxima of the interest operator are detected and merged with their nonsurviving neighbors at low levels of the pyramid. Remaining candidates which correspond to more complex configurations are detected at higher levels of the pyramid where the content of the graph is already simplified hence simplifying the decision process. For example, within a segmentation scheme, the interest operator may detect the center of homogeneous regions which grow at low levels of the pyramid in order to facilitate the aggregation of vertices encoding pixels on the boundary of two regions. Let us additionally note that this asynchronous decimation process is consistent with some psychovisual experiments [11] showing that our brain uses asynchronous processes.

11.3.4 Maximal Independent Edge Set

Within the MIVS framework (Section 11.3.2), two surviving vertices cannot be adjacent. The number of surviving vertices defining a MIVS is thus highly dependent of the connectedness of the graph. Moreover, using the stochastic construction scheme of a MIVS defined by Meer [8] and Montanvert et al. [5] one vertex is defined as a survivor if it corresponds to a local maxima of a random variable. The probability that a vertex survives is thus relative to the size of its neighborhood. Adjacency relationships have thus an influence on:

1. The height of the pyramid

2. The number of iterations required to build each level of the pyramid from the previous one

Several experiments performed by Kropatsch et al. [12, 13] have shown that the mean degree of vertices increases along the pyramid. This increasing connectivity induces the selection of a decreasing number of surviving vertices and hence decreases the decimation ratio along the pyramid.

Such a decrease of the decimation ratio increases the height of the pyramid and thus violates one of the two main properties of regular pyramids that irregular pyramids should preserve (Section 11.3): The logarithmic height of the pyramid according to the size of its base.

In order to correct this important drawback, Kropatsch et al. [12, 13] propose to base the decimation of a pyramid on the construction of a forest $\mathcal{F}$ of $\mathcal{G}$ such that:

- Cond. I. Any vertex of $\mathcal{G}$ belongs to exactly one tree of $\mathcal{F}$
- Cond. II. Each tree is composed of at least two vertices.

Each tree of $\mathcal{F}$ defines a reduction window on which a surviving vertex is selected. Remaining vertices of the tree are considered as nonsurviving vertices. (Cond. I.) insures that each vertex is classified either as a survivor or as a nonsurvivor. (Cond. II.) insures that the number of vertices decreases by a factor at least 2, hence providing a reduction factor at least equal to 2.

The first step toward the construction of the forest $\mathcal{F}$ is based on the notion of *maximal independent edge set* (**MIES**). Given a graph $\mathcal{G} = (\mathcal{V}, \mathcal{E})$, a maximal independent edge set corresponds to a maximal independent vertex set on the *edge graph* $\mathcal{G} = (\mathcal{E}, \mathcal{E}')$ where $\mathcal{E}'$ is defined using the following neighborhood relationship on $\mathcal{E}$:

$$\forall (e_{uv}, e_{xy}) \in \mathcal{E}^2 \ e_{uv} \in \mathcal{N}(e_{xy}) \text{ iff } \{u, v\} \cap \{x, y\} \neq \emptyset \tag{11.10}$$

where $e_{u,v}$ denotes an edge between vertices u and v.

Using equation 11.10, two edges are said to be neighbors if they are incident to a same vertex. The set of surviving edges defined by this decimation process corresponds to a *maximal matching* of the initial graph $\mathcal{G} = (\mathcal{V}, \mathcal{E})$ (Definition 1 and Figure 11.10).

Definition 1
Given a graph $G = (\mathcal{V}, \mathcal{E})$, a subset of edges $\mathcal{C} \subset \mathcal{E}$ is a matching if none of the edges of $\mathcal{C}$ are incident to a same vertex. Let us call the necessary property of $\mathcal{C}$ the matching property.

Maximal matching: *A matching is said to be maximal if no edge may be added to the set without violating the matching property;*

Maximum matching: *A maximum matching is a matching that contains the largest possible number of edges;*

Perfect matching: *A matching $\mathcal{C}$ is perfect if any vertex is incident to an edge of $\mathcal{C}$. Any perfect matching is maximum.*

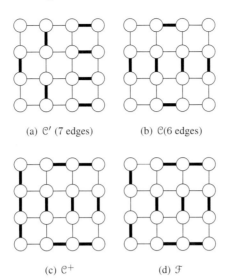

(a) $\mathcal{C}'$ (7 edges)   (b) $\mathcal{C}$(6 edges)

(c) $\mathcal{C}^+$ (d) $\mathcal{F}$

FIGURE 11.10
Two maximal matchings with different numbers of edges (a), (b). Surviving edges are drawn by thick lines. The augmented set $\mathcal{C}^+$ (c) and the final forest $\mathcal{F}$ (d).

A vertex incident to an edge of a matching is said to be saturated. Otherwise, it is unsaturated.

Note that a maximal matching $\mathcal{C}$ does not contain any loop and thus corresponds to a forest of the graph $\mathcal{G}$. However, $\mathcal{C}$ is usually not perfect (Definition 1) and thus violates (Cond. I., Section 11.3.4) since unsaturated vertices do not belong to any tree of our forest. Let us consider such an unsaturated vertex v and a non-self-loop edge $e_{v,w}$ incident to v. The matching $\mathcal{C}$ being maximal, w must be incident to one edge of $\mathcal{C}$. The set $\mathcal{C} \cup \{e_{v,w}\}$ is still a forest of $\mathcal{G}$ and contains the previously unsaturated vertex v. Let us consider the $\mathcal{C}^+$, obtained from $\mathcal{C}$ by adding edges connecting unsaturated vertices to $\mathcal{C}$. The set $\mathcal{C}^+$ is no more a matching. However, by construction, $\mathcal{C}^+$ is a spanning forest of $\mathcal{G}$ composed of trees of depth 1 or 2. Trees of depth 2 may be decomposed in two trees of depth 1 by removing from $\mathcal{C}^+$ edges both end-points of which are incident to some other edges of $\mathcal{C}^+$.

The resulting set $\mathcal{F}$ is a spanning forest of $\mathcal{G}$ composed of trees of depth 1. One vertex may thus be selected in each tree such that the tree rooted on this vertex has a depth 1. Selected vertices correspond to surviving vertices, while remaining vertices are classified as nonsurvivors.

The main motivation of this method is to provide a better stability of the reduction factor. The problem identified within the MIVS framework being connected to the vertex degree, the proposed method uses the notion of maximal independent edge set to avoid the use of vertex's neighborhood. Experiments provided by Haxhimusa et al. [12] (Section 11.3.6) show that the decimation ratio remains close to 2 along the pyramid.

11.3.5   Maximal Independent Directed Edge Set

Using the maximal independent edge set framework (Section 11.3.4), surviving vertices are selected so as to obtain rooted trees of depth one. Such a selection scheme is thus induced by the decomposition of the initial graph into trees and cannot readily incorporate a priori constraints concerning surviving vertices. However, in some applications, the definition of the surviving vertex within each reduction window is an important issue. For example, within the analysis of line drawing framework [14], end points of a line or intersection points must be preserved, and hence selected as survivors, for geometric accuracy reasons.

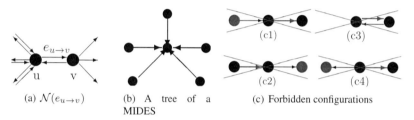

(a) $\mathcal{N}(e_{u \to v})$ (b) A tree of a MIDES   (c) Forbidden configurations

FIGURE 11.11
Directed neighborhood $\mathcal{N}(e_{u \to v})$ of edge $e_{u \to v}$(a). A tree of a MIDES defining unambiguously its surviving vertex (b). Two edges within a same neighborhood cannot be both selected. Hence, red edges in (c) cannot be selected once black edges have been added to the MIDES.

In order to provide a better selection of surviving vertices, Kropatsch et al [12], propose to adapt the MIES framework as follows: The decimation of a graph is based on the construction of a spanning forest $\mathcal{F}$ such that:

- **Cond. I.**   Any vertex of the graph belongs to exactly one tree of $\mathcal{F}$ of depth one

- **Cond. II.**   Each tree is encoded using directed edges. Its surviving vertex is defined either as the only vertex of the tree or as the unique target of all tree's edges (Figure 11.11(b)).

Each tree of $\mathcal{F}$ defining a reduction window, (Cond. I., Section 11.3.5) insures that the set of reduction windows encodes a partition of the vertex set of the graph. Moreover, the surviving vertex within each tree is uniquely determined by (Cond. II., Section 11.3.5). Such a use of directed edges provides a simple way to constraint the selection of surviving vertices by restricting the construction of the forest $\mathcal{F}$ to edges encoding possible selection of surviving and nonsurviving vertices. Such edges are called **preselected edges**.

Note that any undirected graph may be converted into a directed graph by transforming each of its undirected edges $e_{u,v}$ into a pair of directed reverse edges $e_{u \to v}$ and $e_{v \to u}$ with opposite sources and targets. However, the set of preselected edges of such a graph may contain $e_{u \to v}$ without containing $e_{v \to u}$. Let us denote by $\mathcal{G} = (\mathcal{V}, \mathcal{E})$ the directed subgraph induced by the set of preselected edges $\mathcal{E}$. The first step toward the construction of the spanning forest $\mathcal{F}$ consists to build a maximal independent set on $\mathcal{E}$ using the following directed edge's neighborhood relationship (Figure 11.11(a)):

Definition 2
Let $e_{u \to v}$ be a directed edge of $\mathcal{G}$ with $u \neq v$. The directed neighborhood $\mathcal{N}(e_{u \to v})$ is given by all directed edges with the same source u, targeting the source u or emanating from the target v:

$$\mathcal{N}(e_{u \to v}) = \{e_{u \to w} \in E\} \cup \{e_{w \to u} \in E\} \cup \{e_{v \to w} \in E\} \qquad (11.11)$$

Such a maximal independent set on the set of directed edges is called a *maximal independent directed edge set* (**MIDES**). Note that, since two neighbors cannot be simultaneously selected within a maximal independent set, the definition of a neighborhood allows us to specify forbidden configurations (Figure 11.11(c)). Within the MIDES, once an edge $e_{u \to v}$ is selected, the set of edges incident to u or v which may be added to the MIDES is restricted to edges whose target is equal to v. A MIDES thus defines a forest of $\mathcal{G}$ whose trees are composed of edges pointing on a same target node (Figure 11.11(b)). This unique target node is selected as a surviving vertex, while source nodes of each tree are selected as nonsurviving vertices attached to this unique target. Figure 11.11(c) shows some configurations of adjacent edges according to our neighborhood relationship (Def. 2). Since two adjacent elements cannot be both selected within a maximal independent set, such configurations cannot occur within a MIDES and may be interpreted as forbidden configurations. Configurations (c1) and (c2) correspond to trees of depth two which violate Cond. I. (Section 11.3.5). Configurations (c3) and (c4) do not provide an unambiguous designation of the surviving vertex and hence contradict Cond. II. (Section 11.3.5).

Figure 11.12(d) shows the result of a MIDES applied on a graph encoding a 4×4, 4-connected grid. Surviving vertices are marked with an extra filled circle ($\bullet$), while nonsurviving vertices are marked by two concentric circles ($\circledcirc$). As illustrated by this last figure, the resulting forest may not span all the vertices of the graph (vertex $\bigcirc$, on the third row, second column). Such isolated vertices may be interpreted as trees of depth 0 and inserted into the forest in order to obtain a spanning forest $\mathcal{F}$ (Figure 11.12(e)) fulfilling (Cond. I., Section 11.3.5) and (Cond. II., Section 11.3.5). The construction of this spanning forest is thus achieved by the following three steps:

- Step I. Define a maximal independent directed edge set on the directed graph $G = (\mathcal{V}, \mathcal{E})$, where $\mathcal{E}$ encodes a set of preselected edges;

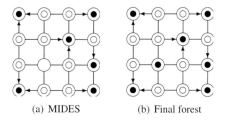

(a) MIDES   (b) Final forest

FIGURE 11.12
A MIDES built on a 4×4 grid.

Algorithms		height	$|\mathcal{V}_l| / |\mathcal{V}_{l+1}|$		iterations	
			μ_{pyr}	σ_{pyr}	μ_{pyr}	σ_{pyr}
MIVS	μ_{data}	20.8	2.0	1.3	3.0	0.8
	σ_{data}	5.2	0.3	1.1	0.2	0.1
MIES	μ_{data}	14.0	2.3	0.2	4.0	1.2
	σ_{data}	0.14	0.01	0.05	0.1	0.1
MIDES	μ_{data}	12.0	2.6	0.3	2.8	1.1
	σ_{data}	0.4	0.1	0.2	0.1	0.1

TABLE 11.1
Statistics on height of the pyramid, decimation ratio, and number of iterations.

- Step II. Define the target of each tree of the resulting forest as a survivor and the sources as nonsurviving vertices attached to this unique target;

- Step III. Complete this forest with isolated vertices.

Note that, contrary to the MIES (Section 11.3.4), the MIDES construction scheme does not require us to split existing trees in order to obtain trees of depth one. Kropatsch et al. [12] have shown that the decimation ratio obtained by a MIDES is at least 2.0 if the MIDES does not produce isolated vertices. Experiments performed in the same paper show that such isolated vertices do not occur frequently. Moreover, each isolated vertex needs only one tree of more edges to keep the balance for a reduction factor of 2.0.

11.3.6   Comparison of MIVS, MIES, and MIDES

Kropatsch et al. [12] have compared the MIVS, MIES, and MIDES construction schemes of a pyramid through several experiments based on a database of pyramids computed on initial graphs encoding 100×100, 150×150, and 200×200, 4-connected planar grids. The outcome of the random variable used within the maximal independent set is repeated with different seeds in order to obtain up to 1000 different pyramids. Statistics of all pyramids built by each specific selection (MIVS, MIES, and MIDES) are calculated to compare the properties of the different strategies by using the following parameters:

- The height of the pyramid (*height*)

- The reduction factor for vertices $\left(\frac{|\mathcal{V}_l|}{|\mathcal{V}_{l+1}|} \right)$

- The number of iterations required to complete a maximal independent set using *repeated maxima selection* of the outcome of a random variable (Section 11.3.2).

Note that the reduction factor and the number of iterations vary within a pyramid. We thus compute the mean (μ_{pyr}) and standard deviation (σ_{pyr}) of these values for each pyramid and further compute the mean (μ_{data}) and standard deviation (σ_{data}) of these global value over our whole dataset of pyramids. The resulting values are displayed in Table 11.1.

As shown by Table 11.1, MIVS provides a mean reduction factor equal to 2.0 but with an important standard deviation within the pyramid (1.3). This important variability of the decimation factor is coherent with the large mean height (20.8) of the pyramid. The maximal height on the dataset obtained using the MIVS being equal to 41. The mean number of iterations is stable ($\sigma_{pyr} = 0.8$) and equal to 3.0. In conclusion, MIVS provide a mean reduction factor of 2.0, but such a mean result hides bad behaviors of the MIVS which have large chances to occur.

Compared to the MIVS, the MIES provides a larger mean decimation ratio (2.3), this decimation ratio being stable along the pyramid ($\sigma_{pyr} = 0.2$). The mean height of the pyramid is also lower (14.0) with a much lower standard deviation (0.14). The maximal height of a pyramid obtained using the MIES construction scheme within these experiments is equal to 15. The mean number of iterations (4.0) is, however, greater than the one of MIVS (3.0) with also a larger standard deviation (1.2). In conclusions, MIES provides a more stable and larger decimation ratio than MIVS. Our observations of this decimation ration was greater than its theoretical bound (2.0) in all experiments.

The mean reduction factor obtained using the MIDES construction scheme (2.6) is the largest of the three methods with a low variation within pyramids ($\sigma_{pyr} = 0.3$). The mean height of the pyramid (12.0) is also lower than the one obtained with MIES and MIVS and remains stable across the database ($\sigma_{data} = 0.4$). The number of iterations needed to complete the maximal independent set is comparable with the one of MIVS. The MIDES thus provides both a large and stable reduction factor and a low number of iterations.

11.4   Irregular Pyramids and Image properties

Different families of irregular pyramids have been proposed [5, 15, 16] to encode image content. These pyramids may vary both according to the method used to build one graph of the pyramid from the graph below and according to the type of graph used to encode each level. The construction scheme of a pyramid, described in Section 11.3, determines its decimation ratio and may be understood as the pyramid's vertical dynamic. Conversely, the choice of a given graph model determines which set of topological and geometrical image properties may be encoded at each level. This last choice may be understood as the determination of the horizontal resolution of the pyramid. Let us additionally note that choices concerning pyramid's construction scheme and graph encoding are usually independent since different pyramid construction schemes may be applied using different graph encoding. The main geometrical and topological properties of an image together with the different graph models used to encode them within the irregular pyramid framework are detailed in the following sections.

11.4.1   Relating Regions

Within an image analysis scheme, a high level image description is usually based on low level image processing steps. The segmentation stage is one of this key step. Segmentation produces an image's partition into *meaningful* connected sets of pixels called *regions*. However, in many applications, the low-level image segmentation stage cannot be readily separated from high-level goals. On the contrary, segmentation algorithms should often extract fine information about a partition in order to guide the segmentation process according to high level objectives. There is thus a need to design graph models which can be both efficiently updated during a segmentation stage and which may be used to extract fine information about the partition.

Region's adjacency also called **meet relationship** constitutes a basic but widely used relationship between regions. Two regions of a partition are said to be adjacent if they share at least one common boundary segment. Figure 11.13(a) shows a partition of an 8×8 image whose boundaries are displayed in Figure 11.13(b) using interpixels elements such as linels (or cracks) and pointels [17, 18]. One may see from these figures that the meet relationship corresponds to different configurations of regions. For example, the central gray region G contains a black region B_2 and shares one common boundary segment with another black region B_1. In the same way, white and black regions, W on the left and B_1 on the right part of the image, respectively, share two connected boundaries.

Finer relationships between regions such as RCC-8 defined by Randel [19] or the relationships defined by K. Shearer et al. [20] in the context of graph matching should thus be used in order to describe fine relationships between regions. Within

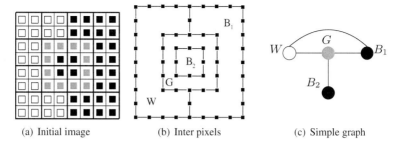

(a) Initial image   (b) Inter pixels   (c) Simple graph

FIGURE 11.13
An image partition (a) whose boundaries are encoded by interpixel elements (b). A simple graph encoding of the resulting partition (c).

the particular context of image segmentation, the following relationships may be defined from these two models:

Meets: The different models used to encode partitions either encode the existence of at least one common boundary segment between two region or create one relationship for each boundary segment between these regions. We denote these two types of encoding *meets_exists* and *meets_each*. A model encoding only the meets_exists relationship is thus not able to distinguish a simple from a multiple adjacency relationship between two regions such as the one between vertices W and B_1 in Figure 11.13. Note that the ability of models to retrieve efficiently a given common boundary segment between two regions is also an important feature of these models;

Contains: The relationship A *contains* B expresses the fact that region B is completely surrounded by region A. For example, the gray region G in Figure 11.13(a) contains the black region B_2;

Inside: The *inside* relationship is the inverse of the *contains* relation: The region B_2 *inside* G is contained in G.

One additional relationship not directly handled by Shearer and Randel's models may be defined within the hierarchical segmentation scheme. Indeed, within such a framework a region defined at a given level of a hierarchy is *composed of* regions defined at levels below.

The following relationships may thus be deduced from the relationships defined by Shearer and Randel and the example provided by Figure 11.13(a) : The *meets_exists, meets_each, contains, inside* and *composed of*. Note that unlike meets relationships, the contains and inside relations are asymmetric. Hence, contains or inside relation between two regions allows us to characterize each of the regions sharing this relation.

11.4.2 Graph models

Different graph models [21] have been proposed to describe the content of an image within the image analysis framework. Each of these graph models encodes a different subset of the relationships defined in Section 11.4.1. In the following sections, we present the main graph models introduced within the irregular pyramid framework together with a description of the relationships between regions encoded by such models.

11.4.2.1 Simple Graph

One of the most common graph data structure, within the segmentation, framework is the *region adjacency graph*. A RAG is defined from a partition by associating one vertex to each region and by creating an edge between two vertices if the associated regions share a common boundary segment (Figure 11.13(c)). A RAG thus encodes the meet_exists relationship. It corresponds to a **simple graph** without any double edge between vertices nor self-loop. The simplification of a RAG is based on the following graph transformations:

Edge removal: The removal of an edge from a graph $\mathcal{G} = (\mathcal{V}, \mathcal{E})$ consists to remove it from $\mathcal{E}$. In order to avoid to disconnect the graph, removed edges should not define a bridge.

Edge contraction: The contraction of an edge consists to collapse the two vertices incident to the edge into a single vertex and to remove the edge. Self-loops are excluded from edge contraction since both end points correspond to a same point which cannot collapse with itself.

The contraction operation is defined for any edge except self-loops and thus for any set of edges which does not contain a cycle (i.e., a forest). Given a graph $\mathcal{G} = (\mathcal{V}, \mathcal{E})$ and a set $\mathcal{N} \subset \mathcal{E}$ defining a forest of $\mathcal{G}$, the contraction of $\mathcal{N}$ in $\mathcal{G}$ is denoted $\mathcal{G}/\mathcal{N}$. In a similar way, given a set $\mathcal{M}$ of edges, the removal of $\mathcal{M}$ in $\mathcal{G}$ is denoted by $\mathcal{G} \setminus \mathcal{M}$. In order to preserve the connectedness of G, $\mathcal{M}$ should not correspond to a cut-set.

Within a non-hierarchical segmentation scheme the RAG model is usually applied as a merging step to overcome the oversegmentation produced by a previous splitting algorithm. Indeed, the existence of an edge between two vertices denotes the existence of at least one common boundary segment between the two associated regions which may thus be merged by removing this segment. Within this framework, edge information may thus be interpreted as a possibility to merge two regions identified by the vertices incident to the edge. This merging operation of two adjacent vertices is encoded within the RAG model by the contraction of the edge encoding their adjacency. This contraction has to be followed by a removal step in order to keep the graph simple by removing all multiple edges or self-loops which may have been created by the contraction operation.

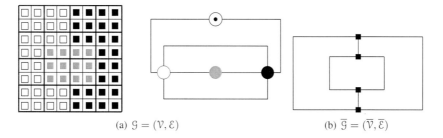

(a) $\mathcal{G} = (\mathcal{V}, \mathcal{E})$ (b) $\overline{\mathcal{G}} = (\overline{\mathcal{V}}, \overline{\mathcal{E}})$

FIGURE 11.14
A nonsimple graph encoding a partition (a) and its dual (b).

The RAG model thus encodes only the existence of a common edge between two regions (the *meets_exists* relationship). Moreover, the existence of a common edge between two vertices does not provide enough information to differentiate a meets relationship from a *contains* or *inside* one. This drawback is illustrated in Figure 11.13(c) where the relationships between G and B_1 is the same than the one between G and B_2 (both B_1 and B_2 are adjacent to G). However, the gray region associated to G contains the black region associated to B_2 but just shares one connected boundary segment with the black region encoded by B_1. These two different configurations of regions are not differentiated by the simple graph model.

11.4.2.2 Dual Graph Model

Edges within a RAG model only encode the meet_exists relationship. One straightforward solution to encode multiple boundaries between regions consists in creating one edge between two vertices for each common boundary segment between the two corresponding regions. The resulting graph is nonsimple. The set of vertices $\mathcal{V}$ and of edges $\mathcal{E}$ of such a graph $\mathcal{G} = (\mathcal{V}, \mathcal{E})$ is defined by the following requirements:

- Req. I.   Each vertex of $\mathcal{V}$ encodes a region of the image, one additional special vertex encodes the background of the image

- Req. II.   Each edge of $\mathcal{E}$ encodes a maximal connected boundary segment between two connected components.

Figure 11.14(a) shows such an encoding where the two connected boundaries between black and white regions are encoded by two edges between the corresponding vertices. Vertex ($\odot$) encodes the background of the image. White and black regions being adjacent to the background, vertices encoding these regions are adjacent to the vertex encoding the background. Such a graph encoding does not allow us to readily perform graph simplifications. This last drawback is illustrated in Figure 11.15(b) whose graph is obtained by contracting the edge between dark gray ($\bullet$) and white

($\bigcirc$) vertices defined in Figure 11.15(a). This edge contraction operation induces the merge of the corresponding regions in Figure 11.15(a) hence producing a partition similar to the one displayed in Figure 11.14(a). As illustrated in Figure 11.15(b), edge contraction between dark gray and white vertices induces the creation of two edges between white and gray vertices. These two edges encode a single boundary segment between the white and gray regions and should thus be merged into a single one. However, the use of a single graph does not allow us to readily distinguish these two edges from the double edges between white and black vertices which encode two nonconnected boundaries and should thus be preserved by any simplification step. Note that the same redundancy problem occurs for the two edges encoding adjacency relationships between the white ($\bigcirc$) and background ($\odot$) vertices.

The *dual graph* model proposed by Kropatsch [22] allows us to perform efficient simplifications of nonsimple graphs after a set of edge contractions. This model assumes that objects to be modeled lie in the 2D plane and may thus be encoded by a connected planar graph. Using this assumption, the dual graph model is defined as a couple of connected dual graphs $(\mathcal{G}, \overline{\mathcal{G}})$, where $\mathcal{G} = (\mathcal{V}, \mathcal{E})$ is a nonsimple graph defined by Req. I (Section 11.4.2.2) and Req. II (Section 11.4.2.2) while $\overline{\mathcal{G}}$ is the dual of $\mathcal{G}$.

The dual graph of $\mathcal{G}$, $\overline{\mathcal{G}} = (\overline{\mathcal{V}}, \overline{\mathcal{E}})$ is defined by creating one vertex (■, Figure 11.14(b)) of $\overline{\mathcal{G}}$ for each face of $\mathcal{G}$ and connecting these vertices such that each edge of $\mathcal{G}$ is crossed by one edge of $\overline{\mathcal{G}}$. Note that $\mathcal{G}$ may encode several edges between two vertices. It is thus a nonsimple graph. Its dual $\overline{\mathcal{G}} = (\overline{\mathcal{V}}, \overline{\mathcal{E}})$ being also nonsimple.

Each dual graph $\overline{\mathcal{G}}$ being deduced automatically from the primal graph $\mathcal{G}$, both graphs $\mathcal{G}$ and $\overline{\mathcal{G}}$ share important properties, which are recalled below:

1. The dual graph operation is idempotent:$\overline{\overline{\mathcal{G}}} = \mathcal{G}$. This important property means that the dual operation does not induce any loss of information since the primal graph may be recovered from its dual. One graph and its dual thus encode the same information differently;

2. Since each edge of $\overline{\mathcal{G}}$ crosses one edge of $\mathcal{G}$, there is a one-to-one correspondence between edges of $\mathcal{G}$ and $\overline{\mathcal{G}}$. Given a set $\mathcal{N}$ of edges, we will denote by $\overline{\mathcal{N}}$ the corresponding set of edges in $\overline{\mathcal{G}}$;

3. The contraction of any forest $\mathcal{N}$ in the primal graph is equivalent to the removal of $\overline{\mathcal{N}}$ in its dual. In other words:

$$\mathcal{G}/\mathcal{N} = \overline{\mathcal{G}} \setminus \overline{\mathcal{N}};\qquad(11.12)$$

Note that since $\mathcal{N}$ does not contain any cycle, $\overline{\mathcal{N}}$ does not define a cut-set. The initial graph $\mathcal{G}$ being connected, both graphs $\mathcal{G}/\mathcal{N}$ and $\overline{\mathcal{G}} \setminus \overline{\mathcal{N}}$ remain connected and dual of each other.

4. Application of the above equation to the dual graph $\overline{\mathcal{G}}$ instead of $\mathcal{G}$ leads to the following equation:

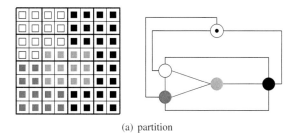

(a) partition

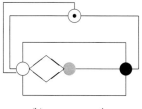

(b) merge operation

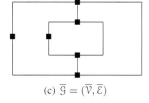

(c) $\overline{\mathcal{G}} = (\overline{\mathcal{V}}, \overline{\mathcal{E}})$

FIGURE 11.15
A partition into four regions and its associated nonsimple graph $\mathcal{G}$(a), the graph $\mathcal{G}$ obtained by the contraction of the edge between white and dark gray vertices. The dual graph of (b), $\overline{\mathcal{G}} = (\overline{\mathcal{V}}, \overline{\mathcal{E}})$ (c).

$$\overline{\mathcal{G}/\mathcal{N}} = \overline{\mathcal{G}} \setminus \overline{\mathcal{N}} \tag{11.13}$$

In other words, the contraction of any forest $\overline{\mathcal{N}}$ of $\overline{\mathcal{G}}$ is equivalent to the removal of $\mathcal{N}$ in $\mathcal{G}$.

As illustrated by Figure 11.15(b), edge contraction may induce the creation of some redundant edges. These edges belong to one of the following categories:

- These edges encode multiple adjacency relationships between two vertices and define degree two faces. They can thus be characterized in the dual graph as degree two dual vertices (Figure 11.15(b and c)). In terms of a partition's encoding these edges correspond to an artificial split of one boundary segment between two regions;

- These edges correspond to a self-loop with an empty inside. These edges thus define degree one faces and are characterized in the dual graph as degree one vertices. Such edges encode artificial inner boundaries of regions.

Both redundant double edges and empty self-loops do not encode relevant topological relations and can be removed without any harm to the involved topology [22].

The resulting pair of dual graphs encodes each connected boundary segment between two regions by one edge between corresponding vertices. The dual graph model thus encodes the *meets_each* relationship through multiple edges between vertices.

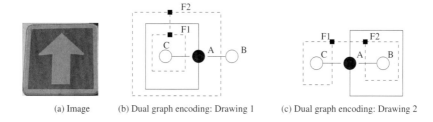

(a) Image   (b) Dual graph encoding: Drawing 1   (c) Dual graph encoding: Drawing 2

FIGURE 11.16
The graph (b) defines the top of a dual graph pyramid encoding an ideal segmentation of (a). The self loop incident to vertex A may surround either vertex B or C without changing the incidence relations between vertices and faces. The dual vertices associated to faces are represented by filled boxes (■). Dual edges are represented by dashed lines.

Within the dual graph model, the encoding of the adjacency between two regions one inside the other is encoded by two edges (Figure 11.16): One edge encoding the common border between the two regions and one self-loop incident to the vertex encoding the surrounding region. One may think to characterize the inside relationship by the fact that the vertex associated to the inside region should be surrounded by the self-loop. However, as shown by Figure 11.16(c), one may exchange the surrounded vertex without modifying the incidence relationships between both vertices and faces. Two dual graphs being defined by these incidence relationships one can thus exchange the surrounded vertex without modifying the encoding of the graphs. This last remark shows that the *inside/contains* relationships cannot be characterized locally within the dual graph framework.

11.4.2.3 Combinatorial Maps

Combinatorial maps and generalized combinatorial maps define a general framework that allows us to encode any subdivision of nD topological spaces orientable or nonorientable with or without boundaries. An exhaustive comparison of combinatorial maps with other boundary representations such as cell-tuples and quad-edges is presented in [23]. Recent trends in combinatorial maps apply this framework to the segmentation of 3D images [24, 25] and the encoding of hierarchies [16, 26, 27].

This section is devoted to $2D$ combinatorial maps that we simply call combinatorial maps. A combinatorial map may be seen as a planar graph encoding explicitly the orientation of edges around a given vertex. Figure 11.17 demonstrates the derivation of a combinatorial map from a plane graph $\mathcal{G} = (\mathcal{V}, \mathcal{E})$ (Figure 11.14(a)). First, edges of $\mathcal{E}$ are split into two half edges called *darts*, each dart having its origin at the vertex it is attached to. The fact that two half edges (darts) stem from the same edge is recorded in the permutation α. A second permutation σ encodes the set of darts encountered when turning counterclockwise around a vertex. A combinatorial map is thus defined by a triplet $\mathcal{G} = (\mathcal{D}, \sigma, \alpha)$, where $\mathcal{D}$ is the set of darts and σ, α are two permutations defined on $\mathcal{D}$ such that α is an involution:

$$\forall d \in \mathcal{D} \quad \alpha^2(d) = d \tag{11.14}$$

If darts are encoded by positive and negative integers, involution α may be implicitly encoded by the sign (Figure 11.17(b)).

Given a dart d and a permutation π, the π-cycle of d denoted by $\pi^*(d)$ is the series of darts $(\pi^i(d))_{i \in \mathbb{N}}$ defined by the successive applications of π on the initial dart d. The σ- and α-cycles of a dart d are thus, respectively, denoted by $\sigma^*(d)$ and $\alpha^*(d)$.

The σ-cycle of one vertex encodes its adjacency relationships with neighboring vertices. For example, let us consider the dart 5 in Figure 11.17(b) attached to the white vertex. Turning counterclockwise around this vertex, we encounter the darts $5, -1, 6$, and -3. We thus have $\sigma(5) = -1$, $\sigma(-1) = 6$, $\sigma(6) = -3$, and $\sigma(-3) = 5$. The σ-cycle of 5 is thus defined as $\sigma^*(5) = (5, -1, 6, -3)$. The whole permutation σ is defined as the composition of its different cycles and is equal to $\sigma = (5, -1, 6, -3)(4, 3)(-5, -4, -6, -2)(1, 2)$. Permutation α being implicitly encoded by the sign in Figure 11.17, we have $\alpha^*(5) = (5, -5)$ and $\alpha = (1, -1)(2, -2) \ldots (6, -6)$.

We can already state one major difference between a combinatorial map and an usual graph encoding of a partition. Indeed, a combinatorial map may be seen as a planar graph with a set of vertices (the cycles of σ) connected by edges (the cycles of α). However, compared to an usual graph encoding, a combinatorial map encodes additionally the local orientation of edges around each vertex, thanks to the order defined within each cycle of σ.

Given a combinatorial map $\mathcal{G} = (\mathcal{D}, \sigma, \alpha)$, its dual is defined by $\overline{\mathcal{G}} = (\mathcal{D}, \varphi, \alpha)$

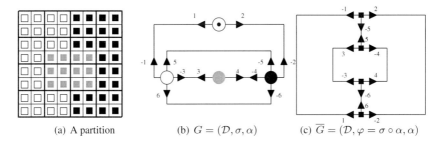

(a) A partition   (b) $G = (\mathcal{D}, \sigma, \alpha)$ (c) $\overline{G} = (\mathcal{D}, \varphi = \sigma \circ \alpha, \alpha)$

FIGURE 11.17
A partition (a) encoded by a combinatorial map (b) and its dual (c).

with $\varphi = \sigma \circ \alpha$. The cycles of the permutation φ encode the sequence of darts encountered when turning around a face of G. Note that, using a counterclockwise orientation for permutation σ, each dart of a φ-cycle has its associated face on its right (see, e.g., the φ-cycle $\varphi^*(5) = (5, -4, 3)$ in Figure 11.17(b)). Cycles of φ may alternatively be interpreted as the sequence of darts encountered when turning clockwise around each dual vertex (Figure 11.17(c)). Permutation φ, on this last figure, being defined as $\varphi = (-1, 2, -5)(5, -4, 3)(4, -6, -3)(1, 6, -2)$.

Each connected boundary segment between two regions is encoded by one edge within the combinatorial map formalism. Models based on combinatorial maps thus encode the *meets_each* relationship. However, within the combinatorial map framework, two connected components S inside R of a partition will be encoded by two combinatorial maps without any information about the respective positioning of S and R. Models based on combinatorial maps have thus designed additional data structure like list of contours [28], inclusion tree [29], or parent-child relationships [30, 18] to encode *contains* and *inside* relationships. These additional data structures should be closely related to the combinatorial map model in order to update both models when a partition is modified.

11.4.3 Simple Graph Pyramids

Simple graph pyramids are defined as a stack of successively reduced simple graphs, each graph being built from the graph below by selecting a set of vertices named surviving vertices and mapping each nonsurviving vertex to a survivor [8, 5]. Using such a framework, the graph $\mathcal{G}_{l+1} = (\mathcal{V}_{l+1}, \mathcal{E}_{l+1})$ defined at level $l + 1$ is deduced from the graph defined at level l by the following steps:

1. Define a partition of $\mathcal{V}_l$ into connected reduction windows and select one surviving vertex within each reduction window. Surviving vertices define the set $\mathcal{V}_{l+1}$ while nonsurviving vertices of each reduction windows are adjacent to the unique survivor of the reduction window (Section 11.3);

2. Define adjacency relationships between vertices of $\mathcal{G}_{l+1}$ in order to define $\mathcal{E}_{l+1}$.

The selection of surviving vertices and the definition of a partition of $\mathcal{V}_l$ into reduction windows is described in Section 11.3. Note that the attachment of each nonsurviving vertex to the unique survivor of its reduction window induces a parent/child relationship between $\mathcal{V}_l$ and $\mathcal{V}_{l+1}$. Each nonsurviving vertex is the child of the unique survivor of its reduction window. Conversely, any surviving vertex is the parent of all nonsurviving vertices within its reduction window.

We are thus here mainly concerned with the second and last step of the decimation process that consists to connect surviving vertices in $\mathcal{G}_{l+1}$ in order to define $\mathcal{E}_{l+1}$. Meer [8] attaches the outcome of a random variable to each vertex and defines a partition of $\mathcal{V}_l$ into reduction windows using the maximal independent vertex set

(MIVS) construction scheme (Section 11.3.2). Figure 11.18(a) (see also Figure 11.9) shows a decomposition of a 4×4, 8-connected grid into 4 reduction windows using such a construction scheme. Surviving vertices are marked by an extra circle, and reduction windows are superimposed to the figure. Given this decomposition of $\mathcal{V}_l$ into reduction windows, Meer joins two parents by an edge if they have adjacent children (e.g., vertices labeled 10 and 7 on the right side of Figure 11.18(a)). Since each child is attached directly to its parent, two adjacent parents at the reduced level are connected in the level below by paths of length less than or equal to three. Let us call these paths *connecting paths*. The set of edges $\mathcal{E}_{l+1}$ induced by connecting paths between surviving vertices in Figure 11.18(a) is shown in Figure 11.18(b).

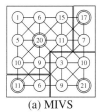 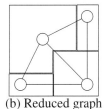

(a) MIVS   (b) Reduced graph

FIGURE 11.18
The reduced graph (b) deduced from the maximal independent vertex set (a) together with reduction windows superimposed to (b).

Two surviving vertices are thus connected in $\mathcal{G}_{l+1}$ if they are connected in $\mathcal{G}_l$ by a path of length lower or equal than 3. Two reduction windows adjacent by more than one such path will thus be connected by a single edge in the reduced graph. The stack of graphs produced by the above decimation process is thus a stack of simple graphs each simple graph encoding only the existence of one common boundary segment between two regions (the *meets_exists* relationship). Moreover, as mentioned in Section 11.4.2, the RAG model that corresponds to a simple graph does not allow us to encode *contains* and *inside* relationships.

Let us finally note that the above construction scheme of the reduced graph $\mathcal{G}_{l+1} = (\mathcal{V}_{l+1}, \mathcal{E}_{l+1})$ from $\mathcal{G}_l = (\mathcal{V}_l, \mathcal{E}_l)$ may be adapted to all parallel reduction schemes detailed in Section 11.3. Using different reduction schemes than the MIVS, it may be interesting to note that the construction of $\mathcal{G}_{l+1}$ from $\mathcal{G}_l$ may be equivalently performed by merging all the vertices of each reduction window into a single vertex (Section 11.4.2.1). This operation may be decomposed as follows:

1. The selection of an unique edge between each nonsurviving vertex and its parent

2. The contraction of the set of edges previously selected

3. The removal of any double edge or self-loop that may have been created by the previous step.

Note, that MIES (Section 11.3.4) and MIDES (Section 11.3.5) construction schemes explicitly connect each nonsurviving vertex to its parent, hence providing the first step of the above process.

11.4.4   Dual Graph Pyramids

As mentioned in Section 11.4.3, the construction of each level of a simple graph pyramid may be decomposed into a contraction step, followed by the removal of any multiple edge or self-loop. Since each edge to be contracted connects a nonsurviving vertex to its surviving parent, the set $\mathcal{CK}$ of contracted edges defines a forest of $\mathcal{CK}$.

11.4.4.1   Contraction Kernel

The different parallel construction schemes of irregular pyramids presented in Section 11.3 allow us to define different forests $\mathcal{F}$ composed of trees of depth at most one hence allowing an efficient parallel contraction of each reduction window.

Kropatsch [3] proposed to abstract the set of edges to be contracted through the notion of *contraction kernel*. A contraction kernel is defined on a graph $\mathcal{G} = (\mathcal{V}, \mathcal{E})$ by a set of surviving vertices $\mathcal{S}$, and a set of nonsurviving edges $\mathcal{CK}$ (shown as bold lines in Figure 11.19) such that:

- $\mathcal{CK}$ is a spanning forest of $\mathcal{G}$;

- Each tree of $\mathcal{CK}$ is rooted by a vertex of $\mathcal{S}$.

Since the set of nonsurviving edges $\mathcal{CK}$ forms a forest of the initial graph, no self-loop may be contracted and the contraction operation is well defined. The decimation of a graph by contraction kernels differs from the one defined in Section 11.4.3 on the two following points:

- First, using contraction kernels, the set of surviving vertices is not required to form a MIVS. Therefore, two surviving vertices may be adjacent in the contracted graph;

- Secondly, using contraction kernels a nonsurviving vertex is not required to be directly linked to its parent but may be connected to it by a branch of a tree (see Figure 11.19(a)). The set of children of a surviving vertex may thus vary from a single vertex to a tree with any depth.

Let us note that the reduction of a graph by a sequence of contraction kernels $\mathcal{CK}_1, \ldots, \mathcal{CK}_n$ composed of trees of depth one is equivalent to the application of one single contraction kernel $\mathcal{CK}$ called the *equivalent contraction kernel* on the initial graph [31]. The set of edges of $\mathcal{CK}$ is defined as the union of all edges of the

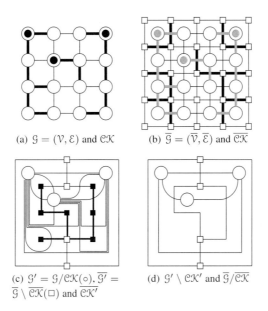

(a) $\mathcal{G} = (\mathcal{V}, \mathcal{E})$ and $\mathcal{CK}$

(b) $\overline{\mathcal{G}} = (\overline{\mathcal{V}}, \overline{\mathcal{E}})$ and $\overline{\mathcal{CK}}$

(c) $\mathcal{G}' = \mathcal{G}/\mathcal{CK}(\circ), \overline{\mathcal{G}}' = \overline{\mathcal{G}} \setminus \overline{\mathcal{CK}}(\square)$ and $\mathcal{CK}'$

(d) $\mathcal{G}' \setminus \mathcal{CK}'$ and $\overline{\mathcal{G}}/\overline{\mathcal{CK}}$

FIGURE 11.19

A contraction kernel $(\mathcal{S}, \mathcal{CK})$ composed of three trees. Vertices belonging to $\mathcal{S}$ are marked with a filled circle inside($\bullet$). Dual vertices incident to redundant dual edges are shown as filled boxes ($\blacksquare$).

contraction kernels $(\mathcal{CK}_i)_{i\in\{1,\ldots,n\}}$. Conversely, any contraction kernel may be decomposed into a sequence of contractions kernels composed of trees of depth one. Contraction kernels should thus be understood as an abstraction of the forest of trees of depth one defined in Section 11.3. Conversely, methods presented in Section 11.3 provide practical heuristics to build contraction kernels fulfilling different properties.

Figure 11.19(a) illustrates a contraction kernel K composed of 3 trees drawn by thick lines (—). The root of each tree being marked by dot circles (●).

This kernel is defined on a graph $\mathcal{G}$ encoding a 4×4 planar sampling grid. The dual of $\mathcal{G}$ is shown in Figure 11.19(b) together with the set of dual edges $\overline{\mathcal{CK}}$. The graph $\mathcal{G}$ is superimposed to $\overline{\mathcal{G}}$ in order to highlight relationships between $\mathcal{CK}$ and $\overline{\mathcal{CK}}$. Contraction of $\mathcal{CK}$ in $\mathcal{G}$ is equivalent to the removal of $\overline{\mathcal{CK}}$ in $\overline{\mathcal{G}}$ (equation 11.12, Section 11.4.2). The resulting pair of reduced graph $(\mathcal{G}/\mathcal{CK}, \overline{\mathcal{G}} \setminus \overline{\mathcal{CK}})$ is shown in Figure 11.19(c).

As mentioned in Section 11.4.2, the contraction of a set of edges may create redundant edges such as double edges surrounding degree 2 faces and self-loops corresponding to degree one faces. These redundant edges may be efficiently characterized in the dual graph, respectively, as degree 2 and degree 1 dual vertices. Such redundant dual vertices are shown in Figure 11.19(c) as filled boxes (■). Removal of degree one dual vertices may be achieved by contracting the only edge incident to such vertices. In the same way, degree 2 dual vertices may be removed by contracting one of their incident edge. The resulting set of edges $\overline{\mathcal{CK}'}$ defines a forest of the dual graph $\overline{\mathcal{G}}$ and is called a *removal kernel* (thick edges (—) in Figure 11.19(c)). Indeed, contraction of $\overline{\mathcal{CK}'}$ in $\overline{\mathcal{G}}$ is equivalent to the removal of $\mathcal{CK}'$ in $\mathcal{G}$ (Section 11.4.2). The reduction of a dual graph $(\mathcal{G}, \overline{\mathcal{G}})$ is thus performed in two steps:

- A first contraction step that computes $(\mathcal{G}', \overline{\mathcal{G}'}) = (\mathcal{G}/\mathcal{CK}, \overline{\mathcal{G}} \setminus \mathcal{CK})$ from $(\mathcal{G}, \overline{\mathcal{G}})$ (Figure 11.19(a-c));

- The definition of a removal kernel $\overline{\mathcal{CK}'}$ from $(\mathcal{G}', \overline{\mathcal{G}'})$ that removes all redundant double edges and empty self-loops of $\mathcal{G}$ by computing $(\mathcal{G}' \setminus \mathcal{CK}', \overline{\mathcal{G}'}/\mathcal{CK}')$ (Figure 11.19(d)).

A *dual graph pyramid* introduced by Kropatsch [3] is defined as a stack of pairs of dual graphs $((\mathcal{G}_0, \overline{\mathcal{G}_0}), \ldots, (\mathcal{G}_n, \overline{\mathcal{G}_n}))$, successively reduced. Each graph $\mathcal{G}_{i+1}$ is deduced from $\mathcal{G}_i$ by a contraction kernel $\mathcal{CK}_i$ followed by the application of a removal kernel removing any redundant double edge or empty self loop that may have been created by the contraction step. The initial graph $\mathcal{G}_0$ and its dual $\overline{\mathcal{G}_0}$ may encode a planar sampling grid or be deduced from an oversegmentation of an image.

11.4.4.2 Dual Graph pyramids and Topological relationships

Given one tree of a contraction kernel, the contraction of its edges collapses all the vertices of the tree into a single vertex and keeps all the connections between the vertices of the tree and the remaining vertices of the graph. The multiple boundaries between the newly created vertex and the remaining vertices of the graph are thus preserved. Each graph of a dual graph pyramid thus encodes the *meets_each* relationships. This property is not modified by the application of a removal kernel that only removes redundant edges.

Moreover, due to the forest requirement, the encoding of the adjacency between two regions one inside the other is encoded by two edges (Figure 11.16): One edge encoding the common border between the two regions and one self-loop incident to the vertex encoding the surrounding region. As mentioned in Section 11.4.2.2 (Figure 11.16), such a couple of edges does not allow to characterize locally contains and inside relationships.

11.4.5 Combinatorial Pyramids

As in the dual graph pyramid scheme [32] (Section 11.4.4), a *combinatorial pyramid* is defined by an initial combinatorial map successively reduced by a sequence of contraction or removal operations. The initial combinatorial map encodes a planar sampling grid or a first segmentation, and the remaining combinatorial maps of a combinatorial pyramid encode a stack of image partitions successively reduced. As mentioned in Section 11.4.2, page 330, a combinatorial map may be understood as a dual graph with an explicit encoding of the orientation of the edges incident to each vertex. This explicit encoding of the orientation is preserved within the combinatorial pyramid framework [33].

Contraction operations are controlled by contraction kernels (CK) defined as an acyclic set of darts encoding the set of edges to be contracted. Given a combinatorial map $\mathcal{G} = (\mathcal{D}, \sigma, \alpha)$, a kernel $\mathcal{CK}$ defined on $\mathcal{G}$ is thus included in $\mathcal{D}$ and symmetric according to α:

$$\forall d \in \mathcal{CK}, \ \alpha(d) \in \mathcal{CK} \tag{11.15}$$

The removal of redundant edges is performed as in the dual graph reduction scheme by a removal kernel. Such kernels are, as contraction kernels, defined as sets of darts to be deleted. Contrary to the dual graph pyramid framework, removal kernels are decomposed in two subkernels: A removal kernel of empty self-loops (RKESL) which contains all darts incident to a degree 1 dual vertex and a removal kernel of empty double edges (RKEDE) which contains all darts incident to a degree 2 dual vertex. These two removal kernels are defined as follows: The removal kernel of empty self-loops RKESL is initialized by all self-loops surrounding a dual vertex of degree 1. RKESL is further expanded by all self-loops that contain only other self-loops already in RKESL until no further expansion is possible. For the removal of empty double edges RKEDE, we ignore all empty self-loops in RKESL in computing the degree of the dual vertex.

Note that the removal of degree two dual vertices by a RKEDE may be achieved

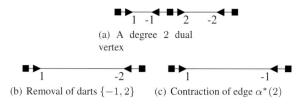

(a) A degree 2 dual
vertex

(b) Removal of darts $\{-1, 2\}$ (c) Contraction of edge $\alpha^*(2)$

FIGURE 11.20

The dual vertex of degree 2, $\varphi^*(-1) = (-1, 2)$ (a) may be suppressed by removing darts -1 and 2 that define the degree 2 dual vertex (b). Permutation α should then be updated as $\alpha'(1) = -2$ and $\alpha'(-2) = 1$. This dual vertex may be alternatively removed by contracting in the dual graph one of the two edges incident to this dual vertex. The contracted edge $(2, -2)$ is removed in (c). Involution α remains unchanged in this case, but the contraction operation has replaced -2 by -1 within the φ cycle encoding the right vertex, hence modifying the dual vertex to which dart -1 is attached to.

using two distinct operations illustrated in Figure 11.20. A first solution (Figure 11.20(b)) consists to remove all darts incident to a degree two vertex. This solution implies to update involution α at each level of the pyramid but simplifies the modification that may affect a dual vertex since a dart remains incident to a same dual vertex until it is removed by some contraction or removal operation. Conversely, the second solution (Figure 11.20(c)) keeps involution α constant but may change the attachment of a dart to a dual vertex. Both encoding of a RKEDE define valid combinatorial pyramids and we choose the first solution for the remaining of this chapter.

The successive application of a RKESL and a RKEDE may be encoded as the application of a single removal kernel. However, this decomposition allows us to distinguish darts encoding boundaries between two regions from inner boundaries within the implicit encoding scheme of combinatorial pyramids (next section). Both contraction and removal operations defined within the combinatorial pyramid framework are thus defined as in the dual graph framework but additionally preserve the orientation of edges around each vertex. Further details about the construction scheme of a combinatorial pyramid may be found in [16].

11.4.5.1   Relationships between regions

Concerning the encoding of relationships between regions, combinatorial pyramids provide, as dual graph pyramids, an encoding of both *meets_each* and *contains/inside* relationships. This last relationship is encoded in both models by self-loops. A region R_1 inside a region R_2 being characterized by a self-loop incident to v_2 (associated to R_2) and surrounding v_1 (encoding R_1). However, using the dual graph framework, such configurations of vertices may be characterized only through global calculus over the whole graph. On the other hand, the explicit encoding of the orientation provided by combinatorial maps allows us to retrieve the set of regions inside a region encoded by a vertex v in a time proportional to the degree of v [27].

One may think to use the additional data structures (Section 11.4.2.3) defined within the combinatorial map framework to encode inside relationships within the combinatorial pyramid model. This solution would break the efficiency of the hierarchical data structure. Indeed, contains and inside relationships may be both created and removed within the pyramid. Therefore, the explicit encoding of these relationships by additional data structures would lead to associate to a pyramid a sequence of data structures whose size do not decrease between two successive levels of a pyramid. The overall size of each level of the pyramid may thus not decrease with a fixed rate or even increase, hence violating the main requirement of a hierarchical data structure (Section 11.3). Note that, conversely, the encoding of contains and inside relationships by self-loops is not adapted to the combinatorial map framework. Indeed, region splitting operations encoded within the combinatorial map model by the addition of edges may lead to cumbersome computations when these additional borders modify the contains/inside relationships encoded by self-loops. Such additions of edges that are not allowed within the combinatorial pyramid model are frequent within the nonhierarchical combinatorial map model. An explicit encoding of contains/inside relationships by additional data structures is thus preferred within this last model.

11.4.5.2 Implicit Encoding of a Combinatorial Pyramid

Let us consider an initial combinatorial map $\mathcal{G}_0 = (\mathcal{D}, \sigma, \alpha)$ and a sequence of kernels $\mathcal{CK}_1, \ldots, \mathcal{CK}_n$ successively applied on $\mathcal{G}_0$ to build the pyramid. Each combinatorial map $\mathcal{G}_i = (\mathcal{SD}_i, \sigma_i, \alpha_i)$ is defined from $\mathcal{G}_{i-1} = (\mathcal{SD}_{i-1}, \sigma_{i-1}, \alpha_{i-1})$ by the application of the kernel $\mathcal{CK}_i$ on $\mathcal{G}_{i-1}$, and the set of surviving darts at level i: $\mathcal{SD}_i$ is equal to $\mathcal{SD}_{i-1} \setminus K_i$. We have thus:

$$\mathcal{SD}_{n+1} \subseteq \mathcal{SD}_n \subseteq \ldots \mathcal{SD}_1 \subseteq \mathcal{D} \tag{11.16}$$

The set of darts of each reduced combinatorial map of a pyramid is thus included in the one of the base level combinatorial map. This last property allows us to define the two following functions:

1. One function state from $\{1, \ldots, n\}$ to the states $\{CK, RKESL, RKEDE\}$ that specifies the type of kernel applied at each level;

2. One function level defined for all darts in $\mathcal{D}$ such that level(d) is equal to the maximal level where d survives:

$$\forall d \in \mathcal{D} \; \text{level}(d) = \max\{i \in \{1, \ldots, n+1\} \mid d \in \mathcal{SD}_{i-1}\} \tag{11.17}$$

a dart d surviving up to the top level has thus a level equal to $n + 1$. Note that if $d \in \mathcal{CK}_i, i \in \{1, \ldots, n\}$, then level$(d) = i$.

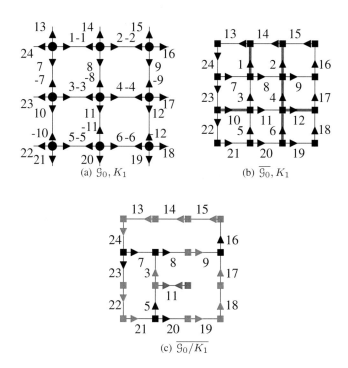

FIGURE 11.21

A combinatorial map (a) and its dual (b) encoding a 3×3 grid with the contraction kernel $K_1 = \alpha^*(1, 2, 4, 12, 6, 10)$ (▬) superimposed to both figures. Only positive darts are represented in (b) and (c) in order to not overload them. The resulting reduced dual combinatorial map (c) should be further simplified by a removal kernel of empty self-loops (RKESL) $K_2 = \alpha^*(11)$ to remove dual vertices of degree 1 (■). Dual vertices of degree 2 (■) are removed by the RKEDE $K_3 = \varphi^*(24, 13, 14, 15, 9, 22, 17, 21, 19, 18,) \cup \{3, -5\}$. Note that the dual vertex $\varphi^*(3) = (3, -5, 11)$ becomes a degree 2 vertex only after the removal of $\alpha^*(11)$.

We have shown [16] that the sequence of reduced combinatorial maps $\mathcal{G}_0, \ldots, \mathcal{G}_{n+1}$ may be encoded without any loss of information using only the base level combinatorial map $\mathcal{G}_0$ and the two functions level and state. Such an encoding is called an implicit encoding of the pyramid. Figure 11.22(a) illustrates the implicit encoding of the pyramid defined in Figure 11.21 by an initial combinatorial map encoding a 3×3 grid and three kernels $K_1, K_2,$ and K_3. These kernels defined at levels 1, 2, and 3 are, respectively, represented by red (▶), green (▶), and blue (▶) darts in Figure 11.22(a).

The encoding of the function *state* being proportional to the height of the pyramid, its memory cost is negligible compared to the cost of the base level combinatorial map. The encoding of the function level requires us to associate one integer to each dart. The memory cost of such a function for a pyramid composed of n levels is thus equal to $\log_2(n) |\mathcal{D}|$. Using an implicit encoding of the involution α, the encoding of permutation σ requires $|\mathcal{D}| \log_2(|\mathcal{D}|)$ bytes. The total memory cost of an implicit pyramid is thus equal to:

$$[\log_2(n) + \log_2(|\mathcal{D}|)] |\mathcal{D}| . \tag{11.18}$$

On the other hand, using an explicit encoding of a combinatorial pyramid, if we suppose that the number of darts decreases by a factor 2 between each level, the total number of darts of the whole pyramid is equal to:

$$|\mathcal{D}| \sum_{i=0}^{p} \frac{1}{2^i} \approx 2 |\mathcal{D}| \tag{11.19}$$

where $|\mathcal{D}|$ is supposed to be a power of 2 ($|\mathcal{D}| = 2^p$). The total memory cost of the explicit encoding of a combinatorial pyramid is thus equal to:

$$2 |\mathcal{D}| \log_2(|\mathcal{D}|). \tag{11.20}$$

The ratio between Equations 11.18 and 11.20 is equal to:

$$\frac{1}{2} \left(1 + \frac{\log_2(n)}{\log_2(|\mathcal{D}|)} \right) \tag{11.21}$$

The value of $\log_2(n)$ being usually much lower than $\log_2(|\mathcal{D}|)$, the implicit encoding provides a lower memory cost than the explicit encoding. Let us additionally note that if we bound the maximal level of a pyramid using, for example, 32 bits integers to store the function level, we obtain a hierarchical encoding which is *in practice* independent of the height of the pyramid.

11.4.5.3 Dart's Embedding and Segments

The *receptive field* of a dart $d \in SD_i$ corresponds to the set of darts reduced to d at level i [16]. Using the implicit encoding of a combinatorial pyramid, the receptive field $RF_i(d)$ of $d \in \mathcal{SD}_i$ is defined as a sequence $d_1 \ldots d_q$ of darts in $\mathcal{D}$ such that $d_1 = d, d_2 = \sigma_0(d)$ and for each j in $\{3, \ldots, q\}$:

$$d_j = \begin{cases} \varphi_0(d_{j-1}) & \text{if state(level}(d_{j-1})) = CK \\ \sigma_0(d_{j-1}) & \text{if state(level}(d_{j-1})) \in \{RKEDE, RKESL\} \end{cases} \tag{11.22}$$

The dart d_q is defined as the last dart of the sequence that has been contracted or removed below the level i. Therefore, the successor of d_q according to Equation 11.22, d_{q+1} satisfies level$(d_{q+1}) > i$. Moreover, we have shown [16] that d, d_q, and d_{q+1} are additionally connected by the two following relationships:

$$\sigma_i(d) = d_{q+1} \text{ and } \alpha_i(d) = \alpha_0(d_q) \tag{11.23}$$

Using a combinatorial map encoding of image content, regions and maximal connected boundaries between regions are, respectively, encoded by σ and α cycles. Therefore, each dart $d \in \mathcal{SD}_i$ encodes a boundary segment between the regions associated to $\sigma_i^*(d)$ and $\sigma_i^*(\alpha_i(d))$. Moreover, in the lower levels of the pyramid, the two vertices of an edge may belong to a same region. We call the corresponding boundary segment an *internal boundary* in contrast to an *external boundary* that separates two different regions. The receptive field of d at level i ($RF_i(d)$) contains both darts corresponding to this boundary segment and additional darts corresponding to internal boundaries. The sequence of external boundary darts contained in $RF_i(d)$ is denoted by $\partial RF_i(d)$ and is called a *segment*. The order on $\partial RF_i(d)$ is deduced from the receptive field $RF_i(d)$. Given a dart $d \in \mathcal{SD}_i$, the sequence $\partial RF_i(d) = d_1, \ldots, d_q$ is retrieved by [34]:

$$d_1 = d \text{ and } \forall j \in \{1, \ldots, q-1\} \, d_{j+1} = \varphi_0^{n_j}(\alpha_0(d_j)) \tag{11.24}$$

The dart d_q is the last dart of $\partial RF_i(d)$ that belongs to a double edge kernel. This dart is thus characterized using equation 11.23 by $d_q = \alpha_0(\alpha_i(d))$. Note that each dart of the base level corresponds to an oriented crack [17, 18] (Section 11.4.2). A segment thus corresponds to a sequence of oriented cracks encoding a connected boundary segment between two regions [34].

The value n_j is defined for each $j \in \{1, \ldots, q-1\}$ by :

$$n_j = \min\{k \in \mathbb{N}^* \mid \text{state(level}(\varphi_0^k(\alpha_0(d_j)))) = RKEDE\}. \tag{11.25}$$

A segment may thus be interpreted as a maximal sequence, according to Equation 11.24, of darts removed as double edges. Such a sequence connects two darts (d and $\alpha_0(d_q) = \alpha_i(d)$) surviving up to level i. The retrieval of the boundaries using equations 11.24 and 11.25 is one of the major reasons that lead us to distinguish empty self-loop and redundant double edges removal kernels. Let us additionally note that if $\mathcal{G}_0$ encodes the 4-connected planar sampling grid, each φ_0 cycle is composed of at most 4 darts (Figure 11.19(b)). Therefore, the computation of d_{j+1} from d_j (Equation 11.24) requires at most 4 iterations and the determination of the whole sequence of cracks composing a boundary segment between two regions is performed in a time proportional to the length of this boundary.

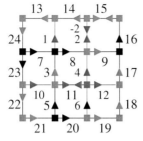

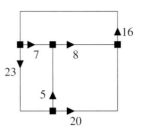

(a) Color encoding of functions state and level

(b) Top of the pyramid defined in Figure 11.21.

FIGURE 11.22
Implicit encoding (a) and top level combinatorial map (b) of the pyramid defined in Figure 11.21. Red, green and blue darts have level equal to 1, 2, and 3 respectively. States associated to these levels are, respectively, equal to $CK, RKESL, RKEDE$.

Let us consider the border of level 3 associated to dart 16 in Figure 11.22. According to Equation 11.24, 16 is the first dart of $\partial RF_3(16)$. Moreover, $\varphi_0(\alpha_0(16)) = \varphi_0(-16) = 15$ belongs the RKEDE K_3 (Figure 11.22(a)). We have thus $n_1 = 1$ (Equation 11.25), and the second dart d_2 of $\partial RF_3(16)$ is equal to 15. Using once again Equation 11.24, d_3 is the first dart encountered from $\varphi_0(\alpha_0(15))$ by iterating permutation φ_0. Since $\varphi_0((\alpha_0(15)) = \varphi_0(-15) = -2$ is a contracted dart, we should iterate permutation φ_0 at least one more time. The dart $\varphi_0^2((\alpha_0(15)) = \varphi_0(-2) = 14$ belongs to the RKEDE K_3, and we have thus $n_2 = 2$ and $d_3 = 14$. Further iterations of Equations 11.24 and 11.25 lead to the sequence:

$$\partial RF_3(16) = 16.15.14.13.24 \qquad (11.26)$$

which encodes the border between the first line of the image and the background.

In conclusion, the implicit encoding of a combinatorial pyramid provides an efficient encoding of the pyramid with a low memory cost and an efficient access to the geometry of segments defining the image's partition.

Figure 11.23 illustrates an application of combinatorial pyramids to image segmentation. The gradient of the original image (Figure 11.23(a)) is encoded by a planar sampling grid encoded by a combinatorial map. Each edge of such a combinatorial map is valuated by a gradient measure between its two incident pixels [35]. The encoding of a watershed transform on this combinatorial map is performed by computing a contraction kernel whose each tree spans a watershed's basin. The application of such a contraction kernel to obtain Figure 11.23(b) may be performed sequentially or using any parallel method defined in Section 11.3. Each edge of this partition is valuated by a relevance measure encoded by gray levels in Figure 11.23(c). This measure of relevance is based on edge's dynamic and is computed using the profile of the gradient along each segment [26]. Such a measure is thus based on the computation of the external boundary of each dart. Finally, further levels of the pyramid are computed from Figure 11.23(c) by merging least relevant edges and updating the remaining edge's relevance at each step. Such a strategy uses the implicit encoding of combinatorial maps in order to reduce memory requirements. The last level of the pyramid before a wrong merge operation is performed on Lena's hat is displayed Figure 11.23(d).

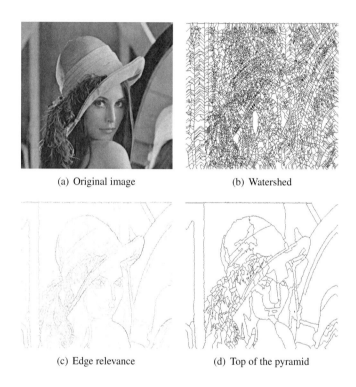

(a) Original image   (b) Watershed

(c) Edge relevance   (d) Top of the pyramid

FIGURE 11.23
An application of combinatorial pyramid to hierarchical image segmentation.

11.5   Conclusion

Graphs have become a representational tool to bridge the gap between the raw information sources provided by sensors and the highly complex structure of the sensed world and the knowledge about it. A typical example is a digital image: it is the projection of the world's structure and geometry into the image plane where some instances and basic objects can be observed together with certain properties like brightness or color and object-to-object relations like adjacency or closeness.

Embedded attributed graphs and combinatorial maps capture such low level visual cues and allow the aggregation of larger pieces of information that needs to be connected with the overall knowledge about the world. Both the sensed data and the high-level knowledge appear in huge amounts, and their processing is often of high complexity. In order to achieve the necessary efficiency hierarchies in different forms can be used.

This chapter introduced both regular and irregular pyramid frameworks connecting the embedded pixel arrangement with mid- and high-level abstraction hierarchies. Both frameworks allow us to efficiently analyze the content of an image using parallel calculus. Compared to non hierarchical methods, regular and irregular pyramids additionally provide a hierarchical description of the image's content. Several drawbacks of regular pyramids, forbidding their use as a general framework for image analysis, lead to the definition of irregular pyramids. Such pyramids may be defined both by their construction scheme and by the graph model used within the pyramid.

We have presented the main parallel reduction schemes of irregular pyramids in Section 11.3. These construction schemes are based on the notion of *independent sets* defined on different parts of a graph according to each construction scheme. As mentioned by Table 11.2 (first column), the MIES and MIDES frameworks provide a theoretical bound on the height of the pyramid, while MIVS and D3P do not. All reduction schemes described in Section 11.3, but the D3P are based on the notion of maximal independent set (Table 11.2, second column). These last methods may be further classified according to the elements on which the maximal independent set is computed (Table 11.2, third column).

	Theoretical bound of pyramid's height	Maximal independent set	Independent set on
MIVS	✗	✓	Vertices
D3P	✗	✗	Vertices
MIES	✓	✓	Undirected edges
MIDES	✓	✓	Directed edges

TABLE 11.2
Main positive (✓) and negative (✗) properties of parallel irregular pyramid construction schemes.

This chapter concludes on a presentation of the main type of graphs used within the irregular pyramid framework. Each graph model determines which type of image features may be encoded at each level of a pyramid. This choice of a graph model thus determines the range of image's features that may be efficiently used by an image analysis or segmentation method and is hence an important parameter of such methods. As indicated in Table 11.3 (first column), all graph models encode the existence of a common boundary between two regions (meets_exists relationship). However, the explicit encoding of connected boundaries between regions is only provided by models based on dual graphs and combinatorial maps (table 11.3, second column). Contains and inside relationships are not encoded by models based on simple graphs. We expect that computing such relationships using the dual graph model requires global calculus on the whole graph. On the other hand, contains and inside relationships are explicitly encoded by additional data structures within the combinatorial map framework and may be efficiently computed by local calculus using combinatorial pyramids (Table 11.3, third column). Finally, the composed of relationship is only encoded by hierarchical models.

	meets_exists	meets_each	contains/ inside	Composed of
RAG	✓	✗	✗	✗
Combinatorial maps	✓	✓	✓	✗
Simple graph pyramids	✓	✗	✗	✓
Dual graph pyramids	✓	✓	Global calculus?	✓
Combinatorial pyramids	✓	✓	Local calculus	✓

TABLE 11.3
Relationships available(✓) or not available (✗) using models described in this chapter.

This study of embedded hierarchical graph structures is probably only the first step to understand the complex strategies enabling humans to understand their environment and to react in real time. Further directions of research consider substructures invariant with respect to certain types of operations: topological structures invariant to continuous deformations [36]; or temporal structures capturing the components of a moving object [37, 38]; or generalizations being invariant to certain less relevant distictions of a class of objects enabling the focus on the main properties of the class.

Bibliography

[1] P. Burt, T.-H. Hong, and A. Rosenfeld, "Segmentation and estimation of image region properties through cooperative hierarchial computation," *IEEE Transactions on Systems, Man and Cybernetics*, vol. 11, no. 12, pp. 802–809, December 1981.

[2] M. Bister, J. Cornelis, and A. Rosenfeld, "A critical view of pyramid segmentation algorithms," *Pattern Recognition Letter*, vol. 11, no. 9, pp. 605–617, September 1990.

[3] W. G. Kropatsch, "From equivalent weighting functions to equivalent contraction kernels," in *Digital Image Processing and Computer Graphics (DIP-97): Applications in Humanities and Natural Sciences*, E. Wenger and L. I. Dimitrov, Eds., vol. 3346. SPIE, October 1997, pp. 310–320.

[4] A. Rosenfeld and C.-Y. Sher, "Detecting image primitives using feature pyramids," *Journal of Information Sciences*, vol. 107, pp. 127–147, june 1998.

[5] A. Montanvert, P. Meer, and A. Rosenfeld, "Hierarchical image analysis using irregular tessellations," *IEEE Transactions on Pattern Analysis and Machine Intelligence*, vol. 13, no. 4, pp. 307–316, April 1991.

[6] L. Brun and W. Kropatsch, "Construction of combinatorial pyramids," in *Graph based Representations in Pattern Recognition*, ser. LNCS, E. Hancock and M. Vento, Eds., vol. 2726. York, UK: IAPR-TC15, June 2003, pp. 1–12.

[7] C. Chevalier and I. Safro, "Comparison of coarsening schemes for multilevel graph partitioning," in *Learning and Intelligent Optimization*, T. Stützle, Ed. Berlin, Heidelberg: Springer-Verlag, 2009, pp. 191–205.

[8] P. Meer, "Stochastic image pyramids," *Computer Vision Graphics Image Processing*, vol. 45, no. 3, pp. 269–294, march 1989.

[9] J. M. Jolion and A. Montanvert, "The adaptative pyramid: A framework for 2d image analysis," *Computer Vision, Graphics, and Image Processing*, vol. 55, no. 3, pp. 339–348, May 1992.

[10] J.-M. Jolion, "Data driven decimation of graphs," in *Proceedings of 3rd IAPR-TC15 Workshop on Graph based Representation in Pattern Recognition*, J.-M. Jolion, W. Kropatsch, and M. Vento, Eds., Ischia-Italy, May 2001, pp. 105–114.

[11] S.Thorpe, D.Fize, and C. Marlot, "Speed of processing in the human visual system," *Nature*, vol. 381, pp. 520–522, june 1996.

[12] W. G. Kropatsch, Y. Haxhimusa, Z. Pizlo, and G. Langs, "Vision pyramids that do not grow too high," *Pattern Recognition Letters*, vol. 26, no. 3, pp. 319 – 337, February 2005, in Memoriam: Azriel Rosenfeld.

[13] Y. Haxhimusa, *The Structurally Optimal Dual Graph Pyramid and its Application in Image Partitioning*. IOS Press and AKA, june 2007, ISBN: 978-1-58603-743-7.

[14] M. Burge and W. Kropatsch, "A minimal line property preserving representation of line images," *Computing*, vol. 62, no. 4, pp. 355–368, 1999.

[15] R. Englert and W. G. Kropatsch, "Image Structure From Monotonic Dual Graph Contraction," in *Applications of Graph Transformations with Industrial Relevance*, ser. LNCS, M. Nagl, A. Schürr, and M. Münch, Eds., vol. Vol. 1799. Kerkrade, Netherlands: Springer, Berlin Heidelberg, New York, September 2000, pp. 550–553.

[16] L. Brun and W. Kropatsch, "Combinatorial pyramids," in *IEEE International conference on Image Processing (ICIP)*, Suvisoft, Ed., vol. II. Barcelona: IEEE, September 2003, pp. 33–37.

[17] V. Kovalevsky, "Finite topology as applied to image analysis," *Computer Vision, Graphics, and Image Processing*, vol. 46, no. 2, pp. 141–161, May 1989.

[18] L. Brun, M. Mokhtari, and J. P. Domenger, "Incremental modifications on segmented image defined by discrete maps," *Journal of Visual Communication and Image Representation*, vol. 14, no. 3, pp. 251–290, September 2003.

[19] D. Randell, C. Z, and A. Cohn, "A spacial logic based on regions and connections," in *Principle of Knowledge Representation and Reasoning: Proceedings 3rd International Conference*, B. Nebel, W. Swartout, and C. Rich, Eds., Cambridge MA, October 1992, pp. 165–176.

[20] K. Shearer, H. Bunke, and S. Venkatesh, "Video indexing and similarity retrieval by largest common subgraph detection using decision trees," *Pattern Recognition*, vol. 34, no. 5, pp. 1075–1091, May 2001.

[21] L. Brun and M. Mokhtari, "Graph based representations in different application domains," in 3rd *IAPR-TC15 Workshop on Graph-based Representations in Pattern Recognition*, J. M. Jolion, W. Kropatsch, and M. Vento, Eds., IAPR-TC15. Ischia Italy: CUEN, May 2001, pp. 115–124, invited conference.

[22] D. Willersinn and W. G. Kropatsch, "Dual graph contraction for irregular pyramids," in *International Conference on Pattern Recogntion D: Parallel Computing.* Jerusalem, Israel: International Association for Pattern Recognition, October 1994, pp. 251–256.

[23] P. Lienhardt, "Topological models for boundary representations: a comparison with n-dimensional generalized maps," *Computer-Aided Design*, vol. 23, no. 1, pp. 59–82, February 1991.

[24] Y. Bertrand, G. Damiand, and C. Fiorio, "Topological map: Minimal encoding of 3d segmented images," in 3rd *Workshop on Graph-based Representations in Pattern Recognition*, J. M. Jolion, W. Kropatsch, and M. Vento, Eds., IAPR-TC15. Ischia(Italy): CUEN, May 2001, pp. 64–73.

[25] S. Fourey and L. Brun, "Efficient encoding of n-D combinatorial pyramids," in *Proceedings of the International Conference on Pattern Recognition (ICPR'2010).* Istanbul, Turkey: IEEE, August 2010, pp. 1036–1039.

[26] L. Brun, M. Mokhtari, and F. Meyer, "Hierarchical watersheds within the combinatorial pyramid framework," in *Proc. of DGCI 2005*, E. Andres, G. Damiand, and P. Lienhardt, Eds., vol. 3429, IAPR-TC18. Poitiers, France: LNCS, April 2005, pp. 34–44.

[27] L. Brun and W. Kropatsch, "Contains and inside relationships within combinatorial pyramids," *Pattern Recognition*, vol. 39, no. 4, pp. 515–526, April 2006.

[28] C. Fiorio, "A topologically consistent representation for image analysis: the frontiers topological graph," in *6th International Conference on Discrete Geometry for Computer Imagery (DGCI'96)*, ser. LNCS, S. Miguet, A. Montanvert, and U. S., Eds., vol. 1176, Lyon, France, November 1996, pp. 151–162.

[29] G. Damiand, Y. Bertrand, and C. Fiorio, "Topological model for two-dimensional image representation: definition and optimal extraction algorithm," *Computer Vision and Image Understanding*, vol. 93, no. 2, pp. 111 – 154, February 2004.

[30] J. P. Braquelaire and L. Brun, "Image segmentation with topological maps and inter-pixel representation," *Journal of Visual Communication and Image representation*, vol. 9, no. 1, pp. 62–79, March 1998.

[31] W. G. Kropatsch, "Equivalent contraction kernels to build dual irregular pyramids," *Advances in Computer Science*, vol. Advances in Computer Vision, pp. 99–107, 1997.

[32] ——, "Building Irregular Pyramids by Dual Graph Contraction," *IEEE-Proc. Vision, Image and Signal Processing*, vol. 142, no. 6, pp. 366–374, December 1995.

[33] L. Brun and W. Kropatsch, "Dual contraction of combinatorial maps," in 2nd *IAPR-TC-15 Workshop on Graph-based Representations*, W. Kropatsch and J.-M. Jolion, Eds., vol. 126. Haindorf, Austria: Österreichische Computer Gesellschaft, May 1999, pp. 145–154.

[34] L. Brun, "Traitement d'images couleur et pyramides combinatoires," Habilitation à diriger des recherches, Université de Reims, 2002.

[35] L. Brun, P. Vautrot, and F. Meyer, "Hierarchical watersheds with inter-pixel boundaries," in *Image Analysis and Recognition: International Conference ICIAR 2004, Part I.* Porto (Portugal): Springer Verlag Heidelberg (LNCS), 2004, pp. 840–847.

[36] R. Gonzalez-Diaz, A. Ion, M. Iglesias-Ham, and W. G. Kropatsch, "Irregular Graph Pyramids and Representative Cocycles of Cohomology Generators," in *Proceedings of the 7th IAPR-TC-15 International Workshop, GbRPR 2009, on Graph Based Representations in Pattern Recognition*, ser. LNCS, A. Torsello, F. Escolano, and M. Vento, Eds., vol. 5534. Berlin Heidelberg: Springer-Verlag, May 2009, pp. 263–272.

[37] W. G. Kropatsch, "When Pyramids Learned Walking," in *The 14th International Congress on Pattern Recognition, CIARP 2009*, ser. LNCS, J. O. E. Eduardo Bayro-Corrochano, Ed., vol. 5856. Berlin Heidelberg: Springer-Verlag, November 2009, pp. 397–414.

[38] N. M. Artner, A. Ion, and W. G. Kropatsch, "Rigid Part Decomposition in a Graph Pyramid," in *The 14th International Congress on Pattern Recognition, CIARP 2009*, ser. LNCS, J. O. E. Eduardo Bayro-Corrochano, Ed., vol. 5856. Berlin Heidelberg: Springer-Verlag, November 2009, pp. 758–765.

12

Graph-Based Dimensionality Reduction

John A. Lee

Center for Molecular Imaging, Radiotherapy, and Oncology (MIRO)
Université catholique de Louvain
Louvain-la-Neuve, Belgium
john.lee@uclouvain.be

Michel Verleysen

Machine Learning Group (MLG), ICTEAM Institute
Université catholique de Louvain
Louvain-la-Neuve, Belgium
Statistique, Analyse, Modélisation Multidisciplinaire (SAMM)
Université Paris I Panthéon-Sorbonne
Paris, France
michel.verleysen@uclouvain.be

CONTENTS

12.1 Summary 352
12.2 Introduction 352
12.3 Classical methods 353
    12.3.1 Principal component analysis 354
    12.3.2 Multidimensional scaling 355
    12.3.3 Nonlinear MDS and Distance Preservation 356
12.4 Nonlinearity through Graphs 357
12.5 Graph-Based Distances 358
12.6 Graph-Based Similarities 361
    12.6.1 Laplacian Eigenmaps 361
    12.6.2 Locally linear embedding 364
12.7 Graph embedding 365
    12.7.1 From LD to HD: Self-Organizing Maps 366
    12.7.2 From HD to LD: Isotop 367
12.8 Examples and comparisons 368
    12.8.1 Quality Assessment 369
    12.8.2 Data Sets 370
    12.8.3 Methods 371
    12.8.4 Results 372
12.9 Conclusions 373
    Bibliography 374

12.1 Summary

Dimensionality reduction aims at representing high-dimensional data into low-dimensionality space. In order to make sense, the low-dimensional representation, or embedding, has to preserve some well-defined structural properties of data. The general idea is that similar data items should be displayed close to each other, whereas longer distances should separate dissimilar ones. This principle applies to data that can be either sampled from a manifold of the data space or distributed among several clusters. In both cases, one generally agrees that rendering the local struture comes prior to reproducing the global arrangement of data. In practice, the structural properties can be pairwise measurements such as dissimilarities (all kinds of distances) or similarities (decreasing functions of distances). Relative proximities such as rank of sorted distances can be used as well. Within this context, graphs can serve several purposes. For instance, they can model the fact that (dis)similarities are missing or overlooked. Even if all pairwise dissimilarities are available, one can consider that only the shortest ones are meaningful and reflect the local data structure. Graphs corresponding to K-ary neighborhoods or ϵ-balls can be built and utilized to compute the length of shortest paths or commute-time distances associated to random walks.

This chapter reviews some of the most prominent dimensionality reduction methods that rely on graphs. These include several techniques based on geodesic distances, such as Isomap and its variants. Spectral methods involving a graph Laplacian are described as well. More biologically inspired techniques, such as the self-organizing map, identify a topographic mapping between a predefined graph and the data manifold. Illustrative experiments focus on image data (scanned handwritten digits) and results are assessed using rank-based quality criteria.

12.2 Introduction

The interpretation of high-dimensional data remains a difficult task, mainly because human vision is not used to deal with spaces having more than three dimensions. Part of this difficulty stems from the curse of dimensionality, a convenient expression that encompasses all weird and unexpected properties of high-dimensional spaces. Dimensionality reduction (DR) aims at constructing an alternative low-dimensional representation of data, in order to improve readability and interpretability. Of course, this low-dimensional representation must be meaningful and faithful to the genuine data. In practice, the representation must preserve important structural properties of the data set, such as relative proximities, similarities, or dissimilarities. The general idea is that dissimilar data items should be represented far from each other, whereas similar ones should appear close to each other. Dimensionality reduction serves other purposes than just data visualization. For instance, DR can be used in data compression and denoising. It can also preprocess data, with the hope that a simplified representation can accelerate any subsequent processing or improve its outcome.

Linear DR is well known, with techniques such as principal component analysis [1] (PCA) and classical metric multidimensional scaling [2, 3] (MDS). The former tries to preserve the covariances in the low-dimensional space, whereas the latter attempts to reproduce the Gram matrix of pairwise dot products. Nonlinear dimensionality reduction [4] (NLDR) emerged later, with nonlinear variants of multidimensional scaling [5, 6, 7] such as Sammon's nonlinear mapping [8] (NLM) and curvilinear component analysis [9, 10] (CCA). Most of these methods are based on the preservation of pairwise distances. Research in NLDR is multidisciplinary and follows many approaches, ranging from artificial neural networks [11, 12, 13, 9, 14] to spectral techniques [15, 16, 17, 18, 19, 20, 21]. If linear DR assumes that data are distributed within or near a linear subspace, NLDR necessitates more complex models.

The most generic framework consists in assuming that data are sampled from a smooth manifold. For this reason, modern NLDR is sometimes referred to as manifold learning [22, 20]. Under this hypothesis, one seeks to re-embed the manifold in a space having the lowest possible dimensionality, without modifying its topological properties. In practice, smooth manifolds are difficult to conciliate with the discrete nature of data. In constrast, graph structures have proven to be very useful, and tight connections between NLDR and graph embedding [23, 24, 25] exist.

Another usual hypothesis is to assume that data are sampled from clusters rather than from a manifold. Dimensionality reduction methods that emphasize clusters are often closely related to spectral clustering [26, 27, 28, 29]. In this domain, graphs are very handy as well, with useful tools such as the Laplacian matrix.

This chapter is organized as follows. Section 12.3 introduces the necessary notations and classical methods such as PCA and MDS. Section 12.4 details how graphs can help introducing nonlinearities in the DR methods. Next, Sections 12.5 and 12.6 review some of the major DR methods that perform manifold learning by using either distances (global methods) or similarities (local methods). Section 12.7 deals with graph embedding, another paradigm used in nonlinear dimensionality reduction. Finally, Section 12.8 compares the methods on a few examples, before Section 12.9 that draws the conclusions and sketches some perspectives for the near future.

12.3   Classical methods

Let $\mathbf{X} = [\mathbf{x}_i]_{1 \le i \le N}$ denotes a multivariate data set, with $\mathbf{x}_i \in \mathbb{R}^M$. Let $\mathbf{Y} = [\mathbf{y}_i]_{1 \le i \le N}$ denote its low-dimensional representation, with $\mathbf{y}_i \in \mathbb{R}^P$ and $P < M$.

Dimensionality reduction aims at finding a tranformation $\mathcal{T}$ from $\mathbb{R}^{M \times N}$ to $\mathbb{R}^{P \times N}$ that minimizes the reconstruction error defined by

$$E = \|\mathbf{X} - \mathcal{T}^{-1}(\mathbf{Y})\|_2^2 , \qquad (12.1)$$

where $\mathbf{Y} = \mathcal{T}(\mathbf{X})$ and $\|\mathbf{U}\|_2^2 = \mathrm{Tr}(\mathbf{U}^T\mathbf{U}) = \mathrm{Tr}(\mathbf{U}\mathbf{U}^T)$ is the Frobenius norm of $\mathbf{U}$. In other words, the encoding-decoding process resulting from the successive application of $\mathcal{T}$ and $\mathcal{T}^{-1}$ should produce minimal distortion.

12.3.1 Principal component analysis

The simplest option for $\mathcal{T}$ is obviously a linear transformation. In principal component analysis [30, 31, 32, 33, 1], this transformation can be written as $\mathcal{T}(\mathbf{X}) = \mathbf{V}_P^T(\mathbf{X} - \mathbf{u}\mathbf{1}^T)$, where $\mathbf{u}$ is an offset and $\mathbf{V}_P$ is an orthogonal matrix ($\mathbf{V}_P^T\mathbf{V}_P = \mathbf{I}$) with P columns. Orthogonality reduces the number of free parameters in $\mathbf{V}_P$ and provides the P-dimensional subspace of $\mathbb{R}^M$ with a basis of orthonormal vectors. With such a linear transformation, the reconstruction error can be rewritten as

$$E(\mathbf{u}, \mathbf{V}_P; \mathbf{X}) = \|(\mathbf{X} - \mathbf{u}\mathbf{1}^T) - \mathbf{V}_P\mathbf{V}_P^T(\mathbf{X} - \mathbf{u}\mathbf{1}^T)\|_2^2 \qquad (12.2)$$

$$= \|(\mathbf{I} - \mathbf{V}_P\mathbf{V}_P^T)(\mathbf{X} - \mathbf{u}\mathbf{1}^T)\|_2^2 \qquad (12.3)$$

$$= \mathrm{Tr}\left((\mathbf{X} - \mathbf{u}\mathbf{1}^T)^T(\mathbf{I} - \mathbf{V}_P\mathbf{V}_P^T)(\mathbf{I} - \mathbf{V}_P\mathbf{V}_P^T)(\mathbf{X} - \mathbf{u}\mathbf{1}^T)\right) \qquad (12.4)$$

$$= \mathrm{Tr}\left((\mathbf{X} - \mathbf{u}\mathbf{1}^T)^T(\mathbf{I} - \mathbf{V}_P\mathbf{V}_P^T)(\mathbf{X} - \mathbf{u}\mathbf{1}^T)\right) . \qquad (12.5)$$

In order to identify $\mathbf{u}$, we can compute the partial derivative of the reconstruction error with respect to $\mathbf{u}$ and equate it with zero. Defining $\mathbf{E} = \mathbf{X} - \mathbf{u}\mathbf{1}^T$, we have

$$\frac{\partial E(\mathbf{u}, \mathbf{V}_P; \mathbf{X})}{\partial \mathbf{u}} = \frac{\partial \,\mathrm{Tr}(\mathbf{E}^T(\mathbf{I} - \mathbf{V}_P\mathbf{V}_P^T)\mathbf{E})}{\partial \mathbf{E}} \frac{\partial \mathbf{E}^T}{\partial \mathbf{u}} = 2(\mathbf{I} - \mathbf{V}_P\mathbf{V}_P^T)\mathbf{E}\mathbf{1} = 0 \qquad (12.6)$$

Assuming that $\mathbf{I} - \mathbf{V}_P\mathbf{V}_P^T$ has full rank, we have $(\mathbf{X} - \mathbf{u}\mathbf{1}^T)\mathbf{1}^T = 0$ and therefore $\mathbf{u} = \mathbf{X}\mathbf{1}^T/N$. The optimal offset is thus the sample mean of $\mathbf{X}$. Knowing this, we can simplify the transformation into $\mathcal{T}(\mathbf{X}) = \mathbf{V}_P^T\mathbf{X}$ if we assume that $\mathbf{X}\mathbf{1} = 0$. The reconstruction error is then rewritten as

$$E(\mathbf{V}_P; \mathbf{X}) = \|(\mathbf{I} - \mathbf{V}_P\mathbf{V}_P^T)\mathbf{X}\|_2^2 = \mathrm{Tr}(\mathbf{X}\mathbf{X}^T) - \mathrm{Tr}(\mathbf{V}_P^T\mathbf{X}\mathbf{X}^T\mathbf{V}_P) , \qquad (12.7)$$

where the first term is constant. The second term can be minimized under the constraint $\mathbf{V}_P^T\mathbf{V}_P = \mathbf{I}$ using Lagrange's technique. The Lagrangian can be written as

$$L(\mathbf{V}_P, \mathbf{\Lambda}; \mathbf{X}) = \mathrm{Tr}(\mathbf{V}_P^T\mathbf{X}\mathbf{X}^T\mathbf{V}_P) + \mathrm{Tr}(\mathbf{\Lambda}(\mathbf{I} - \mathbf{V}_P^T\mathbf{V}_P)) , \qquad (12.8)$$

where $\mathbf{\Lambda}$ is the diagonal matrix containing Lagrange's multipliers. The partial derivative of $L(\mathbf{V}_P, \mathbf{\Lambda}; \mathbf{X})$ with respect to $\mathbf{V}_P$ is given by

$$\frac{\partial L(\mathbf{V}_P, \mathbf{\Lambda}; \mathbf{X})}{\partial \mathbf{V}_P} = 2\mathbf{X}\mathbf{X}^T\mathbf{V}_P - 2\mathbf{\Lambda}\mathbf{V}_P . \qquad (12.9)$$

After equating the partial derivative with zero and rearranging the terms, we obtain $\mathbf{\Lambda V}_P = \mathbf{XX}^T\mathbf{V}_P$, which turns out to be an eigenproblem. Without loss of generality and because $\mathbf{X}$ is centered, the product $\mathbf{XX}^T$ can be replaced with the sample covariance $\mathbf{C}(\mathbf{X}) = \mathbf{XX}^T/N$. The covariance matrix is symmetric and positive semidefinite by construction. Its eigenvalues are therefore larger than or equal to zero. If we write the eigenvalue decomposition as $\mathbf{C}(\mathbf{X}) = \mathbf{V\Lambda V}^T$, then the solution of the maximization problem for a given P is provided by the eigenvectors in $\mathbf{V}$ that are associated with the P largest eigenvalues. These eigenvectors give the columns of $\mathbf{V}_P = [\mathbf{v}_i]_{1\leq i\leq P}$. The covariance in the P-dimensional subspace is given by

$$\mathbf{C}(\mathbf{Y}) = \mathbf{YY}^T/N = \mathbf{V}_P^T\mathbf{XX}^T\mathbf{V}_P/N = \mathbf{V}_P^T\mathbf{C}(\mathbf{X})\mathbf{V}_P = \mathbf{V}_P^T\mathbf{V}_P\mathbf{\Lambda V}_P^T\mathbf{V}_P = \mathbf{\Lambda}_P \ ,$$
(12.10)

where $\mathbf{\Lambda}_P$ denotes the restriction of $\mathbf{\Lambda}$ to its first P rows and columns. As the covariance is diagonal in the subspace, this shows that PCA also decorrelates the data set. Matrix $\mathbf{V}_P$ is also the solution that minimizes $\|\mathbf{C}(\mathbf{X}) - \mathbf{V}_P\mathbf{\Lambda}_P\mathbf{V}_P^T\|_2^2$. This shows that minimal reconstruction error is equivalent to variance preservation. If $\mathbf{V}_P$ maximizes $\mathrm{Tr}(\mathbf{V}_P^T\mathbf{XX}^T\mathbf{V}_P)$, then it also trivially maximizes $\mathrm{Tr}(\mathbf{X}^T\mathbf{V}_P\mathbf{V}_P^T\mathbf{X})$.

12.3.2 Multidimensional scaling

In the previous section, PCA requires the data set to be available in the form of coordinates in $\mathbf{X}$. In contrast, classical metric multidimensional scaling [2, 3](CMMDS) starts from the Gram matrix, which is defined as $\mathbf{G}(\mathbf{X}) = \mathbf{X}^T\mathbf{X}$ and is positive semidefinite by construction. The aim and model of CMMDS are basically the same as those of PCA. The solution remains the same as well, but it is computed differently. To see this, let us use the singular value decomposition of $\mathbf{X}$. It can be written as $\mathbf{X} = \mathbf{VSU}^T$, where both $\mathbf{U}$ and $\mathbf{V}$ are orthogonal matrices and $\mathbf{S}$ contains the singular values on its diagonal in descending order of magnitude. The covariance matrix can be rewritten as $\mathbf{C}(\mathbf{X}) = \mathbf{XX}^T/N = \mathbf{VSU}^T\mathbf{US}^T\mathbf{V}^T/N = \mathbf{V}(\mathbf{SS}^T/N)\mathbf{V}^T$. The last expression is equivalent to the eigenvalue decomposition of the covariance matrix if we rename the product $\mathbf{SS}^T/N$ into $\mathbf{\Lambda}$. Similarly, the Gram matrix can be rewritten as $\mathbf{G}(\mathbf{X}) = \mathbf{X}^T\mathbf{X} = \mathbf{US}^T\mathbf{V}^T\mathbf{VSU}^T = \mathbf{U}(\mathbf{S}^T\mathbf{S})\mathbf{U}^T$. The last expression is equivalent to the eigenvalue decomposition of the Gram matrix and shows the tight connection with the eigenvalue decomposition of the covariance matrix. The coordinates in the P-dimensional subspace can be rewritten as $\mathbf{Y} = \mathbf{V}_P^T\mathbf{X} = \mathbf{V}_P\mathbf{VSU}^T = \mathbf{S}_P\mathbf{U}^T$, where $\mathbf{S}_P$ is the restriction of $\mathbf{S}$ to its P first columns. This shows that the coordinates are given by the P leading eigenvectors of the Gram matrix, scaled by the square root of their associated eigenvalue.

Until here, we have assumed that $\mathbf{X}$ is centered in the Gram matrix. If it is not, then we need to compute

$$\mathbf{G}(\mathbf{X}) = (\mathbf{X} - (\mathbf{X1}/N)\mathbf{1}^T)^T(\mathbf{X} - (\mathbf{X1}/N)\mathbf{1}^T) = (\mathbf{I} - \mathbf{11}^T/N)(\mathbf{X}^T\mathbf{X})(\mathbf{I} - \mathbf{11}^T/N) \ .$$
(12.11)

This shows that the Gram matrix can be centered as such, without explicitly knowing

$\mathbf{X}$. The centering matrix $\mathbf{I} - \mathbf{11}^T/N$ is a powerful tool that can also be used if the data set consists of pairwise squared Euclidean distances. In this case, we write

$$\mathbf{\Delta}^2(\mathbf{X}) = [\|\mathbf{x}_i - \mathbf{x}_j\|_2^2]_{1 \leq i,j \leq N} = \mathrm{diag}(\mathbf{G}(\mathbf{X}))\mathbf{1}^T - 2\mathbf{G}(\mathbf{X}) + \mathbf{1}\,\mathrm{diag}(\mathbf{G}(\mathbf{X}))^T \;, \tag{12.12}$$

where operator diag provides the column vector built from the diagonal entries of its argument. Double centering, namely, multiplying the matrix of squared distances by the centering matrix on both the left and right sides, yields

$$(\mathbf{I} - \mathbf{11}^T/N)\mathbf{\Delta}^2(\mathbf{X})(\mathbf{I} - \mathbf{11}^T/N) = -2(\mathbf{I} - \mathbf{11}^T/N)\mathbf{G}(\mathbf{X})(\mathbf{I} - \mathbf{11}^T/N) \;. \tag{12.13}$$

This results from

$$\mathrm{diag}(\mathbf{G}(\mathbf{X}))\mathbf{1}^T(\mathbf{I} - \mathbf{11}^T/N) = \mathrm{diag}(\mathbf{G}(\mathbf{X}))\mathbf{1}^T - \mathrm{diag}(\mathbf{G}(\mathbf{X}))\mathbf{1}^T\mathbf{11}^T/N \tag{12.14}$$

$$= \mathrm{diag}(\mathbf{G}(\mathbf{X}))\mathbf{1}^T - \mathrm{diag}(\mathbf{G}(\mathbf{X}))\mathbf{1}^T \tag{12.15}$$

$$= \mathbf{00}^T \tag{12.16}$$

and similarly $(\mathbf{I} - \mathbf{11}^T/N)\mathbf{1}\,\mathrm{diag}(\mathbf{G}(\mathbf{X}))^T = \mathbf{00}^T$. Classical metric MDS is thus a flexible method that can be applied to coordinates, pairwise dot products, or pairwise squared Euclidean distances. For a given P, it can be shown that $\mathbf{Y} = \mathbf{S}_P\mathbf{U}^T$ minimizes a cost function called STRAIN [34] and defined as

$$E(\mathbf{Y}; \mathbf{X}) = \|\mathbf{G}(\mathbf{X}) - \mathbf{G}(\mathbf{Y})\|_2^2 = \sum_{i,j} \left(\mathbf{x}_i^T\mathbf{x}_j - \mathbf{y}_i^T\mathbf{y}_j\right)^2 \;. \tag{12.17}$$

In other words, CMMDS tries to preserve dot products in the linear subspace.

12.3.3 Nonlinear MDS and Distance Preservation

Given the close relationship between dot products and squared Euclidean distances, one might extend the CMMDS principle to distance preservation. Unfortunately, distance preservation cannot be achieved with spectral decomposition such as in PCA and CMMDS. It requires more generic optimization tools such as gradient descent or ad hoc algorithms. As an advantage, these tools are quite flexible and allow for more freedom in the definition of the cost function to be minimized. This freedom also means that the P-dimensional coordinates are no longer constrained to a linear transformation of those in the M-dimensional space. For instance, one can consider the STRESS function [6] defined as $E(\mathbf{D}; \mathbf{\Delta}) = \|\mathbf{\Delta}(\mathbf{X}) - \mathbf{\Delta}(\mathbf{Y})\|_2^2$. If $\delta_{ij}(\mathbf{X})$ and $\delta_{ij}(\mathbf{Y})$ denote the pairwise distance in the M- and P-dimensional space, respectively, then a more general form of the STRESS is given by

$$E(\mathbf{Y}; \mathbf{X}) = \sum_{i,j} w_{ij}(\delta_{ij}(\mathbf{X}) - \delta_{ij}(\mathbf{X}))^2 \;, \tag{12.18}$$

where w_{ij} controls the weight of each difference of distances in the cost function. These weights can be chosen arbitrarily or can depend on δ_{ij} and/or d_{ij}. The emphasis put on the preservation of small distances can be reinforced by defining

$w_{ij} = 1/\delta_{ij}$, such as in Sammon's stress function, which is used in his nonlinear mapping (NLM) technique [8]. Another possibility is to define $w_{ij} = f(\delta_{ij})$, where f is a nonincreasing positive function. For instance, in curvilinear component analysis (CCA) [10], we have $w_{ij} = H(\lambda_i - \delta_{ij})$, where H is a step function and λ_i a width parameter.

Yet another cost function for MDS is the squared stress, often shortened as SSTRESS [7], whose generic definition is

$$E(\mathbf{Y}; \mathbf{X}) = \sum_{i,j} w_{ij}(\delta_{ij}^2(\mathbf{X}) - \delta_{ij}^2(\mathbf{Y}))^2 \ . \tag{12.19}$$

As distances are squared, the SSTRESS is actually closer to the STRAIN of CM-MDS.

Various optimization procedures can be used to minimize these cost functions. Gradient descent with a diagonal approximation of the Hessian is used in Sammon's NLM, whereas CCA relies on a stochastic gradient descent. Yet another technique consists of successive majorizations of the STRESS with convex functions whose minimum can be computed analytically [34].

12.4   Nonlinearity through Graphs

The projection of data onto linear subspaces, such as in the methods described in the previous section, might be insufficient in many cases. As an emblematic example, let us consider a manifold looking like the thin layer of jam in a Swiss roll cake (see Section 12.8 and Figure 12.1 for an illustration). This data set is distributed in a three-dimensional space on a surface shaped as a spiral. As a matter of fact, the underlying manifold is a two-dimensional rectangular subspace, which actually corresponds to a latent parameterization of the Swiss roll manifold. No linear projection is able to provide a satisfying representation of this latent space: there is always the risk that pieces of the Swiss roll will be superimposed in the representation. Intuitively, the solution to this problem would be to first unroll the manifold and to perform the projection afterwards. All NLDR methods implement this intuition in some way. They explore the idea sketched at the end of the previous section: local neighborhood relationships should be faithfully rendered in the low-dimensional representation, whereas more distant relationships are less important. Unrolling the Swiss roll puts this principle into practice: close points remain near each other, while the distances between nonneighbors is increased. Graphs provide an efficient way to encode neighborhood relationships. For instance, graphs can represent K-ary neighborhoods (the set of the K-nearest neighbors around each data point) or ϵ-ball neighborhoods (the set of all neighbors lying within a ball of radius ϵ centered on each data point). However, graphs also generate sparsity, which has to be compensated for. The variety of

NLDR methods reflects the different possibilities to intelligently fill the holes left by sparsity.

Three categories of methods can be distinguished:

- Some methods start with pairwise distances, keep those associated with local neighborhoods only, and replace the missing distances with values that facilitate manifold unfolding. Eventually, CMMDS or any of its nonlinear variants can be used to compute the low-dimensional representation. Typical methods in this category are Isomap [16] (and all other methods using graph distances [35, 36, 37, 38, 39]) and maximum variance unfolding [20].

- Some methods define pairwise similarities (or affinities, proximities, vicinities, i.e., any positive quantity that decreases with distance). Nonneighboring data items are generally given a null similarity, which leads to a sparse similarity matrix. Next, the sparse similarity matrix is converted into a distance matrix that can be processed by CMMDS. Typical methods in this category are Laplacian eigenmaps [40, 18] and locally linear embedding [17, 22].

- Some methods rely on what could be called graph placement or graph embedding [24, 25]. These are often ad hoc methods that are driven by mechanical concepts such as spring forces applied to masses. A graph can be used to represent the masses and the springs that connect them. Two subcategories exist. The graph can defined a priori in the low-dimensional space and placed in the high-dimensional one or the graph can be built in a data-dependent way in the high-dimensional and the placement achieved in the low-dimensional one. The self-organizing map [11, 41] belongs to the first subcategory, whereas force-directed placement, Isotop [42], and the exploratory observation machine (XOM) [43], are in the second one.

12.5   Graph-Based Distances

Let us imagine a manifold made of a curved sheet of paper, which is embedded in our physical three-dimensional world. The Swiss roll is an example of such manifold. Let us also consider the Euclidean distance between two relatively distant points of the manifold. Depending on the sheet curvature, the distance will vary, although the sheet keeps the same size (paper does not allow any stretching or shrinking). This shows that distance preservation makes little sense if the manifold to be embedded in a lower-dimensional space needs some unfolding or unrolling. This issue arises because Euclidean distances measure lengths along straight lines, whereas the manifold occupies a nonlinear subspace. The solution to this problem is obviously to compute distances as the ant crawls instead of as the crow flies. In other words, distances should be computed along the manifold or, more accurately, along manifold geodesic curves. In a smooth manifold, the geodesic curve between two points on the manifold is the smooth one-dimensional submanifold with the shortest length. The term geodesic distance refers to this length. In a manifold such as a sheet of paper, geodesic distances are invariant to curvature changes. Therefore, geodesic distances capture the internal structure of the manifold without influence from the way it is embedded.

As a matter of fact, geodesic distances cannot be evaluated if no analytical expression of the manifold is available. However, if at least some points of the manifold are known, for instance, through data set $\mathbf{X}$, then geodesic distances can be approximated by using a graph [44]. Each data vector $\mathbf{x}_i$ is associated with a graph vertex and either K-ary neighborhoods or ϵ-ball neighborhoods provide the edges. Such a graph yields a discrete representation of the manifold. Within this framework, the geodesic distance between $\mathbf{x}_i$ and $\mathbf{x}_j$ can be approximated by the length of the shortest path that connects the two corresponding vertices. Shortest paths and their lengths can be easily computed by Dijkstra's or Floyd's algorithms [45, 46].

At this point, we know that geodesic distances approximated by graph distances can characterize the internal structure of a manifold. But how can we force its unfolding in view of dimensionality reduction? The solution consists in trying to reproduce the graph distances measured in the high-dimensional space with Euclidean distances in the low-dimensional space. In this way, geodesic curves are matched with straight lines. In practice, several methods can perform this hybrid distance preservation. For instance, CMMDS can be applied to the matrix containing all squared pairwise shortest path lengths (instead of squared Euclidean distances). This method is known as Isomap [16]. Although CMMDS is purely linear, Isomap achieves a nonlinear embedding: the computation of the shortest path lengths can be thought of as being equivalent to applying a nonlinear transformation to the data. Nonlinear variants of CMMDS can be used as well, such as Sammon's nonlinear mapping [8] or curvilinear component analysis [10], for instance. This yields geodesic Sammon mapping [37, 38] and curvilinear distance analysis [35, 36].

Graph distances have three main drawbacks. First, let us consider a smooth manifold $\mathcal{M}$ that is isometric to some subset of a Euclidean space. Let us assume that $\mathbf{\Delta}_\mathcal{M}^2(\mathbf{X})$ contains the actual squared geodesic distances for some manifold points stored in $\mathbf{X}$. If we compute $-1/2(\mathbf{I} - \mathbf{1}\mathbf{1}^T/N)\mathbf{\Delta}_\mathcal{M}^2(\mathbf{X})(\mathbf{I} - \mathbf{1}\mathbf{1}^T/N)$, then we should end up with a valid Gram matrix, which is positive semidefinite. Unfortunately, this statement does not hold true if geodesic distances are not exact. This means that if $\mathbf{\Delta}_\mathcal{G}^2(\mathbf{X})$ contains shortest path lengths for some graph $\mathcal{G}$ instead of geodesic distances, the product $-1/2(\mathbf{I} - \mathbf{1}\mathbf{1}^T/N)\mathbf{\Delta}_\mathcal{G}^2(\mathbf{X})(\mathbf{I} - \mathbf{1}\mathbf{1}^T/N)$ will not necessarily be positive semidefinite. The spectral decomposition used in Isomap can thus yield negative eigenvalues. In practice, their magnitude is often negligible and workarounds are described in the literature [16, 47, 48]. This issue matters only for methods relying on spectral decomposition, while it is negligible for methods that rely on a nonlinear variant of CMMDS such as Sammon's NLM or CCA.

A second drawback arises with manifolds that are not isometric to some subset of a Euclidean space. A piece of a spherical surface, for instance, is not isometric to a piece of plane. In this case, distance preservation can only be imperfect. The weighting schemes used in the nonlinear variant of CMMDS can partly address this issue. Isometry is also lost in the case of a nonconvex manifold. For example, a sheet of paper with a hole is not isometric to a piece of plane: some geodesic distances are forced to circumvent the hole in the manifold. Again, a weighted distance preservation giving more importance to short distances can help.

The third shortcoming of graph distances is related to the graph construction. If the value of the parameters K or ϵ is not appropriately chosen, there might be too few or too many edges in the graph. Too few edges typically lead to an overestimation of the actual geodesic distances (paths are zigzagging). In some cases, the graph representation of the manifold can comprise disconnected parts, yielding to infinite distances. Too many edges can cause significant underestimation of some geodesic distances. This could happen if a short circuit edge accidentally connects two points that are not actual neighbors.

If we compare graph distances to the corresponding Euclidean ones, ones sees that the former are longer than or equal to the latter. We have seen that the use of these longer distances in CMMDS can be seen as a way to unfold the manifold. However, Dijkstra's and Floyd's algorithms compute shortest paths in a greedy way, without taking into account the objective of DR. This means that they build a matrix of pairwise distances, but there is no guarantee that this matrix is optimal for DR. For instance, in the case of Isomap (CMMDS with graph distances), there is no explicit optimization of the Gram matrix such that the eigenspectrum energy concentrates into a minimal number of eigenvalues.

This issue [48] is addressed in maximum variance unfolding (MVU) [20]. MVU starts with a neighborhood graph (K-nearest neighbors or ϵ-balls). Just as in Isomap, the idea to achieve manifold unfolding is to stretch distances between nonneighboring data points. For this purpose, one considers a matrix of squared pairwise distances $\mathbf{\Delta}^2(\mathbf{Y}) = [\delta_{ij}^2(\mathbf{Y})]_{1 \le i,j \le N}$ and the simple objective function to be maximized:

$$E(\mathbf{Y}) = \frac{1}{2N}\mathbf{1}^T\mathbf{\Delta}(\mathbf{Y})\mathbf{1} = \frac{1}{2N}\sum_{i,j}\delta_{ij}^2(\mathbf{Y}) \ , \qquad (12.20)$$

subject to $\delta_{ij}^2(\mathbf{Y}) = \delta_{ij}^2(\mathbf{X}) = \|\mathbf{x}_i - \mathbf{x}_j\|_2^2$ if $\mathbf{x}_i \sim \mathbf{x}_j$. Additional constraints stem from the assumption that $\mathbf{\Delta}(\mathbf{Y})$ contains the pairwise Euclidean distances for some P-dimensional representation $\mathbf{Y}$ of data set $\mathbf{X}$. In other words, $\mathbf{\Delta}^2(\mathbf{Y})$ must satisfy the equality $\mathbf{\Delta}^2(\mathbf{Y}) = \text{diag}(\mathbf{G}(\mathbf{Y}))\mathbf{1}^T - 2\mathbf{G}(\mathbf{Y}) + \mathbf{1}\,\text{diag}(\mathbf{GY})^T$, where $\mathbf{GY} = \mathbf{X}^T\mathbf{X}$ is the Gram matrix for some $\mathbf{Y}$. Without loss of generality, one can assume that the mean of $\mathbf{Y}$ is null, i.e., $\mathbf{X}\mathbf{1} = \mathbf{0}$ and $\mathbf{1}^T\mathbf{G}(\mathbf{Y})\mathbf{1} = \mathbf{1}^T\mathbf{Y}^T\mathbf{Y}\mathbf{1} = 0$. This helps to reformulate the problem with only the Gram matrix. Indeed, we can write

$$E(\mathbf{Y}) = \frac{1}{2N}\mathbf{1}^T\mathbf{\Delta}^2(\mathbf{Y})\mathbf{1} \tag{12.21}$$

$$= \frac{1}{2N}\mathbf{1}^T(\mathrm{diag}(\mathbf{G}(\mathbf{Y}))\mathbf{1}^T - 2\mathbf{G}(\mathbf{Y}) + \mathbf{1}\,\mathrm{diag}(\mathbf{G}(\mathbf{Y}))^T)\mathbf{1} \tag{12.22}$$

$$= \frac{1}{2N}\mathbf{1}^T(\mathrm{diag}(\mathbf{G}(\mathbf{Y}))\mathbf{1}^T + \mathbf{1}\,\mathrm{diag}(\mathbf{G}(\mathbf{Y}))^T)\mathbf{1} = \mathrm{Tr}(\mathbf{G}(\mathbf{Y})) \; . \tag{12.23}$$

In order for CMMDS to be applicable to matrix $\mathbf{G}(\mathbf{Y})$, the following constraints must be satisfied:

- To be a Gram matrix, $\mathbf{G}(\mathbf{Y})$ must be positive semidefinite.

- If $\mathbf{G}(\mathbf{Y})$ factorizes into centered coordinates, then the product $\mathbf{1}^T\mathbf{G}(\mathbf{Y})\mathbf{1}$ is equal to 0.

- For all neighbors $\mathbf{x}_i \sim \mathbf{x}_j$, we must have $g_{ii}(\mathbf{Y})-2g_{ij}(\mathbf{Y})+g_{jj}(\mathbf{Y}) = \|\mathbf{x}_i-\mathbf{x}_j\|_2^2$.

Provided the neighborhood graph is fully connected, the last set of constraints also controls the scale of the embedding $\mathbf{Y}$. Such a constrained optimization problem involving a positive semidefinite matrix can be solved using semidefinite programming [49]. Once the optimal $\mathbf{G}(\mathbf{Y})$ is determined, an eigenvalue decomposition such as in CMMDS can factorize it in order to identify $\mathbf{Y}$.

Both Isomap and MVU can be decomposed in a two-step optimization procedure, where the second step is CMMDS applied to a modified distance matrix. The difference resides in the first step of each method. In Isomap, Dijkstra's or Floyd's algorithm minimizes the length of vertex-to-vertex paths. While this objective proves to be useful to unfold a manifold, it is not directly related to DR. In contrast, MVU implements a similar idea (distances must be stretched) in a more principled way, with an objective function that takes into account the needs and constraints of the subsequent CMMDS step.

12.6 Graph-Based Similarities

The previous section has shown that graphs provide a handy framework to build dissimilarities that are more relevant than the Euclidean distances when it comes to nonlinear manifold unfolding. However, graphs can also formalize similarities, which are usually defined as decreasing positive functions of the corresponding distances. Similarities provide, therefore, a very natural way to emphasize the local structure of data, such as neighborhoods around each data point. In contrast, the global structure and large distances are associated with small similarity values. If the latter are neglected, the matrix of pairwise similarities becomes sparse and a graph can efficiently represent it.

12.6.1   Laplacian Eigenmaps

In Laplacian eigenmaps [40, 18], the idea to achieve manifold unfolding is the dual of that of MVU. Instead of stretching the distances between nonneighboring data points, the distances between neighbors are shrunk in the embedding space. For this purpose, the edges of the neighborhood graph are annotated with similarities; nonneighboring points have null similarities. The simplest similarity definition is

$$w_{ij} = \begin{cases} 1 & \text{if } \mathbf{x}_i \sim \mathbf{x}_j \\ 0 & \text{otherwise} \end{cases} . \tag{12.24}$$

As the neighborhood graph is undirected, matrix $\mathbf{W} = [w_{ij}]_{1 \le i,j \le N}$ is symmetric in addition to be sparse. The minimization of the distances between neighbors can be achieved with the cost function

$$E(\mathbf{Y}; \mathbf{W}) = \frac{1}{2} \sum_{i,j} w_{ij} \|\mathbf{y}_i - \mathbf{y}_j\|_2^2 \tag{12.25}$$

$$= \frac{1}{2} \sum_{i,j} w_{ij} (\mathbf{y}_i^T \mathbf{y}_i + \mathbf{y}_j^T \mathbf{y}_j - 2\mathbf{y}_i^T \mathbf{y}_j) \tag{12.26}$$

$$= \sum_i \mathbf{y}_i^T \left(\sum_j w_{ij} \right) \mathbf{y}_i - \sum_{i,j} \mathbf{y}_i^T w_{ij} \mathbf{y}_j \tag{12.27}$$

$$= \text{Tr}(\mathbf{Y}^T \mathbf{D} \mathbf{Y}) - \text{Tr}(\mathbf{Y}^T \mathbf{W} \mathbf{Y}) = \text{Tr}(\mathbf{Y} \mathbf{L} \mathbf{Y}^T) , \tag{12.28}$$

where $\mathbf{D}$ is a diagonal matrix with $d_{ii} = \sum_j w_{ji} = \sum_j w_{ij}$ and $\mathbf{L} = \mathbf{D} - \mathbf{W}$ is the unnormalized Laplacian matrix of the graph whose edges are weighted with $\mathbf{W}$.

The minimization of $E(\mathbf{Y}; \mathbf{W})$ admits the trivial solution $\mathbf{Y} = \mathbf{0}\mathbf{0}^T$. In order to avoid this, we impose the scale constraint $\mathbf{Y}\mathbf{D}\mathbf{Y}^T = \mathbf{I}$. This leads to the Lagrangian

$$L(\mathbf{Y}, \mathbf{\Lambda}; \mathbf{W}) = \text{Tr}(\mathbf{Y}\mathbf{L}\mathbf{Y}^T) - \text{Tr}(\mathbf{\Lambda}(\mathbf{Y}\mathbf{D}\mathbf{Y}^T - \mathbf{I})) = \text{Tr}(\mathbf{Y}\mathbf{L}\mathbf{Y}^T) - \text{Tr}(\mathbf{Y}^T \mathbf{\Lambda}\mathbf{D}\mathbf{Y}) + \text{Tr}(\mathbf{\Lambda}) , \tag{12.29}$$

The partial derivative with respect to $\mathbf{Y}$ is

$$\frac{\partial L(\mathbf{Y}, \mathbf{\Lambda}; \mathbf{W})}{\partial \mathbf{Y}} = 2\mathbf{Y}\mathbf{L} - 2\mathbf{Y}\mathbf{\Lambda}\mathbf{D} . \tag{12.30}$$

Equating the partial derivative to zero and rearranging the terms leads to $\mathbf{L}\mathbf{Y}^T = \mathbf{\Lambda}\mathbf{D}\mathbf{Y}^T$, which is a generalized eigenvalue problem. As $\mathbf{D}$ is diagonal, it is equivalent to $\tilde{\mathbf{L}}\mathbf{Y}^T = \mathbf{\Lambda}\mathbf{Y}^T$, where $\tilde{\mathbf{L}} = \mathbf{D}^{-1/2}\mathbf{L}\mathbf{D}^{-1/2} = \mathbf{I} - \mathbf{D}^{-1/2}\mathbf{W}\mathbf{D}^{-1/2}$ is the normalized Laplacian. Both the unnormalized and normalized Laplacian matrices are positive semidefinite. Notice that they are also singular since by construction $\mathbf{L}\mathbf{1} = \mathbf{D}\mathbf{1} - \mathbf{W}\mathbf{1} = \mathbf{0}$. The multiplicity of the zero eigenvalue is given by the number of connected components in the graph. As we look for a value of $\mathbf{Y}$ with P rows that minimizes $E(\mathbf{Y}; \mathbf{W})$, the solution is provided by the trailing P eigenvectors of $\mathbf{D}^{-1/2}\mathbf{L}\mathbf{D}^{-1/2}$, namely, those associated with the eigenvalues having the lowest nonzero magnitude.

Eventually, if $\tilde{\mathbf{L}} = \mathbf{U\Lambda U}^T$, then $\mathbf{X} = \mathbf{U}_P^T$, where $\mathbf{U}_P$ denotes the restriction of $\mathbf{U}$ to its first P columns.

The embedding found by Laplacian eigenmaps corresponds to a multivariate extension of the solution to a min-cut problem in a graph [50]. There are several variants to Laplacian eigenmaps, depending on the way the graph edges are weighted. Moreover, the unnormalized Laplacian can replace the normalized one. This amounts to changing the scale constraint to $\mathbf{YY}^T = \mathbf{I}$. Let us also consider yet another scaling possibility, namely, $\mathbf{Y} = \mathbf{\Omega}_P^{1/2} \mathbf{U}_P^T$, where $\mathbf{\Omega}_P$ is a diagonal matrix composed of the inverse of the nonzero smallest P eigenvalues λ_i^{-1} in descending order. The corresponding Gram matrix is given by $\mathbf{U}_P \mathbf{\Omega} \mathbf{U}_P^T$. If we assume that the graph consists of a single connected component, then the zero eigenvalue has multiplicity one and $\tilde{\mathbf{L}}^+ = \mathbf{U}_{N-1} \mathbf{\Omega} \mathbf{U}_{N-1}^T$ is the Moore-Penrose pseudo-inverse of $\tilde{\mathbf{L}}$. This means that Laplacian eigenmaps is equivalent to CMMDS applied to a Gram matrix that is the pseudo-inverse of the normalized Laplacian. The computation of the Laplacian and its inversion are equivalent to a nonlinear transformation applied to the data set $\mathbf{X}$ and denoted by $\mathbf{Z} = \phi(\mathbf{X})$. The Gram matrix in this space is given by $\mathbf{G}(\mathbf{Z}) = \tilde{\mathbf{L}}^+$, and the corresponding Euclidean distances are $\mathbf{1} \operatorname{diag}(\mathbf{G}(\mathbf{Z}))^T - 2\mathbf{G}(\mathbf{Z}) + \operatorname{diag}(\mathbf{G}(\mathbf{Z}))\mathbf{1}^T$. These distances are referred to as *commute time distances* [27, 51]. They are closely related to diffusion distances [28] and also to the length (or duration) of random walks in a neighborhood graph whose edges are weighted with transition probabilities. An analogy with electrical networks is possible as well [52], with graph edges being resistances and the commute time distances being the globlal effective resistance between two given vertices through the whole network. This analogy allows establishing a formal relationship between distances (resistances) and similarities that are inversely proportional to distances (conductances); it also highlights the duality between DR methods relying on a dense matrix of distances and those involving a sparse matrix of similarities.

From a more general point of view, the notation $\mathbf{G}(\mathbf{Z})$ used above with $\mathbf{Z} = \phi(\mathbf{X})$ translates the idea that CMMDS can be applied to nonlinearly transformed coordinates. Most of the time, this transformation remains implicit. For instance, commute time distances in Laplacian eigenmaps or geodesic distances in Isomap induce a (hopefully useful) transformation that promotes CMMDS from a linear DR method to a nonlinear one, while keeping many advantages, such as a convex optimization. Maximum variance unfolding goes a step further and actually customizes tranformation ϕ. Along with locally linear embedding described in the next section, all these spectral methods can be cast within the framework of kernel PCA [15]. Kernel PCA is the ancestor of all spectral methods and relies on key properties of Mercer's kernels. Such a kernel is a smooth function $\kappa(\mathbf{x}_i, \mathbf{x}_j)$ that is symmetric with respect to its arguments and induces a scalar product in a so-called feature space $\mathcal{F}$. In other words, we have $\kappa(\mathbf{x}_i, \mathbf{x}_j) = \langle \phi(\mathbf{x}_i), \phi(\mathbf{x}_j) \rangle_{\mathcal{F}}$, where $\phi : \mathbb{R}^M \to \mathcal{F}, \mathbf{x} \mapsto \mathbf{z} = \phi(\mathbf{x})$. Taken the other way round, this property allows ϕ to be induced from a given Mercer kernel κ. In particular, one can build a matrix $[\kappa(\mathbf{x}_i, \mathbf{x}_j)]_{1 \leq i,j \leq N}$ with the guarantee that it is a Gram matrix in some feature space $\mathcal{F}$. Sample mean removal in $\mathcal{F}$ can be performed indirectly as well, just by pre- and postmultiplying the Gram matrix by centering matrix $I - \mathbf{1}\mathbf{1}^T/N$. Somewhat of a misnomer, kernel PCA actually applies CMMDS to this kernelized Gram matrix in order to identify $\mathbf{Z}$ and subsequently a linear P-dimensional projection $\mathbf{Y}$ of $\mathbf{Z}$ with maximum variance. The pioneering work about kernel PCA [15] establishes the theoretical framework of spectral NLDR but provides no clue as to which Mercer kernel performs the best in practice.

12.6.2 Locally linear embedding

The idea behind locally linear embedding [17] (LLE) is to represent each data point as a regularized linear mixture of its neighbors. For points sampled from a smooth manifold, the resulting linear coefficients can be assumed to depend only upon local proximity relationships. Topological operations applied to the manifold, such as unfolding and flattening, have thus a minor impact on these coefficients. Therefore, the same coefficients could be reused to determine a new data embedding in a lower-dimensional space.

In practice, LLE relies on the availability of a neighborhood graph, built with K-ary neighborhoods or ϵ-balls. The first step of LLE is to identify the reconstruction coefficients in the high-dimensional data space. For this purpose, LLE uses a first cost function defined as

$$E(\mathbf{W}; \mathbf{X}) = \frac{1}{2}\sum_i \left\| \mathbf{x}_i - \sum_j w_{ij}\mathbf{x}_j \right\|_2^2 = \frac{1}{2}\sum_i \left\| \sum_j w_{ij}(\mathbf{x}_i - \mathbf{x}_j) \right\|_2^2 , \quad (12.31)$$

where $\mathbf{W}$ is subject to the following constraints: $\mathbf{W1} = \mathbf{1}$, $w_{ii} = 0$, and $w_{ij} = 0$ if and only if $\mathbf{x}_i$ and $\mathbf{x}_j$ are not neighbors. Each row of $\mathbf{W}$ can be identified separately. Let $\mathbf{G}_i$ denote the local Gram-like matrix involving all neighbors of $\mathbf{x}_i$. It can be written as $\mathbf{G}_i = [(\mathbf{x}_k - \mathbf{x}_i)^T(\mathbf{x}_l - \mathbf{x}_i)]_{k,l}$, where $\mathbf{x}_k \sim \mathbf{x}_i$ and $\mathbf{x}_l \sim \mathbf{x}_i$. If vector $\mathbf{w}_i$ contains the nonzero entries of the ith row of $\mathbf{W}$, then we have $\mathbf{w}_i = \min_{\mathbf{w}} \mathbf{w}^T\mathbf{G}_i\mathbf{w}$ with $\mathbf{w}^T\mathbf{1} = 1$. The solution is given by

$$\mathbf{w}_i = \frac{\mathbf{G}_i\mathbf{1}}{\mathbf{1}^T\mathbf{G}_i\mathbf{1}} . \quad (12.32)$$

In order to avoid trivial solutions when the rank of $\mathbf{G}_i$ is lower than the number of neighbors K, it is advised in [17, 22, 53] to replace $\mathbf{G}_i$ with $\mathbf{G}_i + (\Delta^2\,\mathrm{Tr}\,/K)(\mathbf{G}_i)\mathbf{I}$ with $\Delta = 0.1$. This regularization scheme prevents trivial solutions, such as $w_{ij} = 0$ if $\mathbf{x}_i \sim \mathbf{x}_j$. To some extent, $\mathbf{W}$ can be interpreted as a sparse similarity matrix.

Once the reconstruction coefficients are known, a second cost function can be defined, where $\mathbf{W}$ is fixed and coordinates are the unknown:

$$E(\mathbf{Y};\mathbf{W}) = \sum_i \left\| \mathbf{y}_i - \sum_j w_{ij}\mathbf{y}_j \right\|_2^2 \tag{12.33}$$

$$= \sum_i \mathbf{y}_i^T\mathbf{y}_i - 2\sum_i \mathbf{y}_i^T\left(\sum_j w_{ij}\mathbf{y}_j\right) + \sum_{i,j} w_{ij}\mathbf{y}_i^T\mathbf{y}_i w_{ij} \tag{12.34}$$

$$= \mathrm{Tr}(\mathbf{Y}^T\mathbf{Y}) - 2\,\mathrm{Tr}(\mathbf{Y}^T\mathbf{Y}\mathbf{W}^T) + \mathrm{Tr}(\mathbf{W}\mathbf{Y}^T\mathbf{Y}\mathbf{W}^T) \tag{12.35}$$

$$= \mathrm{Tr}(\mathbf{Y}\mathbf{I}\mathbf{Y}^T) - 2\,\mathrm{Tr}(\mathbf{Y}\mathbf{I}\mathbf{W}\mathbf{Y}^T) + \mathrm{Tr}(\mathbf{Y}\mathbf{W}^T\mathbf{W}\mathbf{Y}^T) \tag{12.36}$$

$$= \mathrm{Tr}(\mathbf{Y}(\mathbf{I}-\mathbf{W})^T(\mathbf{I}-\mathbf{W})\mathbf{Y}^T) \ . \tag{12.37}$$

The last equality shows that LLE is similar to Laplacian eigenmaps. The product $\mathbf{M} = (\mathbf{I}-\mathbf{W})^T(\mathbf{I}-\mathbf{W}) = \mathbf{I} - (\mathbf{W}^T + \mathbf{W} - \mathbf{W}^T\mathbf{W})$ is symmetric and positive semidefinite by construction. This product has the same structure as a normalized Laplacian matrix, with the diagonal elements of the first term being equal to the sum of the rows (or columns) of the second term. For the first term, we have $\mathrm{diag}\,\mathbf{I} = \mathbf{1}$, whereas the second term leads to $(\mathbf{W}^T+\mathbf{W}-\mathbf{W}^T\mathbf{W})\mathbf{1} = \mathbf{W}^T\mathbf{1}+\mathbf{1}-\mathbf{W}^T\mathbf{1}$, knowing that $\mathbf{W}\mathbf{1} = \mathbf{1}$. With respect to $\mathbf{W}$, which is generally not symmetric, the product in $\mathbf{M}$ can be seen as a kind of squared Laplacian.

The minimization of LLE's second objective function can be achieved with a spectral decomposition of $\mathbf{M} = \mathbf{U}\mathbf{\Lambda}\mathbf{U}^T$. The matrix $\mathbf{M}$ is singular, as $\mathbf{M}\mathbf{1} = (\mathbf{I}-\mathbf{W})^T(\mathbf{I}-\mathbf{W})\mathbf{1} = \mathbf{0} = 0\mathbf{1}$. This shows that $\mathbf{1}$ is an eigenvector of $\mathbf{M}$ with zero as associated eigenvalue. As in Laplacian eigenmaps, the multiplicity of the null eigenvalue is given by the number of connected components in the neighborhood graph. The solution of the minimization is formed by the transpose of the trailing P eigenvectors in $\mathbf{U}$, namely, those associated with the smallest strictly positive eigenvalues in $\mathbf{\Lambda}$.

As with Laplacian eigenmaps, the solution can be rescaled in various ways. For instance, one might consider $\mathbf{Y} = \mathbf{\Lambda}_P^{-1/2}\mathbf{U}^T$, where $\mathbf{\Lambda}_P$ is a diagonal matrix made of the smallest nonzero P eigenvalues of $\mathbf{\Lambda}$ in ascending order. This particular solution can be cast within the framework of CMMDS and corresponds to a Gram matrix equal to the Moore-Penrose pseudo-inverse of $\mathbf{M}$.

12.7 Graph embedding

This section deals with more heuristic ways to use graphs in dimensionality reduction. The described methods are of two kinds. In the first category, a predefined graph with a fixed planar representation is fitted in the high-dimensional data space. The self-organizing maps are the most widely known example of this kind. The second category works in the opposite direction. A graph is build in the high-dimensional space, according to the data distribution; next, this graph is embedded in a low-dimensional visualization space.

12.7.1   From LD to HD: Self-Organizing Maps

A self-organizing map (SOM) [41] can be seen as a nonlinear generalization of PCA. As detailed in Section 12.3.1, the objective of PCA is to find a linear subspace that minimizes the reconstruction error. Let us assume that this linear subspace is a two-dimensional plane. Intuitively, PCA places this plane amidst the cloud of data points, by minimizing the distances between the points and their projection on the plane. In order to extend PCA to nonlinear subspaces, one could replace the plane by some manifold. The SOM implements this idea by substituting the continuous plane with a discretized representation such as an articulated grid.

Let us assume that a grid is defined by an undirected graph $\mathcal{G} = (\mathcal{V}, \mathcal{E})$. Each vertex v_i is equipped with coordinates $\boldsymbol{\xi}_k$ in the M-dimensional data space and $\boldsymbol{\gamma}_k$ in the P-dimensional grid space. Each edge e_{kl} is weighted with a positive number w_{kl} that indicates the distance between v_k and v_l. If no edge connects v_k and v_l, then the distance is infinite. In order to obtain a simple and readable visualization of data, coordinates $\boldsymbol{\gamma}_k$ are often chosen in such a way that the grid nodes are regularly placed within a rectangle. Direct neighbors of v_k are typically located on a square or a hexagon (honeycomb configuration).

In PCA, the objective function measures the distortion between the genuine data points and their projection on the linear subspace. In a SOM, data points are projected onto the closest grid node. Hence, several data points can be associated with the same grid node, which plays a similar role as a centroid in a vector quantization [54] technique or in K-means-like algorithms [55]. For each data point $\mathbf{x}_i$, the index of the closest grid node is given by $\ell = \arg\min_k \|\mathbf{x}_i - \boldsymbol{\xi}_k\|_2$. Just as in a K-means algorithm, the SOM tries to minimize the distortion between the data and the grid nodes. In the SOM, however, an additional objective must be reached concurrently: grid neighbors must remain as close as possible in the data space. These two concurrent objectives are difficult to formalize into a simple cost function, although some attempts exist [56]. For this reason, an iterative and heuristic procedure is typically used to update the coordinates in $\boldsymbol{\Xi} = [\boldsymbol{\xi}_k]_{1 \leq k \leq Q}$. The simplest procedure is inspired by biological considerations: the SOM works like a neural network that progressively learns a set of patterns. In practice, the data set $\mathbf{X}$ contains the patterns, which are each presented several times to the SOM in random order. Let us assume that $\mathbf{x}_i$ is considered. The first step is to determine the closest grid node $\boldsymbol{\xi}_\ell$ as described above. Next, all grid nodes will be updated according to

$$\boldsymbol{\xi}_k \leftarrow \boldsymbol{\xi}_k + \alpha f(w_{k\ell}/\sigma)(\mathbf{x}_i - \boldsymbol{\xi}_k) \ , \tag{12.38}$$

where α is a learning rate (or step size), f is a positive decreasing function of its argument, and σ is a sort of bandwidth (or neighborhood radius). The learning rate can be decreased either after each presentation of a data vector or only between two complete sweeps of the whole data set. Function f can be a flipped step function or a decreasing exponential. Parameter σ controls the grid elasticity. If σ is large, $f(w_{k\ell}/\sigma)$ is maximal for $k = \ell$ and will not significantly decrease for close neighbors, which will thus closely follow the movement of $\boldsymbol{\xi}_\ell$. On the contrary, if σ is low, $\boldsymbol{\xi}_\ell$ has less influence on its neighbors. If the SOM is to be compared to an assembly of springs and masses, then the lower σ is, the heavier the masses are and the weaker the springs are.

Self-organizing maps are still widely used in various fields such as exploratory data analysis and data visualization. There are many variants of the basic algorithm described above and many ways to display [57] and assess its outcome [58, 59, 60, 61, 62]. There also probabilistic variants such as the generative topographic mapping [63] that relies on a Bayesian approach and maximizes a log-likelihood by an iterative expectation-maximization procedure. The neural gas algorithm [64, 65] works in a similar was as the SOM, although it has no predefined graph structure; hence, the neural gas does not provide a low-dimensional representation.

12.7.2   From HD to LD: Isotop

An SOM, as it is described in the previous section, has several shortcomings. First, it relies on centroids (the grid nodes) like a K-means algorithm. This means that the SOM does not really provide a visualization of each data point. Second, the SOM embeds a predefined graph in the data space, whereas one usually expects a DR method to work the other way round by embedding the data set in a low-dimensional space. There are actually many methods that address those two limitations. Most of these methods are loosely related to the fields of graph embedding [23] and graph drawing [24, 25, 66]. Their general principle is first to build a neighborhood graph in the high-dimensional data space and next to embed this graph in a lower-dimensional visualization space using graph layout algorithms. Most of these algorithms are heuristic and find their inspiration in mechanics. For instance, in a force-based layout, the graph vertices are masses and edges are springs connecting them. The final layout or embedding then results from a free energy minimization that equilibrates the mass-spring system. As an illustrative example, we describe hereafter Isotop [67, 42, 4], a method that is very close to a SOM.

Let us assume that data set $\mathbf{X}$ is a sample drawn from some unknown distribution in a high-dimensional space. In order to obtain a lower-dimensional representation $\mathbf{Y}$ of $\mathbf{X}$, let us further assume that the support of this distribution is a manifold that can be mapped into a low-dimensional space. Starting from this pair of distributions, we can consider a new point with coordinates $\mathbf{x}$ and $\mathbf{y}$ in the high- and low-dimensional spaces, respectively. If we want this point to have the same neighbors in both spaces, we can define the cost function of Isotop as

$$E(\mathbf{Y}; \mathbf{X}, \rho, \sigma) = \mathrm{E}_\mathbf{y} \left[\sum_{i=1}^{N} f\left(\frac{\delta(\mathbf{x}_i, \mathbf{x})}{\rho}\right) \left(1 - \exp\left(-\frac{\|\mathbf{y}_i - \mathbf{y}\|_2^2}{2\sigma^2}\right)\right) \sigma^2 \right] ,$$

(12.39)

where $\mathrm{E}_\mathbf{y}$ is the expectation with respect to $\mathbf{y}$, $\delta(\mathbf{x}_i, \mathbf{x})$ is a distance between $\mathbf{x}_i$ and $\mathbf{x}$, and f is a monotically decreasing function. Parameters ρ and σ are bandwidths that determine the neighborhood radii in each space. The scaled upside-down Gaussian bell $\sigma^2(1 - \exp(-u^2/\sigma^2))$ introduces adjustable saturation into the quadratic factor u^2. The gradient of the cost function can be written as

$$\frac{\partial E(\mathbf{Y}; \mathbf{X}, \rho, \sigma)}{\partial \mathbf{y}_i} = \mathrm{E}_\mathbf{y} \left[f\left(\frac{\delta(\mathbf{x}_i, \mathbf{x})}{\rho}\right) \exp\left(-\frac{\|\mathbf{y}_i - \mathbf{y}\|_2^2}{2\sigma^2}\right) (\mathbf{y}_i - \mathbf{y}) \right] . \quad (12.40)$$

If we adopt a stochastic gradient descent, we can drop the expectation and write the update

$$\mathbf{y}_i \leftarrow \mathbf{y}_i - \alpha f\left(\frac{\delta(\mathbf{x}_i, \mathbf{x})}{\rho}\right) \exp\left(-\frac{\|\mathbf{y}_i - \mathbf{y}\|_2^2}{2\sigma^2}\right) (\mathbf{y}_i - \mathbf{y}) , \quad (12.41)$$

where α is a slowly decreasing step size. In practice, we still need to know the distribution of $\mathbf{y}$ and the value of $\mathbf{x}$ for a given $\mathbf{y}$. We can approximate the distribution with a mixture of unit variance Gaussians, namely,

$$p(\mathbf{y}) = \frac{1}{N} \sum_{i=1}^{N} \frac{1}{\sqrt{2\pi}} \exp(-\|\mathbf{y} - \mathbf{y}_i\|_2^2/2) , \quad (12.42)$$

and draw randomly generated instances of $\mathbf{y}$ from it. The distance between $\mathbf{x}_i$ and $\mathbf{x}$ can be approximated by $\delta(\mathbf{x}_i, \mathbf{x}) \approx \delta(\mathbf{x}_i, \mathbf{x}_j) = \delta_{ij}(\mathbf{X})$, where $j = \arg\min_i \|\mathbf{y} - \mathbf{y}_i\|_2^2$. All distances $\delta_{ij}(\mathbf{X})$ in the high-dimensional space are known and can be Euclidean or correspond to the shortest paths in a neighborhood graph, like in methods described in previous sections. In the case of shortest paths, the difference with a SOM becomes obvious: the graph is established in the high-dimensional data space in a data-driven way, whereas the user arbitrarily defines it in the low-dimensional visualization space. Apart from this key difference, Isotop and a SOM share many common features. In both methods, coordinates are updated in an iterative way, and the amplitude of the update is modulated by a factor that depends on distances in a graph.

As most graph layout techniques are based on heuristic approaches, there are of course many possible variants, such as the exploratory observation machine (XOM) [43].

12.8 Examples and comparisons

12.8.1 Quality Assessment

Dimensionality reduction aims at providing a faithful representation of data. The algorithms described in this chapter, among others, provide ways to obtain such a representation. However, it is not obvious for the user to decide which method best suits the problem. An objective evaluation of the data representation quality is necessary. Nevertheless, using the cost function of a specific method as quality criterion obviously biases any comparison result toward the chosen specific method. There is thus a need for criteria that are as independent as possible from all methods.

Intuitively, a good representation should preserve neighborhood relationships: close points in the data space should remain near each other in the representation, while distant points should remain far from each other in the representation. This idea could be implemented very simply with a distance preservation criterion, such as Sammon's stress [8]. However, distance criteria prove to be too strict, since neighborhood relationships can be preserved even if distances are stretched or shrunk. For this reason, modern quality criteria [68, 69, 70, 71, 72] involve sorted distances and ranked neighbors.

Let us define the rank of $\mathbf{x}_j$ with respect to $\mathbf{x}_i$ in the high-dimensional space as $r_{ij}(\mathbf{X}) = |\{k : \delta_{ik}(\mathbf{X}) < \delta_{ij}(\mathbf{X})$ or $(\delta_{ik}(\mathbf{X}) = \delta_{ij}(\mathbf{X})$ and $1 \leq k < j \leq N)\}|$, where $|\mathcal{A}|$ denotes the cardinality of set $\mathcal{A}$. Intuitively, the rank counts the number of points that are closer neighbors of $\mathbf{x}_i$ than $\mathbf{x}_j$, including $\mathbf{x}_i$ itself. The ranking is established according to some given distances (Euclidean in the following experiments), and ties are circumvented by switching to point indices. Similarly, the rank of $\mathbf{y}_j$ with respect to $\mathbf{y}_i$ in the low-dimensional space is $r_{ij}(\mathbf{Y}) = |\{k : \delta_{ik}(\mathbf{Y}) < \delta_{ij}(\mathbf{Y})$ or $(\delta_{ik}(\mathbf{Y}) = \delta_{ij}(\mathbf{Y})$ and $1 \leq k < j \leq N)\}|$. Hence, reflexive ranks are set to zero $(r_{ii}(\mathbf{X}) = r_{ii}(\mathbf{Y}) = 0)$ and ranks are unique, i.e., there are no *ex aequo* ranks: $r_{ij}(\mathbf{X}) \neq r_{ik}(\mathbf{X})$ for $k \neq j$, even if $\delta_{ij}(\mathbf{X}) = \delta_{ik}(\mathbf{X})$. This means that nonreflexive ranks belong to $\{1, \ldots, N - 1\}$. The nonreflexive K-ary neighborhoods of $\mathbf{x}_i$ and $\mathbf{y}_i$ are denoted by $\mathcal{N}_i^K(\mathbf{X}) = \{j : 1 \leq r_{ij}(\mathbf{X}) \leq K\}$ and $\mathcal{N}_i^K(\mathbf{Y}) = \{j : 1 \leq r_{ij}(\mathbf{Y}) \leq K\}$, respectively.

The co-ranking matrix [70] can then be defined as

$$\mathbf{Q} = [q_{kl}]_{1 \leq k, l \leq N-1} \quad \text{with} \quad q_{kl} = |\{(i, j) : r_{ij}(\mathbf{X}) = k \text{ and } r_{ij}(\mathbf{Y}) = l\}| \ .$$
$$(12.43)$$

The co-ranking matrix is the joint histogram of the ranks and is actually a sum of N permutation matrices of size $N - 1$. With an appropriate gray scale, the co-ranking matrix can also be displayed and interpreted in a similar way as a Shepard diagram [5]. Historically, this scatterplot has often been used to assess results of multidimensional scaling and related methods [10]; it shows the distances $\delta_{ij}(\mathbf{X})$ with respect to the corresponding distances $\delta_{ij}(\mathbf{Y})$, for all pairs (i, j), with $i \neq j$. The analogy between the co-ranking matrix and Shepard's diagram suggests that meaningful criteria should focus on the upper and lower triangle of the co-ranking matrix $\mathbf{Q}$. This can be done by considering the pair of criteria proposed in [71, 72]. They are defined as

$$Q_{\text{NX}}(K) = \frac{1}{KN} \sum_{k=1}^{K} \sum_{l=1}^{K} q_{kl} \qquad (12.44)$$

and

$$B_{\text{NX}}(K) = \frac{1}{KN} \sum_{k=1}^{K} \left(\sum_{l=k+1}^{K} q_{kl} - \sum_{l=1}^{k-1} q_{kl} \right) . \qquad (12.45)$$

The first criterion assesses the overall quality of the embedding and varies between 0 and 1. It measures the average agreement between the corresponding neighborhoods in the high- and low-dimensional spaces. To see this, $Q_{\text{NX}}(K)$ can be rewritten as

$$Q_{\text{NX}}(K) = \frac{1}{KN} \sum_{i=1}^{N} |\mathcal{N}_i^K(\mathbf{X}) \cap \mathcal{N}_i^K(\mathbf{Y})| . \qquad (12.46)$$

The second criterion assesses the balance between the upper and lower triangles of the upper left K-by-K block of the co-ranking matrix. This criterion is useful to distinguish two different types of errors: distant points in the data space that become erroneously neighbors in the visualization space and neighbors that are mistakenly represented too far away. In the case of a nonlinear manifold, the first error type occurs, e.g., if two distant patches are superimposed in the visualization ($B_{\text{NX}}(K)$ will be positive), whereas the second error type happens, e.g., with torn patches ($B_{\text{NX}}(K)$ will be negative).

12.8.2 Data Sets

Two typical data sets illustrate the methods described in the previous sections. The first one is an academic example based on the so-called Swiss roll [16], which is a 2D manifold embedded in a 3D space. The Swiss roll basically looks like a rectangular piece of plane curved as a spiral. The parametric equations are

$$\mathbf{x} = \left[\sqrt{u} \cos(3\pi\sqrt{u}), \sqrt{u} \sin(3\pi\sqrt{u}), \pi v \right]^T , \qquad (12.47)$$

where u and v are uniformly distributed between 0 and 1. The distribution of $\mathbf{x}$ on the Swiss roll is uniform as well. To add some difficulty to the exercise, a disc in the center of the curved rectangle is removed, as shown in Figure 12.1 (left). There are approximately 950 points in the data set.

The second data set includes about 1000 images of handwritten digits. They are randomly drawn from the MNIST data base [73]. Typical images are shown in Figure 12.2. Each image contains 28^2 pixels with a gray level ranging from 0 to 1. All images are converted into 784-dimensional vectors. Classical metric MDS is used to

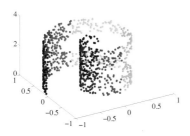 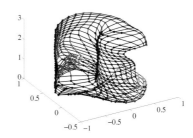

FIGURE 12.1
Left: About 950 points sampled from a Swiss roll with a hole. The color varies with respect to the spiral radius. Right: a 20-by-30 SOM learned on the Swiss roll sample.

FIGURE 12.2
Examples of scanned handwritten digits randomly drawn from the MNIST database. Each image is 28 pixels wide and 28 pixels high. Gray levels range from 0 to 1.

preprocess the data set: 97.5% of the total variance is preserved, leading to a reduced dimensionality about 200, depending on the random sample. This allows a first drastic reduction of the dimensionality with almost no loss of information or structure, which is intended to accelerate all subsequent computations.

For both data sets, two-dimensional representations are computed with NLDR methods. In the case of the Swiss roll, the embedding dimensionality is thus in agreement with the intrinsic dimensionality of the underlying manifold. In contrast, the degrees of freedom in the MNIST data base are expected to be much more numerous, and a good 2D visualization is much more difficult to obtain.

12.8.3 Methods

The compared methods are CMMDS, Sammon's NLM, CCA, Isotop, a 20-by-30 SOM, LLE, Laplacian eigenmaps, and MVU. The first four methods are used successively with Euclidean distances and geodesic distances. The latter are computed with K-ary neighborhoods, where $K = 7$. All data points outside of the largest connected component are discarded. The same value of K is used for other methods involving K-ary neighborhoods (LLE and Laplacian eigenmaps). In order to reduce the huge computational cost of MVU, we used the faster variant described in [74], with $K = 5$ and 5% of data points randomly chosen to be landmarks. All methods run with default parameter values if not otherwise specified.

12.8.4 Results

Figure 12.3 shows a visual representation of the two-dimensional embeddings of the Swiss roll obtained with the twelve considered methods. In the particular case of the SOM, only grid nodes have coordinates both in the high- and low-dimensional space. As a simple workaround to this limitation, data points are given the same coordinates as their closest grid nodes in the high-dimensional space. Grid nodes that are never selected as the closest one are shown as a blank cell. The configuration of the SOM in the data space is visible in Figure 12.1 (right).

The quality assessment curves are shown in Figure 12.4, for a neighborhood size K ranging from 1 to 300. The solid and dashed lines refer to $Q_{\mathrm{NX}}(K)$ (above, with baseline $K/(N-1)$) and $B_{\mathrm{NX}}(K)$ (below, with baseline 0), respectively. As can be seen, for distance-based methods, the use of graph distances yield a better unfolding of the Swiss roll than Euclidean ones. Among these methods, CCA performs the best, followed by Sammon's NLM and finally CMMDS. This shows that a global minimum found by a spectral optimization technique such as in CMMDS does not systematically outperform methods relying on gradient descent. Other spectral methods, such as LLE, Laplacian eigenmaps, and MVU, work well though the quality of their results is slightly lower. In particular, MVU suffers from convergence issues in its semidefinite programming step (the constraints are apparently too tight and no solution is found). The SOM achieves an intermediate performance; due to the inherent vector quantization in this method, small neighborhoods ($K < 20$) are not well preserved (rank information is lost for all data points sharing the same closest grid node). Back to Figure 12.3, we see that methods showing a negative value for $B_{\mathrm{NX}}(K)$ are those that are able to actually unfold the Swiss roll. Among those methods, CCA with graph distance is the only method able to faithfully render the true latent space of the Swiss roll: a rectangle with a circular hole. Other distance-based methods tend to stretch the hole; this effect is caused by an overestimation of the distances, as geodesic paths are forced to circumvent the hole.

As to the handwritten digit images, the 2D visualizations computed with the various NLDR methods are shown in Fiure. 12.5. Each visualization consists of an array of 20-by-20 bins decorated with thumbnail images. Each thumbnail is the average of all images falling in the corresponding bin. If dissimilar images are gathered in the same bin, the thumbnail looks blurred. Empty bins are left blank. The shape of the embedding provided by the SOM trivially depends on the chosen grid. As can be seen, methods relying on (stochastic) gradient descent, such as Sammon's NLM, CCA, and Isotop, yield disc-shaped embeddings. In contrast, spectral methods involving a sparse similarity matrix such as LLE and Laplacian eigenmaps tend to produce spiky embeddings. Knowing that the data set is likely to be clustered (one cluster for each of the ten digits), the objective functions of these methods tend to increase the distances between the clusters. In Laplacian eigenmaps, this effect results from minimizing the distances between neighbors, while keeping a fixed variance for all points; this remains valid for LLE, due to their close relationship. Maximal distances from one cluster to all others are obtained in a hyper-pyramidal configuration; for ten clusters, such a configuration spans at least nine dimensions. The linear projection of a hyper-pyramid look exactly like the spiky embeddings observed in Figure 12.5, with three corners correctly represented in two dimensions and all others collapsed in the center. This reasoning is indirectly confirmed by looking at the eigenvalue spectrum involved in these methods, showing that variance actually spreads out across many dimensions. This is a fundamental difference between spectral and nonspectral methods: the former prune dimensions only after nonlinear transformation of data, whereas the latter are directly forced to work in a low-dimensional space.

Quality assessment curves are displayed in Figure 12.6. As expected, the overall performance level is lower than for the Swiss roll, as the intrinsic dimensionality of data is much higher than two. For small values of K, hardly half of the neighbors are preserved, as indicated by $Q_{\mathrm{NX}}(K)$. The best two methods are CCA with graph distances and the SOM; they outperform all others by far. For small neighborhood sizes, the SOM is hindered by its inherent vector quantization. As with the Swiss roll, the methods with the highest values of $Q_{\mathrm{NX}}(K)$ are also those with the lower $B_{\mathrm{NX}}(K)$. The third method is Isotop with graph distances. In this example, Isotop is much faster than the SOM, because the computation of closest points is much cheaper in the low-dimensional visualization space than in the high-dimensional data space. Spectral methods yield lower performances. The semidefinite programming step in MVU does not succeed in finding a satisfying solution. Sammon's NLM with Euclidean distances performs the worst and also hardly converges.

12.9   Conclusions

Dimensionality reduction proves to be a powerful tool for data visualization and exploratory analysis. From early linear techniques such as PCA and CMMDS to modern nonlinear methods, more than a century has passed. As shown in this chapter, the use of graphs is an important breakthrough in the domain and largely contributes to a significant performance leap. Methods relying on graphs can be categorized with respect to mainly two criteria, namely, the data properties they consider (dissimilarities or similarities) and their optimization technique (spectral or nonspectral). For instance, Isomap and MVU involve graph distances (shortest paths and specifically optimized distances, respectively) and a spectral optimization (CMMDS). Sammon's nonlinear mapping and CCA can use shortest paths as well, but they work with a pseudo-Newton optimization and stochastic gradient descent, respectively. Spectral methods based on similarities are, for instance, Laplacian eigenmaps, LLE, and their variants. Isotop and the SOM also entail similarities, but they rely on heuristic optimization schemes. Spectral methods provide strong theoretical guarantees, such as the capability to find a global minimum. In practice, however, nonspectral techniques often outperform them, thanks to a greater flexibility and the possibility to handle more relevant or more complex cost functions.

If distance preservation has been extensively investigated for quite some time, the use of similarities remains largely unexplored. Spectral methods that uses sparse matrices of pairwise similarities, such as Laplacian eigenmaps and LLE, eventually seem to be more suited to clustering problems than to dimensionality reduction. On the other hand, nonspectral methods like Isotop and the SOM lack sound theoretical foundations or are impeded by their inherent vector quantization. In the near future, progress is likely to stem from more elaborate similarity preservation schemes, such as those developed in [75, 76, 77]. There is a hope that while naturally enforcing the preservation of the local structure, carefully designed similarities [78] can also address important issues such as the phenomenon of distance concentration [79].

Bibliography

[1] I. Jolliffe, *Principal Component Analysis.* New York, NY: Springer-Verlag, 1986.

[2] G. Young and A. Householder, "Discussion of a set of points in terms of their mutual distances," *Psychometrika*, vol. 3, pp. 19–22, 1938.

[3] W. Torgerson, "Multidimensional scaling, I: Theory and method," *Psychometrika*, vol. 17, pp. 401–419, 1952.

[4] J. Lee and M. Verleysen, *Nonlinear dimensionality reduction.* Springer, 2007.

[5] R. Shepard, "The analysis of proximities: Multidimensional scaling with an unknown distance function (parts 1 and 2)," *Psychometrika*, vol. 27, pp. 125–140, 219–249, 1962.

[6] J. Kruskal, "Multidimensional scaling by optimizing goodness of fit to a nonmetric hypothesis," *Psychometrika*, vol. 29, pp. 1–28, 1964.

[7] Y. Takane, F. Young, and J. de Leeuw, "Nonmetric individual differences multidimensional scaling: an alternating least squares method with optimal scaling features," *Psychometrika*, vol. 42, pp. 7–67, 1977.

[8] J. Sammon, "A nonlinear mapping algorithm for data structure analysis," *IEEE Transactions on Computers*, vol. CC-18, no. 5, pp. 401–409, 1969.

[9] P. Demartines and J. Hérault, "Vector quantization and projection neural network," ser. Lecture Notes in Computer Science, A. Prieto, J. Mira, and J. Cabestany, Eds. New York: Springer-Verlag, 1993, vol. 686, pp. 328–333.

[10] ——, "Curvilinear component analysis: A self-organizing neural network for nonlinear mapping of data sets," *IEEE Transactions on Neural Networks*, vol. 8, no. 1, pp. 148–154, Jan. 1997.

[11] T. Kohonen, "Self-organization of topologically correct feature maps," *Biological Cybernetics*, vol. 43, pp. 59–69, 1982.

[12] M. Kramer, "Nonlinear principal component analysis using autoassociative neural networks," *AIChE Journal*, vol. 37, no. 2, pp. 233–243, 1991.

[13] E. Oja, "Data compression, feature extraction, and autoassociation in feedforward neural networks," in *Artificial Neural Networks*, T. Kohonen, K. Mäkisara, O. Simula, and J. Kangas, Eds.   North-Holland: Elsevier Science Publishers, B.V., 1991, vol. 1, pp. 737–745.

[14] J. Mao and A. Jain, "Artificial neural networks for feature extraction and multivariate data projection," *IEEE Transactions on Neural Networks*, vol. 6, no. 2, pp. 296–317, Mar. 1995.

[15] B. Schölkopf, A. Smola, and K.-R. Müller, "Nonlinear component analysis as a kernel eigenvalue problem," *Neural Computation*, vol. 10, pp. 1299–1319, 1998.

[16] J. Tenenbaum, V. de Silva, and J. Langford, "A global geometric framework for nonlinear dimensionality reduction," *Science*, vol. 290, no. 5500, pp. 2319–2323, Dec. 2000.

[17] S. Roweis and L. Saul, "Nonlinear dimensionality reduction by locally linear embedding," *Science*, vol. 290, no. 5500, pp. 2323–2326, 2000.

[18] M. Belkin and P. Niyogi, "Laplacian eigenmaps for dimensionality reduction and data representation," *Neural Computation*, vol. 15, no. 6, pp. 1373–1396, June 2003.

[19] D. Donoho and C. Grimes, "Hessian eigenmaps: Locally linear embedding techniques for high-dimensional data," in *Proceedings of the National Academy of Arts and Sciences*, vol. 100, 2003, pp. 5591–5596.

[20] K. Weinberger and L. Saul, "Unsupervised learning of image manifolds by semidefinite programming," *International Journal of Computer Vision*, vol. 70, no. 1, pp. 77–90, 2006.

[21] L. Xiao, J. Sun, and S. Boyd, "A duality view of spectral methods for dimensionality reduction," in *Proceedings of the 23rd International Conference on Machine Learning*, Pittsburg, PA, 2006, pp. 1041–1048.

[22] L. Saul and S. Roweis, "Think globally, fit locally: Unsupervised learning of nonlinear manifolds," *Journal of Machine Learning Research*, vol. 4, pp. 119–155, June 2003.

[23] N. Linial, E. London, and Y. Rabinovich, "The geometry of graphs and some of its algorithmic applications," *Combinatorica*, vol. 15, no. 2, pp. 215–245, 1995.

[24] G. Di Battista, P. Eades, R. Tamassia, and I. Tollis, *Graph Drawing: Algorithms for the Visualization of Graphs*.   Prentice Hall, 1998.

[25] I. Herman, G. Melançon, and M. Marshall, "Graph visualization and navigation in information visualization: A survey," *IEEE Transactions on Visualization and Computer Graphics*, vol. 6, pp. 24–43, 2000.

[26] Y. Bengio, P. Vincent, J.-F. Paiement, O. Delalleau, M. Ouimet, and N. Le Roux, "Spectral clustering and kernel PCA are learning eigenfunctions," Département d'Informatique et Recherche Opérationnelle, Université de Montréal, Montréal, Tech. Rep. 1239, July 2003.

[27] M. Saerens, F. Fouss, L. Yen, and P. Dupont, "The principal components analysis of a graph, and its relationships to spectral clustering," in *Proceedings of the 15th European Conference on Machine Learning (ECML 2004)*, 2004, pp. 371–383.

[28] B. Nadler, S. Lafon, R. Coifman, and I. Kevrekidis, "Diffusion maps, spectral clustering and eigenfunction of Fokker-Planck operators," in *Advances in Neural Information Processing Systems (NIPS 2005)*, Y. Weiss, B. Schölkopf, and J. Platt, Eds.   Cambridge, MA: MIT Press, 2006, vol. 18.

[29] M. Brand and K. Huang, "A unifying theorem for spectral embedding and clustering," in *Proceedings of International Workshop on Artificial Intelligence and Statistics (AISTATS'03)*, C. Bishop and B. Frey, Eds., Jan. 2003.

[30] K. Pearson, "On lines and planes of closest fit to systems of points in space," *Philosophical Magazine*, vol. 2, pp. 559–572, 1901.

[31] H. Hotelling, "Analysis of a complex of statistical variables into principal components," *Journal of Educational Psychology*, vol. 24, pp. 417–441, 1933.

[32] K. Karhunen, "Zur Spektraltheorie stochastischer Prozesse," *Ann. Acad. Sci. Fennicae*, vol. 34, 1946.

[33] M. Loève, "Fonctions aléatoire du second ordre," in *Processus stochastiques et mouvement Brownien*, P. Lévy, Ed. Paris: Gauthier-Villars, 1948, p. 299.

[34] A. Kearsley, R. Tapia, and M. Trosset, "The solution of the metric STRESS and SSTRESS problems in multidimensional scaling using newton's method," *Computational Statistics*, vol. 13, no. 3, pp. 369–396, 1998.

[35] J. Lee, A. Lendasse, N. Donckers, and M. Verleysen, "A robust nonlinear projection method," in *Proceedings of ESANN 2000, 8th European Symposium on Artificial Neural Networks*, M. Verleysen, Ed. Bruges, Belgium: D-Facto public., Apr. 2000, pp. 13–20.

[36] J. Lee and M. Verleysen, "Curvilinear distance analysis versus isomap," *Neurocomputing*, vol. 57, pp. 49–76, Mar. 2004.

[37] J. Peltonen, A. Klami, and S. Kaski, "Learning metrics for information visualisation," in *Proceedings of the 4th Workshop on Self-Organizing Maps (WSOM'03)*, Hibikino, Kitakyushu, Japan, Sept. 2003, pp. 213–218.

[38] L. Yang, "Sammon's nonlinear mapping using geodesic distances," in *Proc. 17th International Conference on Pattern Recognition (ICPR'04)*, 2004, vol. 2.

[39] P. Estévez and A. Chong, "Geodesic nonlinear mapping using the neural gas network," in *Proceedings of IJCNN 2006*, 2006, in press.

[40] M. Belkin and P. Niyogi, "Laplacian eigenmaps and spectral techniques for embedding and clustering," in *Advances in Neural Information Processing Systems (NIPS 2001)*, T. Dietterich, S. Becker, and Z. Ghahramani, Eds. MIT Press, 2002, vol. 14.

[41] T. Kohonen, *Self-Organizing Maps*, 2nd ed. Heidelberg: Springer, 1995.

[42] J. Lee, C. Archambeau, and M. Verleysen, "Locally linear embedding versus Isotop," in *Proceedings of ESANN 2003, 11th European Symposium on Artificial Neural Networks*, M. Verleysen, Ed. Bruges, Belgium: d-side, Apr. 2003, pp. 527–534.

[43] A. Wismüller, "The exploration machine - a novel method for data visualization," in *Lecture Notes in Computer Science. Advances in Self-Organizing Maps*, 2009, pp. 344–352.

[44] M. Bernstein, V. de Silva, J. Langford, and J. Tenenbaum, "Graph approximations to geodesics on embedded manifolds," Stanford University, Palo Alto, CA, Tech. Rep., Dec. 2000.

[45] E. Dijkstra, "A note on two problems in connection with graphs," *Numerical Mathematics*, vol. 1, pp. 269–271, 1959.

[46] M. Fredman and R. Tarjan, "Fibonacci heaps and their uses in improved network optimization algorithms," *Journal of the ACM*, vol. 34, pp. 596–615, 1987.

[47] F. Shang, L. Jiao, J. Shi, and J. Chai, "Robust positive semidefinite l-isomap ensemble," *Pattern Recognition Letters*, vol. 32, no. 4, pp. 640–649, 2011.

[48] H. Choi and S. Choi, "Robust kernel isomap," *Pattern Recognition*, vol. 40, no. 3, pp. 853–862, 2007.

[49] K. Weinberger and L. Saul, "Unsupervised learning of image manifolds by semidefinite programming," in *Proceedings of the IEEE Conference on Computer Vision and Pattern Recognition (CVPR-04)*, vol. 2, Washington, DC, 2004, pp. 988–995.

[50] J. Shi and J. Malik, "Normalized cuts and image segmentation," *IEEE Transactions on Pattern Analysis and Machine Intelligence*, vol. 22, no. 8, pp. 888–905, 2000.

[51] L. Yen, D. Vanvyve, F. Wouters, F. Fouss, M. Verleysen, and M. Saerens, "Clustering using a random-walk based distance measure," in *Proceedings of ESANN 2005, 13th European Symposium on Artificial Neural Networks*, M. Verleysen, Ed. Bruges, Belgium: d-side, Apr. 2005, pp. 317–324.

[52] L. Grady and J. Polimeni, *Discrete Calculus: Applied Analysis on Graphs for Computational Science*. New York: Springer, 2010.

[53] G. Daza-Santacoloma, C. Acosta-Medina, and G. Castellanos-Dominguez, "Regularization parameter choice in locally linear embedding," *Neurocomputing*, vol. 73, no. 10–12, pp. 1595–1605, 2010.

[54] Y. Linde, A. Buzo, and R. Gray, "An algorithm for vector quantizer design," *IEEE Transactions on Communications*, vol. 28, pp. 84–95, 1980.

[55] J. MacQueen, "Some methods for classification and analysis of multivariate observations," in *Proceedings of the Fifth Berkeley Symposium on Mathematical Statistics and Probability. Volume I: Statistics*, L. Le Cam and J. Neyman, Eds.   Berkeley and Los Angeles, CA: University of California Press, 1967, pp. 281–297.

[56] E. Erwin, K. Obermayer, and K. Schulten, "Self-organizing maps: ordering, convergence properties and energy functions," *Biological Cybernetics*, vol. 67, pp. 47–55, 1992.

[57] A. Ultsch, "Maps for the visualization of high-dimensional data spaces," in *Proc. Workshop on Self-Organizing Maps (WSOM 2003)*, Kyushu, Japan, 2003, pp. 225–230.

[58] H.-U. Bauer, M. Herrmann, and T. Villmann, "Neural maps and topographic vector quantization," *Neural Networks*, vol. 12, pp. 659–676, 1999.

[59] G. Goodhill and T. Sejnowski, "Quantifying neighbourhood preservation in topographic mappings," in *Proceedings of the Third Joint Symposium on Neural Computation*.   University of California, Pasadena, CA: California Institute of Technology, 1996, pp. 61–82.

[60] H.-U. Bauer and K. Pawelzik, "Quantifying the neighborhood preservation of self-organizing maps," *IEEE Transactions on Neural Networks*, vol. 3, pp. 570–579, 1992.

[61] M. de Bodt, E. Cottrell and M. Verleysen, "Statistical tools to assess the reliability of self-organizing maps," *Neural Networks*, vol. 15, no. 8–9, pp. 967–978, 2002.

[62] K. Kiviluoto, "Topology preservation in self-organizing maps," in *Proc. Int. Conf. on Neural Networks, ICNN'96*, I. N. N. Council, Ed., vol. 1, Piscataway, NJ, 1996, pp. 294–299, also available as technical report A29 of the Helsinki University of Technology.

[63] C. Bishop, M. Svensén, and K. Williams, "GTM: A principled alternative to the self-organizing map," *Neural Computation*, vol. 10, no. 1, pp. 215–234, 1998.

[64] T. Martinetz and K. Schulten, "A "neural-gas" network learns topologies," in *Artificial Neural Networks*, T. Kohonen, K. Mäkisara, O. Simula, and J. Kangas, Eds.   Amsterdam: Elsevier, 1991, vol. 1, pp. 397–402.

[65] ——, "Topology representing networks," *Neural Networks*, vol. 7, no. 3, pp. 507–522, 1994.

[66] M. Jünger and P. Mutzel, *Graph Drawing Software*.   Springer-Verlag, 2004.

[67] J. Lee and M. Verleysen, "Nonlinear projection with the Isotop method," in *LNCS 2415: Artificial Neural Networks, Proceedings of ICANN 2002*, J. Dorronsoro, Ed.   Madrid (Spain): Springer, Aug. 2002, pp. 933–938.

[68] J. Venna and S. Kaski, "Neighborhood preservation in nonlinear projection methods: An experimental study," in *Proceedings of ICANN 2001*, G. Dorffner, H. Bischof, and K. Hornik, Eds.   Berlin: Springer, 2001, pp. 485–491.

[69] L. Chen and A. Buja, "Local multidimensional scaling for nonlinear dimension reduction, graph drawing, and proximity analysis," *Journal of the American Statistical Association*, vol. 101, no. 485, pp. 209–219, 2009.

[70] J. Lee and M. Verleysen, "Rank-based quality assessment of nonlinear dimensionality reduction," in *Proceedings of ESANN 2008, 16th European Symposium on Artificial Neural Networks*, M. Verleysen, Ed.   Bruges: d-side, Apr. 2008, pp. 49–54.

[71] ——, "Quality assessment of nonlinear dimensionality reduction based on k-ary neighborhoods," in *JMLR Workshop and Conference Proceedings (New challenges for feature selection in data mining and knowledge discovery)*, Y. Saeys, H. Liu, I. Inza, L. Wehenkel, and Y. Van de Peer, Eds., Sept. 2008, vol. 4, pp. 21–35.

[72] ——, "Quality assessment of dimensionality reduction: Rank-based criteria," *Neurocomputing*, vol. 72, no. 7–9, pp. 1431–1443, 2009.

[73] Y. LeCun, L. Bottou, Y. Bengio, and P. Haffner, "Gradient-based learning applied to document recognition," *Proceedings of the IEEE*, vol. 86, no. 11, pp. 2278–2324, Nov. 1998.

[74] K. Weinberger, B. Packer, and L. Saul, "Nonlinear dimensionality reduction by semidefinite programming and kernel matrix factorization," in *Proceedings of the Tenth International Workshop on Artificial Intelligence and Statistics (AISTATS 2005)*, R. Cowell and Z. Ghahramani, Eds. Bardados: Society for Artificial Intelligence and Statistics, Jan. 2005, pp. 381–388.

[75] G. Hinton and S. Roweis, "Stochastic neighbor embedding," in *Advances in Neural Information Processing Systems (NIPS 2002)*, S. Becker, S. Thrun, and K. Obermayer, Eds. MIT Press, 2003, vol. 15, pp. 833–840.

[76] L. van der Maaten and G. Hinton, "Visualizing data using t-SNE," *Journal of Machine Learning Research*, vol. 9, pp. 2579–2605, 2008.

[77] J. Venna, J. Peltonen, K. Nybo, H. Aidos, and S. Kaski, "Information retrieval perspective to nonlinear dimensionality reduction for data visualization," *Journal of Machine Learning Research*, vol. 11, pp. 451–490, 2010.

[78] J. A. Lee and M. Verleysen, "Shift-invariant similarities circumvent distance concentration in stochastic neighbor embedding and variants," in *Proc. International Conference on Computational Science (ICCS 2011)*, Singapore, 2011.

[79] D. François, V. Wertz, and M. Verleysen, "The concentration of fractional distances," *IEEE Transactions on Knowledge and Data Engineering*, vol. 19, no. 7, pp. 873–886, July 2007.

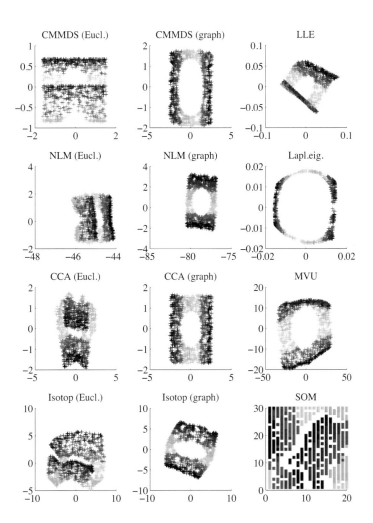

FIGURE 12.3
Two-dimensional embeddings of the Swiss roll with various NLDR methods.
Distance-based methods use either Euclidean norm ('Eucl.') or shortest paths in a
neighborhood graph ('Graph').

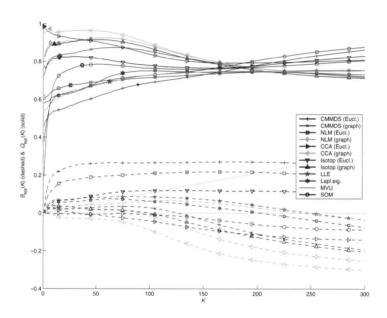

FIGURE 12.4

Quality assessment curves for the two-dimensional embeddings of the Swiss roll with various NLDR methods.

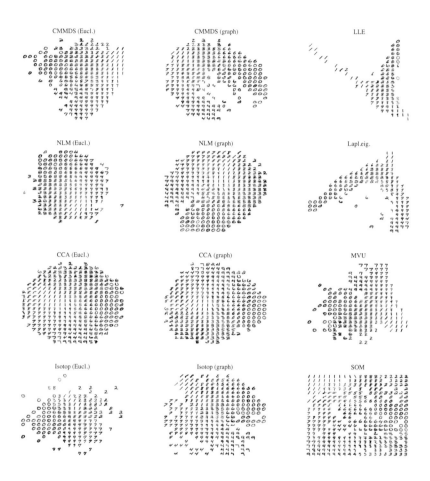

FIGURE 12.5
Two-dimensional embeddings of the handwritten digits with various NLDR methods. Distance-based methods use either Euclidean norm ('Eucl.') or shortest paths in a neighborhood graph ('Graph').

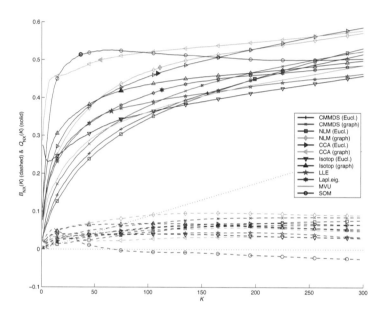

FIGURE 12.6
Quality assessment curves for the two-dimensional embeddings of the handwritten
digits with various NLDR methods.

13

Graph Edit Distance—Theory, Algorithms, and Applications

Miquel Ferrer

Institut de Robòtica i Informàtica Industrial
Universitat Politècnica de Catalunya - CSIC
Barcelona, Spain
Email: mferrer@iri.upc.edu

Horst Bunke

NICTA, Victoria Research Laboratory
The University of Melbourne, Parkville
Victoria, Australia
Email: bunke@iam.unibe.ch

CONTENTS

13.1 Introduction 384
13.2 Definitions and Graph Matching 386
    13.2.1 Exact Graph Matching 387
        13.2.1.1 Graph Similarity Measures Based on the MACS and the MICS 391
    13.2.2 Error-Tolerant Graph Matching 393
    13.2.3 Graph Edit Distance 395
13.3 Theoretical Aspects of GED 396
    13.3.1 Edit Costs 396
        13.3.1.1 Conditions on Edit Costs 397
        13.3.1.2 Example of Edit Costs 398
    13.3.2 Automatic Learning of Edit Costs 399
        13.3.2.1 Learning Probabilistic Edit Costs 399
        13.3.2.2 Self-Organizing Edit Costs 400
13.4 GED Computation 401
    13.4.1 Optimal Algorithms 401
    13.4.2 Suboptimal Algorithms 403
    13.4.3 GED Computation by Means of Bipartite Graph Matching 404
13.5 Applications of GED 407
    13.5.1 Weighted Mean of a Pair of Graphs 407
    13.5.2 Graph Embedding into Vector Spaces 409
    13.5.3 Median Graph Computation 411
13.6 Conclusions 417
    Bibliography 417

Graphs provide us with a powerful and flexible representation formalism which can be employed in various fields of intelligent information processing. Graph matching refers to the process of evaluating the similarity between graphs. Two approaches to this task have been proposed, namely, exact and inexact (or error-tolerant) graph matching. While the former approach aims at finding a strict correspondence between two graphs to be matched, the latter is able to cope with errors and measures the difference of two graphs in a broader sense. As a matter of fact, inexact graph matching has been a highly significant research topic in the area of pattern analysis for a long time. The graph edit distance (GED) is a foundation of inexact graph matching, and has emerged as an important way to measure the similarity between pairs of graphs in an error-tolerant fashion.

13.1   Introduction

One of the basic objectives in pattern recognition is to develop systems for the analysis or classification of objects [1, 2]. In principle, these objects or patterns can be of any kind. For instance, they can include images taken from a digital camera, speech signals captured with a microphone, or words written with a pen or a tablet PC, to name a few. A first issue to be addressed in any pattern recognition system is how to represent these objects. Feature vectors is one of the most common and widely used data representations. That is, for each object, a set of relevant properties, or features, are extracted and arranged in a vector. Then, a classifier can be trained to recognize the unknown objects. The main advantage of this representation is that a large number of algorithms for pattern analysis and classification become immediately available [1]. This is mainly due to the fact that vectors are simple structures that have many interesting and useful mathematical properties.

However, some disadvantages arise from the simple structure of feature vectors. Regardless of the complexity of the object, within one particular application, feature vectors have always to have the same length and structure (a simple list of pre-determined components). Therefore, for the representation of complex objects where the relations between their parts become important for their analysis or classification, graphs appear as an appealing alternative. One of the main advantages of graphs over feature vectors is that graphs can explicitly model relations between different parts of an object, whereas feature vectors are only able to describe an object as an aggregation of numerical properties. In addition, graphs permit us to associate any kind of label (not only numbers) to both edges and nodes, extending in this way the spectrum of properties that can be represented. Furthermore, the dimensionality of graphs, that is, the number of nodes and edges, can be different for every object, even for objects of the same class. Thus, the more complex an object is, the larger the number of nodes and edges can be.

Although these visible differences between graphs and vectors exist, one can see from the literature that vector and graph representations are not mutually exclusive. For instance, there is a substantial literature in complex networks that characterizes the graph structure by means of feature vectors using different measures on graphs, such as connectivity and average distance. Two surveys on that field are [3, 4]. Recently, an extensive work comparing the representational power of feature vectors and graphs without restrictions in the node labels under the context of Web content mining has been presented in [5]. Experimental results consistently show an improvement in the accuracy of the graph-based approaches over comparable vector-based methods. In addition, in some cases the experiments even showed an improvement in execution time over the vector model.

Actually, graphs have been used to solve computer vision problems for decades. Some examples include recognition of graphical symbols [6, 7], character recognition [8, 9], shape analysis [10, 11], 3D-object recognition [12, 13], and video and image database indexing [14]. However, in spite of the strong mathematical foundation underlying graphs and their high power of representation, working with graphs is usually harder and more challenging than working with feature vectors.

Graph matching is the specific process of evaluating the structural similarity of two graphs[1,2]. It can be split into two categories, namely exact and error-tolerant graph matching. In exact graph matching, the basic objective is to decide whether two graphs or parts of them are identical in terms of their structure and labels. Methods for exact graph matching include *graph isomorphism, subgraph isomorphism* [15], *maximum common subgraph* [16], and *minimum common supergraph* [17]. The main advantage of exact graph matching methods is their stringent definition and solid mathematical foundation. Nevertheless, it is still imperative for the underlying graphs to share isomorphic parts to be considered similar. This means that the two graphs under comparison must be identical to a large extent in terms of structure and labels to produce high similarity values. In practice, node and edge labels used to describe the properties of the underlying objects (represented by the graphs) are often of continuous nature. In this situation, at least two problems arise. First, it is not sufficient to check whether two labels are identical or not, but one has also to evaluate their similarity. In addition, under all the matching procedures mentioned so far, two labels will be considered different regardless of the degree of difference between them. This may lead to considering two graphs with similar, but not identical labels completely dissimilar even if they have identical structure. It is therefore clear that a more sophisticated method is needed to measure the dissimilarity between graphs, taking into account such limitations. This leads to the definition of inexact, or error-tolerant, graph matching.

In error-tolerant graph matching, the idea is to evaluate the similarity of two graphs in a broader sense that better reflects the intuitive understanding of graph sim-

---

[1] As it is straightforward to transform a similarity measure into a dissimilarity measure and vice versa, we use the term *similarity* and *dissimilarity* interchangeably in this chapter.

[2] We defined the graph matching process as simply the way to ontain a similarity measure between graphs. However, one can see the graph matching a richer process where more valuable information, such as the mapping between nodes and edges can be obtained.

ilarity. In fact, error-tolerant graph matching methods need not be defined by means of a graph similarity measure at all, but can be formulated in entirely different terms. In the ideal case, two graphs representing the same class of objects are identical. That is, a graph extraction process turning objects into graphs should always result in exactly the same graph for objects of the same class. In practice, of course, the graph extraction process suffers from noise and various distortions. In graph matching, error-tolerance means that the matching algorithm is able to cope with the structural difference between the actually extracted graph and the ideal graph.

In this chapter, we review a number of exact and error-tolerant graph distance measures. The main focus is the graph edit distance (GED), which is recognized as one of the most flexible and universal error-tolerant matching paradigm. Graph edit distance is not constrained to particular classes of graphs and can be tailored to specific applications. The main restriction is that the computation of edit distance is inefficient, particularly for large graphs. Recently, however, faster approximate solutions have become available (see Section 13.4) This chapter is devoted to a detailed description of graph edit distance. We review the graph edit distance from three different points of view, namely, theory, algorithms, and applications. In the first part of the chapter, fundamental concepts related to GED are introduced. Special emphasis will be put on the theory underlying the edit distance cost function, giving some examples and introducing the fundamental problem of automatic edit cost learning. The second part of the chapter gives an introduction to GED computation. We review in detail exact and approximate methods for GED computation and finish by introducing one of the most recent approximate algorithms for GED computation, which is based on bipartite graph matching. The last part of the chapter is devoted to GED applications. We concentrate on three basic applications: the weighted mean of a pair of graphs, graph embedding into vector spaces, and the median graph computation problem. We finish the chapter by giving some conclusions and pointing to future directions in GED theory and practice.

13.2 Definitions and Graph Matching

Definition 13.2.1
Graph: *Given L, a finite or infinite set of labels for nodes and edges, a graph $\mathcal{G}$ is defined by the tuple $\mathcal{G} = (\mathcal{V}, \mathcal{E})$ where,*

- $\mathcal{V}$ *is a finite set of nodes*

- $\mathcal{E} \subseteq \mathcal{V} \times \mathcal{V}$ *is the set of edges*[3]

Without loss of generality, Definition 13.2.1 can be seen as the definition of a

---

[3]Unless otherwise stated, in this work we will suppose graphs with directed edges.

labeled graph. Notice that there is not any restriction concerning the nature of the labels of nodes and edges. That is, the label alphabet is not constrained at all. L may be defined as a vector space (i.e. $L = \mathbb{R}^n$) or simply as a finite or infinite set of discrete labels (i.e., $L = \{\alpha, \beta, \gamma, \cdots\}$). The set of labels L can also include the *null* label (often represented by ε). If all the nodes and edges are labeled with the same *null* label, the graph is considered as *unlabeled*. A *weighted graph* is a special type of labeled graph in which each node is labeled with the null label and each edge (v_i, v_j) is labeled with a real number or *weight* w_{ij}, usually belonging, but not restricted, to the interval $[0, 1]$. An *unweighted graph* can be seen as a particular instance of a *weighted graph* where $\forall (v_i, v_j) \in \mathcal{E}, w_{ij} = 1$. The label of an element $\bullet$ (where $\bullet$ can be a node or an edge) is denoted by $L(\bullet)$.

In the more general case, vertices and edges may represent more complex information. That is, they can contain information of a different nature at the same time. For instance, a complex attribute for a node representing a region of an image could be composed of the color histogram of the region, a description of its shape and symbolic information relating this region with its adjacent regions. In this case graphs are called *attributed graphs* or simply AG. Notice that labeled, weighted, and unweighted graphs are particular instances of attributed graphs, where the attributes are simple labels or numbers.

Definition 13.2.2
Subgraph: Let $\mathcal{G}_1 = (\mathcal{V}_1, \mathcal{E}_1)$, and $\mathcal{G}_2 = (\mathcal{V}_2, \mathcal{E}_2)$ be two graphs. Graph $\mathcal{G}_1$ is a subgraph of $\mathcal{G}_2$, denoted by $\mathcal{G}_1 \subseteq \mathcal{G}_2$, if

- $\mathcal{V}_1 \subseteq \mathcal{V}_2$
- $\mathcal{E}_1 = \mathcal{E}_2 \cap (\mathcal{V}_1 \times \mathcal{V}_1)$

From Definition 13.2.2 it follows that, given a graph $\mathcal{G} = (\mathcal{V}, \mathcal{E})$, a subset $\mathcal{V}' \subseteq \mathcal{V}$ of its vertices uniquely defines a subgraph, called the subgraph *induced* by $\mathcal{V}'$. That is, an induced subgraph of $\mathcal{G}$ can be obtained by removing some of its nodes $(\mathcal{V} - \mathcal{V}')$ and all their adjacent edges. However, if the second condition of Definition 13.2.2 is replaced by $\mathcal{E}_1 \subseteq \mathcal{E}_2$ then the resulting subgraph is called *noninduced*. A noninduced subgraph of $\mathcal{G}$ is obtained by removing some of its nodes $(\mathcal{V} - \mathcal{V}')$ and all their adjacent edges plus some additional edges. An example of a graph $\mathcal{G}$, an induced, and a noninduced subgraph of $\mathcal{G}$ is given in Figure 13.1 (note that node labels are represented using different shades of grey). Of course, given a graph $\mathcal{G}$, an induced subgraph of $\mathcal{G}$ is also a noninduced subgraph of $\mathcal{G}$.

13.2.1 Exact Graph Matching

The operation of comparing two graphs is commonly referred to as *graph matching*. The aim of exact graph matching is to determine whether two graphs, or parts of two graphs, are identical in terms of their structure and labels. The equality of two graphs

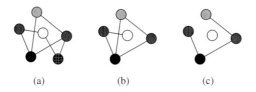

(a)   (b)   (c)

FIGURE 13.1
Original model graph $\mathcal{G}$ (*a*), an induced subgraph of $\mathcal{G}$ (*b*) and a non-induced sub-graph of $\mathcal{G}$ (*c*).

can be tested by means of a bijective function, called *graph isomorphism*, defined as follows:

Definition 13.2.3
Graph isomorphism: *Let* $\mathcal{G}_1 = (\mathcal{V}_1, \mathcal{E}_1)$ *and* $\mathcal{G}_2 = (\mathcal{V}_2, \mathcal{E}_2)$ *be two graphs. A* graph isomorphism *between* $\mathcal{G}_1$ *and* $\mathcal{G}_2$ *is a bijective mapping* $f : \mathcal{V}_1 \longrightarrow \mathcal{V}_2$ *such that*

- $L(v) = L(f(v))$ *for all nodes* $v \in \mathcal{V}_1$

- *for each edge* $e_1 = (v_i, v_j) \in \mathcal{E}_1$, *there exists an edge* $e_2 = (f(v_i), f(v_j)) \in \mathcal{E}_2$ *such that* $L(e_1) = L(e_2)$

- *for each edge* $e_2 = (v_i, v_j) \in \mathcal{E}_2$, *there exists an edge* $e_1 = (f^{-1}(v_i), f^{-1}(v_j)) \in \mathcal{E}_1$ *such that* $L(e_1) = L(e_2)$

It is clear from this definition that isomorphic graphs are identical in terms of structure and labels. To check whether two graphs are isomorphic or not, we have to find a function mapping every node of the first graph to a node of the second graph in such a way that the edge structure of both graphs is preserved, and the labels for the nodes and the edges are consistent. Graph isomorphism is an equivalence relation on graphs, since it satisfies the conditions of reflexivity, symmetry, and transitivity.

Related to graph isomorphism is the concept of subgraph isomorphism. It permits one to check whether a part (subgraph) of one graph is identical to another graph. A subgraph isomorphism exists between two given graphs $\mathcal{G}_1$ and $\mathcal{G}_2$ if there is a graph isomorphism between the smaller graph and a subgraph of the larger graph, i.e., if the smaller graph is contained in the larger graph. Formally, subgraph isomorphism is defined as follows:

Definition 13.2.4
Subgraph Isomorphism: *Let* $\mathcal{G}_1 = (\mathcal{V}_1, \mathcal{E}_1)$ *and* $\mathcal{G}_2 = (\mathcal{V}_2, \mathcal{E}_2)$ *be two graphs. An injective function* $f : \mathcal{V}_1 \longrightarrow \mathcal{V}_2$ *is called a subgraph isomorphism from* $\mathcal{G}_1$ *to* $\mathcal{G}_2$ *if there exists a subgraph* $\mathcal{G} \subseteq \mathcal{G}_2$, *such that* f *is a graph isomorphism between* $\mathcal{G}_1$ *and* $\mathcal{G}$.

Most of the algorithms for *graph isomorphism* and *subgraph isomorphism* are based on some form of tree search with backtracking. The main idea is to iteratively expand a partial match (initially empty) by adding new pairs of nodes satisfying some constraints imposed by the matching method with respect to the previously matched pairs of nodes. These methods usually apply some heuristic conditions to prune useless search paths as early as possible. Eventually, either the algorithm finds a complete match or reaches a point where the partial match cannot be further expanded because of the matching constraints. In the latter case, the algorithm backtracks until it finds a partial match for which another alternative expansion is possible. The algorithm halts when all the possible mappings that satisfy the constraints have been tried.

One of the most important algorithms based on this approach is described in [15]. It addresses both the graph and subgraph isomorphism problem. To prune unfruitful paths at an early stage, the author proposes a refinement procedure that drops pairs of nodes that are inconsistent with the partial match being explored. The branches of this partial match leading to these incompatible matches are not expanded. A similar strategy is used in [18], where the authors include an additional preprocessing step that creates an initial partition of the nodes of the graph based on a distance matrix to reduce the size of the search space. More recent approaches are the VF [19] and the VF2 [20] algorithms. In these works, the authors define a heuristic based on the analysis of the nodes adjacent to the nodes of the partial mapping. This procedure is fast to compute and leads in many cases to an improvement over the approach of [15]. Another recent approach which combines explicit search in state-space and the use of energy minimization is [21]. The basic heuristic of this algorithm can be interpreted as a greedy algorithm to form maximal cliques in an association graph. In addition, the authors allow for vertex swaps during the clique creation. This last characteristic makes this algorithm faster than a PBH (pivoting-based heuristic).

Probably, the most important approach to graph isomorphism testing that is not based on tree search is described in [22]. It uses concepts from group theory. First, the automorphism group of each input graph is constructed. After that, a canonical labelling is derived and isomorphism is checked by simply verifying the equality of the canonical forms.

It is still an open question whether the graph isomorphism problem belongs to the NP class or not. Polynomial algorithms have been developed for special classes of graphs such as *bounded valence graphs* [23] (i.e., graphs where the maximum number of edges adjacent to a node is bounded by a constant); *planar graphs* [24] (i.e., graphs that can be drawn on a plane without graph edges crossing) and *trees* [11] (i.e., graphs with no cycles). But no polynomial algorithms are known for the general case. Conversely, the subgraph isomorphism problem is proven to be NP-complete [25].

Matching graphs by means of graph and subgraph isomorphism is limited in the sense that an exact correspondence must exist between two graphs or between a graph and part of another graph. But let us consider the situation of Figure 13.2(a).

It is clear that the two graphs are similar, since most of their nodes and edges are identical. But it is also clear that none of them is related to the other one by (sub)graph isomorphism. Therefore, under the (sub)graph isomorphism paradigm, they will be considered completely different graphs. Consequently, in order to overcome such a drawbacks of the (sub)graph isomorphism, to establish a measure of partial similarity between any two graphs, and to relax this rather stringent condition, the concept of the largest common part of two graphs is introduced.

Definition 13.2.5
Maximum common subgraph (MACS): *Let* $\mathcal{G}_1 = (\mathcal{V}_1, \mathcal{E}_1)$ *and* $\mathcal{G}_2 = (\mathcal{V}_2, \mathcal{E}_2)$ *be two graphs. A graph* $\mathcal{G}$ *is called a common subgraph* (cs) *of* $\mathcal{G}_1$ *and* $\mathcal{G}_2$ *if there exists a subgraph isomorphism from* $\mathcal{G}$ *to* $\mathcal{G}_1$ *and from* $\mathcal{G}$ *to* $\mathcal{G}_2$. *A common subgraph of* $\mathcal{G}_1$ *and* $\mathcal{G}_2$ *is called maximum common subgraph (MACS) if there exists no other common subgraph of* $\mathcal{G}_1$ *and* $\mathcal{G}_2$ *with more nodes than* $\mathcal{G}$.

The notion of maximum common subgraph of two graphs can be interpreted as an intersection. Intuitively, it is the largest part of either graph that is identical in terms of structure and labels. It is clear that the larger the maximum common subgraph is, the more similar the two graphs are.

Macs computation has been intensively investigated. There are two major approaches in the literature. In [16, 26, 27] a backtracking search is used. A different strategy is proposed in [28, 29], where the authors reduce the MACS problem to the problem of finding a maximum clique (i.e., a completely connected subgraph) of a suitably constructed *association graph* [30]. A comparison of some of these methods on large databases of graphs is given in [31, 32].

It is well known that the MACS and the maximum clique problems are NP-complete. Therefore, some approximate algorithms have been developed. A survey of these approximate approaches and the study of their computational complexity is given in [33]. Nevertheless, in [34] it is shown that when graphs have unique node labels, the computation of the MACS can be done in polynomial time. This important result has been exploited in [5] to perform data mining on Web pages based on their contents. Other applications of the MACS include comparison of molecular structures [35, 36] and matching of 3-D graph structures [37].

Dual to the definition of the maximum common subgraph (interpreted as an intersection operation) is the concept of the minimum common supergraph.

Definition 13.2.6
(Minimum common supergraph (MICS)): *Let* $\mathcal{G}_1 = (\mathcal{V}_1, \mathcal{E}_1)$ *and* $\mathcal{G}_2 = (\mathcal{V}_2, \mathcal{E}_2)$ *be two graphs. A graph* $\mathcal{G}$ *is called a common supergraph* (CS) *of* $\mathcal{G}_1$ *and* $\mathcal{G}_2$ *if there exists a subgraph isomorphism from* $\mathcal{G}_1$ *to* $\mathcal{G}$ *and from* $\mathcal{G}_2$ *to* $\mathcal{G}$. *A common supergraph of* $\mathcal{G}_1$ *and* $\mathcal{G}_2$ *is called minimum common supergraph (MICS) if there exists no other common supergraph of* $\mathcal{G}_1$ *and* $\mathcal{G}_2$ *with less nodes than* $\mathcal{G}$.

The minimum common supergraph of two graphs can be seen as a graph with the minimum required structure so that both graphs are contained in it as subgraphs. In

[17], it is demonstrated that the computation of the minimum common supergraph can be reduced to the computation of the maximum common subgraph. This result has not only relevant theoretical implications, but also practical consequences in the sense that any algorithm for the MACS computation, such as all algorithms described before, can also be used to compute the MICS.

An example of the concepts of MACS and MICS is given in Figure 13.2. Notice that, in general, neither $MACS(\mathcal{G}_1, \mathcal{G}_2)$ nor $MICS(\mathcal{G}_1, \mathcal{G}_2)$ are uniquely defined for two given graphs $\mathcal{G}_1$ and $\mathcal{G}_2$.

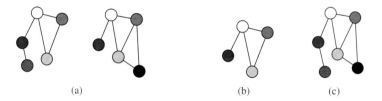

(a)　　　　　　　　　　　　　　(b)　　　　　(c)

FIGURE 13.2
Two graphs $\mathcal{G}_1$ and $\mathcal{G}_2$ (*a*), $MACS(\mathcal{G}_1, \mathcal{G}_2)$ (*b*) and $MICS(\mathcal{G}_1, \mathcal{G}_2)$ (*c*).

13.2.1.1 Graph Similarity Measures Based on the MACS and the MICS

The concepts of maximum common subgraph and minimum common supergraph can be used to measure the similarity of two graphs. The basic idea is the intuitive observation that the larger the common part of two graphs is, the higher their similarity. In the following, several distance measures between graphs based on the concepts of the MACS and the MICS will be presented.

For instance, in [38] a distance metric between two graphs is defined in the following way:

$$d_1(\mathcal{G}_1, \mathcal{G}_2) = |\mathcal{G}_1| + |\mathcal{G}_2| - 2|MACS(\mathcal{G}_1, \mathcal{G}_2)| \tag{13.1}$$

It is clear in this definition that, if two graphs are similar, they will have a $MACS(\mathcal{G}_1, \mathcal{G}_2)$ similar to both of them. Then, the term $|MACS(\mathcal{G}_1, \mathcal{G}_2)|$ will be close to $|\mathcal{G}_1|$ and $|\mathcal{G}_2|$ and therefore the distance close to 0. Conversely, if the graphs are dissimilar, the term $|MACS(\mathcal{G}_1, \mathcal{G}_2)|$ will tend to 0 and the distance will be large.

Another distance measure based on the MACS [39] is

$$d_2(\mathcal{G}_1, \mathcal{G}_2) = 1 - \frac{|MACS(\mathcal{G}_1, \mathcal{G}_2)|}{\max(|\mathcal{G}_1|, |\mathcal{G}_2|)} \tag{13.2}$$

In this case, if the two graphs are very similar, then their maximum common subgraph will obviously be almost as large as one of the two graphs. Then, the value of the fraction will therefore tend to 1 and the distance will be close to 0. For two dissimilar graphs, their maximum common subgraph will be small, and the ratio will then be close to 0 and the distance close to 1. Clearly, the distance metric d_2 is bounded between 0 and 1, and the more similar two graphs are the smaller is d_2.

A similar distance is defined in [40], but the union graph is used as a normalization factor instead of the size of the larger graph,

$$d_3(\mathcal{G}_1, \mathcal{G}_2) = 1 - \frac{|\text{MACS}(\mathcal{G}_1, \mathcal{G}_2)|}{|\mathcal{G}_1| + |\mathcal{G}_2| - |\text{MACS}(\mathcal{G}_1, \mathcal{G}_2)|} \qquad (13.3)$$

By "graph union", the authors of [40] mean that the denominator represents the size of the union graph in the set theory point of view. As a matter of fact, this denominator is equal to $|\text{MICS}(\mathcal{G}_1, \mathcal{G}_2)|$. The behaviour of distance d_3 is similar to that of d_2. The use of the graph union is motivated by the fact that changes in the size of the smaller graph that keep the $\text{MACS}(\mathcal{G}_1, \mathcal{G}_2)$ constant are not taken into account in d_2, whereas distance d_3 does take this variations into account. This measure was also demonstrated to be a metric and gives a distance value in the interval $[0, 1]$.

Another approach based on the difference between the minimum common supergraph and the maximum common subgraph was proposed in [41]. In this approach, the distance is given by

$$d_4(\mathcal{G}_1, \mathcal{G}_2) = |\text{MICS}(\mathcal{G}_1, \mathcal{G}_2)| - |\text{MACS}(\mathcal{G}_1, \mathcal{G}_2)| \qquad (13.4)$$

The basic idea is that for similar graphs, the size of the maximum common subgraph and the minimum common supergraph will be similar, and the resulting distance will be small. On the other hand, if the graphs are dissimilar, the two terms will be significantly different, resulting in larger distance values. It is important to notice that also d_4 is a metric. In fact, it can be easily shown that this distance is equal to d_1.

Obviously, the similarity measures d_1 to d_4 admit a certain amount of error-tolerance. That is, with the use of these similarity measures, two graphs do not need to be directly related in terms of (sub)graph isomorphism to be successfully matched. Nevertheless, it is still imperative for these graphs to share isomorphic parts to be considered similar. This means that the two graphs must be identical to a large extent in terms of structure and labels to produce large values of the similarity measures. In many practical applications, however, node and edge labels used to describe the properties of the underlying objects (represented by graphs) are of continuous nature. In this situation, at least two problems emerge. First, it is not sufficient to evaluate whether two labels are identical or not, but one has to evaluate their similarity. In addition, using all the distances given so far, two labels will be considered different, regardless of the actual degree of dissimilarity between them. This may lead to the problem of considering two graphs with similar, but not identical, labels completely different, even if they have identical structure. It is therefore clear that there is a need of a more sophisticated method to measure the dissimilarity between graphs, which takes into account such limitations. This leads to the definition of *inexact* or *error-tolerant graph matching*.

13.2.2 Error-Tolerant Graph Matching

The methods for exact graph matching outlined above have solid mathematical foundation. Nevertheless, their stringent conditions restrict their applicability to a very small range of real-world problems only. In many practical applications, when objects are encoded using graph-based representations, some degree of distortion may occur due to multiple reasons. For instance, there may exist some form of noise in the acquisition process, nondeterministic elements in some of the processing steps, etc. Hence, graph representations of the same object may differ somewhat, and two identical objects may not have an exact match when transformed into their graph-based representations. Therefore, it is necessary to introduce some degree of error tolerance into the matching process, so that structural differences between the models can be taken into account. For this reason, a number of error-tolerant, or inexact, graph matching methods have been proposed. They deal with a more general graph matching problem than (sub)graph isomorphism, the MACS, and the MICS problem.

The key idea of error-tolerant graph matching is to measure the similarity between two given graphs instead of simply giving an indication of whether they are identical or not. Usually, in these algorithms the matching between two different nodes that do not preserve the edge compatibility is not forbidden. But a cost is introduced to penalize the difference. The task of any related algorithm is then to find the mapping that minimizes the matching cost. For instance, an error-tolerant graph matching algorithm is expected to discover not only that the graphs in the Figure 13.2(a) share some parts of their structure but also that they are rather similar despite the difference in their structure.

As in the case of exact matching, techniques based on tree search can also be used for inexact graph matching. Differently from exact graph matching where only identical nodes are matched, in this case, the search is usually controlled by the partial matching cost obtained so far and a heuristic function to estimate the cost of matching the remaining nodes. The heuristic function is used to prune unsuccessful paths in a *depth-first* search approach or to determine the order in which the search tree has to be traversed by an A* algorithm. For instance, the use of an A* strategy is proposed in [42] to obtain an optimal algorithm to compute the graph distance. In [43] an A*-based approach is used to cast the matching between two graphs as a bipartite matching problem. A different approach is used in [44], where the authors present an inexact graph matching method that also exploits some form of contextual information, defining a distance between function described graphs (FDG). Finally, a parallel branch and bound algorithm has been presented in [45] to compute a distance between two graphs with the same number of nodes.

Another class of approaches to inexact graph matching are those based on genetic algorithms [46, 47, 48, 49, 50]. The main advantage of genetic algorithms is that they are able to cope with huge search spaces. In genetic algorithms, the possible solutions are encoded as chromosomes. These chromosomes are generated randomly following some operators inspired by evolution theory, such as mutation and crossover. To determine how good a solution (chromosome) is, a fitness function is defined. The search space is explored with a combination of the fitness function and the randomness of the biologically inspired operators. The algorithm tends to favour well-performing candidates with a high value of the fitness function. The two main drawbacks of genetic algorithms are that they are nondeterministic algorithms and that the final output is dependent on the initialization phase. On the other hand, these algorithms are able to cope with difficult optimization problems and they have been widely used to solve NP-complete problems [51]. For instance, in [48] the matching between two given graphs is encoded in a vector form as a set of node-to-node correspondences. Then an underlying distortion model (similar to the idea of cost function presented in the next section) is used to evaluate the quality of the match by simply accumulating individual costs.

Spectral methods [52, 53, 54, 55, 56, 57, 58] have also been used for inexact graph matching. The basic idea of these methods is to represent graphs by the eigendecomposition of their adjacency or Laplacian matrix. Among the pioneering works of the spectral graph theory applied to graph matching problems is [56]. This work proposes an algorithm for the weighted graph isomorphism problem, where the objective is to match a subset of nodes of the first graph to a subset of nodes of the second graph, usually by means of a matching matrix M. One of the major limitations of this approach is that the graphs must have the same number of nodes. A more recent paper [58] proposes a solution to the same problem by combining the use of eigenvalues and eigenvectors with continuous optimization techniques. In [59] some spectral features are used to cluster sets of nodes that are likely to be matched in an optimal correspondence. This method does not suffer from the limitation of [56] that all graphs must have the same number of nodes. Another approach combining clustering and spectral graph theory has been presented in [60]. In this approach the nodes are embedded into a so-called graph eigenspace using the eigenvectors of the adjacency matrix. Then, a clustering algorithm is used to find nodes of the two graphs that are to be put in correspondence. A work that is partly related to spectral techniques has been proposed in [61, 55]. Here, the authors assign a *Topological Signature Vector*, or TSV, to each nonterminal node of a directed acyclic graph (DAG). The TSV associated to a node is based on the sum of the eigenvalues of its descendant DAGs. It can be seen as a signature to represent the shape of an object and can be used both for indexing graphs in a database and for graph matching [62].

A completely different approach is to cast the inexact graph matching problem as a nonlinear optimization problem. The first family of methods based on this idea is *relaxation labelling* [63, 64]. The main idea is to formulate the matching of the two graphs under consideration as a labeling problem. That is, each node of one of the graphs is to be assigned to one label out of all the possible labels. This label determines the corresponding node of the other graph. To model the compatibility of the node labeling, a Gaussian probability distribution is used. In principle, only the labels attached to the nodes or the nodes together with their surrounding edges can be taken into consideration in this probability distribution. Then, the labelling is iteratively refined until a sufficiently accurate labelling is found. Any node assignment established in this way implies an assignment of the edges between the two graphs. In another approach described in [65], the nodes of the input graph are seen as the observed data while the nodes of the model graph act as hidden random variables. The iterative matching process is handled by the expectation-maximization algorithm [66]. Finally, in another classical approach [67], the problem of weighted graph matching is tackled using a technique called graduated assignment which makes it possible to avoid poor local optima.

In the last paragraphs we have described some of the most relevant algorithms for error-tolerant graph matching. Relaxation methods, graduate assignment, spectral method,s etc. can be used to establish a mapping between the nodes and edges of the two graphs, but do not directly give us any distances or dissimilarities as they are required by a classifier or clustering algorithm. In the next section we review graph edit distance, which is one of the most widely used error-tolerant graph matching methods. Differently from other methods, edit distance is suitable to compute graph distances. As far as the computational complexity is concerned, any of the graph matching problems, subgraph isomorphism, maximum common subgraph, edit distance, etc. are NP complete. Hence computational tractability is an issue for all of them. However, no experimental comparison has been conducted to directly compare all of these methods with respect to performance and efficiency. An excellent review of algorithms for both exact and error-tolerant graph matching can be found in [68].

13.2.3   Graph Edit Distance

The basic idea behind the graph edit distance is to define a dissimilarity measure between two graphs by the minimum amount of distortion required to transform one graph into the other [69, 70, 71, 72]. To this end, a number of distortions or edit operations ed, consisting of the insertion, deletion, and substitution (replacement of a label by another one) of both nodes and edges are defined. Then, for every pair of graphs, $\mathcal{G}_1$ and $\mathcal{G}_2$, there exists a sequence of edit operations, or edit path $\partial(\mathcal{G}_1, \mathcal{G}_2) = (ed_1, \ldots, ed_k)$ (where each ed_i denotes an edit operation), that transform one graph into the other. In Figure 13.3, an example of an edit path between two given graphs $\mathcal{G}_1$ and $\mathcal{G}_2$ is shown. In this example, the edit path consists of one edge deletion, one node substitution, one node insertion, and two edge insertions.

In general, several edit paths may exist between two given graphs. This set of edit paths is denoted by $\wp(\mathcal{G}_1, \mathcal{G}_2)$. To quantitatively evaluate which is the best edit path, an edit cost function is introduced. The basic idea is to assign a cost c to each edit operation according to the amount of distortion that it introduces in the transformation. The edit distance between two graphs $\mathcal{G}_1$ and $\mathcal{G}_2$, denoted by $d(\mathcal{G}_1, \mathcal{G}_2)$, is defined by the minimum cost edit path that transforms one graph into the other.

Definition 13.2.7
Graph Edit Distance: *Given two graphs* $\mathcal{G}_1 = (\mathcal{V}_1, \mathcal{E}_1)$, *and* $\mathcal{G}_2 = (\mathcal{V}_2, \mathcal{E}_2)$, *the graph edit distance between* $\mathcal{G}_1$ *and* $\mathcal{G}_2$ *is defined by:*

$$d(\mathcal{G}_1, \mathcal{G}_2) = \min_{(ed_1, \dots, ed_k) \in \wp(\mathcal{G}_1, \mathcal{G}_2)} \sum_{i=1}^{k} c(ed_i) \tag{13.5}$$

where $\wp(\mathcal{G}_1, \mathcal{G}_2)$ *denotes the set of edit paths that transform* $\mathcal{G}_1$ *into* $\mathcal{G}_2$ *and* $c(ed)$ *denotes the cost of an edit operation* ed.

g_1 $\qquad\qquad\qquad\qquad\qquad\qquad\qquad\qquad$ g_2

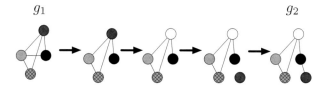

FIGURE 13.3
A possible edit path between two graphs $\mathcal{G}_1$ and $\mathcal{G}_2$.

13.3 Theoretical Aspects of GED

In this section we introduce some theoretical aspects concerning the graph edit distance. We first introduce some important aspects related to the edit costs that may affect the result with respect to the underlying application. Then, we briefly show how these edit costs can be learnt automatically.

13.3.1 Edit Costs

One of the key components of the graph edit distance is the costs of the edit operations. Edit operation costs specify how expensive it is to apply a node deletion and a node insertion instead of a node substitution, and if node operations are more important than edge operations. In this context, it is clear that the underlying graph representation must be taken into account in order to define edit costs in an appropriate way. For some representations, nodes and their labels may be more important than edges. For other representations, for example, in a network monitoring application, where a missing edge represents a broken physical link, it may be crucial whether or not two nodes are linked by an edge. Thus, in order to obtain a suitable graph edit distance measure, it is of key importance to define the edit costs in such a way that the structural variations of graphs are modeled according to the given domain. In other words, the definition of the edit costs strongly depends on the underlying application.

13.3.1.1   Conditions on Edit Costs

In the definition of edit distance given above, the distance $d(\mathcal{G}_1, \mathcal{G}_2)$ between two graphs has been defined over the set $\wp(\mathcal{G}_1, \mathcal{G}_2)$ formed by all the edit paths from $\mathcal{G}_1$ to $\mathcal{G}_2$. Theoretically, one could take a valid edit path ∂ from $\wp$ and construct an infinite number of edit paths from ∂ by arbitrarily inserting and deleting a single node, making the cardinality of $\wp$ infinite. However, as we will show, if the edit cost function satisfies a few weak conditions, it is enough to evaluate a finite subset of edit paths from $\wp$ to find one with minimum cost among all edit paths. The first condition we require on the edit costs is the *nonnegativity*,

$$c(ed) \geq 0 \tag{13.6}$$

for all node and edge edit operations ed.

This condition is certainly reasonable since edit costs are usually seen as distortions or penalty costs associated with their edit operations. The next conditions account for unnecessary substitutions. They state that removing unnecessary substitutions from an edit path does not increase the sum of the edit operation cost,

$$c(u \rightarrow w) \leq c(u \rightarrow v) + c(v \rightarrow w) \tag{13.7}$$
$$c(u \rightarrow \epsilon) \leq c(u \rightarrow v) + c(v \rightarrow \epsilon) \tag{13.8}$$
$$c(\epsilon \rightarrow v) \leq c(\epsilon \rightarrow u) + c(u \rightarrow v) \tag{13.9}$$

where ϵ denotes the null element.

For instance, instead of substituting u with v and then substituting v with w, one could safely replace these operations (right-hand side of line 1) by the operation $u \rightarrow w$ (left-hand side) and will never miss out a minimum cost edit path. It is important to note that the analogous triangle inequalities must also be satisfied for edge operations. Edit paths belonging to $\wp(\mathcal{G}_1, \mathcal{G}_2)$ and containing unnecessary substitutions $u \rightarrow w$ according to the conditions above, can therefore be replaced by a shorter edit path that is possibly less, and definitely not more, expensive. After removing those irrelevant edit paths from $\wp(\mathcal{G}_1, \mathcal{G}_2)$ containing unnecessary substitutions, one can finally eliminate from the remaining ones those edit paths that contain unnecessary insertions of nodes followed by their deletion, which is reasonable due to the nonnegativity of edit costs.

Provided that these conditions are satisfied, it is clear that only the deletions, insertions, and substitutions of nodes and edges of the two involved graphs have to be considered. Since the objective is to edit the first graph $\mathcal{G}_1 = (\mathcal{V}_1, \mathcal{E}_1)$ into the second graph $\mathcal{G}_2 = (\mathcal{V}_2, \mathcal{E}_2)$, we are only interested in how to map nodes and edges from $\mathcal{G}_1$ to $\mathcal{G}_2$. Hence, in Definition 13.2.7, the infinite set of edit paths $\wp(\mathcal{G}_1, \mathcal{G}_2)$ can be reduced to the finite set of edit paths containing edit operations of this kind only. In the remainder of this chapter, only edit cost functions satisfying the conditions stated above will be considered.

Note that the resulting graph edit distance measure need not be metric. For instance, the edit distance is not symmetric in the general case, that is $d(\mathcal{G}_1, \mathcal{G}_2) = d(\mathcal{G}_2, \mathcal{G}_1)$ may not hold. However, the graph edit distance can be turned into a metric by requiring the underlying cost functions on edit operations to satisfy the metric conditions of positive definiteness, symmetry, and the triangle inequality [69]. That is, if the conditions

$$c(u \to v) = 0 \text{ if and only if } u = v \tag{13.10}$$
$$c(p \to q) = 0 \text{ if and only if } p = q \tag{13.11}$$
$$c(u \to v) = c(v \to u) \tag{13.12}$$
$$c(p \to q) = c(q \to p) \tag{13.13}$$
$$c(u \to \epsilon) > 0 \tag{13.14}$$
$$c(\epsilon \to v) > 0 \tag{13.15}$$
$$c(p \to \epsilon) > 0 \tag{13.16}$$
$$c(\epsilon \to q) > 0 \tag{13.17}$$

are fulfilled, then the edit distance is a metric. Note that u and v represent nodes, and p and q edges.

13.3.1.2 Example of Edit Costs

For numerical node and edge labels, it is common to measure the dissimilarity of labels by the Euclidean distance and assign constant costs to insertions and deletions. For two graphs $\mathcal{G}_1 = (\mathcal{V}_1, \mathcal{E}_1)$ and $\mathcal{G}_2 = (\mathcal{V}_2, \mathcal{E}_2)$ and nonnegative parameters $\alpha, \beta, \gamma, \theta \in \mathbb{R}^+ \cup \{0\}$, this cost function is defined for all nodes $u \in \mathcal{V}_1, v \in \mathcal{V}_2$ and for all edges $p \in \mathcal{E}_1, q \in \mathcal{E}_2$ by

$$c(u \to \epsilon) = \gamma \tag{13.18}$$
$$c(\epsilon \to v) = \gamma \tag{13.19}$$
$$c(u \to v) = \alpha \cdot \|L(u) - L(v)\| \tag{13.20}$$
$$c(p \to \epsilon) = \theta \tag{13.21}$$
$$c(\epsilon \to q) = \theta \tag{13.22}$$
$$c(p \to q) = \beta \cdot \|L(u) - L(v)\| \tag{13.23}$$

where $L(\cdot)$ denotes the label of the element $\cdot$.

Note that edit costs are defined with respect to labels of nodes and edges, rather than the node and edge identifiers themselves. The cost function above defines substitution costs proportional to the Euclidean distance of the respective labels. The basic idea is that the further away two labels are, the stronger is the distortion associated with the corresponding substitution. Insertions and deletions are assigned constant costs, so that any node substitution having higher costs than a fixed threshold $\gamma + \gamma$ will be replaced by a node deletion and insertion operation. This behavior reflects the intuitive understanding that the substitution of nodes and edges should be favored to a certain degree over insertions and deletions. The parameters α and β weight the importance of the node and edge substitutions against each other, and relative to node and edge deletions and insertions. The edit cost conditions from the previous section are obviously fulfilled, and if each of the four parameters is greater than zero, we obtain a metric graph edit distance. The main advantage of the cost function above is that its definition is very straightforward.

13.3.2 Automatic Learning of Edit Costs

As we have already seen in the previous sections, the definition of the edit costs is a crucial aspect, and it is also application dependent. It means that, on one hand, given an application different edit costs may lead to very different results and, on the other hand, the same edit costs in two different applications will give in general different results. In this context, it is therefore of key importance to be able to learn adequate edit costs for a given application, so that the results are in coherence with the underlying graph representation. Consequently, the question of how to determine the best edit costs for a given situation arises. Such a task, however, may be extremely complex and therefore automatic procedures for edit cost learning would be desirable. In this section we give a sketch of two different approaches to learn edit costs in an automatic way.

13.3.2.1 Learning Probabilistic Edit Costs

A probabilistic model for edit costs was introduced in [73]. Given a space of labels L, the basic idea is to model the distribution of edit operations in the label space.

To this end, the edit costs are defined in such a way that they can be learned from a labeled sample set of graphs. The probability distribution in the label space is then adapted so that those edit paths corresponding to pairs of graphs in the same class have low costs, while higher costs are assigned if a pair of graphs belong to different classes. The process of editing an edit path into another, which is seen as a stochastic process, is motivated by the cost learning method for string edit distance presented in [74].

The method works as follows. In the first step, a probability distribution on edit operations is defined. To this end, a system of Gaussian mixtures is used to approximate the unknown underlying distribution. Under this model, assuming that the edit operations are statistically independent, we obtain a probability measure on edit paths. To derive the cost model, pairs of similar graphs are required during the training phase. The objective of the learning algorithm is to derive the most likely edit paths between the training graphs and adapt the model in such a way that the probability of the optimal edit path is increased. If we denote a multivariate Gaussian density with mean μ and covariance matrix $\sum$ as $G(.|\mu, \sum)$, the probability of an edit path $f = (ed_1, ed_2, \ldots, ed_n)$ is given by

$$p(f) = p(ed_1, ed_2, \ldots, ed_n) = \prod_{j=1}^{n} \beta_{t_j} \sum_{i=1}^{m_{t_j}} \alpha_{t_j}^i G(ed_j | \mu_{t_j}^i, \sum_{t_j}^i) \qquad (13.24)$$

where t_j denotes the type of edit operation ed_j. Every type of edit operation is additionally provided with a weight β_{t_j}, a number of mixture components m_{t_j}, and a mixture weight $\alpha_{t_j}^i$ for each component $i \in 1, 2, \ldots, m_{t_j}$. Then, the probability distribution between two graphs is obtained by considering the probability of all edit paths between them. Thus, given two graphs $\mathcal{G}_1$ and $\mathcal{G}_2$ and the set of edit paths $\wp = f_1, f_2, \ldots, f_m$ from $\mathcal{G}_1$ to $\mathcal{G}_2$, the probability of $\mathcal{G}_1$ and $\mathcal{G}_2$ is formally given by

$$p_{\text{graph}}(\mathcal{G}_1, \mathcal{G}_2) = \int_{i \in [1, \ldots, m]} p(f_i | \Phi) \, dp(f_i | \Phi) \qquad (13.25)$$

where Φ denotes the parameters of the edit path distribution.

This learning procedure leads to a distribution that assigns high probabilities to graphs belonging to the same class.

13.3.2.2 Self-Organizing Edit Costs

One disadvantage of the probabilistic edit model described before is that a certain number of graph samples are required to make an accurate estimation of the distribution parameters. The second approach, which is based on self-organizing maps (SOMs) [75], is specially useful when only a few samples are available.

In a similar way as the probabilistic method described above, SOMs can be used to model the distribution of edit operations. In this approach, the label space is represented by means of a SOM. As in the probabilistic case, a sample set of edit operations is derived from pairs of graphs belonging to the same class. The initial regular grid of the competitive layer of the SOM is initially transformed into a deformed grid. From this deformed grid, a distance measure for substitution costs and a density estimation for insertion and deletion costs are obtained. To derive the distance measure for substitutions, the SOM is trained with pairs of labels, one of them belonging to the source node (or edge) and the other belonging to the target node (or edge). The competitive layer of the SOM is then adapted so as to draw the two regions corresponding to the source and the target label closer to each other. The edit cost of node (edge) substitution is defined to be proportional to the distance in the trained SOM. In the case of deletions and insertions, the SOM is adapted so as to draw neurons closer to the deleted or inserted label. The edit costs of these operations are then defined according to the competitive neural density at the respective position. That is, the cost of deletion or insertion is lower when the number of neurons at a certain position in the competitive layer increases. Consequently, this self-organizing training procedure leads to lower costs for those pairs of graphs that are in the training set and belong to the same class. For more details the reader is referred to [76].

13.4 GED Computation

As we have already seen, one of the major advantages of the graph edit distance over other exact graph matching methods is that it allows one to potentially match every node of a graph to every node of another graph. On the one hand, this high degree of flexibility makes graph edit distance particularly appropriate for noisy data. But, on the other hand, it is usually computationally more expensive than other graph matching approaches. For the graph edit distance computation, one can distinguish between two different approaches, namely optimal and approximate or suboptimal algorithms. In the former group, the best edit path (i.e., the edit path with minimum cost) between two given graphs is always found. However, these methods suffer from an exponential time and space complexity, which makes their application only feasible for small graphs. The latter computation paradigm only ensures to find a local minimum cost among all edit paths. It has been shown that in some cases, this local minimum is not far from the global one, but this property cannot always be guaranteed. Nevertheless, the time complexity becomes usually polynomial [68].

13.4.1 Optimal Algorithms

Optimal algorithms for the graph edit distance computation are typically implemented by means of a tree search strategy which explores the space of all possible mappings of the nodes and edges of the first graph to the nodes and edges of the second graph. One of the most often used techniques is based on the well-known A* algorithm [77], which follows a best-first search strategy. Under this approach, the underlying search space is represented as an ordered tree. Thus, at the first level there is the root node, which represents the starting point of the search procedure. Inner nodes of the search tree correspond to partial solutions (i.e., partial edit paths), while leaf nodes represent complete—not necessarily optimal—solutions (i.e., complete edit paths). The tree is constructed dynamically during the algorithm execution. In order to determine the most promising node in the current search tree, i.e., the node which will be used for further expansion of the desired mapping in the next iteration, a heuristic function is usually used. Given a node n in the search tree, $g(n)$ is used to denote the cost of the optimal path from the root node to the current node n. The estimation of the cost of the remaining edit path from n to a leaf node is denoted by $h(n)$. The sum $g(n) + h(n)$ gives the total cost assigned to an open node in the search tree. One can show that, given that the estimation of the future cost $h(n)$ is lower than, or equal to, the real cost, the algorithm is admissible, i.e., an optimal path from the root node to a leaf node is guaranteed to be found [77]. For the sake of completeness, the pseudo code of an optimal algorithm for the graph edit distance computation is presented in the following.

Algorithm 2: Optimal graph edit distance algorithm.

Input : Two graphs $\mathcal{G}_1 = \{\mathcal{V}_1, \mathcal{E}_1\}$ and $\mathcal{G}_1 = \{\mathcal{V}_2, \mathcal{E}_2\}$, where
$\mathcal{V}_1 = \{u_1, \ldots, u_{|\mathcal{V}_1|}\}$ and $\mathcal{V}_2 = \{v_1, \ldots, v_{|\mathcal{V}_2|}\}$
Output: An optimal edit path $\partial_{\min} = \{u_i \to v_p, u_j \to v_q, u_k \to \epsilon, \ldots\}$ from
$\mathcal{G}_1$ to $\mathcal{G}_2$

begin

1   Init OPEN to the empty set $\{\}$
2   For each node $v \in \mathcal{V}_2$ insert the substitution $\{u_1 \to v\}$ into OPEN
3   Insert the deletion $\{u_1 \to \epsilon\}$ into OPEN
4   Remove $\partial_{\min} = \mathrm{argmin}_{n \in \text{OPEN}}\{g(\partial) + h(\partial)\}$ from OPEN
5   **while** $\partial_{\min}$ *is not a complete edit path* **do**
6   Let $\partial_{\min} = \{u_1 \to v_{i_1}, \ldots, u_k \to v_{i_k}\}$
7   **if** $k < |\mathcal{V}_1|$ **then**
8   For each node $v \in \mathcal{V}_2$ insert, $p_{\min} \cup \{u_{k+1} \to v\}$ into OPEN
9   Insert $\partial_{\min} \cup \{u_{k+1} \to \epsilon\}$ into OPEN
10   **else**
11   Insert $\partial_{\min} \cup_{v \in \mathcal{V}_2 \setminus \{v_{i_1}, \ldots, v_{i_k}\}} \{\epsilon \to v\}$ into OPEN
12   Remove $\partial_{\min} = \mathrm{argmin}_{p \in \text{OPEN}}\{g(\partial) + h(\partial)\}$ from OPEN
13   Return $\partial_{\min}$ as the solution

In Algorithm 2, the A*-based method for optimal graph edit distance computation is used. The nodes of the source graph $\mathcal{G}_1$ are processed in the order $(u_1, u_2, \ldots)$. While there are unprocessed nodes of the first graph, substitutions (line 8) and insertions (line 9) of a node are considered. If all nodes of the first graph have been processed, the remaining nodes of the second graph are inserted in a single step (line 11). These three operations produce a number of successor nodes in the search tree. The set OPEN of partial edit paths contains the search tree nodes to be processed in the next steps. The most promising partial edit path $\partial \in$ OPEN, i.e., the one that minimizes $g(\partial)+h(\partial)$, is always chosen first (lines 4 and 12). This procedure guarantees that the first complete edit path found by the algorithm is always optimal, i.e., has minimal costs among all possible competing paths (line 13).

Note that edit operations on edges are implied by edit operations on their adjacent nodes. Thus, insertions, deletions, and substitutions of edges depend on the edit operations performed on the adjacent nodes. Obviously, the implied edge operations can be derived from every partial or complete edit path during the search procedure given in Algorithm 2. The costs of these implied edge operations are dynamically added to the corresponding paths in OPEN.

The use of heuristics has been proposed in [77]. This allows one to integrate more knowledge about partial solutions in the search tree and possibly reduce the number of partial solutions that need to be evaluated to find the solution. The main objective in the introduction of heuristics is the estimation of a lower bound $h(\partial)$ on the future costs. The simplest scenario is to set this lower bound estimation $h(\partial)$ equal to zero for all ∂, which is equivalent to using no heuristic information about the present situation at all. On the other hand, the most complex scenario would be to perform a complete edit distance computation for each node of the search tree to a leaf node. It is easy to see, however, that in this case the function $h(\partial)$ is not a lower bound, but the exact value of the optimal costs. Of course, the computation of such a perfect heuristic is both unreasonable and intractable. Somewhere in between these two extremes, one can define a function $h(\partial)$ evaluating how many edit operations have to be performed in a complete edit path at least [69].

13.4.2 Suboptimal Algorithms

The method described so far finds an optimal edit path between two graphs. Unfortunately, regardless of the use of a heuristic function $h(\partial)$, the computational complexity of the edit distance algorithm is exponential in the number of nodes of the involved graphs. This restricts the applicability of the optimal algorithm since the running time and space complexity may be huge even for rather small graphs.

For this reason, a number of methods addressing the high computational complexity of graph edit distance computation have been proposed in recent years. One direction to tackle this problem and to make graph matching more efficient is to restrict considerations to special classes of graphs. Examples include the classes of planar graphs [24], bounded-valence graphs [23], trees [78], and graphs with unique node labels [34]. Another approach, which has been proposed recently [79], requires the nodes of graphs to be planarly embedded, which is satisfied in many computer vision applications of graph matching. Another approach is to perform a local search to solve the graph matching problem. The main idea is to optimize local criteria instead of global, or optimal, ones [80]. In [81], for instance, a linear programming method for computing the edit distance of graphs with unlabeled edges is proposed. The method can be used to derive lower and upper edit distance bounds in polynomial time. A number of graph matching methods based on genetic algorithms have been proposed [47]. Genetic algorithms offer an efficient way to cope with large search spaces, but are nondeterministic. In [82] a simple variant of an edit distance algorithm based on A* together with a heuristic is proposed. Instead of expanding all successor nodes in the search tree, only a fixed number s of nodes to be processed are kept in the OPEN set. Whenever a new partial edit path is added to OPEN in Algorithm 1, only the s partial edit paths ∂ with the lowest costs $g(\partial) + h(\partial)$ are kept, and the remaining partial edit paths in OPEN are removed. Obviously, this procedure corresponds to a pruning of the search tree during the search procedure, i.e., not the full search space is explored, but only those nodes are expanded that belong to the most promising partial matches. Another approach to solve the problem of graph edit distance computation by means of bipartite graph matching is introduced in [83]. This approach will be explained in detail in the next section since we will use it in the last part of the chapter. In a recent approach [84], graph edit distance by means of an HMM-based approach is used to perform image indexing. For more information about the graph edit distance, the reader is referrer to a recent survey [85].

13.4.3   GED Computation by Means of Bipartite Graph Matching

In this approach the graph edit distance is approximated by finding an optimal match between nodes of two graphs together with their local structure. The computation of graph edit distance is actually reduced to the *assignment problem.*

The assignment problem considers the task of finding an optimal assignment of the elements of a set A to the elements of a set B, where A and B have the same cardinality. Assuming that numerical costs c are given for each assignment pair, an optimal assignment is one which minimizes the sum of the individual assignment costs. Usually, given two sets A and B with $|A| = |B| = n$, a cost matrix $C \in M_{n \times n}$ containing real numbers is defined. The matrix elements C_{ij} correspond to the costs of assigning the i-th element of A to the j-th element of B. The assignment problem is then reduced to find a permutation $p = p_1, \ldots, p_n$ of the integers $1, 2, \ldots, n$ that minimizes $\sum_{i=1}^{n} C_{ip_i}$.

The assignment problem can be reformulated as finding an optimal matching in a complete bipartite graph and is therefore also referred to as the bipartite graph matching problem. Solving the assignment problem in a brute force manner by enumerating all permutations and selecting the one that minimizes the objective function leads to an exponential complexity which is unreasonable, of course. However, there exists an algorithm which is known as Munkres' algorithm [86] that solves the bipartite matching problem in polynomial time. The assignment cost matrix C defined above is the input to the algorithm, and the output corresponds to the optimal permutation, i.e., the assignment pairs resulting in the minimum cost. In the worst case, the maximum number of operations needed by the algorithm is $O(n^3)$. Note that the $O(n^3)$ complexity is much smaller than the $O(n!)$ complexity required by a brute force algorithm.

The assignment problem described above can be used to find the optimal correspondence between the nodes of one graph and the nodes of another graph. Let us assume that a source graph $\mathcal{G}_1 = \{\mathcal{V}_1, \mathcal{E}_1\}$ and a target graph $\mathcal{G}_2 = \{\mathcal{V}_2, \mathcal{E}_2\}$ of equal size are given. If we set $A = \mathcal{V}_1$ and $B = \mathcal{V}_2$, then one can use Munkres' algorithm in order to map the nodes of $\mathcal{V}_1$ to the nodes of $\mathcal{V}_2$ such that the resulting node substitution costs are minimal. In that approach the cost matrix C is defined such that the entry C_{ij} corresponds to the cost of substituting the i-th node of $\mathcal{V}_1$ with the j-th node of $\mathcal{V}_2$. That is, $C_{ij} = c(u_i \rightarrow v_j)$, where $u_i \in \mathcal{V}_1$ and $v_j \in \mathcal{V}_2$, for $i, j = 1, \ldots, |\mathcal{V}_1|$.

Note that the above approach only allows the matching of graphs with the same number of nodes. Such a constraint is too restrictive in practice since it cannot be expected that all graphs in a specific problem domain always have the same number of nodes. However, this lack can be overcome by defining a quadratic cost matrix C which is more general in the sense that it allows insertions and/or deletions to occur in both graphs under consideration. Thus, given two graphs $\mathcal{G}_1 = \{\mathcal{V}_1, \mathcal{E}_1\}$ and $\mathcal{G}_2 = \{\mathcal{V}_2, \mathcal{E}_2\}$ with $\mathcal{V}_1 = \{u_1, \ldots, u_m\}$ and $\mathcal{V}_2 = \{v_1, \ldots, v_n\}$, and $|\mathcal{V}_1|$ not necessarily being equal to $|\mathcal{V}_2|$, the new cost matrix C is defined as

$$C = \begin{bmatrix} c_{1,1} & c_{1,2} & \cdots & c_{1,m} & c_{1,\epsilon} & \infty & \cdots & \infty \\ c_{2,1} & c_{2,2} & \cdots & c_{2,m} & \infty & c_{2,\epsilon} & \cdots & \infty \\ \vdots & \vdots & \ddots & \vdots & \vdots & \ddots & \ddots & \infty \\ c_{n,1} & c_{n,2} & \cdots & c_{n,m} & \infty & \infty & \cdots & c_{n,\epsilon} \\ c_{\epsilon,1} & \infty & \cdots & \infty & 0 & 0 & 0 & 0 \\ \infty & c_{\epsilon,2} & \cdots & \infty & 0 & 0 & \ddots & \vdots \\ \vdots & \ddots & \ddots & \infty & \vdots & \ddots & \ddots & 0 \\ \infty & \infty & \cdots & c_{\epsilon,m} & 0 & \cdots & 0 & 0 \end{bmatrix} \quad (13.26)$$

where $c_{\epsilon j}$ denotes the costs of a node insertion $c(\epsilon \rightarrow v_j)$, $c_{i\epsilon}$ denotes the cost of a node deletion $c(u_i \rightarrow \epsilon)$, and c_{ij} denotes the cost of a node substitution $c(u_i \rightarrow v_j)$.

In the matrix C, the diagonal of the bottom left corner represents the costs of all possible node insertions. In the same way, the diagonal of the right upper corner represents the costs of all possible node deletions. Finally, the left upper corner of the cost matrix represents the costs of all possible node substitutions. Since each node can be deleted or inserted at most once, any nondiagonal element of the right-upper and left-lower part is set to ∞. The bottom right corner of the cost matrix is set to zero since substitutions of the form $c(\epsilon \rightarrow \epsilon)$ do not cause any costs.

Munkres' algorithm [86] can be executed using the new cost matrix C as input. The algorithm finds the optimal permutation $\hbar = \hbar_1, \ldots, \hbar_{n+m}$ of the integers

$1, 2, \ldots, n + m$ that minimizes $\sum_{i=1}^{n+m} C_{i\hbar}$, i.e., the minimum cost. Obviously, this is equivalent to the minimum cost assignment of the nodes of $\mathcal{G}_1$ to the nodes of $\mathcal{G}_2$. Thus, the answer of the Munkres' algorithm (each row and each column contains only one assignment), ensures that each node of graph $\mathcal{G}_1$ is either uniquely assigned to a node of $\mathcal{G}_2$ (substitution), or to the node ϵ (deletion). In the same way, each node of graph $\mathcal{G}_2$ is either uniquely assigned to a node of $\mathcal{G}_1$ (substitution), or to the node ϵ (insertion). The remaining ϵ nodes corresponding to rows $n + 1, \ldots, n + m$ and columns $m + 1, \ldots, m + n$ in C that are not used cancel each other out without any costs.

Note that, up to now, only the nodes have been taken into account, and no information about the edges has been considered. This could lead to obtaining a poor approximation of the real cost of the optimal edit path. Therefore, in order to achieve a better approximation of the true edit distance, it is highly desirable to involve edge operations and their costs in the node assignment process as well. In order to achieve this goal, the authors propose an extension of the cost matrix [87]. Thus, to each entry $c_{i,j}$ in the cost matrix C, i.e. to each node substitution $c(u_i \rightarrow v_j)$, the minimum sum of edge edit operation costs implied by the substitution is added. For instance, imagine that $\mathcal{E}_{u_i}$ denotes the set of edges adjacent to node u_i and $\mathcal{E}_{v_j}$ denotes the set of edges adjacent to node v_j. We can define a cost matrix D similarly to Equation (13.26) with the elements of $\mathcal{E}_{u_i}$ and $\mathcal{E}_{v_j}$. Then, applying the Munkres' algorithm over this matrix D one can find the optimal assignment of the elements $\mathcal{E}_{u_i}$ to the elements of $\mathcal{E}_{v_j}$. Clearly, this procedure leads to the minimum sum of edge edit costs implied by the given node substitution $u_i \rightarrow v_j$. Then, these edge edit costs are added to the entry $c_{i,j}$. For deletions and insertions the procedure is slightly different. For the entries $c_{i,\epsilon}$, which denote the cost of a node deletion, the cost of the deletion of all adjacent edges of u_i is added, and for the entries $c_{\epsilon,j}$, which denote the cost of a node insertion, the cost of all insertions of the adjacent edges of v_j is added.

Regardless of the improved handling of the edges, note that Munkres' algorithm used in its original form is optimal for solving the assignment problem, but it provides us with a suboptimal solution for the graph edit distance problem only. This is due to the fact that each node edit operation is considered individually (considering the local structure only), such that no implied operations on the edges can be inferred dynamically. The result returned by Munkres' algorithm corresponds to the minimum cost mapping, according to matrix C, of the nodes of $\mathcal{G}_1$ to the nodes of $\mathcal{G}_2$. Given this mapping, the implied edit operations of the edges are inferred, and the accumulated costs of the individual edit operations on both nodes and edges can be computed. The approximate edit distance values obtained by this procedure are equal to, or larger than, the exact distance values, since the bipartite matching algorithm finds an optimal solution in a subspace of the complete search space. Although the method provides only approximate solutions to the graph edit distance, it has been shown that, in some cases, the provided results are very close to the optimal solutions [83]. This fact in conjunction with the polynomial time complexity makes this method particularly useful. In a recent paper [88], an even faster version of this method has been reported, where the assignment algorithm of Volgenant and Jonker

[89] rather than the Munkres' algorithm is used. An application of bipartite graph matching to approximately solving the graph isomorphism problem is described in [90].

13.5   Applications of GED

In this final part of the chapter we will show three applications where the graph edit distance is used. The first two applications, the weighted mean of a pair of graphs and the graph embedding into vector spaces, are general procedures that can be widely used in the context of other applications. An example of the general nature of weighted mean and graph embedding is the third application, the median graph computation, where the first two procedures are used to solve this hard problem approximately.

13.5.1   Weighted Mean of a Pair of Graphs

Consider two points in the n-dimensional real space, $x, y \in \mathbb{R}^n$, where $n \geq 1$. Their weighted mean can be defined as a point z such that

$$z = \gamma x + (1 - \gamma)y, \ \ 0 \leq \gamma \leq 1 \tag{13.27}$$

Clearly, if $\gamma = \frac{1}{2}$, then z is the (normal) mean of x and y. If z is defined according to Equation 13.27 then $z - x = (1 - \gamma)(y - x)$ and $y - z = (1 - \gamma)(y - x)$. In other words, z is a point on the line segment in n dimensions that connects x and y, and the distance between z and both x and y is controlled via the parameter γ.

In [91] the authors described the same concept but applied it to the domain of graphs, which resembles the weighted mean as described by Equation 13.27. Formally speaking, the weighted mean of a pair of graphs can be defined as follows.

Definition 13.5.1
Weighted mean: *Let $\mathcal{G}$ and $\mathcal{G}'$ be graphs with labels from the alphabet L. Let U be the set of graphs that can be constructed using labels from L and let*

$$I = \{h \in U \mid d(\mathcal{G}, \mathcal{G}') = d(\mathcal{G}, h) + d(h, \mathcal{G}')\},$$

be the set of intermediate graphs. Given $0 \leq a \leq d(\mathcal{G}, \mathcal{G}')$, the weighted mean of $\mathcal{G}$ and $\mathcal{G}'$ is a graph

$$g'' = WM(\mathcal{G}, \mathcal{G}', a) = \arg \min_{h \in I} |d(\mathcal{G}, h) - a|. \tag{13.28}$$

Intuitively, given two graphs, $\mathcal{G}$ and $\mathcal{G}'$, and a parameter a, the weighted mean is an intermediate graph, not necessarily unique, whose distance to $\mathcal{G}$ is as similar as possible to a. Consequently, its distance to $\mathcal{G}'$ is also the closest to $d(\mathcal{G}, \mathcal{G}') - a$. Again, although the definition is valid for any distance function, we let d be the graph edit distance. For this distance function, an efficient computation of the weighted mean is given in [91]. Figure 13.4 shows an example of the weighted mean of a pair of graphs where the distance is the graph edit distance.

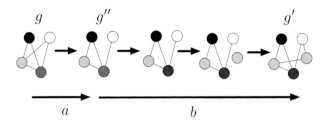

FIGURE 13.4
Example of the weighted mean of a pair of graphs. In this case a coincides with the cost of deletion of the edge between the white and the light grey nodes. Therefore, the graph $\mathcal{G}''$ is a weighted mean for which $|d(\mathcal{G}, \mathcal{G}'') - a| = 0$.

Note that the so called *error*, $\epsilon(a) = |d(\mathcal{G}, \mathcal{G}'') - a|$, is not necessarily null. This fact, regardless of the exactness of the computation, depends on the properties of the search space U.

As a practical example of the use of the weighted mean, consider graph representations of drawings consisting of straight line segments. Figure 13.5(a) shows an example of these kind of drawings. In our representation, a node in the graph corresponds to an endpoint of a line segment or a point of intersection between two line segments, and edges represent lines that connect two adjacent intersections or end points which each other. The label of a node represents the location of that node in the image. There are no edge labels. Figure 13.5(b) shows an example of such a representation. The cost of the edit operations under this representation is as follows:

- Node insertion and deletion cost $c(u \rightarrow \epsilon) = c(\epsilon \rightarrow u) = \beta_1$

- Node substitution cost $c(u, v) = c((u_1, u_2), (v_1, v_2)) = \sqrt{(u_1 - v_1)^2 + (u_2 - v_2)^2}$

- Cost of deleting or inserting an edge e is $c(e \rightarrow \epsilon) = c(\epsilon \rightarrow e) = \beta_2$

The cost of a node substitution is equal to the Euclidean distance of the locations of the two considered nodes in the image, while the cost of node deletions and insertions, as well as edge deletions and insertions, are user defined constants, β_1 and β_2. As there are no edge labels, no edge substitutions need to be considered. Figure 13.6

FIGURE 13.5
Example of graph-based representation.

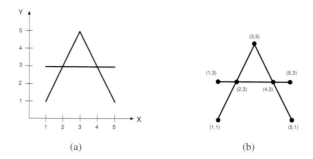

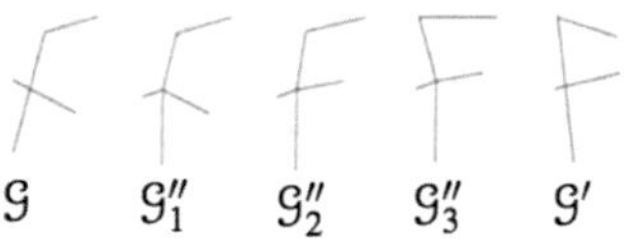

FIGURE 13.6
Example of the weighted mean of a pair of graphs. $\mathcal{G}$ and $\mathcal{G}'$ represent the original graphs while the intermediate graphs $\mathcal{G}_1''$ to $\mathcal{G}_3''$ represents graphs obtained with different values of a.

shows an example of the result of applying the weighted mean of two graphs with different values of a.

From the domain of line drawing analysis, a number of practical examples of weighted mean graph were given in the literature. These examples demonstrate that weighted mean is a useful concept to synthesize patterns $\mathcal{G}''$ that have certain degrees of similarity to given patterns $\mathcal{G}$ and $\mathcal{G}'$. In the examples the intuitive notion of shape similarity occurs to correspond well with the formal concept of graph edit distance.

13.5.2 Graph Embedding into Vector Spaces

Graph embedding [92] aims at converting graphs into real vectors, and then operate in the associated space to make easier some typical graph-based tasks, such as matching and clustering [93, 94]. To this end, different graph embedding procedures have been proposed in the literature. Some of them are based on spectral graph theory. Others take advantage of typical similarity measures to perform the embedding task. For instance, a relatively early approach based on the adjacency matrix of a graph is proposed in [53]. In this work, graphs are converted into a vector representation using some spectral features extracted from the adjacency matrix of a graph. Then, these vectors are embedded into eigenspaces with the use of the eigenvectors of the covariance matrix of the vectors. This approach is then used to perform graph clustering. Another approach has been presented in [95]. This work is similar to the previous one, but in this case the authors use the coefficients of some symmetric polynomials, constructed from the spectral features of the Laplacian matrix, to represent the graphs in a vectorial form. Finally, in a recent approach [96], the idea is to embed the nodes of a graph into a metric space and view the graph edge set as geodesics between pairs of points on a Riemannian manifold. Then, the problem of matching the nodes of a pair of graphs is viewed as the alignment of the embedded point sets. In this section we will describe a new class of graph embedding procedures based on the selection of some prototypes and graph edit distance computation. The basic intuition of this work is that the description of regularities in the observation of classes and objects is the basis to perform pattern classification. Thus, based on the selection of concrete prototypes, each point is embedded into a vector space by taking its distance to all these prototypes. Assuming these prototypes have been chosen appropriately, each class will form a compact zone in the vector space.

General Embedding Procedure Using Dissimilarities

The idea of this graph embedding framework [87] stems from the seminal work done by Duin and Pekalska [97] where dissimilarities for pattern representation are used for the first time. Later, this method was extended so as to map string representations into vector spaces [98]. The following approach further extends and generalizes methods described in [97, 98] to the domain of graphs. The key idea of this approach is to use the distances of an input graph to a number of training graphs, termed *prototype graphs*, as a vectorial description of the graph. That is, we use the dissimilarity representation for pattern recognition rather than the original graph based representation.

Assume we have a set of training graphs $T = \{\mathcal{G}_1, \mathcal{G}_2, \dots, \mathcal{G}_n\}$ and a graph dissimilarity measure $d(\mathcal{G}_i, \mathcal{G}_j)$ $(i, j = 1 \dots n; \mathcal{G}_i, \mathcal{G}_j \in T)$. Then, a set $\Omega = \{\rho_1, \dots, \rho_m\} \subseteq T$ of m prototypes is selected from T (with $m \leq n$). After that, the dissimilarity between a given graph of $\mathcal{G} \in T$ and every prototype $\rho \in \Omega$ is computed. This leads to m dissimilarity values, $d_1, \dots, d_m$ where $d_k = d(\mathcal{G}, \rho_k)$. These dissimilarities can be arranged in a vector $(d_1, \dots, d_m)$. In this way, we can transform any graph of the training set T into an m-dimensional vector using the prototype set P. More formally this embedding procedure can be defined as follows:

Definition 13.5.2
Graph embedding: *Given a set of training graphs* $T = \{\mathcal{G}_1, \mathcal{G}_2, \dots, \mathcal{G}_n\}$ *and a set of prototypes* $\Omega = \{\rho_1, \dots, \rho_m\} \subseteq T$, *the graph embedding*

$$\psi : T \longrightarrow \mathbb{R}^m \tag{13.29}$$

is defined as the function

$$\psi(\mathcal{G}) \longrightarrow (d(\mathcal{G}, \rho_1), d(\mathcal{G}, \rho_2), \dots, d(\mathcal{G}, \rho_m)) \tag{13.30}$$

where $\mathcal{G} \in T$, and $d(\mathcal{G}, \rho_i)$ is a graph dissimilarity measure between the graph $\mathcal{G}$ and the i-th prototype ρ_i.

Obviously, by means of this definition a vector space is obtained where each axis is associated with a prototype graph $\rho_i \in \Omega$. The coordinate values of an embedded graph $\mathcal{G}$ are the distances of $\mathcal{G}$ to the elements in P. In this way one can transform any graph $\mathcal{G}$ from the training set T as well as any other graph set S (for instance, a validation or a test set of a classification problem), into a vector of real numbers. In other words, the graph set to be embedded can be arbitrarily extended. Training graphs which have been selected as prototypes before have a zero entry in their corresponding graph map. Although any dissimilarity measure can be employed to embed the graphs, the graph edit distance has been used to perform graph classification with very good results in [87].

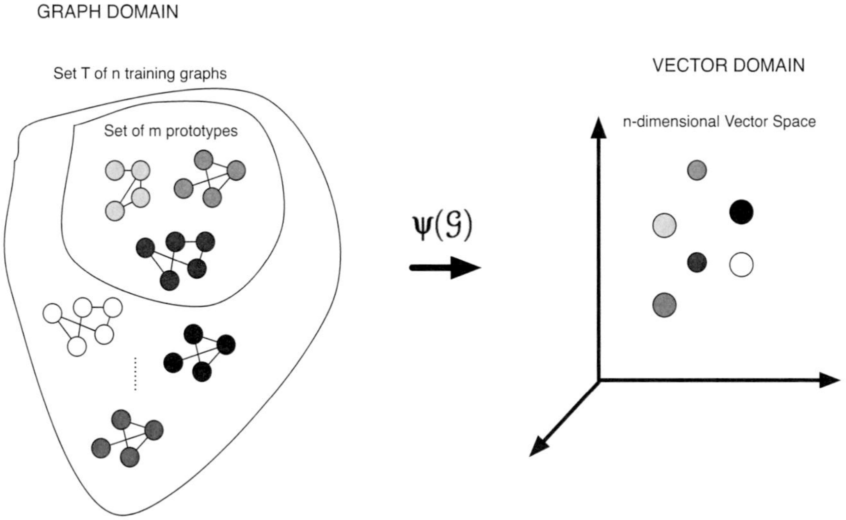

FIGURE 13.7
Overview of graph embedding.

13.5.3   Median Graph Computation

In some machine learning algorithms a representative of a set of objects is needed. For instance, in the classical k-means clustering algorithm, a representative of each cluster is computed and used at the next iteration to reorganize the clusters. In classification tasks using the k-nearest neighbour classifier, a representative of a class could be useful to reduce the number of comparisons needed to assign the unknown input pattern to its closest class. While the computation of a representative of a set is a relatively simple task in the vector domain (it can be computed by means of the mean or the median vector), it is not clear how to obtain a representative of a set of graphs. One of the proposed solutions to this problem is the median graph.

In this section we present a novel approach to the computation of the median graph that is faster and more accurate than previous approximate algorithms. It is based on the graph embedding procedure presented in the last section.

Introduction to the Median Graph Problem

The generalized median graph has been proposed to represent a set of graphs.

Definition 13.5.3
Median Graph: *Let U be the set of graphs that can be constructed using labels from L. Given $S = \{\mathcal{G}_1, \mathcal{G}_2, ..., \mathcal{G}_n\} \subseteq U$, the generalized median graph $\bar{\mathcal{G}}$ of S is defined as:*

$$\bar{\mathcal{G}} = \operatorname*{argmin}_{\mathcal{G} \in U} \sum_{\mathcal{G}_i \in S} d(\mathcal{G}, \mathcal{G}_i) \qquad (13.31)$$

That is, the generalized median graph $\bar{\mathcal{G}}$ of S is a graph $\mathcal{G} \in U$ that minimizes the sum of distances (SOD) to all the graphs in S. Notice that $\bar{\mathcal{G}}$ is usually not a member of S, and in general more than one generalized median graph may exist for a given set S. Graph $\bar{\mathcal{G}}$ can be seen as a representative of the set. Consequently, it can be potentially used by any graph-based algorithm where a representative of a set of graphs in needed, for example, in graph-based clustering.

However, the computation of the median graph is extremely complex. As implied by Equation (13.31), a distance measure $d(\mathcal{G}, \mathcal{G}_i)$ between the candidate median $\mathcal{G}$ and every graph $\mathcal{G}_i \in S$ must be computed. Therefore, since the computation of the graph distance is a well-known NP-complete problem, the computation of the generalized median graph can only be done in exponential time, both in the number of graphs in S and their size (even in the special case of strings, the time required is exponential in the number of input strings [99]). As a consequence, in real applications we are forced to use suboptimal methods in order to obtain approximate solutions for the generalized median graph in reasonable time. Such approximate methods [100, 101, 102, 103] apply some heuristics in order to reduce the complexity of the graph distance computation and the size of the search space. Recent works [104, 105] rely on graph embedding in vector spaces.

Computation via Embedding

Given a set $S = \{\mathcal{G}_1, \mathcal{G}_2, \ldots, \mathcal{G}_n\}$ of n graphs, the first step is to embed every graph of S into the n-dimensional space of real numbers, i.e., each graph becomes a point in $\mathbb{R}^n$ (Figure 13.7). The second step consists in computing the median using the vectors obtained in the previous step. Finally, we go from the vector space back

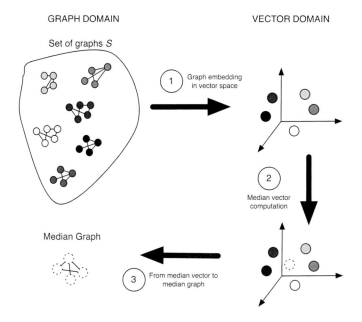

FIGURE 13.8
Overview of the approximate procedure for median graph computation.

to the graph domain, converting the median vector into a graph. The resulting graph is taken as the median graph of S. These three steps are depicted in Figure 13.8. In the following subsections, these three main steps will be further explained.

- **Step I. Graph Embedding in Vector Space:** The embedding procedure we use in this paper follows the procedure proposed in [106] (Section 13.5.2) but we let the training set T and the prototype set P be the same, i.e, the set S for which the median graph is to be computed. So, we compute the graph edit distance between every pair of graphs in set S. These distances are arranged in a distance matrix. Each row or column of the matrix can be seen as an n-dimensional vector. Since each row and column of the distance matrix is assigned to one graph, such an n-dimensional vector is the vectorial representation of the corresponding graph (Figure 13.9).

- **Step II. Computation of the Median Vector:** Once all graphs have been embedded in the vector space, the median vector is computed. To this end we use the concept of *Euclidean median*.

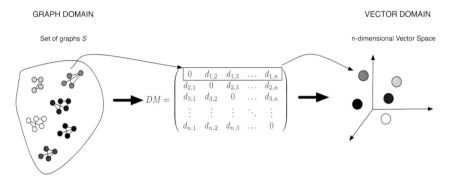

FIGURE 13.9
Step I. Graph embedding.

Definition 13.5.4
Euclidean median: *Given a set* $X = \{x_1, x_2, \ldots, x_m\}$ *of* m *points with* $x_i \in \mathbb{R}^n$ *for* $i = 1 \ldots m$, *the geometric median is defined as*

$$\text{Euclidean median} = \underset{y \in \mathbb{R}^n}{\operatorname{argmin}} \sum_{i=1}^{m} ||x_i - y||$$

where $||x_i - y||$ *denotes the Euclidean distance between the points* $x_i, y \in \mathbb{R}^n$.

That is, the *Euclidean median* is a point $y \in \mathbb{R}^n$ that minimizes the sum of the Euclidean distances between itself and all the points in X. It corresponds with the definition of the median graph, but in the vector domain. The Euclidean median cannot be calculated in a straightforward way. The exact location of the Euclidean median can not always be found when the number of elements in X is greater than 5 [107]. No algorithm in polynomial time is known, nor has the problem been shown to be NP-hard [108]. In this work, we will use the most common approximate algorithm for the computation of the Euclidean median, which is the Weiszfeld's algorithm [109]. It is an iterative procedure that converges to the Euclidean median. To this end, the algorithm first chooses an initial estimate solution y (this initial solution is often chosen randomly). Then, the algorithm defines a set of weights that are inversely proportional to the distances from the current estimate to the samples, and creates a new estimate that is the weighted average of the samples according to these weights. The algorithm may finish either when a predefined number of iterations is reached, or under some other criteria, for instance, when the difference between the current estimate and the previous one is less than a threshold.

- **Step III. Back to Graph Domain:** This step is similar to the so called preimage problem in spectral analysis where one wants to go back from the spectral to the original domain. In our particular case, the last step in median graph computation is to transform the Euclidean median into a graph. Such a graph will be considered as an approximation of the median graph of the set S. To this end we will use a triangulation procedure based on the weighted mean of a pair of graphs [91] and the edit path between two given graphs.

This triangulation procedure, illustrated in Figure 13.10, works as follows. Given the n-dimensional points representing every graph in S (the white dots in Figure 13.10(a)), and the Euclidean Median vector v_m (the grey dot in Figure 13.10(a)), we first select the three closest points to the Euclidean median (v_1 to v_3 in Figure 13.10(a)). Notice that we know the corresponding graph of each of these points (in Figure 13.10(a) we have indicated this fact by labelling them with the pair $v_j, \mathcal{G}_j$ with $j = 1 \ldots 3$). Then, we compute the median vector v'_m of these three points (represented as a black dot in Figure 13.10(a)). Notice that v'_m is in the plane formed by v_1, v_2 and v_3. With v_1 to v_3 and v'_m at hand (Figure 13.10(b)), we arbitrarily choose two out of these three points (without loss of generality we can assume that we select v_1 and v_2), and we project the remaining point (v_3) onto the line joining v_1 and v_2. In this way, we obtain a point v_i in between v_1 and v_2 (Figure 13.10(c)). With this point at hand, we can compute the percentage of the distance in between v_1 and v_2 where v_i is located (Figure 13.10(d)). As we know the corresponding graphs of the points v_1 and v_2 we can obtain the graph $\mathcal{G}_i$ corresponding to v_i by applying the weighted mean procedure (Figure 13.10(e)). Once $\mathcal{G}_i$ is known, we can obtain the percentage of distance in between v_i and v_3 where v'_m is located and obtain $\mathcal{G}'_m$ applying again the weighted mean procedure (Figure 13.10(f)). Finally, $\mathcal{G}'_m$ is chosen as the approximation for the generalized median of the set S.

Discussion of the Approximations

The proposed approximate embedding procedure is composed of three steps: the graph embedding into a vector space, the median vector computation and the return to the graph domain. Each of these steps introduces some kind of approximation to the final solution. In the first step, in order to deal with large graphs, an approximate edit distance algorithm is normally used. Thus, each vector representing a graph includes small errors in its coordinates with respect to the optimal distance between two graphs. Nevertheless, the two approximate methods we used for the edit distance computation [82, 83] provide correlation scatter plots showing a high accuracy with respect to the exact distance computation. Also, the median vector computation introduces a certain amount of error, since the Weiszfeld method obtains approximations for the median vector. This factor may lead to choosing three points that might not be the best points to go back to the graph domain. In addition, small errors may be introduced when choosing v'_m instead of v_m to obtain the weighted mean of a pair of graphs. Finally, when the weighted mean between two points is computed, the graph edit path is composed of a set of discrete steps, each having its own cost. When returning to the graph domain, the percentage of distance needed to obtain the weighted mean of a pair of graphs may fall in between two of these edit operations. Since we choose only one of them, small errors may also be introduced in this step.

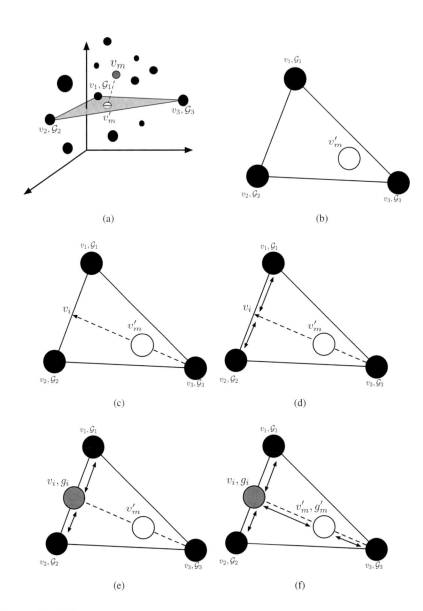

FIGURE 13.10
Illustration of the triangulation procedure.

Despite of these possible errors, the proposed method has shown very good results and compares favourably to competing algorithms [105].

13.6 Conclusions

In this chapter we have reviewed the graph edit distance (GED) from three points of view. It is a method that is recognized as one of the most flexible and universal approaches to error-tolerant graph matching. In the first part of the chapter, we introduced fundamental concepts related to GED, such as graph similarity to finally introduce the concept of graph edit distance. Then we focused on the theory underlying the graph edit distance, putting special emphasis on the cost function, giving some examples, and introducing the fundamental problem of automatic edit cost learning. In the second part of the chapter we gave an introduction to GED computation. We reviewed in detail exact and approximate methods for GED computation. We finished this second part by introducing one of the most recent approximate algorithms for GED computation, which is based on bipartite graph matching. Finally, the last part of the chapter is devoted to GED applications. We concentrated on three basic applications: the weighted mean of a pair of graphs, graph embedding into vector spaces, and the median graph computation problem. All of them demonstrate that graph edit distance is a truly versatile and flexible similarity measure that is potentially beneficial to various fields where graph representations are being used.

Bibliography

[1] R. Duda, P. Hart, and D. Stork, *Pattern Classification*. Wiley Interscience. 2nd Edition, 2000.

[2] M. Friedman and A. Kandel, *Introduction to Pattern Recognition*. World Scientific, 1999.

[3] L. da F. Costa, F. A. Rodrigues, G. Travieso, and P. V. Boas, "Characterization of complex networks: A survey of measurements," *Advances in Physics*, vol. 56, no. 1, pp. 167–242, 2007.

[4] L. Grady and J. R. Polimeni, *Discrete Calculus: Applied Analysis on Graphs for Computational Science*. Springer, 2010.

[5] A. Schenker, H. Bunke, M. Last, and A. Kandel, *Graph-Theoretic Techniques for Web Content Mining*. River Edge, NJ: World Scientific Publishing Co., Inc., 2005.

[6] S. Lee, J. Kim, and F. Groen, "Translation-, rotation-, and scale invariant recognition of hand-drawn symbols in schematic diagrams," *International Journal of Pattern Recognition and Artificial Intelligence*, vol. 4, pp. 1–15, 1990.

[7] J. Lladós, E. Martí, and J. J. Villanueva, "Symbol recognition by error-tolerant subgraph matching between region adjacency graphs," *IEEE Trans. Pattern Anal. Mach. Intell.*, vol. 23, no. 10, pp. 1137–1143, 2001.

[8] S. W. Lu, Y. Ren, and C. Y. Suen, "Hierarchical attributed graph representation and recognition of handwritten chinese characters," *Pattern Recognition*, vol. 24, no. 7, pp. 617–632, 1991.

[9] J. Rocha and T. Pavlidis, "A shape analysis model with applications to a character recognition system," *IEEE Trans. Pattern Anal. Mach. Intell.*, vol. 16, no. 4, pp. 393–404, 1994.

[10] V. Cantoni, L. Cinque, C. Guerra, S. Levialdi, and L. Lombardi, "2-D object recognition by multi-scale tree matching," *Pattern Recognition*, vol. 31, no. 10, pp. 1443–1454, 1998.

[11] M. Pelillo, K. Siddiqi, and S. W. Zucker, "Matching hierarchical structures using association graphs," *IEEE Trans. Pattern Anal. Mach. Intell.*, vol. 21, no. 11, pp. 1105–1120, 1999.

[12] F. Serratosa, R. Alquézar, and A. Sanfeliu, "Function-described graphs for modelling objects represented by sets of attributed graphs," *Pattern Recognition*, vol. 36, no. 3, pp. 781–798, 2003.

[13] E. K. Wong, "Model matching in robot vision by subgraph isomorphism," *Pattern Recognition*, vol. 25, no. 3, pp. 287–303, 1992.

[14] K. Shearer, H. Bunke, and S. Venkatesh, "Video indexing and similarity retrieval by largest common subgraph detection using decision trees," *Pattern Recognition*, vol. 34, no. 5, pp. 1075–1091, 2001.

[15] J. R. Ullman, "An algorithm for subgraph isomorphism," *Journal of ACM*, vol. 23, no. 1, pp. 31–42, Jan. 1976.

[16] J. J. McGregor, "Backtrack search algorithms and the maximal common subgraph problem," *Software - Practice and Experience*, vol. 12, no. 1, pp. 23–24, 1982.

[17] H. Bunke, X. Jiang, and A. Kandel, "On the minimum common supergraph of two graphs," *Computing*, vol. 65, no. 1, pp. 13–25, 2000.

[18] D. C. Schmidt and L. E. Druffel, "A fast backtracking algorithm to test directed graphs for isomorphism using distance matrices," *J. ACM*, vol. 23, no. 3, pp. 433–445, 1976.

[19] L. P. Cordella, P. Foggia, C. Sansone, and M. Vento, "Fast graph matching for detecting cad image components," in *15th International Conference on Pattern Recognition*, 2000, pp. 6034–6037.

[20] C. S. L.P Cordella, P. Foggia and M. Vento, "An improved algorithm for matching large graphs," in *Proc. 3rd IAPR Workshop Graph-Based Representations in Pattern Recognition*, 2001, pp. 149–159.

[21] P. Fosser, R. Glantz, M. Locatelli, and M. Pelillo, "Swap strategies for graph matching," in *GbRPR*, ser. Lecture Notes in Computer Science, E. R. Hancock and M. Vento, Eds., vol. 2726. Springer, 2003, pp. 142–153.

[22] B. McKay, "Practical graph isomorphism," *Congressus Numerantum*, vol. 30, pp. 45–87, 1981.

[23] E. Luks, "Isomorphism of graphs of bounded valence can be tested in polynomial time," *Journal of Computer and System Sciences*, vol. 25, pp. 42–65, 1982.

[24] J. E. Hopcroft and J. K. Wong, "Linear time algorithm for isomorphism of planar graphs (preliminary report)," in *STOC '74: Proceedings of the sixth annual ACM symposium on Theory of computing.* New York: ACM Press, 1974, pp. 172–184.

[25] M. R. Garey and D. S. Johnson, *Computers and Intractability: A Guide to the Theory of NP-Completeness.* New York: W. H. Freeman & Co., 1979.

[26] E. B. Krissinel and K. Henrick, "Common subgraph isomorphism detection by backtracking search," *Softw., Pract. Exper.*, vol. 34, no. 6, pp. 591–607, 2004.

[27] Y. Wang and C. Maple, "A novel efficient algorithm for determining maximum common subgraphs," in *9th International Conference on Information Visualisation, IV 2005, 6-8 July 2005, London, UK.* IEEE Computer Society, 2005, pp. 657–663.

[28] E. Balas and C. S. Yu, "Finding a maximum clique in an arbitrary graph," *SIAM J. Comput.*, vol. 15, no. 4, pp. 1054–1068, 1986.

[29] P. J. Durand, R. Pasari, J. W. Baker, and C. che Tsai, "An efficient algorithm for similarity analysis of molecules," *Internet Journal of Chemistry*, vol. 2, no. 17, 1999.

[30] C. Bron and J. Kerbosch, "Finding all the cliques in an undirected graph," *Communication of the ACM*, vol. 16, pp. 189–201, 1973.

[31] H. Bunke, P. Foggia, C. Guidobaldi, C. Sansone, and M. Vento, "A comparison of algorithms for maximum common subgraph on randomly connected graphs," in *Structural, Syntactic, and Statistical Pattern Recognition, Joint IAPR International Workshops SSPR 2002 and SPR 2002, Windsor, Ontario, Canada, August 6-9, 2002, Proceedings. Lecture Notes in Computer Science Vol. 2396*, 2002, pp. 123–132.

[32] D. Conte, P. Foggia, and M. Vento, "Challenging complexity of maximum common subgraph detection algorithms: A performance analysis of three algorithms on a wide database of graphs," *Journal of Graph Algorithms and Applications*, vol. 11, no. 1, pp. 99–143, 2007.

[33] I. M. Bomze, M. Budunuch., P. M. Paralos, and M. Pelillo, *The Maximum Clique Problem.* Dordrecht: Kluwer Academic Publisher, 1999.

[34] P. J. Dickinson, H. Bunke, A. Dadej, and M. Kraetzl, "Matching graphs with unique node labels," *Pattern Anal. Appl.*, vol. 7, no. 3, pp. 243–254, 2004.

[35] J. W. Raymond and P. Willett, "Maximum common subgraph isomorphism algorithms for the matching of chemical structures," *Journal of Computer-Aided Molecular Design*, vol. 16, no. 7, pp. 521–533, 2002.

[36] Y. Takahashi, Y. Satoh, H. Suzuki, and S. Sasaki, "Recognition of largest common structural fragment among a variety of chemical structures," *Analytical Sciences*, vol. 3, no. 1, pp. 23–28, 1987.

[37] H. Y. Sumio Masuda and E. Tanaka, "Algorithm for finding one of the largest common subgraphs of two three-dimensional graph structures," *Electronics and Communications in Japan (Part III: Fundamental Electronic Science)*, vol. 81, no. 9, pp. 48–53, 1998.

[38] H. Bunke, "On a relation between graph edit distance and maximum common subgraph," *Pattern Recognition Letters*, vol. 18, no. 8, pp. 689–694, 1997.

[39] H. Bunke and K. Shearer, "A graph distance metric based on the maximal common subgraph," *Pattern Recognition Letters*, vol. 19, no. 3-4, pp. 255–259, 1998.

[40] W. D. Wallis, P. Shoubridge, M. Kraetz, and D. Ray, "Graph distances using graph union," *Pattern Recognition Letters*, vol. 22, no. 6/7, pp. 701–704, 2001.

[41] M.-L. Fernández and G. Valiente, "A graph distance metric combining maximum common subgraph and minimum common supergraph," *Pattern Recognition Letters*, vol. 22, no. 6/7, pp. 753–758, 2001.

[42] A. Dumay, R. van der Geest, J. Gerbrands, E. Jansen, and J. Reiber, "Consistent inexact graph matching applied to labelling coronary segments in arteriograms," in *11th International Confenrence on Pattern Recognition*, 1992, pp. III:439–442.

[43] S. Berretti, A. D. Bimbo, and E. Vicario, "Efficient matching and indexing of graph models in content-based retrieval," *IEEE Trans. Pattern Anal. Mach. Intell.*, vol. 23, no. 10, pp. 1089–1105, 2001.

[44] F. Serratosa, R. Alquezar, and A. Sanfeliu, "Function-described graphs: a fast algorithm to compute a sub-optimal matching measure," in *Proceedings of the 2nd IAPR-TC15 Workshop on Graph-based Representations in Pattern Recognition*, 1999, pp. 71–77.

[45] R. Allen, L. Cinque, S. L. Tanimoto, L. G. Shapiro, and D. Yasuda, "A parallel algorithm for graph matching and its maspar implementation," *IEEE Trans. Parallel Distrib. Syst.*, vol. 8, no. 5, pp. 490–501, 1997.

[46] S. Auwatanamongkol, "Inexact graph matching using a genetic algorithm for image recognition," *Pattern Recognition Letters*, vol. 28, no. 12, pp. 1428–1437, 2007.

[47] A. D. J. Cross, R. C. Wilson, and E. R. Hancock, "Inexact graph matching using genetic search," *Pattern Recognition*, vol. 30, no. 6, pp. 953–970, 1997.

[48] M. Singh, A. Chatterjee, and S. Chaudhury, "Matching structural shape descriptions using genetic algorithms," *Pattern Recognition*, vol. 30, no. 9, pp. 1451–1462, September 1997.

[49] P. N. Suganthan, "Structural pattern recognition using genetic algorithms," *Pattern Recognition*, vol. 35, no. 9, pp. 1883–1893, 2002.

[50] Y.-K. Wang, K.-C. Fan, and J.-T. Horng, "Genetic-based search for error-correcting graph isomorphism," *IEEE Transactions on Systems, Man, and Cybernetics, Part B*, vol. 27, no. 4, pp. 588–597, 1997.

[51] K. A. D. Jong and W. M. Spears, "Using genetic algorithms to solve NP-complete problems," in *ICGA*, J. D. Schaffer, Ed. Morgan Kaufmann, 1989, pp. 124–132.

[52] T. Caelli and S. Kosinov, "An eigenspace projection clustering method for inexact graph matching," *IEEE Trans. Pattern Anal. Mach. Intell.*, vol. 26, no. 4, pp. 515–519, 2004.

[53] B. Luo, R. C. Wilson, and E. R. Hancock, "Spectral embedding of graphs," *Pattern Recognition*, vol. 36, no. 10, pp. 2213–2230, 2003.

[54] A. Robles-Kelly and E. R. Hancock, "Graph edit distance from spectral seriation," *IEEE Trans. Pattern Anal. Mach. Intell.*, vol. 27, no. 3, pp. 365–378, 2005.

[55] A. Shokoufandeh, D. Macrini, S. J. Dickinson, K. Siddiqi, and S. W. Zucker, "Indexing hierarchical structures using graph spectra," *IEEE Trans. Pattern Anal. Mach. Intell.*, vol. 27, no. 7, pp. 1125–1140, 2005.

[56] S. Umeyama, "An eigendecomposition approach to weighted graph matching problems," *IEEE Transactions on Pattern Analysis and Machine Intelligence*, vol. 10, no. 5, pp. 695–703, September 1988.

[57] R. C. Wilson and E. R. Hancock, "Levenshtein distance for graph spectral features," in *17th Internarional Conference on Pattern Recognition*, 2004, pp. 489–492.

[58] L. Xu and I. King, "A PCA approach for fast retrieval of structural patterns in attributed graphs," *IEEE Transactions on Systems, Man and Cybernetics - Part B*, vol. 31, no. 5, pp. 812–817, October 2001.

[59] M. Carcassoni and E. R. Hancock, "Weighted graph-matching using modal clusters," in *CAIP '01: Proceedings of the 9th International Conference on Computer Analysis of Images and Patterns*. London, UK: Springer-Verlag, 2001, pp. 142–151.

[60] S. Kosinov and T. Caelli, "Inexact multisubgraph matching using graph eigenspace and clustering models," in *Structural, Syntactic, and Statistical Pattern Recognition, Joint IAPR International Workshops SSPR 2002 and SPR 2002, Windsor, Ontario, Canada, August 6-9, 2002, Proceedings. Lecture Notes in Computer Science Vol. 2396*, 2002, pp. 133–142.

[61] A. Shokoufandeh and S. J. Dickinson, "A unified framework for indexing and matching hierarchical shape structures," in *IWVF*, ser. Lecture Notes in Computer Science, C. Arcelli, L. P. Cordella, and G. S. di Baja, Eds., vol. 2059.   Springer, 2001, pp. 67–84.

[62] W.-J. Lee, R. P. W. Duin, and H. Bunke, "Selecting structural base classifiers for graph-based multiple classifier systems," in *MCS*, ser. Lecture Notes in Computer Science, N. E. Gayar, J. Kittler, and F. Roli, Eds., vol. 5997.   Springer, 2010, pp. 155–164.

[63] W. J. Christmas, J. Kittler, and M. Petrou, "Structural matching in computer vision using probabilistic relaxation," *IEEE Trans. Pattern Anal. Mach. Intell.*, vol. 17, no. 8, pp. 749–764, 1995.

[64] R. C. Wilson and E. R. Hancock, "Structural matching by discrete relaxation," *IEEE Trans. Pattern Anal. Mach. Intell.*, vol. 19, no. 6, pp. 634–648, 1997.

[65] B. Luo and E. R. Hancock, "Structural graph matching using the EM algorithm and singular value decomposition," *IEEE Trans. Pattern Anal. Mach. Intell.*, vol. 23, no. 10, pp. 1120–1136, 2001.

[66] A. P. Dempster, N. M. Laird, and D. B. Rubin, "Maximum likelihood from incomplete data via the EM algorithm," *Journal of the Royal Statistical Society*, vol. 39, no. 1, pp. 1–38, 1977. [Online]. Available: http://links.jstor.org/sici?sici=0035-9246\%281977\%2939\%3A1\%3C1\%3AMLFIDV\%3E2.0.CO\%3B2-Z

[67] S. Gold and A. Rangarajan, "A graduated assignment algorithm for graph matching," *IEEE Trans. Pattern Anal. Mach. Intell.*, vol. 18, no. 4, pp. 377–388, 1996.

[68] D. Conte, P. Foggia, C. Sansone, and M. Vento, "Thirty years of graph matching in pattern recognition," *International Journal of Pattern Recognition and Artificial Intelligence*, vol. 18, no. 3, pp. 265–298, 2004.

[69] H. Bunke and G. Allerman, "Inexact graph matching for structural pattern recognition," *Pattern Recognition Letters*, vol. 1, no. 4, pp. 245–253, 1983.

[70] M. A. Eshera and K. S. Fu, "A graph distance measure for image analysis," *IEEE Transactions on Systems, Man and Cybernetics*, vol. 14, pp. 398–408, 1984.

[71] A. Sanfeliu and K. Fu, "A distance measure between attributed relational graphs for pattern recognition," *IEEE Transactions on Systems, Man and Cybernetics*, vol. 13, no. 3, pp. 353–362, May 1983.

[72] W. H. Tsai and K. S. Fu, "Error-correcting isomorphisms of attributed relational graphs for pattern analysis," *IEEE Transactions on Systems, Man and Cybernetics*, vol. 9, pp. 757–768, 1979.

[73] M. Neuhaus and H. Bunke, "Automatic learning of cost functions for graph edit distance," *Inf. Sci.*, vol. 177, no. 1, pp. 239–247, 2007.

[74] E. S. Ristad and P. N. Yianilos, "Learning string-edit distance," *IEEE Trans. Pattern Anal. Mach. Intell.*, vol. 20, no. 5, pp. 522–532, 1998.

[75] T. Kohonen, *Self-Organizing Maps*. Berlin: Springer, 1995.

[76] M. Neuhaus and H. Bunke, "Self-organizing maps for learning the edit costs in graph matching," *IEEE Transactions on Systems, Man, and Cybernetics, Part B*, vol. 35, no. 3, pp. 503–514, 2005.

[77] P. E. Hart, N. J. Nilsson, and B. Raphael, "A formal basis for the heuristic determination of minimum costs paths," *IEEE Transactions of Systems, Science, and Cybernetics*, vol. 4, no. 2, pp. 100–107, 1968.

[78] A. Torsello, D. H. Rowe, and M. Pelillo, "Polynomial-time metrics for attributed trees," *IEEE Trans. Pattern Anal. Mach. Intell.*, vol. 27, no. 7, pp. 1087–1099, 2005.

[79] M. Neuhaus and H. Bunke, "An error-tolerant approximate matching algorithm for attributed planar graphs and its application to fingerprint classification," in *SSPR/SPR*, ser. Lecture Notes in Computer Science, A. L. N. Fred, T. Caelli, R. P. W. Duin, A. C. Campilho, and D. de Ridder, Eds., vol. 3138. Springer, 2004, pp. 180–189.

[80] M. C. Boeres, C. C. Ribeiro, and I. Bloch, "A randomized heuristic for scene recognition by graph matching," in *WEA*, ser. Lecture Notes in Computer Science, C. C. Ribeiro and S. L. Martins, Eds., vol. 3059. Springer, 2004, pp. 100–113.

[81] D. Justice and A. O. Hero, "A binary linear programming formulation of the graph edit distance," *IEEE Trans. Pattern Anal. Mach. Intell.*, vol. 28, no. 8, pp. 1200–1214, 2006.

[82] M. Neuhaus, K. Riesen, and H. Bunke, "Fast suboptimal algorithms for the computation of graph edit distance," in *Joint IAPR International Workshops, SSPR and SPR 2006. Lecture Notes in Computer Science 4109*, 2006, pp. 163–172.

[83] K. Riesen and H. Bunke, "Approximate graph edit distance computation by means of bipartite graph matching," *Image and Vision Computing*, vol. 27, no. 7, pp. 950–959, 2009.

[84] B. Xiao, X. Gao, D. Tao, and X. Li, "HMM-based graph edit distance for image indexing," *International Journal of Imaging Systems and Technology - Multimedia Information Retrieval*, vol. 8, no. 2-3, pp. 209–218, 2008.

[85] X. Gao, B. Xiao, D. Tao, and X. Li, "A survey of graph edit distance," *Pattern Analysis and Applications*, vol. 13, pp. 113–129, 2010.

[86] J. Munkres, "Algorithms for the assignment and transportation problems," *Journal of the Society of Industrial and Applied Mathematics*, vol. 5, no. 1, pp. 32–38, March 1957.

[87] K. Riesen and H. Bunke, *Graph Classification and Clustering Based on Vector Space Embedding*. World Scientific, 2010.

[88] S. Fankhauser, K. Riesen, and H. Bunke, "Speeding up graph edit distance computation through fast bipartite matching," in *GbRPR*, ser. Lecture Notes in Computer Science, X. Jiang, M. Ferrer, and A. Torsello, Eds., vol. 6658. Springer, 2011, pp. 102–111.

[89] R. Jonker and T. Volgenant, "A shortest augmenting path algorithm for dense and sparse linear assignment problems," *Computing*, vol. 38, pp. 325–340, 1987.

[90] K. Riesen, S. Fankhauser, H. Bunke, and P. J. Dickinson, "Efficient suboptimal graph isomorphism," in *GbRPR*, ser. Lecture Notes in Computer Science, A. Torsello, F. Escolano, and L. Brun, Eds., vol. 5534. Springer, 2009, pp. 124–133.

[91] H. Bunke and S. Günter, "Weighted mean of a pair of graphs," *Computing*, vol. 67, no. 3, pp. 209–224, 2001.

[92] P. Indyk, "Algorithmic applications of low-distortion geometric embeddings," in *IEEE Symposium on Foundations of Computer Science*, 2001, pp. 10–33.

[93] K. Grauman and T. Darrell, "Fast contour matching using approximate earth mover's distance," in *Proceedings of the 2004 IEEE Computer Society Conference on Computer Vision and Pattern Recognition*, 2004, pp. 220–227.

[94] M. F. Demirci, A. Shokoufandeh, Y. Keselman, L. Bretzner, and S. J. Dickinson, "Object recognition as many-to-many feature matching," *International Journal of Computer Vision*, vol. 69, no. 2, pp. 203–222, 2006.

[95] R. C. Wilson, E. R. Hancock, and B. Luo, "Pattern vectors from algebraic graph theory," *IEEE Trans. Pattern Anal. Mach. Intell.*, vol. 27, no. 7, pp. 1112–1124, 2005.

[96] A. Robles-Kelly and E. R. Hancock, "A Riemannian approach to graph embedding," *Pattern Recognition*, vol. 40, no. 3, pp. 1042–1056, 2007.

[97] E. Pekalska and R. Duin, *The Dissimilarity Representation for Pattern Recognition - Foundations and Applications.* World Scientific, 2005.

[98] B. Spillmann, M. Neuhaus, H. Bunke, E. Pekalska, and R. P. W. Duin, "Transforming strings to vector spaces using prototype selection," in *Structural, Syntactic, and Statistical Pattern Recognition, Joint IAPR International Workshops, SSPR 2006 and SPR 2006, Hong Kong, China, August 17-19, 2006, Proceedings*, 2006, pp. 287–296.

[99] C. de la Higuera and F. Casacuberta, "Topology of strings: Median string is NP-complete," *Theor. Comput. Sci.*, vol. 230, no. 1-2, pp. 39–48, 2000.

[100] M. Ferrer, F. Serratosa, and A. Sanfeliu, "Synthesis of median spectral graph," in *Second Iberian Conference of Pattern Recognition and Image Analysis. Volume 3523 LNCS*, 2005, pp. 139–146.

[101] A. Hlaoui and S. Wang, "Median graph computation for graph clustering," *Soft Comput.*, vol. 10, no. 1, pp. 47–53, 2006.

[102] X. Jiang, A. Münger, and H. Bunke, "On median graphs: Properties, algorithms, and applications," *IEEE Trans. Pattern Anal. Mach. Intell.*, vol. 23, no. 10, pp. 1144–1151, 2001.

[103] D. White and R. C. Wilson, "Mixing spectral representations of graphs," in *18th International Conference on Pattern Recognition (ICPR 2006), 20-24 August 2006, Hong Kong, China.* IEEE Computer Society, 2006, pp. 140–144.

[104] M. Ferrer, E. Valveny, F. Serratosa, I. Bardají, and H. Bunke, "Graph-based k-means clustering: A comparison of the set median versus the generalized median graph," in *CAIP*, ser. Lecture Notes in Computer Science, X. Jiang and N. Petkov, Eds., vol. 5702. Springer, 2009, pp. 342–350.

[105] M. Ferrer, E. Valveny, F. Serratosa, K. Riesen, and H. Bunke, "Generalized median graph computation by means of graph embedding in vector spaces," *Pattern Recognition*, vol. 43, no. 4, pp. 1642–1655, 2010.

[106] K. Riesen, M. Neuhaus, and H. Bunke, "Graph embedding in vector spaces by means of prototype selection," in *6th IAPR-TC-15 International Workshop, GbRPR 2007*, ser. Lecture Notes in Computer Science, vol. 4538. Springer, 2007, pp. 383–393.

[107] C. Bajaj, "The algebraic degree of geometric optimization problems," *Discrete Comput. Geom.*, vol. 3, no. 2, pp. 177–191, 1988.

[108] S. L. Hakimi, *Location Theory.* CRC Press., 2000.

[109] E. Weiszfeld, "Sur le point pour lequel la somme des distances de n points donnés est minimum," *Tohoku Math. Journal*, no. 43, pp. 355– 386, 1937.

14

The Role of Graphs in Matching Shapes and in Categorization

Benjamin Kimia

Brown University
School of Engineering
Providence, Rhode Island, USA
Email: kimia@lems.brown.edu

CONTENTS

14.1 Introduction 423
14.2 Using Shock Graphs for Shape Matching 425
14.3 Using Proximity Graphs for Categorization 428
    14.3.1 Indexing: Euclidean versus Metric Spaces 430
    14.3.2 Metric-based Indexing Methods 431
    14.3.3 Capturing the Local Topology 433
14.4 Conclusion 436
14.5 Acknowledgment 437
    Bibliography 437

14.1 Introduction

This chapter explores the value of graphs in computer vision from the perspective of representing both structural quantum variations as well as metric changes, as applied to the matching and categorization problems. It briefly overviews our particular approach in using shock graphs in matching and recognition, and proximity graphs in categorization. The key concept is that graphs can be viewed as capturing a notion of *discrete topology*, whether it is related to the shape space or to the category space. Specifically, we view both shape matching and categorization as establishing first a similarity space. Subsequently, the cost of a "geodesic" from one point to another is computed as the internal of dissimilarity over this path. The technology for shape matching uses shock graphs, a form of the medial axis, while the technology for categorization uses proximity graphs.

Graphs are ubiquitous in computer vision, as evidenced by the large body of work that is presented and referenced in this book. Some insight into why this is the case can be obtained by observing that the set of images arising from an equivalence class of objects, e.g., horses, are affected by a vast variety of visual transformations such as changes in illumination, pose, and viewing distance, occlusion, articulation, within category type variation, camera gain control, interaction with background clutter, and others. The sorting out of commonalities from among such an incredible variety of variations to establish an equivalence class is certainly an incredible feat, one that our visual system solves on a regular basis.

The regularity of objects, in contrast, places them in a low-dimensional space embedded in a very high-dimensional image space. Even within this low-dimensional space the distribution of object representations is not uniform. Rather, in the *similarity space*, objects from the same class are clustered and arranged as a cluster in a neighborhood of similar categories, see; Figure 14.1. This implies that in an arbitrary path traversing an exemplar from one category to an exemplar from another category, say from a giraffe to a horse, two distinct types of events can be observed: initially a giraffe's deformation may lead us to other giraffe exemplars, but eventually as the deformation varies in type and magnitude we venture into noncategory exemplars before we finally arrive at a horse exemplar. This typifies the *dual nature* of our observations in general in computer vision, one *quantitative* and *metric*, and one *structural* and in *quantum* steps. This dual nature is one of the key challenges underlying the success of computer vision algorithms: **Optimizing a functional is metric and usually assumes a uniform structure, while what is required is optimizing over both quantum changes and metric variations.** Herein lies the intrinsic value of an attributed graph as a representation: it embodies both these aspects.

FIGURE 14.1
(Left) A cartoon view of the space of observable of images, shapes, categories, etc. as lying on a low-dimensional manifold embedded in a very high-dimensional space. The structure of these abstract points when arranged in an appropriate similarity space is inherently clustered, reflecting the nature of objects and categories, as highlighted for the space of categories on the right. (Right) The similarity space of categories naturally reflects the topology of categories at the basic, super- and sub-levels of categorization, and clustering of exemplar categories.

Jan Koenderink observed a fundamental point about the use of representations for shape matching, that representations of shape can be either *static* of *dynamic* [1]. In a *static* representation, each shape maps to a point in some abstract space, e.g., a high-dimensional Euclidean feature space. Two shapes are then matched by measuring the distance between the two representation, e.g., using the Euclidean distance, in the representation space. In contrast, in a *dynamic* representation, the matching of two points is not done by measuring distance in the representation space, but rather it is based on the cost of deforming one representation to another in an intrinsic manner[1]. The cost itself is computed by summing up the cost of deforming each shape to a neighboring shape, thus requiring a basic, atomic cost for comparing two infinitesimally close shapes. The deformation cost can then be viewed as the cost of the *geodesic* between two shapes in an underlying similarity space. This is equally true for a category space. The critical advantage of a dynamic representation is that it only allows for the transformations which are legal, in contrast to infeasible transformations implicitly represented in a static space. Furthermore, the straight line in the embedding space may not be the appropriate geodesic when we consider the requirements of matching. For example, it may not traverse or correspond to "legal" shapes.

The framework presented is based on these ideas and is fairly general. The basic ingredients are visual constructs such as shapes, categories, image fragments, *etc.* These constructs are assumed to have an inherent sense of similarity which can then be used to define a continuous space as above. The next step is to find the optimal geodesic connecting two points whose cost defines a global sense of similarity. Such an approach, however, must consider that for any practical implementation the similarity space must be discretized so that the geodesic path costs can be computed in a realistic manner. Such a discretization requires three steps:

1. Definition of an *equivalence class* for the points (shapes, categories, *etc.*) in the underlying space;

2. Definition of which equivalence classes are immediate neighbors;

3. Definition of an atomic costs for transitioning from one equivalence class to another.

This structure converts the original continuous space to a discrete one, one which can be effectively represented by an *attributed graph*: each equivalence class is a node of the graph, while each immediate neighbor defines a link or an edge in the graph. Attributing each graph link with a cost allows for the computation of costs for each geodesic path by summation. We now illustrate the construction of this structure in the context of two examples, one for matching shapes for recognition and one for representing categorization for indexing.

---

[1]This is akin to establishing a Riemannian metric in the tangent space via an inner product that is fully locally defined, yet it is capable of defining global geodesic cost. The critical point is that this is done *without* requiring an embedding.

14.2   Using Shock Graphs for Shape Matching

While the word *shape* or *form* has a strongly intuitive meaning to us, Figure 14.2, the construction of such a concept in computer vision has been elusive. Shape is multifaceted [2]: *(1)* it simultaneously involves a representation of the shape *silhouette* as well as the *shape interior*. Concepts such as the contour tangents are explicit in one and implicit in the other, and vice versa for a "neck"; *(2)* shape involves *local attributes* as well as *global attributes*; *(3)* shape can be thought of being made as *hierarchy of parts* or as whole; *(4)* shape can be perceived at various *scales*. A representation of such a complex notion by a simple collection of features embedded in a high-dimensional Euclidean space would be hard-pressed to capture the subtly of shape.

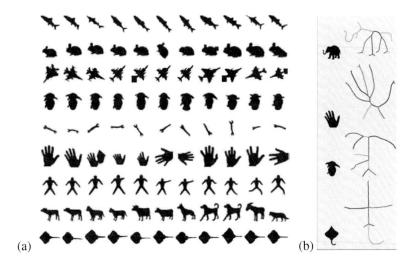

(a)   (b)

FIGURE 14.2
a) A database of shapes which has been popular in gauging the performance of shape matching algorithms. b) The shock graph of a few shapes is illustrated partially by showing the decomposition of shock segments, but not the shock dynamics.

Our approach to shape is based the *shock-graph*, a variant of the *medial axis*. Formally, the medial axis is the closure of centers of circles with at least two points of tangency to the contour of the shape. A classic construction of the medial axis is as the quench points of "grassfire", when the shape is considered as field of grass with fire initiated at the shape boundary [3]. In this wave propagation analogy, the medial axis can also be viewed as the shock points of the evolving front [4, 2]. In this case, each shock point has a dynamic associated with it that allows a finer classification than that allowed by the static view of the medial axis; some points are sources of flow, others are sinks of flow, and some other are junctures of flow, with continuously flowing segments connecting them. This leads to a formal definition of a shock graph as a directed graph with nodes representing flow sources, sinks, and junctions, while links are continuous segments connecting these. See [5] for a formal classification of the local form of shock points, [6] for its use in 2D recognition, [7] for 3D recognition, and [8] for top-down model-based segmentation; examples are illustrated in Figure 14.2b. Observe that this dynamic view of a shock graph allows for a geometrical decomposition of medial axis into finer branches, leading to a qualitatively more informative description. Given a medial axis sketch one gets a coarse idea of the equivalence class of shapes that gave rise to it, but given a shock graph this class is much better defined. That the topology of the shock graph, without any metric information, captures the categorical structure of the shape is a critical element in its success in shape recognition. The actual dynamics of shock propagation on the shock graph then provides the metric structure for its reconstruction [5]. The shock graph topology provides the answer to the first of the three steps described in the introduction:

Definition 14.2.1
Two shapes are said to be equivalent *if they share the same shock graph topology. A* shape cell *is the set of all shapes that share the same shock graph topology.*

This *shape cell* provides a *discretization* of the shape space into equivalence classes of shapes.

The second step in the process is based on the instabilities of the representation. Any representation involves such *instabilities* or *transitions*, which are defined as points where the shape undergoes an infinitesimally small smooth perturbation and the representation undergoes a *structural change*. This is unstable because a small change in the shape causes a large change in the representation. Every representation experiences such instabilities, e.g., representing the boundary of a shape with a polygon, etc. These instabilities define the neighborhood structure of a shape cell, Figure 14.3(a); the shapes in a shape cell as they deform near the instability may venture into a different distinct shape cell. Thus two shape cells are neighbors if the shapes in one can be transformed to shapes in the other by one single transition, aside from continuous deformations. The equivalence class of paths that goes through the same shape cells can then provide a recipe for the second discretization step:

Definition 14.2.2
Two deformations paths are equivalent *if they traverse the same ordered set of shape cells, Figure 14.3(b).*

The third component of our approach is an atomic cost for going from one shape cell to another. The cost of deforming two very close shapes when structural correspondences are not an issue is easier to define. For example one can use the *Hausdorff metric* or a variant. We have in the past used a measure which penalized differences in the geometric and dynamic attributes of the shock graph [6]. The search over all

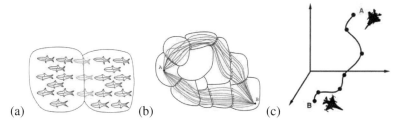

FIGURE 14.3
From [6]: (a) Each of the two sets of shapes define a shape cell. The two shapes cells are immediate neighbors. (b) The set of paths from shape A to shape B can be grouped into equivalent classes of paths, where two paths are equivalent if they are composed of the same sequence of transitions. c) The geodesic between two shapes is then the path of least cost, and this cost defines the dissimilarity between two shapes.

such paths leads to the geodesic, the path of least cost, which defines the metric of dissimilarity between two shapes, Figure 14.3(c).

With such a three-step discretization of the shape space, all that is left to do is to examine all equivalent classes of deformation paths from one shape to another. One more important aspect that keeps the problem finite without any loss of generality is to avoid considering paths which unnecessarily venture into complexity and then back towards simplicity. This can be done by splitting the full path between two shapes into two paths of nonincreasing complexity, as in Figure 14.4(a). The full space is therefore structured as two finite trees with each having one of the two shapes at the root, as in Figure 14.4(b); the two trees have many nodes which are equivalent. Our task is to search for the least cost path, Figure 14.4(c), the geodesic, by searching through this space, which can be graphically represented.

An approach to the efficient search of this space is by dynamic programming, which requires an adaptation into the domain of shock graphs. Specifically, each instability or transition of the shock graph can be thought of as an *edit* in the graph domain. We have therefore developed an edit distance for shock graphs that runs in polynomial time and refer the read to the following papers for detail [9, 10, 6]. The solution generates the optimal path of deformation and a similarity based cost which we regard as the distance between two shapes, as in Figure 14.5. This distance has been effectively used for shape matching as illustrated on multiple datasets [6]. Figure 14.6 shows shape recognition performance on the Kimia-99 and Kimia-216 shape databases [11].

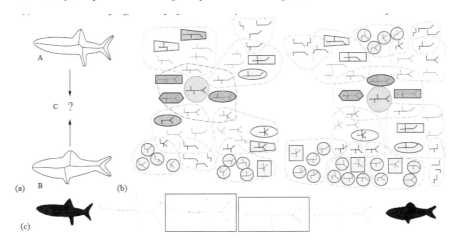

FIGURE 14.4

From [6]: (a) Two shapes and their shock graphs. (b) The set of possible equivalent classes of deformation paths from one fish to another is illustrated. (c) The optimal path involves one transition for each shape in this case. In general, however, the geodesic involves many more transitions.

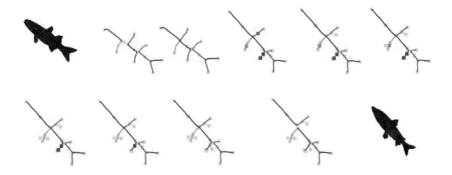

FIGURE 14.5

The sequence of edits which optimally transform one fish into another through operations on the shock graph.

14.3 Using Proximity Graphs for Categorization

A second example that illustrates the use of graphs for capturing the local topology of a similarity space is the categorization task [2]. The successful experience in building

---

[2]This work is based on Maruthi Narayanan and Benjamin Kimia, Proximity Graphs: Applications to Shape Categorization, June 2010 Technical Report LEMS-211 Division of Engineering, Brown University.

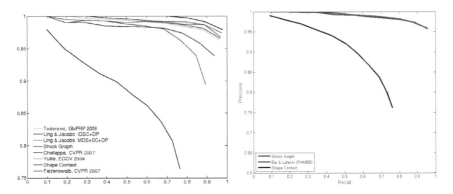

FIGURE 14.6

The recognition performance of the shock-graph matching based on the approach described above for the Kimia-99 and Kimia-216 shape databases [11].

an explicit structure for the shape space by using a local similarity and neighborhood structure, as described in the previous section, motivates the construction of a similar structure for the spaces of abstract categories, denoted as space X. Our assumption is that we can define a metric of dissimilarity d for any pair of points in X, expecting it to be meaningful when the categories are fairly similar, but not necessarily meaningful when they are not. Given a query q, the proximity search task is to find its closest neighbors or neighbors within a given range. Typically, the computation of the metric is expensive so that the $O(N)$ computation of $d(q, x_i)$, where x_i is one of N sample points in the metric space X, is not feasible without some form of *index structure* [12, 13, 14, 15]. The key insight here is to represent a metric space via a *proximity graph* and use this *discrete topology* to construct a *hierarchical index* to allow for *sublinear search*, while allowing for *dynamic insertion and deletion* of points for incremental learning.

14.3.1   Indexing: Euclidean versus Metric Spaces

There is a significant difference between *indexing* into a Euclidean space and a general metric space, especially when the latter is intrinsically high-dimensional. In general, indexing into Euclidean spaces relies on classical spatial access methods such as KD-tree [16], R-tree [17], and Quadtree [18], with a long history of use in computer vision, e.g., [19, 20]. However, when objects cannot be represented as points in a Euclidean space, a *low-distortion embedding* is often used, e.g., the Karhunen-Loeve transform [21], PCA [22], MDS [23, 24], random projections [25, 26, 27], to approximate a metric space. Unfortunately, a low-distortion embedding for the shape space requires a very high-dimensional global space: using the shape similarity metric described earlier, the hammer (square) and bird (diamond) categories are well separated in R^2, but the addition of keys (triangle) now requires three dimensions, Figure 14.7. This trend continues with more categories so that the embedding becomes practical beyond a few categories. It is well-known that the search becomes intractable beyond 10-20 dimensions due to the *curse of dimensionality* [28, 29]. Thus, Euclidean spatial access methods are not appropriate for our applications.

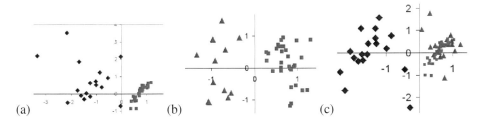

FIGURE 14.7

The required embedding dimension for many categories is high: Two dimensions suffice for embedding the categories of hammer and bird (a) (this is done by standard MDS), or hammer and keys (b), but the three categories of hammer, keys, and birds require an additional dimension. This increase in dimensions with the addition of new categories continues at a sub-linear rate, but the resulting dimensions is very high.

14.3.2   Metric-based Indexing Methods

Metric-based indexing methods go back to Burkhard and Keller [30], and typically use the triangle inequality and the classical divide-and-conquer approach [31, 32, 33, 34, 35, 36, 37, 32, 30, 38, 33]; see [39] for a review. *Metric-based indexing* methods are typically either based on pivots or on clustering. *Pivot-based* algorithms use the distance from query to a set of distinguished elements (pivots or prototypes).The triangle inequality is used to discard elements represented by the pivot without necessarily computing a distance to query. These algorithms generally improve as more pivots are added, although the space requirements of the indexes increase as well.

Clustering or *compact partitioning* algorithms divide the set into spatial zones as compact as possible, either by Voronoi type regions delimited by hyperplanes [33, 34, 35] or a covering radius [36, 37] or both [32]. They are able to discard complete zones by performing few distance evaluations between the query q and a centroid of the zone. The partition into zones can be *hierarchical*, but the indexes use a fixed amount of memory and do not improve by having more space. As shown in [14], compact partitioning algorithms deal better with high dimensional metric spaces, with pivot-based system becoming impractical beyond dimension 20.

We have explored the use of exact metric-based methods and have shown that they have poor indexing efficiency for our application [41] where the best indexing efficiency among the *VP-tree* [33, 38], AESA [42], unbalanced BK-tree [43], GH-tree [33], and GNAT [32] on a database of shapes achieved savings of 39% of the computations for AESA, not nearly enough to tackle much larger databases. A key issue underlying poor indexing is the high-dimensionality of data, indicated by a

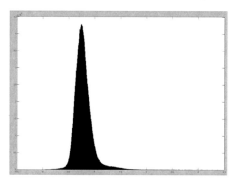

FIGURE 14.8

The intrinsic dimensionality of a shape dataset is revealed by the "peakiness" of the histogram of distances between pairs of shape. Specifically [31, 40] defines intrinsic dimensionality of a metric space as $\rho = \frac{\mu^2}{2\sigma^2}$, where μ and σ^2 are the mean and variance of its histogram of distances. The histogram of pairwise distanced for the MPEG7 shape dataset gives an intrinsic dimensionality of 22. This number is quite high for such a relatively small dataset of 70 categories and 20 exemplars per category, when compared to a realistic situation.

peaky histogram, Figure 14.8. As the intrinsic data dimensionality increases, indexes lose their efficiency since all the elements are more or less at the same distance from each other. Intuitively, pivot-based methods define "rings" of points at the same distance around pivots. The intersections of rings define the zones in which the space is partitioned, which need to remain spatially compact as the dimension grows, requiring numerous rings and therefore pivots. Compact partitioning by construction defines compact regions and requires no additional memory; It is easier to prove that a compact region is far enough from the query than to do the same for a sparse region. This may explain why AESA had the best performance in our applications.

Another class of recent methods, *nonlinear dimensionality reduction (NDR)*, applies when the data lives in a high-dimensional space but lies on a low-dimensional manifold [44, 45, 46]. In [44] the "geodesic" distance between two points is estimated by limiting the interaction of each point to its $k$nearest neighbors, and then MDS is applied. In [45, 46] each point is described as a weighted combination of its k-nearest neighbors. Both techniques assume a uniform distribution over the manifold and suffer when this assumption is violated. Indeed, our preliminary results using the ISOMAP [44] on our 1032 shape database showed that while the idea of limiting the interaction of each shape to a local neighborhood is useful, such demands should not be made globally.

We have earlier explored [41] a *wave propagation approach* to use a kNN graph connecting each shape to its k-nearest neighbors (see a summary of our approach in [47], pages 637-641) and used a number of exemplars as initial seeds for a wave propagation-based search. The multiply connected front propagated only from its closest point to the query. This rather simple method *doubled* the performance from

39% to 78% [41]. While this is encouraging and significant, this level of savings is not yet nearly sufficient for practical indexing into millions of fragments and thousands of categories. What is lacking is a method to capture local topology as described below.

14.3.3 Capturing the Local Topology

Exact metric search methods have focused their attention on the triangle inequality, which is weak in a high-dimensional space for partitioning space for access. Our spaces, however, enjoy an additional underlying structure which can be exploited, namely, a "manifold" structure, speaking in intuitive terms. Visual objects and categories are not randomly distributed, but are highly clustered and have lower dimensionality than their embedding spaces. As in NDR methods above, capturing local topology is a first step towards defining such a structure. Specifically, with an explicit local topology, we can define the notion of propagation and wavefront. This allows us to define Euclidean-like concepts such as *direction*. Given a point and a front around this point, *the ordering of points along the front defines a sense of direction*. This sense of direction is very important in Euclidean spaces as it helps avoid further search in a direction that takes us away from the query. Metric spaces which do not have a sense of direction are at a great loss, having to rely on the much weaker triangle inequality.

The need for a local topology arises for other reasons as well. Second, the metric of similarity we define, e.g., between fragments and categories, only makes sense when the two points are close to each other: we can define the distance between a horse and a donkey and compare it in a meaningful way to the distance between a horse and a zebra. But how can the distance between a horse and a clock be meaningfully related to the distance between a horse and a scorpion? We argue that a *proper notion of distance* between two points should be built as the geodesic distance between the two points built up from and grounded on distances between *nearby* points. This approach has paid off in our experience with shape recognition as discussed in the previous section. Third, an idea from the cognitive science community, "categorical perception", argues that the similarity space is warped so as to optimally differentiate between neighboring categories [48]. The local representation of topology would allow a selective "stretching" of the underlying space! Fourth, our experiments also reveal that in cases where close by shapes from two distinct categories were clustered, e.g., a bird and a flying dinosaur, consideration of distances to third parties could reveal the distinction. This idea of using higher-order local structures by maintaining k-tuple structure is akin to the concept of volume preserving embedding which suggests that not only pairwise distances but also volumes must be preserved [49]. This is applicable when local topology is explicitly represented.

The kNN approach in defining a neighborhood in NDR is successful when data is uniformly distributed, but fails in clustered data. This was clearly demonstrated by the barbell configuration shown in [46]. What is lacking in the kNN approach is that *(i)* the space is often disconnected, and *(ii)* not all local directions are captured such that there are "holes" in the kNN's idea of a neighborhood.

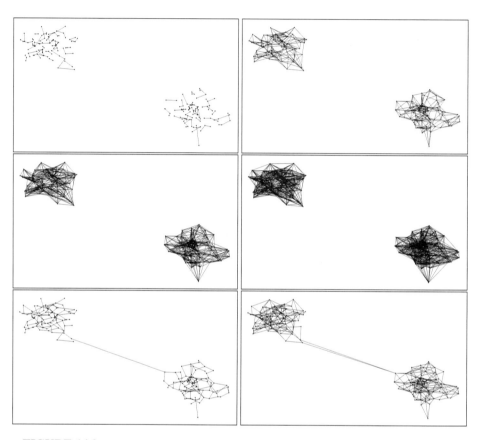

FIGURE 14.9
In the first two rows, the kNN graph is shown for reasonable values of $k = 1, 5, 10, 15$, which are all disconnected. In contrast, the bottom row shows the RNG and GG graphs which are connected between the two clusters. Of course for very high values of k the kNN graph is connected, but at this point the kNN graph is close to a complete graph, thus defeating any efficiency gains.

Capturing the local topology with proximity graphs. Alternatively, we suggest that the proper method to capture local topology is to use **proximity graphs** [50, 51], in particular the β-*skeleton* and the *Gabriel graphs*, which capture the spatial distribution of points rather than their distance-based ranking. These graphs are connected and avoid the drawback with the use of k-nearest neighbor as a local neighborhood that does not fully "fill the space" around each point when it is close to a category boundary (e.g., key versus hammer) because the nearest neighbors are all on one side, as in Figure 14.9. The geodesic paths on the Gabriel graph, in particular, have minimal distortion. Another advantage of proximity graphs is that they allow for incremental update and deletion since the construction is by nature local.

Proximity graphs are a class of graphs that represent the neighborhood relationship among a set of N exemplar points, $X^* = \{x_i, i = 1, \cdots, N\}$ in a metric space (X, d). In the following, we consider graphs whose nodes (vertices) are x_i and where the links between nodes establish a neighborhood of *proximity* relationship between the nodes. For this reason, these graphs can be described as *proximity graphs* or sometimes *geometric graphs*. In this class fall well-known graphs such as the nearest neighbor graph, Delaunay triangulation (DT) graphs, minimal spanning tree (MST), and also the less well-known graphs such as the relative neighborhood graph (RNG), the Gabriel graph, and $\beta-$skeleton graphs.

Definition 14.3.1
The nearest neighbor (NN) *of a point x_i is the point $x_j, i \neq j$, that minimizes $d(x_i, x_j)$. When two points ore more have equal minimal distances to x_i, the point with the maximum index is chosen to break the tie. The nearest neighbor is denoted as $NN(x_i)$. The* nearest neighborhood graph (NNG) *is a directed graph where each point x_i is connected to its nearest neighbor [52]. The* k-nearest neighborhood graph (kNN) *is a directed graph where each point x_i is connected to the k closest points to it.*

Definition 14.3.2
The relative neighborhood graph (RNG) *[53] of a set of points X^* is a graph such that vertices $x_i \in X^*$ and $x_j \in X^*$ are connected with an edge if and only if there exists no points $x_k \in X^*$ for which*

$$d(x_i, x_k) \leq d(x_i, x_j) \quad \text{and} \quad d(x_j, x_k) \leq d(x_i, x_j). \tag{14.1}$$

Definition 14.3.3
The Gabriel graph (GG) *[54] of a set of points X^* is a graph such that two vertices $x_i \in X^*$ and $x_j \in X^*$ are connected with an edge if and only if for all other points $x_k \in X^*$ we have*

$$d^2(x_i, x_k) + d^2(x_j, x_k) \geq d^2(x_i, x_j). \tag{14.2}$$

Definition 14.3.4
The β-skeleton graph (BSG) *[55] of a set of points X^* is a graph such that two vertices $x_i \in X^*$ and $x_j \in X^*$ are connected with an edge if and only if for all other points $x_k \in X^*$ we have*

$$d^2(x_i, x_k) + d^2(x_j, x_k) + 2\sqrt{1 - \beta^2}\, d(x_i, x_k)d(x_j, x_k) \geq d^2(x_i, x_j), \qquad \beta \leq 1. \tag{14.3}$$

$$\max((\frac{2}{\beta}-1)d^2(x_i, x_k)+d(x_j, x_k)^2, d^2(x_i, x_k)+(\frac{2}{\beta}-1)d(x_j, x_k)^2) \geq d^2(x_i, x_j),\ \beta \geq 1. \tag{14.4}$$

Note that the radius of the circles that make up the lune are $R = \frac{d}{2\beta}$.

β−skeletons were first defined by Kirkpatrick and Radke (1985) [55] as a scale-invariant extension of the alpha-shapes of [56]. The behavior of the β−skeleton graph as β varies continuously from 0 to ∞ is to form a sequence of graphs extending from the complete graph to the empty graph. There are also special cases: $\beta = 1$ leads to the Gabriel graph, which is known to contain the Euclidean minimum spanning tree in the plane. $\beta = 2$ gives the RNG. A generalization to γ−graphs can be found in [57].

There is an interesting interpretation of the β−skeletons to the support vector machines (SVM) in the context of solving a geometric classification problem [58, 59].

Observation: It should be noted that $NNG \subset RNG \subset GG \subset DT$.

Also note that the Gabriel graph is a subgraph of the Delaunay triangulation; it can be found in linear time if the Delaunay triangulation is given [60]. The Gabriel graph contains as a subgraph the Euclidean minimum spanning tree and the nearest neighbor graph. It is an instance of a beta-skeleton.

Figure 14.10 shows the comparison of proximity graphs to the kNN graphs in this context. We expect that proximity graphs will see increasing use in computer vision as a way of capturing the local topology.

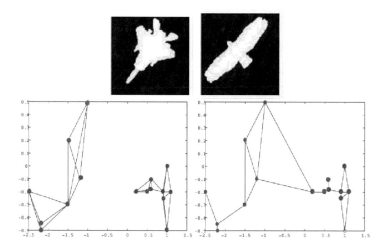

FIGURE 14.10
(top) A difficult case for capturing the local topology using the locally linear embedding from [46]. (bottom) The kNN versus proximity graph connectivity.

14.4   Conclusion

Graphs are ideal structures to capture the local topology among points in a space. We have covered two examples. The first example uses shock graphs, which capture the local topology of shape parts. The second example uses proximity graphs, which capture the local topology of a similarity space. The work also provides a template for generalizing this work to any space where the notions of local similarity and of equivalence class make sense.

14.5   Acknowledgment

The author gratefully acknowledges the support of NSF grant IIS-0083231 in funding research leading to this article.

Bibliography

[1] J. J. Koenderink and A. J. van Doorn, "Dynamic shape," *Biological Cybernetics*, vol. 53, pp. 383–396, 1986.

[2] B. B. Kimia, A. R. Tannenbaum, and S. W. Zucker, "Shapes, shocks, and deformations, I: The components of shape and the reaction-diffusion space," *International Journal of Computer Vision*, vol. 15, no. 3, pp. 189–224, 1995.

[3] H. Blum, "Biological shape and visual science," *J. Theor. Biol.*, vol. 38, pp. 205–287, 1973.

[4] B. B. Kimia, A. R. Tannenbaum, and S. W. Zucker, "Toward a computational theory of shape: An overview," in *European Conference on Computer Vision*, 1990, pp. 402–407.

[5] P. J. Giblin and B. B. Kimia, "On the intrinsic reconstruction of shape from its symmetries," *IEEE Trans. on Pattern Anal. Mach. Intell.*, vol. 25, no. 7, pp. 895–911, July 2003.

[6] T. Sebastian, P. Klein, and B. Kimia, "Recognition of shapes by editing their shock graphs," *IEEE Trans. on Pattern Anal. Mach. Intell.*, vol. 26, pp. 551–571, May 2004.

[7] C. M. Cyr and B. B. Kimia, "A similarity-based aspect-graph approach to 3D object recognition," *International Journal of Computer Vision*, vol. 57, no. 1, pp. 5–22, April 2004.

[8] N. H. Trinh and B. B. Kimia, "Skeleton search: Category-specific object recognition and segmentation using a skeletal shape model," *International Journal of Computer Vision*, vol. 94, no. 2, pp. 215–240, 2011.

[9] P. Klein, S. Tirthapura, D. Sharvit, and B. Kimia, "A tree-edit distance algorithm for comparing simple, closed shapes," in *Tenth Annual ACM-SIAM Symposium on Discrete Algorithms (SODA)*, San Francisco, California, January 9-11 2000, pp. 696–704.

[10] P. Klein, T. Sebastian, and B. Kimia, "Shape matching using edit-distance: an implementation," in *Twelfth Annual ACM-SIAM Symposium on Discrete Algorithms (SODA)*, Washington, D.C., January 7-9 2001, pp. 781–790.

[11] B. Kimia. (2011) Shape databases. [Online]. Available: http://vision.lems.brown.edu/content/available-software-and-databases

[12] G. R. Hjaltason and H. Samet, "Contractive embedding methods for similarity searching in metric spaces," CS Department, Univ. of Maryland, Tech. Rep. TR-4102, 2000.

[13] ——, "Properties of embedding methods for similarity searching in metric spaces," *IEEE Trans. Pattern Anal. Mach. Intell.*, vol. 25, no. 5, pp. 530–549, 2003.

[14] E. Chavez, G. Navarro, R. Baeza-Yates, and J. L. Marroquín, "Searching in metric spaces," *ACM Computing Surveys*, vol. 33, no. 3, pp. 273 – 321, September 2001.

[15] G. Navarro, "Analyzing metric space indexes: What for?" in *Proc. 2nd International Workshop on Similarity Search and Applications (SISAP)*. IEEE CS Press, 2009, pp. 3–10, invited paper.

[16] J. L. Bentley and J. H. Friedman, "Data structures for range searching," *ACM Computing Surveys*, vol. 11, no. 4, pp. 397–409, Decmber 1979.

[17] A. Guttman, "R-tree: A dynamic index structure for spatial searching," in *Proceedings of the 1984 ACM SIGMOD International Conference on Management of Data*, 1984, pp. 47–57.

[18] H. Samet, "The quadtree and related hierarchical data structures," *ACM Computing Surveys*, vol. 16, no. 2, pp. 187–260, 1984.

[19] O. Boiman, E. Shechtman, and M. Irani, "In defense of nearest-neighbor based image classification," in *CVPR*, 2008.

[20] N. Kumar, L. Zhang, and S. K. Nayar, "What is a good nearest neighbors algorithm for finding similar patches in images?" in *European Conference on Computer Vision*, 2008, pp. 364–378.

[21] D. Cremers, S. Osher, and S. Soatto, "Kernel density estimation and intrinsic alignment for shape priors in level set segmentation." *International Journal of Computer Vision*, vol. 69, no. 3, pp. 335–351, 2006.

[22] D. Comaniciu and P. Meer, "Mean shift: a robust approach toward feature space analysis," *IEEE Trans. on Pattern Anal. Mach. Intell.*, vol. 24, no. 5, pp. 603–619, 2002.

[23] T. F. Cox and M. A. A. Cox, *Multidimensional Scaling*. Chapman and Hall, 1994.

[24] J. B. Kruskal and M. Wish, *Multidimensional Scaling*. Beverly Hills, CA: Sage Publications, 1978.

[25] D. Achlioptas, "Database-friendly random projections," in *Symposium on Principles of Database Systems*, 2001. [Online]. Available: citeseer.nj.nec.com/achlioptas01databasefriendly.html

[26] E. Bingham and H. Mannila, "Random projection in dimensionality reduction: Applications to image and text data," in *Proceedings of the ACM SIGKDD International Conference on Knowledge Discovery and Data Mining*, San Francisco, CA, Aug. 2001, pp. 245–250.

[27] N. Gershenfeld, *The Nature of Mathematical Modelling*. Cambridge: Cambridge University Press, 1999.

[28] R. Weber, "Similarity search in high-dimensional data spaces," in *Grundlagen von Datenbanken*, 1998, pp. 138–142. [Online]. Available: citeseer.nj.nec.com/111967.html

[29] C. Böhm, S. Berchtold, and D. A. Keim, "Searching in high-dimensional spaces: Index structures for improving the performance of multimedia databases," *ACM Comput. Surv.*, vol. 33, no. 3, pp. 322–373, 2001.

[30] W. Burkhard and R. Keller, "Some approaches to best-match file searching," *Communications of the ACM*, vol. 16, no. 4, pp. 230–236, 1973.

[31] E. Chavez, G. Navarro, R. Baeza-Yates, and J. Marroquin, "Searching in metric spaces," *ACM Computing Surveys*, vol. 33, no. 3, pp. 273–321, September 2001.

[32] S. Brin, "Near neighbor search in large metric spaces," in *Proc. Intl. Conf. on Very Large Databases (VLDB)*, 1995, pp. 574–584. [Online]. Available: citeseer.nj.nec.com/brin95near.html

[33] J. Uhlmann, "Satisfying general proximity/similarity queries with metric trees," *Information Processing Letters*, vol. 40, pp. 175–179, 1991.

[34] H. Nolteimer, K. Verbarg, and C. Zirkelbach, "Monotonous bisector trees – a tool for efficient partitioning of complex scenes of geometric objects." in *Data Structures and Efficient Algorithms, Lecture Notes in Computer Science*, vol. 594, 1992, pp. 186–203.

[35] F. Dehne and H. Nolteimer, "Voronoi trees and clustering problems." *Information Systems*, vol. 12, no. 2, pp. 171–175, 1987.

[36] E. Chavez and G. Navarro, "An effective clustering algorithm to index high dimensional metric spaces." in *Proc. 7th International Symposium on String Processing and Information Retrieval (SPIRE'00)*. IEEE CS Press, 2000, pp. 75–86.

[37] P. Ciaccia, M. Patella, and P. Zezula, "M-tree: an efficient access method for similarity search in metric spaces." in *Proc. 23rd Conference on Very Large Databases (VLDB'97)*, 1997, pp. 426–435.

[38] P. Yianilos, "Data structures and algorithms for nearest neighbor search in general metric spaces," in *ACM-SIAM Symposium on Discrete Algorithms*, 1993, pp. 311–321.

[39] H. Samet, *The design and analysis of Spatial Data Structures*. Addison-Wesley, MA, 1990.

[40] E. Chávez and G. Navarro, "A compact space decomposition for effective metric indexing," *Pattern Recognition Letters*, vol. 26, no. 9, pp. 1363–1376, 2005.

[41] T. B. Sebastian and B. B. Kimia, "Metric-based shape retrieval in large databases," in *International Conference on Pattern Recognition*, vol. 3, 2002, pp. 30 291–30 296.

[42] E. Vidal, "An algorithm for finding nearest neighbors in (approximately) constant average time," *Pattern Recognition Letters*, vol. 4, pp. 145–157, 1986.

[43] E. Chavez and G. Navarro, "Unbalancing: The key to index high dimensional metric spaces," Universidad Michoacana, Tech. Rep., 1999.

[44] J. B. Tenenbaum, V. de Silva, and J. C. Langford, "A global geometric framework for nonlinear dimensionality reduction," *Science*, vol. 22, no. 290, pp. 2319–2323, December 2000.

[45] S. T. Roweis and L. K. Saul, "Nonlinear dimensionality reduction by locally linear embedding," *Science*, vol. 290, pp. 2323–2326, December 2000.

[46] L. K. Saul and S. T. Roweis, "Think globally, fit locally: unsupervised learning of low dimensional manifolds," *J. Mach. Learn. Res.*, vol. 4, pp. 119–155, 2003.

[47] H. Samet, *Foundations of Multidimensional and Metric Data Structures (The Morgan Kaufmann Series in Computer Graphics and Geometric Modeling)*. San Francisco, CA, USA: Morgan Kaufmann Publishers Inc., 2005.

[48] R. Pevtzow and S. Harnad, "Warping similarity space in category learning by human subjects: The role of task difficulty," *Proceedings Proceedings of SimCat1997:Interdisciplinary Workshop on Similarity and Categorization*, pp. 263–269, 1997.

[49] U. Feige, "Approximating the bandwidth via volume respecting embeddings," in *in Proc. 30th ACM Symposium on the Theory of Computing*, 1998, pp. 90–99. [Online]. Available: citeseer.nj.nec.com/feige99approximating.html

[50] G. Toussaint, "Proximity graphs for nearest neighbor decision rules: Recent progress," 2002. [Online]. Available: citeseer.nj.nec.com/594932.html

[51] J. Jaromczyk and G. Toussaint, "Relative neighborhood graphs and their relatives," *Proceedings of the IEEE*, vol. 80, pp. 1502–1517, 1992. [Online]. Available: citeseer.nj.nec.com/jaromczyk92relative.html

[52] D. Eppstein, M. S. Paterson, and F. F. Yao, "On nearest neighbor graphs," *Discrete & Computational Geometry*, vol. 17, no. 3, pp. 263–282, 1997.

[53] G. T. Toussaint, "The relative neighbourhood graph of a finite planar set," *Pattern Recognition*, vol. 12, no. 261–268, 1980.

[54] R. K. Gabriel and R. R. Sokal, "A new statistical approach to geographic variation analysis," *Systematic Zoology*, vol. 18, no. 3, pp. 259–278, 1969.

[55] D. Kirkpatrick and J. Radke, "A framework for computational morphology," *Computational Geometry*, pp. 217–248, 1985.

[56] H. Edelsbrunner, D. Kirkpatrick, and R. Seidel, "On the shape of a set of points in the plane," *IEEE Trans. on Information Theory*, vol. 29, pp. 551–559, 1983.

[57] R. C. Veltkamp, "The γ-neighborhood graph," *Computational Geometry*, vol. 1, no. 4, pp. 227–246, 1992.

[58] W. Zhang and I. King, "Locating support vectors via β-skeleton technique," in *Proceedings of the 9th International Conference on Neural Information Processing*, vol. 3, 2002, pp. 1423–1427.

[59] ——, "A study of the relationship between support vector machine and gabriel graph," in *International Joint Conference on Neural Networks*, vol. 1, 2002, pp. 239–244.

[60] D. Matula and R. Sokal, "Properties of gabriel graphs relevant to geographic variation research and the clustering of points in the plane," *Geographical Analysis*, vol. 12, no. 3, pp. 205–222, 1980.

15

3D Shape Registration Using Spectral Graph Embedding and Probabilistic Matching

Avinash Sharma

INRIA Grenoble Rhône-Alpes
655 avenue de l'Europe
38330 Montbonnot Saint-Martin, France
Avinash.Sharma@inrialpes.fr

Radu Horaud

INRIA Grenoble Rhône-Alpes
655 avenue de l'Europe
38330 Montbonnot Saint-Martin, France
Radu.Horaud@inrialpes.fr

Diana Mateus

Institut für Informatik
Technische Universität München
Garching b. München Germany
mateus@cs.tum.de

CONTENTS

15.1 Introduction 442
15.2 Graph Matrices 444
    15.2.1 Variants of the Graph Laplacian Matrix 445
15.3 Spectral Graph Isomorphism 445
    15.3.1 An Exact Spectral Solution 446
    15.3.2 The Hoffman–Wielandt Theorem 447
    15.3.3 Umeyama's Method 450
15.4 Graph Embedding and Dimensionality Reduction 452
    15.4.1 Spectral Properties of the Graph Laplacian 452
    15.4.2 Principal Component Analysis of a Graph Embedding 454
    15.4.3 Choosing the Dimension of the Embedding 456
    15.4.4 Unit Hyper-Sphere Normalization 457
15.5 Spectral Shape Matching 458
    15.5.1 Maximum Subgraph Matching and Point Registration 458
    15.5.2 Aligning Two Embeddings Based on Eigensignatures 461
    15.5.3 An EM Algorithm for Shape Matching 462
15.6 Experiments and Results 464
15.7 Discussion 468
15.8 Appendix: Permutation and Doubly- stochastic Matrices 469

15.9 Appendix: The Frobenius Norm 470
15.10Appendix: Spectral Properties of the Normalized Laplacian 471
     Bibliography 472

15.1 Introduction

In this chapter, we discuss the problem of 3D shape matching. Recent advancements in shape acquisition technology has led to the capture of large amounts of 3D data. Existing real-time multi-camera 3D acquisition methods provide a framewise reliable visual-hull or mesh representations for real 3D animation sequences [1, 2, 3, 4, 5, 6]. The task of 3D shape analysis involves tracking, recognition, registration, etc. Analyzing 3D data in a single framework is still a challenging task considering the large variability of the data gathered with different acquisition devices. 3D shape registration is one such challenging shape analysis task. The major difficulties in shape registration arise due to: (1) variation in the shape acquisition techniques, (2) local deformations in nonrigid shapes, (3) large acquisition discrepancies (e.g., holes, topology change, surface acquisition noise), (4) local scale change.

Most of the previous attempts of shape matching can be broadly categorized as *extrinsic* or *intrinsic* approaches depending on how they analyze the properties of the underlying manifold. Extrinsic approaches mainly focus on finding a global or local rigid transformation between two 3D shapes.

There is a large set of approaches based on variations of the iterative closest point (ICP) algorithm [7, 8, 9] that falls in the category of extrinsic approaches. However, the majority of these approaches compute rigid transformations for shape registration and are not directly applicable to nonrigid shapes. Intrinsic approaches are a natural choice for finding dense correspondences between articulated shapes, as they embed the shape in some canonical domain which preserves some important properties of the manifold, e.g., geodesics and angles. Intrinsic approaches are preferable to extrinsic approaches as they provide a global representation which is invariant to nonrigid deformations that are common in the real-world 3D shapes.

Interestingly, mesh representation also enables the adaptation of well-established graph-matching algorithms that use eigenvalues and eigenvectors of graph matrices, and are theoretically well investigated in the framework of *spectral graph theory* (SGT), e.g., [10, 11]. Existing methods in SGT are mainly theoretical results applied to small graphs and under the premise that eigenvalues can be computed exactly. However, spectral graph matching does not easily generalize to very large graphs due to the following reasons: (1) eigenvalues are approximately computed using eigensolvers, (2) eigenvalue multiplicity and hence ordering change are not well studied, and (3) exact matching is intractable for very large graphs. It is important to note that these methods mainly focus on exact graph matching, while the majority of the real-world graph matching applications involve graphs with different cardinality and for which only a subgraph isomorphism can be sought.

The main contribution of this work is to extend the spectral graph methods to very large graphs by combining spectral graph matching with *Laplacian embedding*. Since the embedded representation of a graph is obtained by dimensionality reduction, we claim that the existing SGT methods (e.g., [10]) are not easily applicable. The major contributions of this work are the following: (1) we discuss solutions for the exact and inexact graph isomorphism problems and recall the main spectral properties of the combinatorial graph Laplacian, (2) we provide a novel analysis of the commute-time embedding that allows us to interpret the latter in terms of the PCA of a graph, and to select the appropriate dimension of the associated embedded metric space, (3) we derive a unit hyper-sphere normalization for the commute-time embedding that allows us to register two shapes with different samplings, (4) we propose a novel method to find the eigenvalue-eigenvector ordering and the eigenvector signs using the eigensignatures (histograms) that are invariant to the isometric shape deformations and which fits well in the spectral graph matching framework, and (5) we present a probabilistic shape matching formulation using an *expectation maximization* (EM) framework for implementing a point registration algorithm which alternates between aligning the eigenbases and finding a vertex-to-vertex assignment.

The existing graph-matching methods that use intrinsic representations are [12, 13, 14, 15, 16, 17, 18, 19]. There is another class of methods that allows intrinsic (geodesics) and extrinsic (appearance) features to be combined and which were previously successfully applied for matching features in pairs of images [20, 21, 22, 23, 24, 25, 26, 27, 28, 29]. Some recent approaches apply hierarchical matching to find dense correspondences [30, 31, 32]. However, many of these graph-matching algorithms suffer from the problem of either computational intractability or a lack of a proper metric as the Euclidean metric is not directly applicable while computing distances on nonrigid shapes. A recent benchmarking of shape-matching methods was performed in [33]. Recently, a few methods proposed a diffusion framework for the task of shape registration [34, 35, 36].

In this chapter, we present an intrinsic approach for unsupervised 3D shape registration first proposed in [16, 37]. In the first step, dimensionality reduction is performed using the graph Laplacian, which allows us to embed a 3D shape in an isometric subspace invariant to nonrigid deformations. This leads to an embedded point cloud representation where each vertex of the underlying graph is mapped to a point in a K-dimensional metric space. Thus, the problem of nonrigid 3D shape registration is transformed into a K-dimensional point registration task. However, before point registration, the two eigen spaces need to be correctly aligned. This alignment is critical for the spectral matching methods because the two eigen spaces are defined up to the signs and the ordering of the eigenvectors of their Laplacian matrices. This is achieved by a novel matching method that uses histograms of eigenvectors as eigensignatures. In the final step, a point registration method based on a variant of the expectation-maximization (EM) algorithm [38] is applied in order to register two sets of points associated with the Laplacian embeddings of the two shapes. The pro-

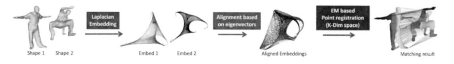

FIGURE 15.1

Overview of the proposed method. First, a Laplacian embedding is obtained for each shape. Next, these embeddings are aligned to handle the issue of sign flip and ordering change using the histogram matching. Finally, an Expectation-Maximization based point registration is performed to obtain dense probabilistic matching between two shapes.

posed algorithm alternates between the estimation of an orthogonal transformation matrix associated with the alignment of the two eigen spaces and the computation of probabilistic vertex-to-vertex assignment. Figure 15.1 presents an overview of the proposed method. According to the results summarized in [33], this method is one among the best-performing unsupervised shape-matching algorithms.

Chapter Overview

Graph matrices are introduced in Section 15.2. The problem of exact graph isomorphism and existing solutions are discussed in Section 15.3. Section 15.4 deals with dimensionality reduction using the graph Laplacian in order to obtain embedded representations for 3D shapes. In the same section, we discuss the PCA of graph embeddings and propose a unit hyper-sphere normalization for these embeddings along with a method to choose the embedding dimension. Section 15.5 introduces the formulation of maximum subgraph isomorphism before presenting a two-step method for 3D shape registration. In the first step, Laplacian embeddings are aligned using histogram matching, while in the second step we briefly discuss an EM point registration method to obtain probabilistic shape registration. Finally, we present shape-matching results in Section 15.6 and conclude with a brief discussion in Section 15.7.

15.2   Graph Matrices

A shape can be treated as a connected *undirected weighted graph* $g = \{\mathcal{V}, \mathcal{E}\}$, where $\mathcal{V}(g) = \{v_1, \ldots, v_n\}$ is the vertex set, and $\mathcal{E}(g) = \{e_{ij}\}$ is the edge set. Let $\mathbf{W}$ be the weighted adjacency matrix of this graph. Each $(i, j)^{\text{th}}$ entry of $\mathbf{W}$ matrix stores weight w_{ij} whenever there is an edge $e_{ij} \in \epsilon(g)$ between graph vertices v_i and v_j and 0 otherwise with all the diagonal elements set to 0 . We use the following notations:

The degree d_i of a graph vertex $d_i = \sum_{i \sim j} w_{ij}$ ($i \sim j$ denotes the set of vertices v_j that are adjacent to v_i), the *degree matrix* $\mathbf{D} = \text{diag}[d_1 \dots d_i \dots d_n]$, the $n \times 1$ vector $\mathbb{1} = (1 \dots 1)^\top$ (the constant vector), the $n \times 1$ *degree vector* $\mathbf{d} = \mathbf{D}\mathbb{1}$, and the *graph volume* $\text{Vol}(g) = \sum_i d_i$.

In spectral graph theory, it is common [39, 40] to use the following expression for the edge weights:

$$w_{ij} = e^{-\frac{\text{dist}^2(v_i, v_j)}{\sigma^2}}, \tag{15.1}$$

where $\text{dist}(v_i, v_j)$ denotes any distance metric between two vertices and σ is a free parameter. In the case of a *fully connected graph*, matrix $\mathbf{W}$ is also referred to as the *similarity matrix*. The *normalized weighted adjacency matrix* writes $\tilde{\mathbf{W}} = \mathbf{D}^{-1/2}\mathbf{W}\mathbf{D}^{-1/2}$. The *transition* matrix of the nonsymmetric reversible Markov chain associated with the graph is $\tilde{\mathbf{W}}_R = \mathbf{D}^{-1}\mathbf{W} = \mathbf{D}^{-1/2}\tilde{\mathbf{W}}\mathbf{D}^{1/2}$.

15.2.1 Variants of the Graph Laplacian Matrix

We can now build the concept of the *graph Laplacian operator*. We consider the following variants of the Laplacian matrix [41, 40, 42]:

- The unnormalized *Laplacian*, which is also referred to as the combinatorial *Laplacian* $\mathbf{L}$,

- The normalized *Laplacian* $\tilde{\mathbf{L}}$, and

- The random-walk *Laplacian* $\tilde{\mathbf{L}}_R$ also referred to as the *discrete Laplace operator*.

In more detail, we have:

$$\mathbf{L} = \mathbf{D} - \mathbf{W} \tag{15.2}$$
$$\tilde{\mathbf{L}} = \mathbf{D}^{-1/2}\mathbf{L}\mathbf{D}^{-1/2} = \mathbf{I} - \tilde{\mathbf{W}} \tag{15.3}$$
$$\tilde{\mathbf{L}}_R = \mathbf{D}^{-1}\mathbf{L} = \mathbf{I} - \tilde{\mathbf{W}}_R \tag{15.4}$$

with the following relations between these matrices:

$$\mathbf{L} = \mathbf{D}^{1/2}\tilde{\mathbf{L}}\mathbf{D}^{1/2} = \mathbf{D}\tilde{\mathbf{L}}_R \tag{15.5}$$
$$\tilde{\mathbf{L}} = \mathbf{D}^{-1/2}\mathbf{L}\mathbf{D}^{-1/2} = \mathbf{D}^{1/2}\tilde{\mathbf{L}}_R\mathbf{D}^{-1/2} \tag{15.6}$$
$$\tilde{\mathbf{L}}_R = \mathbf{D}^{-1/2}\tilde{\mathbf{L}}\mathbf{D}^{1/2} = \mathbf{D}^{-1}\mathbf{L}. \tag{15.7}$$

15.3 Spectral Graph Isomorphism

Let g_A and g_B be two *undirected weighted graphs* with the same number of nodes, n, and let $\mathbf{W}_A$ and $\mathbf{W}_B$ be their adjacency matrices. They are real-symmetric matrices. In the general case, the number r of distinct eigenvalues of these matrices is smaller than n. The standard spectral methods only apply to those graphs whose adjacency matrices have n distinct eigenvalues (each eigenvalue has multiplicity one), which implies that the eigenvalues can be ordered.

Graph isomorphism [43] can be written as the following minimization problem:

$$\mathbf{P}^\star = \arg\min_{\mathbf{P}} \|\mathbf{W}_A - \mathbf{P}\mathbf{W}_B\mathbf{P}^\top\|_F^2 \tag{15.8}$$

where $\mathbf{P}$ is an $n \times n$ permutation matrix (see appendix 15.8) with $\mathbf{P}^\star$ as the desired vertex-to-vertex permutation matrix and $\|\bullet\|_F$ is the Frobenius norm defined by (see Appendix 15.9):

$$\|\mathbf{W}\|_F^2 = \langle \mathbf{W}, \mathbf{W} \rangle = \sum_{i=1}^{n} \sum_{j=1}^{n} w_{ij}^2 = \mathrm{tr}(\mathbf{W}^\top \mathbf{W}) \tag{15.9}$$

Let:

$$\mathbf{W}_A = \mathbf{U}_A \mathbf{\Lambda}_A \mathbf{U}_A^\top \tag{15.10}$$
$$\mathbf{W}_B = \mathbf{U}_B \mathbf{\Lambda}_B \mathbf{U}_B^\top \tag{15.11}$$

be the eigen-decompositions of the two matrices with n eigenvalues $\mathbf{\Lambda}_A = \mathrm{diag}[\alpha_i]$ and $\mathbf{\Lambda}_B = \mathrm{diag}[\beta_i]$ and n orthonormal eigenvectors, the column vectors of $\mathbf{U}_A$ and $\mathbf{U}_B$.

15.3.1 An Exact Spectral Solution

If there exists a vertex-to-vertex correspondence that makes (15.8) equal to 0, we have

$$\mathbf{W}_A = \mathbf{P}^\star \mathbf{W}_B \mathbf{P}^{\star\top}. \tag{15.12}$$

This implies that the adjacency matrices of the two graphs should have the same eigenvalues. Moreover, if the eigenvalues are non-null and, the matrices $\mathbf{U}_A$ and $\mathbf{U}_B$ have full rank and are uniquely defined by their n orthonormal column vectors (which are the eigenvectors of $\mathbf{W}_A$ and $\mathbf{W}_B$), then $\alpha_i = \beta_i, \forall i, 1 \leq i \leq n$ and $\mathbf{\Lambda}_A = \mathbf{\Lambda}_B$. From (15.12) and using the eigen-decompositions of the two graph matrices, we obtain

$$\mathbf{\Lambda}_A = \mathbf{U}_A^\top \mathbf{P}^\star \breve{\mathbf{U}}_B \mathbf{\Lambda}_B \breve{\mathbf{U}}_B^\top \mathbf{P}^{\star\top} \mathbf{U}_A = \mathbf{\Lambda}_B, \tag{15.13}$$

where the matrix $\breve{\mathbf{U}}_B$ is defined by

$$\breve{\mathbf{U}}_B = \mathbf{U}_B \mathbf{S}. \tag{15.14}$$

Matrix $\mathbf{S} = \mathrm{diag}[s_i]$, with $s_i = \pm 1$, is referred to as a sign matrix with the property $\mathbf{S}^2 = \mathbf{I}$. Post multiplication of $\mathbf{U}_B$ with a sign matrix takes into account the fact that the eigenvectors (the column vectors of $\mathbf{U}_B$) are only defined up to a sign. Finally, we obtain the following permutation matrix:

$$\mathbf{P}^\star = \mathbf{U}_B \mathbf{S} \mathbf{U}_A^\top. \tag{15.15}$$

Therefore, one may notice that there are as many solutions as the cardinality of the set of matrices $\mathbf{S}_n$, i.e., $|\mathbf{S}_n| = 2^n$, and that *not all of these solutions correspond to a permutation matrix*. This means that there exist some matrices $\mathbf{S}^\star$ that exactly make $\mathbf{P}^\star$ a permutation matrix. Hence, all those permutation matrices that satisfy (15.15) are solutions of the exact graph isomorphism problem. Notice that once the permutation has been estimated, one can write that the rows of $\mathbf{U}_B$ can be aligned with the rows of $\mathbf{U}_A$:

$$\mathbf{U}_A = \mathbf{P}^\star \mathbf{U}_B \mathbf{S}^\star. \tag{15.16}$$

The rows of $\mathbf{U}_A$ and of $\mathbf{U}_B$ can be interpreted as isometric embeddings of the two graph vertices: A vertex v_i of g_A has as coordinates the ith row of $\mathbf{U}_A$. This means that the spectral graph isomorphism problem becomes a point registration problem, where graph vertices are represented by points in $\mathbb{R}^n$. To conclude, the exact graph isomorphism problem has a spectral solution based on (i) the eigen-decomposition of the two graph matrices, (ii) the ordering of their eigenvalues, and (iii) the choice of a sign for each eigenvector.

15.3.2 The Hoffman–Wielandt Theorem

The Hoffman–Wielandt theorem [44, 45] is the fundamental building block of spectral graph isomorphism. The theorem holds for normal matrices; Here, we restrict the analysis to real symmetric matrices, although the generalization to Hermitian matrices is straightforward.

Theorem 15.3.1
(Hoffman and Wielandt) If $\mathbf{W}_A$ and $\mathbf{W}_B$ are real-symmetric matrices, and if α_i and β_i are their eigenvalues arranged in increasing order, $\alpha_1 \leq \dots \leq \alpha_i \leq \dots \leq \alpha_n$ and $\beta_1 \leq \dots \leq \beta_i \leq \dots \leq \beta_n$, then

$$\sum_{i=1}^{n} (\alpha_i - \beta_i)^2 \leq \|\mathbf{W}_A - \mathbf{W}_B\|_F^2. \tag{15.17}$$

Proof: The proof is derived from [11, 46]. Consider the eigen-decompositions of matrices $\mathbf{W}_A$ and $\mathbf{W}_B$, (15.10), (15.11). Notice that for the time being we are free to prescribe the ordering of the eigenvalues α_i and β_i and hence the ordering of the column vectors of matrices $\mathbf{U}_A$ and $\mathbf{U}_B$. By combining (15.10) and (15.11), we write:

$$\mathbf{U}_A \mathbf{\Lambda}_A \mathbf{U}_A^\top - \mathbf{U}_B \mathbf{\Lambda}_B \mathbf{U}_B^\top = \mathbf{W}_A - \mathbf{W}_B \tag{15.18}$$

or, equivalently

$$\mathbf{\Lambda}_A \mathbf{U}_A^\top \mathbf{U}_B - \mathbf{U}_A^\top \mathbf{U}_B \mathbf{\Lambda}_B = \mathbf{U}_A^\top (\mathbf{W}_A - \mathbf{W}_B) \mathbf{U}_B. \tag{15.19}$$

By the unitary-invariance of the Frobenius norm (see Appendix 15.9) and with the notation $\mathbf{Z} = \mathbf{U}_A^\top \mathbf{U}_B$ we obtain

$$\|\mathbf{\Lambda}_A \mathbf{Z} - \mathbf{Z} \mathbf{\Lambda}_B\|_F^2 = \|\mathbf{W}_A - \mathbf{W}_B\|_F^2, \tag{15.20}$$

which is equivalent to

$$\sum_{i=1}^{n} \sum_{j=1}^{n} (\alpha_i - \beta_j)^2 z_{ij}^2 = \|\mathbf{W}_A - \mathbf{W}_B\|_F^2. \tag{15.21}$$

The coefficients $x_{ij} = z_{ij}^2$ can be viewed as the entries of a doubly-stochastic matrix $\mathbf{X}$: $x_{ij} \geq 0, \sum_{i=1}^{n} x_{ij} = 1, \sum_{j=1}^{n} x_{ij} = 1$. Using these properties, we obtain

$$
\begin{aligned}
\sum_{i=1}^{n} \sum_{j=1}^{n} (\alpha_i - \beta_j)^2 z_{ij}^2 &= \sum_{i=1}^{n} \alpha_i^2 + \sum_{j=1}^{n} \beta_j^2 - 2 \sum_{i=1}^{n} \sum_{j=1}^{n} z_{ij}^2 \alpha_i \beta_j \\
&\geq \sum_{i=1}^{n} \alpha_i^2 + \sum_{j=1}^{n} \beta_j^2 \qquad -2 \max_{Z} \left\{ \sum_{i=1}^{n} \sum_{j=1}^{n} z_{ij}^2 \alpha_i \beta_j \right\}.
\end{aligned}
\tag{15.22}
$$

Hence, the minimization of (15.21) is equivalent to the maximization of the last term in (15.22). We can modify our maximization problem to admit all the doubly-stochastic matrices. In this way we seek an extremum over a convex compact set. The maximum over this compact set is larger than or equal to our maximum:

$$\max_{Z \in \mathcal{O}_n} \left\{ \sum_{i=1}^{n} \sum_{j=1}^{n} z_{ij}^2 \alpha_i \beta_j \right\} \leq \max_{X \in \mathcal{D}_n} \left\{ \sum_{i=1}^{n} \sum_{j=1}^{n} x_{ij} \alpha_i \beta_j \right\} \tag{15.23}$$

where $\mathcal{O}_n$ is the set of orthogonal matrices and $\mathcal{D}_n$ is the set of doubly stochastic matrices (see Appendix 15.8). Let $c_{ij} = \alpha_i \beta_j$ and hence one can write that the right term in the equation above as the dot-product of two matrices:

$$\langle \mathbf{X}, \mathbf{C} \rangle = \operatorname{tr}(\mathbf{X}\mathbf{C}) = \sum_{i=1}^{n} \sum_{j=1}^{n} x_{ij} c_{ij}. \tag{15.24}$$

Therefore, this expression can be interpreted as the projection of $\mathbf{X}$ onto $\mathbf{C}$; see Figure 15.2. The Birkhoff theorem (Appendix 15.8) tells us that the set $\mathcal{D}_n$ of doubly stochastic matrices is a compact convex set. We obtain that the extrema (minimum

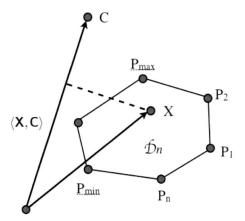

FIGURE 15.2
This figure illustrates the maximization of the dot-product $\langle \mathbf{X}, \mathbf{C} \rangle$. The two matrices can be viewed as vectors of dimension n^2. Matrix $\mathbf{X}$ belongs to a compact convex set whose extreme points are the permutation matrices $\mathbf{P}_1, \mathbf{P}_2, \ldots, \mathbf{P}_n$. Therefore, the projection of this set (i.e., $\mathcal{D}_n$) onto $\mathbf{C}$ has projected permutation matrices at its extremes, namely $\langle \mathbf{P}_{\min}, \mathbf{X} \rangle$ and $\langle \mathbf{P}_{\max}, \mathbf{X} \rangle$ in this example.

and maximum) of the projection of $\mathbf{X}$ onto $\mathbf{C}$ occur at the projections of one of the extreme points of this convex set, which correspond to permutation matrices. Hence, the maximum of $\langle \mathbf{X}, \mathbf{C} \rangle$ is $\langle \mathbf{P}_{\max}, \mathbf{X} \rangle$, and we obtain

$$\max_{X \in \mathcal{D}_n} \left\{ \sum_{i=1}^{n} \sum_{j=1}^{n} x_{ij} \alpha_i \beta_j \right\} = \sum_{i=1}^{n} \alpha_i \beta_{\pi(i)}. \tag{15.25}$$

By substitution in (15.22) we obtain

$$\sum_{i=1}^{n} \sum_{j=1}^{n} (\alpha_i - \beta_j)^2 z_{ij}^2 \geq \sum_{i=1}^{n} (\alpha_i - \beta_{\pi(i)})^2. \tag{15.26}$$

If the eigenvalues are in increasing order, then the permutation that satisfies Theorem 15.17 is the identity matrix, i.e., $\pi(i) = i$. Indeed, let's assume that for some indices k and $k+1$ we have $\pi(k) = k+1$ and $\pi(k+1) = k$. Since $\alpha_k \leq \alpha_{k+1}$ and $\beta_k \leq \beta_{k+1}$, the following inequality holds:

$$(\alpha_k - \beta_k)^2 + (\alpha_{k+1} - \beta_{k+1})^2 \leq (\alpha_k - \beta_{k+1})^2 + (\alpha_{k+1} - \beta_k)^2 \tag{15.27}$$

and hence (15.17) holds. ∎

Corollary 15.3.2
The inequality (15.17) becomes an equality when the eigenvectors of $\mathbf{W}_A$ are aligned

with the eigenvectors of $\mathbf{W}_B$ *up to a sign ambiguity:*

$$\mathbf{U}_B = \mathbf{U}_A \mathbf{S}. \tag{15.28}$$

Proof: Since the minimum of (15.21) is achieved for $\mathbf{X} = \mathbf{I}$ and since the entries of $\mathbf{X}$ are z_{ij}^2, we have that $z_{ii} = \pm 1$, which corresponds to $\mathbf{Z} = \mathbf{S}$. ∎

Corollary 15.3.3
If $\mathbf{Q}$ *is an orthogonal matrix, then*

$$\sum_{i=1}^{n}(\alpha_i - \beta_i)^2 \leq \|\mathbf{W}_A - \mathbf{Q}\mathbf{W}_B\mathbf{Q}^\top\|_F^2. \tag{15.29}$$

Proof: Since the eigen-decomposition of matrix $\mathbf{Q}\mathbf{W}_B\mathbf{Q}^\top$ is $(\mathbf{Q}\mathbf{U}_B)\mathbf{\Lambda}_B(\mathbf{Q}\mathbf{U}_B)^\top$ and since it has the same eigenvalues as $\mathbf{W}_B$, the inequality (15.29) holds and hence Corollary 15.3.3. ∎

These corollaries will be useful in the case of spectral graph-matching methods presented below.

15.3.3 Umeyama's Method

The exact spectral matching solution presented in section 15.3.1 finds a permutation matrix satisfying (15.15). This requires an exhaustive search over the space of all possible 2^n matrices. Umeyama's method presented in [10] proposes a relaxed solution to this problem as outlined below.

Umeyama [10] addresses the problem of *weighted graph matching* within the framework of spectral graph theory. He proposes two methods, the first for *undirected weighted graphs* and the second for *directed weighted graphs*. The adjacency matrix is used in both cases. Let's consider the case of undirected graphs. The eigenvalues are (possibly with multiplicities):

$$\mathbf{W}_A : \quad \alpha_1 \leq \ldots \leq \alpha_i \leq \ldots \leq \alpha_n \tag{15.30}$$
$$\mathbf{W}_B : \quad \beta_1 \leq \ldots \leq \beta_i \leq \ldots \leq \beta_n. \tag{15.31}$$

Theorem 15.3.4
(Umeyama) If $\mathbf{W}_A$ *and* $\mathbf{W}_B$ *are real-symmetric matrices with* n *distinct eigenvalues (that can be ordered),* $\alpha_1 < \ldots < \alpha_i < \ldots < \alpha_n$ *and* $\beta_1 < \ldots < \beta_i < \ldots < \beta_n$, *the minimum of*

$$J(\mathbf{Q}) = \|\mathbf{W}_A - \mathbf{Q}\mathbf{W}_B\mathbf{Q}^\top\|_F^2 \tag{15.32}$$

is achieved for

$$\mathbf{Q}^\star = \mathbf{U}_A \mathbf{S} \mathbf{U}_B^\top \tag{15.33}$$

and hence (15.29) becomes an equality:

$$\sum_{i=1}^{n}(\alpha_i - \beta_i)^2 = \|\mathbf{W}_A - \mathbf{Q}^\star \mathbf{W}_B \mathbf{Q}^{\star\top}\|_F^2. \tag{15.34}$$

Proof: The proof is straightforward. By corollary 15.3.3, the Hoffman–Wielandt theorem applies to matrices $\mathbf{W}_A$ and $\mathbf{Q}\mathbf{W}_B\mathbf{Q}^\top$. By Corollary 15.3.2, the equality (15.34) is achieved for

$$\mathbf{Z} = \mathbf{U}_A^\top \mathbf{Q}^\star \mathbf{U}_B = \mathbf{S} \tag{15.35}$$

and hence (15.33) holds. ∎

Notice that (15.33) can be written as

$$\mathbf{U}_A = \mathbf{Q}^\star \mathbf{U}_B \mathbf{S} \tag{15.36}$$

which is a *relaxed* version of (15.16): The permutation matrix in the exact isomorphism case is replaced by an orthogonal matrix.

A Heuristic for Spectral Graph Matching

Let us consider again the exact solution outlined in Section 15.3.1. Umeyama suggests a heuristic in order to avoid exhaustive search over all possible 2^n matrices that satisfy (15.15). One may easily notice that

$$\|\mathbf{P} - \mathbf{U}_A\mathbf{S}\mathbf{U}_B^\top\|_F^2 = 2n - 2\mathrm{tr}(\mathbf{U}_A\mathbf{S}(\mathbf{P}\mathbf{U}_B)^\top). \tag{15.37}$$

Using Umeyama's notations, $\bar{\mathbf{U}}_A = [|u_{ij}|], \bar{\mathbf{U}}_B = [|v_{ij}|]$ (the entries of $\bar{\mathbf{U}}_A$ are the absolute values of the entries of $\mathbf{U}_A$), one may further notice that

$$\mathrm{tr}(\mathbf{U}_A\mathbf{S}(\mathbf{P}\mathbf{U}_B)^\top) = \sum_{i=1}^{n}\sum_{j=1}^{n} s_j u_{ij} v_{\pi(i)j} \leq \sum_{i=1}^{n}\sum_{j=1}^{n} |u_{ij}||v_{\pi(i)j}| = \mathrm{tr}(\bar{\mathbf{U}}_A\bar{\mathbf{U}}_B^\top\mathbf{P}^\top).$$
$$\tag{15.38}$$

The minimization of (15.37) is equivalent to the maximization of (15.38) and the maximal value that can be attained by the latter is n. Using the fact that both $\mathbf{U}_A$ and $\mathbf{U}_B$ are orthogonal matrices, one can easily conclude that

$$\mathrm{tr}(\bar{\mathbf{U}}_A\bar{\mathbf{U}}_B^\top\mathbf{P}^\top) \leq n. \tag{15.39}$$

Umeyama concludes that when the two graphs are isomorphic, the optimum permutation matrix maximizes $\mathrm{tr}(\bar{\mathbf{U}}_A\bar{\mathbf{U}}_B^\top\mathbf{P}^\top)$, and this can be solved by the Hungarian algorithm [47].

When the two graphs are not exactly isomorphic, theorems 15.3.1 and 15.3.4 allow us to relax the permutation matrices to the group of orthogonal matrices. Therefore, with similar arguments as above, we obtain

$$\mathrm{tr}(\mathbf{U}_A\mathbf{S}\mathbf{U}_B^\top\mathbf{Q}^\top) \leq \mathrm{tr}(\bar{\mathbf{U}}_A\bar{\mathbf{U}}_B^\top\mathbf{Q}^\top) \leq n. \tag{15.40}$$

The permutation matrix obtained with the Hungarian algorithm can be used as an initial solution that can then be improved by some hill-climbing or relaxation technique [10].

The spectral matching solution presented in this section is not directly applicable to large graphs. In the next section, we introduce the notion of dimensionality reduction for graphs, which will lead to a tractable graph-matching solution.

15.4 Graph Embedding and Dimensionality Reduction

For *large* and *sparse* graphs, the results of Section 15.3 and Umeyama's method (Section 15.3.3) hold only *weakly*. Indeed, one cannot guarantee that all the eigenvalues have multiplicity equal to one: the presence of symmetries causes some of eigenvalues to have an algebraic multiplicity greater than one. Under these circumstances and due to numerical approximations, it might not be possible to properly order the eigenvalues. Moreover, for very large graphs with thousands of vertices it is not practical to compute all its eigenvalue–eigenvector pairs. This means that one has to devise a method that is able to match shapes using a small set of eigenvalues and eigenvectors.

One elegant way to overcome this problem, is to reduce the dimension of the eigenspace, along the line of spectral dimensionality reduction techniques. The eigen-decomposition of graph Laplacian matrices (introduced in Section 15.2.1) is a popular choice for the purpose of dimensionality reduction [39].

15.4.1 Spectral Properties of the Graph Laplacian

The spectral properties of the Laplacian matrices introduced in Section 15.2.1 have been thoroughly studied. They are summarized in table 15.1. We derive some subtle properties of the combinatorial Laplacian that will be useful for the task of shape registration. In particular, we will show that the eigenvectors of the combinatorial Laplacian can be interpreted as directions of maximum variance (principal components) of the associated embedded shape representation. We note that the embeddings of the normalized and random-walk Laplacians have different spectral properties which make them less interesting for shape registration, i.e., Appendix 15.10.

Laplacian	Null space	Eigenvalues	Eigenvectors
$\mathbf{L} = \mathbf{U}\boldsymbol{\Lambda}\mathbf{U}^\top$	$\mathbf{u}_1 = \mathbb{1}$	$0 = \lambda_1 < \lambda_2 \leq \ldots \leq \lambda_n$	$\mathbf{u}_{i>1}^\top \mathbb{1} = 0$ $\mathbf{u}_i^\top \mathbf{u}_j = \delta_{ij}$
$\tilde{\mathbf{L}} = \tilde{\mathbf{U}}\boldsymbol{\Gamma}\tilde{\mathbf{U}}^\top$	$\tilde{\mathbf{u}}_1 = \mathbf{D}^{1/2}\mathbb{1}$	$0 = \gamma_1 < \gamma_2 \leq \ldots \leq \gamma_n$	$\tilde{\mathbf{u}}_{i>1}^\top \mathbf{D}^{1/2}\mathbb{1} = 0$ $\tilde{\mathbf{u}}_i^\top \tilde{\mathbf{u}}_j = \delta_{ij}$
$\tilde{\mathbf{L}}_R = \mathbf{T}\boldsymbol{\Gamma}\mathbf{T}^{-1}$, $\mathbf{T} = \mathbf{D}^{-1/2}\tilde{\mathbf{U}}$	$\mathbf{t}_1 = \mathbb{1}$	$0 = \gamma_1 < \gamma_2 \leq \ldots \leq \gamma_n$	$\mathbf{t}_{i>1}^\top \mathbf{D}\mathbb{1} = 0, \mathbf{t}_i^\top \mathbf{D}\mathbf{t}_j = \delta_{ij}$

TABLE 15.1
Summary of the spectral properties of the Laplacian matrices. Assuming a connected graph, the null eigenvalue (λ_1, γ_1) has multiplicity one. The first non-null eigenvalue (λ_2, γ_2) is known as the Fiedler value, and its multiplicity is, in general, equal to one. The associated eigenvector is denoted the Fiedler vector [41].

The combinatorial Laplacian.

Let $\mathbf{L} = \mathbf{U}\boldsymbol{\Lambda}\mathbf{U}^\top$ be the spectral decomposition of the combinatorial Laplacian with $\mathbf{U}\mathbf{U}^\top = \mathbf{I}$. Let $\mathbf{U}$ be written as

$$\mathbf{U} = \begin{bmatrix} u_{11} & \cdots & u_{1k} & \cdots & u_{1n} \\ \vdots & & \vdots & & \vdots \\ u_{n1} & \cdots & u_{nk} & \cdots & u_{nn} \end{bmatrix} \tag{15.41}$$

Each column of $\mathbf{U}$, $\mathbf{u}_k = (u_{1k} \ldots u_{ik} \ldots u_{nk})^\top$ is an eigenvector associated with the eigenvalue λ_k. From the definition of $\mathbf{L}$ in (15.2) (see [39]), one can easily see that $\lambda_1 = 0$ and that $\mathbf{u}_1 = \mathbb{1}$ (a constant vector). Hence, $\mathbf{u}_{k\geq2}^\top \mathbb{1} = 0$ and by combining this with $\mathbf{u}_k^\top \mathbf{u}_k = 1$, we derive the following proposition.

Proposition 15.4.1
The components of the nonconstant eigenvectors of the combinatorial Laplacian satisfy the following constraints:

$$\sum_{i=1}^{n} u_{ik} = 0, \quad \forall k, 2 \leq k \leq n \tag{15.42}$$
$$-1 < u_{ik} < 1, \quad \forall i, k, 1 \leq i \leq n, 2 \leq k \leq n. \tag{15.43}$$

Assuming a connected graph, λ_1 has multiplicity equal to one [40]. Let's organize the eigenvalues of $\mathbf{L}$ in increasing order: $0 = \lambda_1 < \lambda_2 \leq \ldots \leq \lambda_n$. We prove the following proposition [41].

Proposition 15.4.2
For all $k \leq n$, we have $\lambda_k \leq 2\max_i(d_i)$, where d_i is the degree of vertex i.

Proof: The largest eigenvalue of $\mathbf{L}$ corresponds to the maximization of the Rayleigh quotient, or

$$\lambda_n = \max_{\mathbf{u}} \frac{\mathbf{u}^\top \mathbf{L}\mathbf{u}}{\mathbf{u}^\top \mathbf{u}}. \tag{15.44}$$

We have $\mathbf{u}^\top \mathbf{L}\mathbf{u} = \sum_{e_{ij}} w_{ij}(u_i - u_j)^2$. From the inequality $(a-b)^2 \le 2(a^2 + b^2)$ we obtain

$$\lambda_n \le \frac{2\sum_{e_{ij}} w_{ij}(u_i^2 + u_j^2)}{\sum_i u_i^2} = \frac{2\sum_i d_i u_i^2}{\sum_i u_i^2} \le 2\max_i(d_i). \quad \blacksquare \qquad (15.45)$$

This ensures an upper limit on the eigenvalues of $\mathbf{L}$. By omitting the zero eigenvalue and associated eigenvector, we can rewrite $\mathbf{L}$ as

$$\mathbf{L} = \sum_{k=2}^n \lambda_k \mathbf{u}_k \mathbf{u}_k^\top . \qquad (15.46)$$

Each entry u_{ik} of an eigenvector $\mathbf{u}_k$ can be interpreted as a real-valued function that projects a graph vertex v_i onto that vector. The mean and variance of the set $\{u_{ik}\}_{i=1}^n$ are therefore a measure of how the graph *spreads* when projected onto the k-th eigenvector. This is clarified by the following result.

Proposition 15.4.3
The mean $\bar{u}_k$ and the variance σ_{u_k} of an eigenvector $\mathbf{u}_k$. For $2 \le k \le n$, and $1 \le i \le n$ we have

$$\bar{u}_k = \sum_{i=1}^n u_{ik} = 0 \qquad (15.47)$$

$$\sigma_{u_k} = \frac{1}{n}\sum_{i=1}^n (u_{ik} - \bar{u}_k)^2 = \frac{1}{n} \qquad (15.48)$$

Proof: These results can be easily obtained from $\mathbf{u}_{k\ge2}^\top \mathbb{1} = 0$ and $\mathbf{u}_k^\top \mathbf{u}_k = 1$.
$\blacksquare$

These properties will be useful while aligning two Laplacian embeddings and thus registering two 3D shapes.

15.4.2   Principal Component Analysis of a Graph Embedding

The Moore–Penrose pseudo-inverse of the Laplacian can be written as

$$\begin{aligned}
\mathbf{L}^\dagger &= \mathbf{U}\mathbf{\Lambda}^{-1}\mathbf{U}^\top \\
&= (\mathbf{\Lambda}^{-\frac{1}{2}}\mathbf{U}^\top)^\top (\mathbf{\Lambda}^{-\frac{1}{2}}\mathbf{U}^\top) \\
&= \mathbf{X}^\top \mathbf{X} \qquad (15.49)
\end{aligned}$$

where $\mathbf{\Lambda}^{-1} = \mathrm{diag}(0, 1/\lambda_2, \ldots, 1/\lambda_n)$.

The symmetric semi-definite positive matrix $\mathbf{L}^\dagger$ is a *Gram* matrix with the same eigenvectors as those of the graph Laplacian. When omitting the null eigenvalue and associated constant eigenvector, $\mathbf{X}$ becomes a $(n-1) \times n$ matrix whose columns are the coordinates of the graph's vertices in an *embedded (or feature) space*, i.e., $\mathbf{X} = [\mathbf{x}_1 \dots \mathbf{x}_j \dots \mathbf{x}_n]$. It is interesting to note that the entries of $\mathbf{L}^\dagger$ may be viewed as *kernel* dot-products, or a Gram matrix [48]. The Gram-matrix representation allows us to embed the graph in an Euclidean feature-space where each vertex v_j of the graph is a feature point represented as $\mathbf{x}_j$.

The left pseudo-inverse operator of the Laplacian $\mathbf{L}$, satisfying $\mathbf{L}^\dagger \mathbf{L} \mathbf{u} = \mathbf{u}$ for any $\mathbf{u} \perp \text{null}(\mathbf{L})$, is also called the *Green function* of the heat equation. Under the assumption that the graph is connected and thus $\mathbf{L}$ has an eigenvalue $\lambda_1 = 0$ with multiplicity 1, we obtain

$$\mathbf{L}^\dagger = \sum_{k=2}^{n} \frac{1}{\lambda_k} \mathbf{u}_k \mathbf{u}_k^\top . \tag{15.50}$$

The Green function is intimately related to random walks on graphs, and can be interpreted probabilistically as follows. Given a Markov chain such that each graph vertex is the state, and the transition from vertex v_i is possible to any adjacent vertex $v_j \sim v_i$ with probability w_{ij}/d_i, the expected number of steps required to reach vertex v_j from v_i, called the *access* or *hitting time* $O(v_i, v_j)$. The expected number of steps in a round trip from v_i to v_j is called the *commute-time distance*: $\text{CTD}^2(v_i, v_j) = O(v_i, v_j) + O(v_j, v_i)$. The commute-time distance [49] can be expressed in terms of the entries of $\mathbf{L}^\dagger$:

$$
\begin{aligned}
\text{CTD}^2(v_i, v_j) &= \text{Vol}(g)(\mathbf{L}^\dagger(i,i) + \mathbf{L}^\dagger(j,j) - 2\mathbf{L}^\dagger(i,j)) \\
&= \text{Vol}(g) \left(\sum_{k=2}^{n} \frac{1}{\lambda_k} \mathbf{u}_{ik}^2 + \sum_{k=2}^{n} \frac{1}{\lambda_k} \mathbf{u}_{jk}^2 - 2 \sum_{k=2}^{n} \frac{1}{\lambda_k} \mathbf{u}_{ik} \mathbf{u}_{jk} \right) \\
&= \text{Vol}(g) \sum_{k=2}^{n} \left(\lambda_k^{-1/2} (\mathbf{u}_{ik} - \mathbf{u}_{jk}) \right)^2 \\
&= \text{Vol}(g) \|\mathbf{x}_i - \mathbf{x}_j\|^2, \tag{15.51}
\end{aligned}
$$

where the volume of the graph, $\text{Vol}(g)$ is the sum of the degrees of all the graph vertices. The CTD function is positive-definite and sub-additive, thus defining a *metric* between the graph vertices, referred to as *commute-time* (or *resistance*) *distance* [50]. The CTD is inversely related to the number and length of paths connecting two vertices. Unlike the shortest-path (geodesic) distance, CTD captures the connectivity structure of the graph volume rather than a single path between the two vertices. The great advantage of the commute-time distance over the shortest geodesic path is that it is robust to topological changes and therefore is well suited for characterizing complex shapes. Since the volume is a graph constant, we obtain

$$\text{CTD}^2(v_i, v_j) \propto \|\mathbf{x}_i - \mathbf{x}_j\|^2. \tag{15.52}$$

Hence, the Euclidean distance between any two feature points $\mathbf{x}_i$ and $\mathbf{x}_j$ is the commute time distance between the graph vertex v_i and v_j.

Using the first K non-null eigenvalue-eigenvector pairs of the Laplacian $\mathbf{L}$, the *commute-time embedding* of the graph's nodes corresponds to the column vectors of the $K \times n$ matrix $\mathbf{X}$:

$$\mathbf{X}_{K \times n} = \mathbf{\Lambda}_K^{-1/2} (\mathbf{U}_{n \times K})^\top = [\mathbf{x}_1 \ldots \mathbf{x}_j \ldots \mathbf{x}_n]. \tag{15.53}$$

From (15.43) and (15.53) one can easily infer lower and upper bounds for the i-th coordinate of $\mathbf{x}_j$:

$$-\lambda_i^{-1/2} < x_{ji} < \lambda_i^{-1/2}. \tag{15.54}$$

The last equation implies that the graph embedding stretches along the eigenvectors with a factor that is inversely proportional to the square root of the eigenvalues. Theorem 15.4.4 below characterizes the smallest non-null K eigenvalue-eigenvector pairs of $\mathbf{L}$ as the directions of maximum variance (the principal components) of the commute-time embedding.

Theorem 15.4.4
The largest eigenvalue–eigenvector pairs of the pseudo-inverse of the combinatorial Laplacian matrix are the principal components of the commute-time embedding, i.e., the points $\mathbf{X}$ are zero-centered and have a diagonal covariance matrix.

Proof: Indeed, from (15.47) we obtain a zero-mean while from (15.53) we obtain a diagonal covariance matrix:

$$\bar{\mathbf{x}} = \frac{1}{n} \sum_{i=1}^n \mathbf{x}_i = \frac{1}{n} \mathbf{\Lambda}^{-\frac{1}{2}} \begin{pmatrix} \sum_{i=1}^n \mathbf{u}_{i2} \\ \vdots \\ \sum_{i=1}^n \mathbf{u}_{ik+1} \end{pmatrix} = \begin{pmatrix} 0 \\ \vdots \\ 0 \end{pmatrix} \tag{15.55}$$

$$\mathbf{\Sigma}_X = \frac{1}{n} \mathbf{X}\mathbf{X}^\top = \frac{1}{n} \mathbf{\Lambda}^{-\frac{1}{2}} \mathbf{U}^\top \mathbf{U} \mathbf{\Lambda}^{-\frac{1}{2}} = \frac{1}{n} \mathbf{\Lambda}^{-1} \tag{15.56}$$

∎.

Figure 15.3 shows the projection of graph (in this case 3D shape represented as meshes) vertices on eigenvectors.

15.4.3 Choosing the Dimension of the Embedding

A direct consequence of theorem 15.4.4 is that the embedded graph representation is centered, and the eigenvectors of the combinatorial Laplacian are the directions of maximum variance. The *principal* eigenvectors correspond to the eigenvectors associated with the K *largest* eigenvalues of the $\mathbf{L}^\dagger$, i.e., $\lambda_2^{-1} \geq \lambda_3^{-1} \geq \ldots \geq \lambda_K^{-1}$. The variance along vector $\mathbf{u}_k$ is λ_k^{-1}/n. Therefore, the total variance can be computed from the trace of the $\mathbf{L}^\dagger$ matrix :

$$\text{tr}(\mathbf{\Sigma}_X) = \frac{1}{n} \text{tr}(\mathbf{L}^\dagger). \tag{15.57}$$

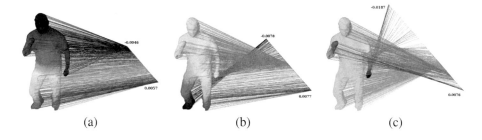

(a)   (b)   (c)

FIGURE 15.3

This is an illustration of the concept of the PCA of a graph embedding. The graph's vertices are projected onto the second, third and fourth eigenvectors of the Laplacian matrix. These eigenvectors can be viewed as the principal directions of the shape.

A standard way of choosing the principal components is to use the *scree diagram*:

$$\theta(K) = \frac{\sum_{k=2}^{K+1} \lambda_k^{-1}}{\sum_{k=2}^{n} \lambda_k^{-1}}. \tag{15.58}$$

The selection of the first K principal eigenvectors therefore depends on the spectral fall-off of the inverses of the eigenvalues. In spectral graph theory, the dimension K is chosen on the basis of the existence of an eigengap, such that $\lambda_{K+2} - \lambda_{K+1} > t$ with $t > 0$. In practice, it is extremely difficult to find such an eigengap, in particular in the case of sparse graphs that correspond to a discretized manifold. Instead, we propose to select the dimension of the embedding in the following way. Notice that (15.58) can be written as $\theta(K) = A/(A+B)$ with $A = \sum_{k=2}^{K+1} \lambda_k^{-1}$ and $B = \sum_{k=K+2}^{n} \lambda_k^{-1}$. Moreover, from the fact that the λ_k's are arranged in increasing order, we obtain $B \leq (n - K - 1)\lambda_{K+1}^{-1}$. Hence:

$$\theta_{\min} \leq \theta(K) \leq 1, \tag{15.59}$$

with

$$\theta_{\min} = \frac{\sum_{k=2}^{K+1} \lambda_k^{-1}}{\sum_{k=2}^{K} \lambda_k^{-1} + (n - K)\lambda_{K+1}^{-1}}. \tag{15.60}$$

This lower bound can be computed from the K smallest non-null eigenvalues of the combinatorial Laplacian matrix. Hence, one can choose K such that the sum of the first K eigenvalues of the $\mathbf{L}^{\dagger}$ matrix is a good approximation of the total variance, e.g., $\theta_{\min} = 0.95$.

15.4.4 Unit Hyper-Sphere Normalization

One disadvantage of the standard embeddings is that when two shapes have large difference in sampling, the embeddings will differ by a significant scale factor. In order to avoid this, we can *normalize* the embedding such that the vertex coordinates lie on a unit sphere of dimension K, which yields

$$\hat{\mathbf{x}}_i = \frac{\mathbf{x}_i}{\|\mathbf{x}_i\|}. \tag{15.61}$$

In more detail, the k-th coordinate of $\hat{\mathbf{x}}_i$ writes as

$$\hat{\mathbf{x}}_{ik} = \frac{\lambda_k^{-\frac{1}{2}} u_{ik}}{\left(\sum_{l=2}^{K+1} \lambda_l^{-\frac{1}{2}} u_{il}^2\right)^{1/2}}. \tag{15.62}$$

15.5   Spectral Shape Matching

In the previous sections we discussed solutions for the exact and inexact graph isomorphism problems, we recalled the main spectral properties of the combinatorial graph Laplacian, and we provided a novel analysis of the commute-time embedding that allows to interpret the latter in terms of the PCA of a graph, and to select the appropriate dimension $K \ll n$ of the associated embedded metric space. In this section, we address the problem of 3D shape registration, and we illustrate how the material developed above can be exploited in order to build a robust algorithm for spectral shape matching.

Let's consider two shapes described by two graphs, g_A and g_B, where $|\mathcal{V}_A| = n$ and $|\mathcal{V}_B| = m$. Let $\mathbf{L}_A$ and $\mathbf{L}_B$ be their corresponding graph Laplacians. Without loss of generality, one can choose the same dimension $K \ll \min(n, m)$ for the two embeddings. This yields the following eigen decompositions:

$$\mathbf{L}_A = \mathbf{U}_{n \times K} \boldsymbol{\Lambda}_K (\mathbf{U}_{n \times K})^\top \tag{15.63}$$
$$\mathbf{L}_B = \mathbf{U}'_{m \times K} \boldsymbol{\Lambda}'_K (\mathbf{U}'_{m \times K})^\top. \tag{15.64}$$

For each one of these graphs, one can build two *isomorphic* embedded representations, as follows:

- An *unnormalized Laplacian embedding* that uses the K rows of $\mathbf{U}_{n \times K}$ as the Euclidean coordinates of the vertices of g_A (as well as the K rows of $\mathbf{U}'_{m \times K}$ as the Euclidean coordinates of the vertices of g_B), and

- A *normalized commute-time embedding* given by (15.61), i.e., $\hat{\mathbf{X}}_A = [\hat{\mathbf{x}}_1 \dots \hat{\mathbf{x}}_j \dots \hat{\mathbf{x}}_n]$ (as well as $\hat{\mathbf{X}}_B = [\hat{\mathbf{x}}'_1 \dots \hat{\mathbf{x}}'_j \dots \hat{\mathbf{x}}'_m]$). We recall that each column $\hat{\mathbf{x}}_j$ (and respectively $\hat{\mathbf{x}}'_j$) is a K-dimensional vector corresponding to a vertex v_j of g_A (and respectively v'_j of g_B).

15.5.1 Maximum Subgraph Matching and Point Registration

Let's apply the graph isomorphism framework of Section 15.3 to the two graphs. They are embedded into two congruent spaces of dimension $\mathbb{R}^K$. If the smallest K non-null eigenvalues associated with the two embeddings are distinct and can be ordered, i.e.,

$$\lambda_2 < \ldots < \lambda_k < \ldots < \lambda_{K+1} \tag{15.65}$$

$$\lambda'_2 < \ldots < \lambda'_k < \ldots < \lambda'_{K+1} \tag{15.66}$$

then, the Umeyama method could be applied. If one uses the unnormalized Laplacian embeddings just defined, (15.33) becomes

$$\mathbf{Q}^\star = \mathbf{U}_{n \times K} \mathbf{S}_K (\mathbf{U}'_{m \times K})^\top \tag{15.67}$$

Notice that here the sign matrix $\mathbf{S}$ defined in 15.33 became a $K \times K$ matrix denoted by $\mathbf{S}_K$. We now assume that the eigenvalues $\{\lambda_2, \ldots, \lambda_{K+1}\}$ and $\{\lambda'_2, \ldots, \lambda'_{K+1}\}$ *cannot be reliably ordered*. This can be modeled by multiplication with a $K \times K$ permutation matrix $\mathbf{P}_K$:

$$\mathbf{Q} = \mathbf{U}_{n \times K} \mathbf{S}_K \mathbf{P}_K (\mathbf{U}'_{m \times K})^\top \tag{15.68}$$

Premultiplication of $(\mathbf{U}'_{m \times K})^\top$ with $\mathbf{P}_K$ permutes its rows such that $\mathbf{u}'_k \to \mathbf{u}'_{\pi(k)}$. Each entry q_{ij} of the $n \times m$ matrix $\mathbf{Q}$ can therefore be written as

$$q_{ij} = \sum_{k=2}^{K+1} s_k u_{ik} u'_{j\pi(k)} \tag{15.69}$$

Since both $\mathbf{U}_{n \times K}$ and $\mathbf{U}'_{m \times K}$ are column-orthonormal matrices, the dot-product defined by (15.69) is equivalent to the cosine of the angle between two K-dimensional vectors. This means that each entry of $\mathbf{Q}$ is such that $-1 \le q_{ij} \le +1$ and that two vertices v_i and v'_j are matched if q_{ij} is close to 1.

One can also use the normalized commute-time coordinates and define an equivalent expression as above:

$$\hat{\mathbf{Q}} = \hat{\mathbf{X}}^\top \mathbf{S}_K \mathbf{P}_K \hat{\mathbf{X}}' \tag{15.70}$$

with:

$$\hat{q}_{ij} = \sum_{k=2}^{K+1} s_k \hat{x}_{ik} \hat{x}'_{j\pi(k)} \tag{15.71}$$

Because both sets of points $\hat{\mathbf{X}}$ and $\hat{\mathbf{X}}'$ lie on a K-dimensional unit hyper-sphere, we also have $-1 \le \hat{q}_{ij} \le +1$.

It should however be emphasized that the rank of the $n \times m$ matrices $\mathbf{Q}, \hat{\mathbf{Q}}$ is equal to K. Therefore, these matrices cannot be viewed as *relaxed permutation matrices* between the two graphs. In fact, they define many-to-many correspondences between the vertices of the first graph and the vertices of the second graph, this being due to the fact that the graphs are embedded on a low-dimensional space. This is one of the main differences between our method proposed in the next section and the Umeyama method, as well as many other subsequent methods, that use all eigenvectors of the graph. As it will be explained below, our formulation leads to a shape-matching method that will alternate between aligning their eigenbases and finding a vertex-to-vertex assignment.

It is possible to extract a one-to-one assignment matrix from $\mathbf{Q}$ (or from $\hat{\mathbf{Q}}$) using either dynamic programming or an assignment method technique such as the Hungarian algorithm. Notice that this assignment is conditioned by the choice of a sign matrix $\mathbf{S}_K$ and of a permutation matrix $\mathbf{P}_K$, i.e., $2^K K!$ possibilities, and that not all these choices correspond to a valid subisomorphism between the two graphs. Let's consider the case of the normalized commute-time embedding; there is an equivalent formulation for the unnormalized Laplacian embedding. The two graphs are described by two sets of points, $\hat{\mathbf{X}}$ and $\hat{\mathbf{X}}'$, both lying onto the K-dimensional unity hyper-sphere. The $K \times K$ matrix $\mathbf{S}_K \mathbf{P}_K$ transforms one graph embedding onto the other graph embedding. Hence, one can write $\hat{\mathbf{x}}_i = \mathbf{S}_K \mathbf{P}_K \hat{\mathbf{x}}'_j$ if vertex v_i matches v_j. More generally, let $\mathbf{R}_K = \mathbf{S}_K \mathbf{P}_K$, and let's extend the domain of $\mathbf{R}_K$ to all possible orthogonal matrices of size $K \times K$, namely $\mathbf{R}_K \in \mathcal{O}_K$ or the *orthogonal group* of dimension K. We can now write the following criterion whose minimization over $\mathbf{R}_K$ guarantees an optimal solution for registering the vertices of the first graph with the vertices of the second graph:

$$\min_{R_K} \sum_{i=1}^{n} \sum_{j=1}^{m} \hat{q}_{ij} \|\hat{\mathbf{x}}_i - \mathbf{R}_K \hat{\mathbf{x}}'_j\|^2 \qquad (15.72)$$

One way to solve minimization problems such as (15.72) is to use a point registration algorithm that alternates between (i) estimating the $K \times K$ orthogonal transformation $\mathbf{R}_K$, which aligns the K-dimensional coordinates associated with the two embeddings, and (ii) updating the assignment variables $\hat{q}_{ij}$. This can be done using either ICP-like methods (the $\hat{q}_{ij}$'s are binary variables), or EM-like methods (the $\hat{q}_{ij}$'s are posterior probabilities of assignment variables). As we just outlined above, matrix $\mathbf{R}_K$ belongs to the orthogonal group $\mathcal{O}_K$. Therefore, this framework differs from standard implementations of ICP and EM algorithms that usually estimate a 2-D or 3-D *rotation* matrix which belong to the *special orthogonal group*.

It is well established that ICP algorithms are easily trapped in local minima. The EM algorithm recently proposed in [38] is able to converge to a good solution starting with a rough initial guess and is robust to the presence of outliers. Nevertheless, the algorithm proposed in [38] performs well under *rigid transformations* (rotation and translation), whereas in our case we have to estimate a more general orthogonal transformation that incorporates both rotations and reflections. Therefore, before describing in detail an EM algorithm well suited for solving the problem at hand, we discuss the issue of estimating an initialization for the transformation aligning the K eigenvectors of the first embedding with those of the second embedding, and we propose a practical method for initializing this transformation (namely, matrices $\mathbf{S}_K$ and $\mathbf{P}_K$ in (15.70)) based on comparing the histograms of these eigenvectors, or *eigensignatures*.

15.5.2 Aligning Two Embeddings Based on Eigensignatures

Both the unnormalized Laplacian embedding and the normalized commute-time embedding of a graph are represented in a metric space spanned by the eigenvectors of the Laplacian matrix, namely the n-dimensional vectors $\{\mathbf{u}_2, \ldots, \mathbf{u}_k, \ldots, \mathbf{u}_{K+1}\}$, where n is the number of graph vertices. They correspond to *eigenfunctions*, and each such eigenfunction maps the graph's vertices onto the real line. More precisely, the k-th eigenfunction maps a vertex v_i onto u_{ik}. Propositions 15.4.1 and 15.4.3 revealed interesting statistics of the sets $\{u_{1k}, \ldots, u_{ik}, \ldots, u_{nk}\}_{k=2}^{K+1}$. Moreover, Theorem 15.4.4 provided an interpretation of the eigenvectors in terms of principal directions of the embedded shape. One can therefore conclude that the probability distribution of the components of an eigenvector have interesting properties that make them suitable for comparing two shapes, namely $-1 < u_{ik} < +1$, $\bar{u}_k = 1/n \sum_{i=1}^{n} u_{ik} = 0$, and $\sigma_k = 1/n \sum_{i=1}^{n} u_{ik}^2 = 1/n$. This means that one can build a histogram for each eigenvector and that all these histograms share the same bin width w and the same number of bins b [51]:

$$w_k = \frac{3.5\sigma_k}{n^{1/3}} = \frac{3.5}{n^{4/3}} \tag{15.73}$$

$$b_k = \frac{\sup_i u_{ik} - \inf_i u_{ik}}{w_k} \approx \frac{n^{4/3}}{2}. \tag{15.74}$$

We claim that these histograms are eigenvector signatures that are invariant under graph isomorphism. Indeed, let's consider the Laplacian $\mathbf{L}$ of a shape and we apply the isomorphic transformation $\mathbf{PLP}^\top$ to this shape, where $\mathbf{P}$ is a permutation matrix. If $\mathbf{u}$ is an eigenvector of $\mathbf{L}$, it follows that $\mathbf{Pu}$ is an eigenvector of $\mathbf{PLP}^\top$ and therefore, while the order of the components of $\mathbf{u}$ are affected by this transformation, their frequency and hence their probability distributions remain the same. Hence, one may conclude that such a histogram may well be viewed as an *eigensignature*.

We denote by $H\{\mathbf{u}\}$ the histogram formed with the components of $\mathbf{u}$, and let $C(H\{\mathbf{u}\}, H\{\mathbf{u}'\})$ be a similarity measure between two histograms. From the eigenvector properties just outlined, it is straightforward to notice that $H\{\mathbf{u}\} \neq H\{-\mathbf{u}\}$: These two histograms are mirror symmetric. Hence, the histogram is not invariant to the sign of an eigenvector. Therefore, one can use the eigenvectors' histograms to estimate both the permutation matrix $\mathbf{P}_K$ and the sign matrix $\mathbf{S}_K$ in (15.70). The problem of finding one-to-one assignments $\{\mathbf{u}_k \leftrightarrow s_k \mathbf{u}'_{\pi(k)}\}_{k=2}^{K+1}$ between the two sets of eigenvectors associated with the two shapes is therefore equivalent to the problem of finding one-to-one assignments between their histograms.

Let $\mathbf{A}_K$ be an assignment matrix between the histograms of the first shape and the histograms of the second shape. Each entry of this matrix is defined by

$$a_{kl} = \sup[C(H\{\mathbf{u}_k\}, H\{\mathbf{u}'_l\}); C(H\{\mathbf{u}_k\}, H\{-\mathbf{u}'_l\})] \qquad (15.75)$$

Similarly, we define a matrix $\mathbf{B}_K$ that accounts for the *sign assignments*:

$$b_{kl} = \begin{cases} +1 & \text{if} \quad C(H\{\mathbf{u}_k\}, H\{\mathbf{u}'_l\}) \geq C(H\{\mathbf{u}_k\}, H\{-\mathbf{u}'_l\}) \\ -1 & \text{if} \quad C(H\{\mathbf{u}_k\}, H\{\mathbf{u}'_l\}) < C(H\{\mathbf{u}_k\}, H\{-\mathbf{u}'_l\}) \end{cases} \qquad (15.76)$$

Extracting a permutation matrix $\mathbf{P}_K$ from $\mathbf{A}_K$ is an instance of the bipartite maximum matching problem, and the Hungarian algorithm is known to provide an optimal solution to this assignment problem [47]. Moreover, one can use the estimated $\mathbf{P}_K$ to extract a sign matrix $\mathbf{S}_K$ from $\mathbf{B}_K$. Algorithm 3 estimates an alignment between two embeddings.

Algorithm 3: *Alignment of Two Laplacian Embeddings*

---

input : Histograms associated with eigenvectors $\{\mathbf{u}_k\}_{k=2}^{K+1}$ and $\{\mathbf{u}'_k\}_{k=2}^{K+1}$.
output : A permutation matrix $\mathbf{P}_K$ and a sign matrix $\mathbf{S}_K$.
  1: Compute the assignment matrices $\mathbf{A}_K$ and $\mathbf{B}_K$.
  2: Compute $\mathbf{P}_K$ from $\mathbf{A}_K$ using the Hungarian algorithm.
  3: Compute the sign matrix $\mathbf{S}_K$ using $\mathbf{P}_K$ and $\mathbf{B}_K$.

Figure 15.4 illustrates the utility of the histogram of eigenvectors as eigensignatures for solving the problem of sign flip and change in eigenvector ordering by computing histogram matching. It is interesting to observe that a threshold on the histogram matching score (15.75) allows us to discard the eigenvectors with low similarity cost. Hence, starting with large K obtained using (15.60), we can limit the number of eigenvectors to just a few, which will be suitable for EM-based point registration algorithm proposed in the next section.

15.5.3 An EM Algorithm for Shape Matching

As explained in section 15.5.1, the maximum subgraph matching problem reduces to a point registration problem in K-dimensional metric space spanned by the eigenvectors of graph Laplacian where two shapes are represented as point clouds. The initial alignment of Laplacian embeddings can be obtained by matching the histogram of eigenvectors as described in the previous section. In this section, we propose an EM algorithm for 3D shape matching that computes a probabilistic vertex-to-vertex assignment between two shapes. The proposed method alternates between the step to estimate an orthogonal transformation matrix associated with the alignment of the two shape embeddings and the step to compute a point-to-point probabilistic assignment variable.

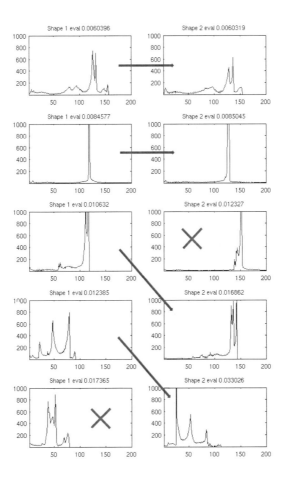

FIGURE 15.4

An illustration of applicability of eigenvector histogram as eigensignature to detect sign flip and eigenvector ordering change. The blue line shows matched eigenvector pairs, and the red cross depicts discarded eigenvectors.

The method is based on a parametric probabilistic model, namely maximum likelihood with missing data. Let us consider the Laplacian embedding of two shapes, i.e., (15.53) : $\hat{\mathbf{X}} = \{\hat{\mathbf{x}}_i\}_{i=1}^n$, $\hat{\mathbf{X}}' = \{\hat{\mathbf{x}}'_j\}_{j=1}^m$, with $\hat{\mathbf{X}}, \hat{\mathbf{X}}' \subset \mathbb{R}^K$. Without loss of generality, we assume that the points in the first set, $\hat{\mathbf{X}}$ are cluster centers of a Gaussian mixture model (GMM) with n clusters and an additional uniform component that accounts for outliers and unmatched data. The matching $\hat{\mathbf{X}} \leftrightarrow \hat{\mathbf{X}}'$ will consist in fitting the Gaussian mixture to the set $\hat{\mathbf{X}}'$.

Let this Gaussian mixture undergo a $K \times K$ transformation $\mathbf{R}$ (for simplicity, we omit the index K) with $\mathbf{R}^{\top}\mathbf{R} = \mathbf{I}_K, \det(\mathbf{R}) = \pm 1$, more precisely $\mathbf{R} \in \mathcal{O}_K$, the group of orthogonal matrices acting on $\mathbb{R}^K$. Hence, each cluster in the mixture is parametrized by a prior p_i, a cluster mean $\boldsymbol{\mu}_i = \mathbf{R}\hat{\mathbf{x}}_i$, and a covariance matrix $\boldsymbol{\Sigma}_i$. It will be assumed that all the clusters in the mixture have the same priors, $\{p_i = \pi_{\text{in}}\}_{i=1}^n$, and the same isotropic covariance matrix, $\{\boldsymbol{\Sigma}_i = \sigma\mathbf{I}_K\}_{i=1}^n$. This parametrization leads to the following *observed-data log-likelihood* (with $\pi_{\text{out}} = 1 - n\pi_{\text{in}}$ and $\mathcal{U}$ is the uniform distribution):

$$\log P(\hat{\mathbf{X}}') = \sum_{j=1}^{m} \log \left(\sum_{i=1}^{n} \pi_{\text{in}}\mathcal{N}(\hat{\mathbf{x}}'_j|\boldsymbol{\mu}_i,\sigma) + \pi_{\text{out}}\mathcal{U} \right) \qquad (15.77)$$

It is well known that the direct maximization of (15.77) is not tractable, and it is more practical to maximize the *expected complete-data log-likelihood* using the EM algorithm, where "complete-data" refers to both the observed data (the points $\hat{\mathbf{X}}'$) and the missing data (the data-to-cluster assignments). In our case, the above expectation writes (see [38] for details)

$$\mathcal{E}(\mathbf{R},\sigma) = -\frac{1}{2} \sum_{j=1}^{m} \sum_{i=1}^{n} \alpha_{ji}(\|\hat{\mathbf{x}}'_j - \mathbf{R}\hat{\mathbf{x}}_i\|^2 + k\log\sigma), \qquad (15.78)$$

where α_{ji} denotes the posterior probability of an assignment $\hat{\mathbf{x}}'_j \leftrightarrow \hat{\mathbf{x}}_i$:

$$\alpha_{ji} = \frac{\exp(-\|\hat{\mathbf{x}}'_j - \mathbf{R}\hat{\mathbf{x}}_i\|^2/2\sigma)}{\sum_{q=1}^{n} \exp(-\|\hat{\mathbf{x}}'_j - \mathbf{R}\hat{\mathbf{x}}_q\|^2/2\sigma) + \emptyset\sigma^{k/2}}, \qquad (15.79)$$

where $\emptyset$ is a constant term associated with the uniform distribution $\mathcal{U}$. Notice that one easily obtains the posterior probability of a data point to remain unmatched, $\alpha_{jn+1} = 1 - \sum_{i=1}^{n} \alpha_{ij}$. This leads to the shape-matching procedure outlined in Algorithm 4.

15.6 Experiments and Results

We have performed several 3D shape registration experiments to evaluate the proposed method. In the first experiment, 3D shape registration is performed on 138 high-resolution (10K-50K vertices) triangular meshes from the publicly available TOSCA dataset [33]. The dataset includes 3 shape classes (human, dog, horse) with simulated transformations. Transformations are split into 9 classes (isometry, topology, small and big holes, global and local scaling, noise, shot noise, sampling). Each

---

Algorithm 4: *EM for shape matching*

---

input : Two embedded shapes $\hat{\mathbf{X}}$ and $\hat{\mathbf{X}}'$;

output : Dense correspondences $\hat{\mathbf{X}} \leftrightarrow \hat{\mathbf{X}}'$ between the two shapes;

1: *Initialization:* Set $\mathbf{R}^{(0)} = \mathbf{S}_K \mathbf{P}_K$ choose a large value for the variance $\sigma^{(0)}$;

2: *E-step:* Compute the current posteriors $\alpha_{ij}^{(q)}$ from the current parameters using (15.79);

3: *M-step:* Compute the new transformation $\mathbf{R}^{(q+1)}$ and the new variance $\sigma^{(q+1)}$ using the current posteriors:

$$\mathbf{R}^{(q+1)} = \arg\min_{\mathbf{R}} \sum_{i,j} \alpha_{ij}^{(q)} \|\boldsymbol{x}_j' - \mathbf{R}\boldsymbol{x}_i\|^2$$

$$\sigma^{(q+1)} = \sum_{i,j} \alpha_{ij}^{(q)} \|\hat{\mathbf{x}}_j' - \mathbf{R}^{(q+1)}\hat{\mathbf{x}}_i\|^2 / k \sum_{i,j} \alpha_{ij}^{(q)}$$

4: *MAP:* Accept the assignment $\hat{\mathbf{x}}_j' \leftrightarrow \hat{\mathbf{x}}_i$ if $\max_i \alpha_{ij}^{(q)} > 0.5$.

---

transformation class appears in five different strength levels. An estimate of average geodesic distance to ground truth correspondence was computed for performance evaluation (see [33] for details).

We evaluate our method in two settings. In the first setting SM1, we use the commute-time embedding (15.53) while in the second setting SM2, we use the unit hyper-sphere normalized embedding (15.61).

Table 15.2 shows the error estimates for dense shape matching using the proposed spectral matching method. In the case of some transforms, the proposed method yields zero error because the two meshes had identical triangulations. Figure 15.5 shows some matching results. The colors emphasize the correct matching of body parts, while we show only 5% of matches for better visualization. In Figure 15.5(e), the two shapes have large difference in the sampling rate. In this case, the matching near the shoulders is not fully correct since we used the commute-time embedding.

Table 15.3 summarizes the comparison of proposed spectral matching method (SM1 and SM2) with generalized multidimensional scaling (GMDS) based matching algorithm introduced in [19] and the Laplace–Beltrami matching algorithm proposed in [12] with two settings LB1 (uses graph Laplacian) and LB2 (uses cotangent weights). GMDS computes the correspondence between two shapes by trying to embed one shape into another with minimum distortion. LB1 and LB2 algorithms combines the surface descriptors based on the eigendecomposition of the Laplace–Beltrami operator and the geodesic distances measured on the shapes when calculating the correspondence quality. The above results in a quadratic optimization problem formulation for correspondence detection, and its minimizer is the best possible correspondence. The proposed method clearly outperform the other two methods

Transform	Strength									
	1		≤ 2		≤ 3		≤ 4		≤ 5	
	SM1	SM2	SM1	SM2	SM1	SM2	SM1	SM2	SM1	SM2
Isometry	0.00	0.00	0.00	0.00	0.00	0.00	0.00	0.00	0.00	0.00
Topology	6.89	5.96	7.92	6.76	7.92	7.14	8.04	7.55	8.41	8.13
Holes	7.32	5.17	8.39	5.55	9.34	6.05	9.47	6.44	12.47	10.32
Micro holes	0.37	0.68	0.39	0.70	0.44	0.79	0.45	0.79	0.49	0.83
Scale	0.00	0.00	0.00	0.00	0.00	0.00	0.00	0.00	0.00	0.00
Local scale	0.00	0.00	0.00	0.00	0.00	0.00	0.00	0.00	0.00	0.00
Sampling	11.43	10.51	13.32	12.08	15.70	13.65	18.76	15.58	22.63	19.17
Noise	0.00	0.00	0.00	0.00	0.00	0.00	0.00	0.00	0.00	0.00
Shot noise	0.00	0.00	0.00	0.00	0.00	0.00	0.00	0.00	0.00	0.00
Average	2.88	2.48	3.34	2.79	3.71	3.07	4.08	3.37	4.89	4.27

TABLE 15.2

3D shape registration error estimates (average geodesic distance to ground truth correspondences) using proposed spectral matching method with commute-time embedding (SM1) and unit hyper-sphere normalized embedding (SM2).

	Strength				
Method	1	≤ 2	≤ 3	≤ 4	≤ 5
LB1	10.61	15.48	19.01	23.22	23.88
LB2	15.51	18.21	22.99	25.26	28.69
GMDS	39.92	36.77	35.24	37.40	39.10
SM1	2.88	3.34	3.71	4.08	4.89
SM2	2.48	2.79	3.07	3.37	4.27

TABLE 15.3

Average shape registration error estimates over all transforms (average geodesic distance to ground truth correspondences) computed using proposed methods (SM1 and SM2), GMDS [19] and LB1, LB2 [12].

| | **Strength** | | |
Transform	1	≤ 3	≤ 5
Isometry	SM1,SM2	SM1,SM2	SM1,SM2
Topology	SM2	SM2	SM2
Holes	SM2	SM2	SM2
Micro holes	SM1	SM1	SM1
Scale	SM1,SM2	SM1,SM2	SM1,SM2
Local scale	SM1,SM2	SM1,SM2	SM1,SM2
Sampling	LB1	SM2	LB2
Noise	SM1,SM2	SM1,SM2	SM1,SM2
Shot noise	SM1,SM2	SM1,SM2	SM1,SM2
Average	SM1,SM2	SM1,SM2	SM1,SM2

TABLE 15.4
3D shape registration performance comparison: The proposed methods (SM1 and SM2) performed best by providing minimum average shape registration error over all the transformation classes with different strength as compare to GMDS [19] and LB1, LB2 [12] methods.

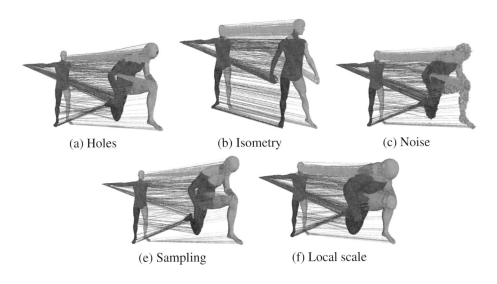

(a) Holes   (b) Isometry   (c) Noise

(e) Sampling   (f) Local scale

FIGURE 15.5
3D shape registration in the presence of different transforms.

with minimum average error estimate computed over all the transformations in the dataset.

In Table 15.4, we show a detailed comparison of the proposed method with other methods. For a detailed quantitative comparison refer to [33]. The proposed method inherently uses diffusion geometry as opposed to geodesic metric used by other two methods and hence outperform them.

In the second experiment, we perform shape registration on two different shapes with similar topology. In Figure 15.6, results of shape registration on different shapes is presented. Figure 15.6(a),(c) shows the initialization step of EM algorithm while Figure 15.6(b),(d) shows the dense matching obtained after EM convergence.

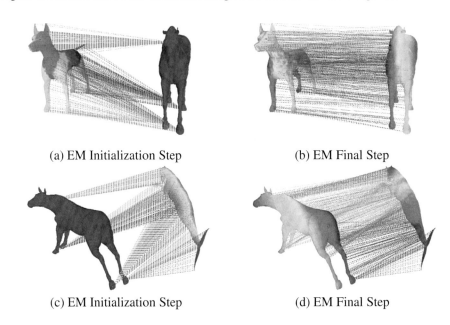

        (a) EM Initialization Step   (b) EM Final Step

        (c) EM Initialization Step   (d) EM Final Step

FIGURE 15.6
3D shape registration performed on different shapes with similar topology.

Finally, we show shape-matching results on two different human meshes captured with a multi-camera system at MIT [5] and University of Surrey [2] in Figure 15.7

15.7   Discussion

This chapter describes a 3D shape registration approach that computes dense correspondences between two articulated objects. We address the problem using spectral matching and unsupervised point registration method. We formally introduce graph isomorphism using the Laplacian matrix, and we provide an analysis of the matching problem when the number of nodes in the graph is very large, that is, of the order of

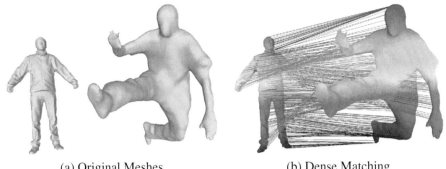

(a) Original Meshes   (b) Dense Matching

FIGURE 15.7
3D shape registration performed on two real meshes captured from different sequence.

$O(10^4)$. We show that there is a simple equivalence between graph isomorphism and point registration under the group of orthogonal transformations, when the dimension of the embedding space is much smaller than the cardinality of the point-sets.

The eigenvalues of a large sparse Laplacian cannot be reliably ordered. We propose an elegant alternative to eigenvalue ordering, using eigenvector histograms and alignment based on comparing these histograms. The point registration that results from eigenvector alignment yields an excellent initialization for the EM algorithm, subsequently used only to refine the registration.

However, the method is susceptible to large topology changes that might occur in the multi-camera shape acquisition setup due to self-occlusion (originated from complex kinematics poses) and shadow effects. This is because Laplacian embedding is a global representation and any major topology change will lead to large changes in embeddings causing failure of this method. Recently, a new shape registration method proposed in [36] provides robustness to the large topological changes using the heat kernel framework.

15.8 Appendix: Permutation and Doubly- stochastic Matrices

A matrix $\mathbf{P}$ is called a *permutation* matrix if exactly one entry in each row and column is equal to 1, and all other entries are 0. Left multiplication of a matrix $\mathbf{A}$ by a permutation matrix $\mathbf{P}$ permutes the *rows* of $\mathbf{A}$, while right multiplication permutes the *columns* of $\mathbf{A}$.

Permutation matrices have the following properties: $\det(\mathbf{P}) = \pm 1$, $\mathbf{P}^\top = \mathbf{P}^{-1}$, the identity is a permutation matrix, and the product of two permutation matrices is a permutation matrix. Hence, the set of permutation matrices $\mathbf{P} \in \mathcal{P}_n$ constitute a subgroup of the subgroup of orthogonal matrices, denoted by $\mathcal{O}_n$, and $\mathcal{P}_n$ has finite cardinality $n!$.

A non-negative matrix $\mathbf{A}$ is a matrix such that all its entries are non-negative. A non-negative matrix with the property that all its row sums are $+1$ is said to be a *(row) stochastic matrix*. A *column stochastic matrix* is the transpose of a row stochastic matrix. A stochastic matrix $\mathbf{A}$ with the property that $\mathbf{A}^\top$ is also stochastic is said to be *doubly stochastic*: all row and column sums are $+1$ and $a_{ij} \geq 0$. The set of stochastic matrices is a compact convex set with the simple and important property that $\mathbf{A}$ is stochastic if and only if $\mathbf{A}\mathbb{1} = \mathbb{1}$, where $\mathbb{1}$ is the vector with all components equal to $+1$.

Permutation matrices are doubly stochastic matrices. If we denote by $\mathcal{D}_n$ the set of doubly stochastic matrices, it can be proved that $\mathcal{P}_n = \mathcal{O}_n \cap \mathcal{D}_n$ [52]. The permutation matrices are the fundamental and prototypical doubly stochastic matrices, for Birkhoff's theorem states that any doubly stochastic matrix is a linear convex combination of finitely many permutation matrices [46]

Theorem 15.8.1
(Birkhoff) A matrix $\mathbf{A}$ is a doubly stochastic matrix if and only if for some $N < \infty$ there are permutation matrices $\mathbf{P}_1, \ldots, \mathbf{P}_N$ and positive scalars $s_1, \ldots, s_N$ such that $s_1 + \ldots + s_N = 1$ and $\mathbf{A} = s_1\mathbf{P}_1 + \ldots + s_N\mathbf{P}_N$.

A complete proof of this theorem is to be found in [46][pages 526–528]. The proof relies on the fact that $\mathcal{D}_n$ is a compact convex set, and every point in such a set is a convex combination of the extreme points of the set. First, it is proved that every permutation matrix is an extreme point of $\mathcal{D}_n$ and, second, it is shown that a given matrix is an extreme point of $\mathcal{D}_n$ if an only if it is a permutation matrix.

15.9   Appendix: The Frobenius Norm

The Frobenius (or Euclidean) norm of a matrix $\mathbf{A}_{n \times n}$ is an *entry-wise* norm that treats the matrix as a vector of size $1 \times nn$. The standard norm properties hold: $\|\mathbf{A}\|_F > 0 \Leftrightarrow \mathbf{A} \neq 0$, $\|\mathbf{A}\|_F = 0 \Leftrightarrow \mathbf{A} = 0$, $\|c\mathbf{A}\|_F = c\|\mathbf{A}\|_F$, and $\|\mathbf{A} + \mathbf{B}\|_F \leq \|\mathbf{A}\|_F + \|\mathbf{B}\|_F$. Additionally, the Frobenius norm is *sub-multiplicative*:

$$\|\mathbf{AB}\|_F \leq \|\mathbf{A}\|_F \|\mathbf{B}\|_F \tag{15.80}$$

as well as *unitarily invariant*. This means that for any two orthogonal matrices $\mathbf{U}$ and $\mathbf{V}$:

$$\|\mathbf{UAV}\|_F = \|\mathbf{A}\|_F. \tag{15.81}$$

It immediately follows the following equalities:

$$\|\mathbf{UAU}^\top\|_F = \|\mathbf{UA}\|_F = \|\mathbf{AU}\|_F = \|\mathbf{A}\|_F. \tag{15.82}$$

15.10 Appendix: Spectral Properties of the Normalized Laplacian

The Normalized Laplacian

Let $\tilde{\mathbf{u}}_k$ and γ_k denote the eigenvectors and eigenvalues of $\tilde{\mathbf{L}}$; The spectral decomposition is $\tilde{\mathbf{L}} = \tilde{\mathbf{U}}\Gamma\tilde{\mathbf{U}}^\top$ with $\tilde{\mathbf{U}}\tilde{\mathbf{U}}^\top = \mathbf{I}$. The smallest eigenvalue and associated eigenvector are $\gamma_1 = 0$ and $\tilde{\mathbf{u}}_1 = \mathbf{D}^{1/2}\mathbb{1}$.

We obtain the following equivalent relations:

$$\sum_{i=1}^{n} d_i^{1/2}\tilde{u}_{ik} = 0, \quad 2 \leq k \leq n \tag{15.83}$$

$$d_i^{1/2}|\tilde{u}_{ik}| < 1, \quad 1 \leq i \leq n, 2 \leq k \leq n. \tag{15.84}$$

Using (15.5) we obtain a useful expression for the combinatorial Laplacian in terms of the spectral decomposition of the normalized Laplacian. Notice, however, that the expression below is NOT a spectral decomposition of the combinatorial Laplacian:

$$\mathbf{L} = (\mathbf{D}^{1/2}\tilde{\mathbf{U}}\Gamma^{1/2})(\mathbf{D}^{1/2}\tilde{\mathbf{U}}\Gamma^{1/2})^\top. \tag{15.85}$$

For a connected graph γ_1 has multiplicity 1: $0 = \gamma_1 < \gamma_2 \leq \ldots \leq \gamma_n$. As in the case of the combinatorial Laplacian, there is an upper bound on the eigenvalues (see [41] for a proof):

Proposition 15.10.1
For all $k \leq n$, we have $\mu_k \leq 2$.

We obtain the following spectral decomposition for the normalized Laplacian:

$$\tilde{\mathbf{L}} = \sum_{k=2}^{n} \gamma_k \tilde{\mathbf{u}}_k \tilde{\mathbf{u}}_k^\top. \tag{15.86}$$

The spread of the graph along the k-th normalized Laplacian eigenvector is given by $\forall(k, i), 2 \leq k \leq n, 1 \leq i \leq n$:

$$\bar{\tilde{u}}_k = \frac{1}{n}\sum_{i=1}^{n} \tilde{u}_{ik} \tag{15.87}$$

$$\sigma_{u_k} = \frac{1}{n} - \bar{\tilde{u}}_k^2. \tag{15.88}$$

Therefore, the projection of the graph onto an eigenvector $\tilde{\mathbf{u}}_k$ is not centered. By combining (15.5) and (15.86), we obtain an alternative representation of the combinatorial Laplacian in terms of the the spectrum of the normalized Laplacian, namely:

$$\mathbf{L} = \sum_{k=2}^{n} \gamma_k (\mathbf{D}^{1/2}\tilde{\mathbf{u}}_k)(\mathbf{D}^{1/2}\tilde{\mathbf{u}}_k)^{\top}. \tag{15.89}$$

Hence, an alternative is to project the graph onto the vectors $\mathbf{t}_k = \mathbf{D}^{1/2}\tilde{\mathbf{u}}_k$. From $\tilde{\mathbf{u}}_{k\geq 2}^{\top}\tilde{\mathbf{u}}_1 = 0$ we get that $\mathbf{t}_{k\geq 2}^{\top}\mathbb{1} = 0$. Therefore, the spread of the graph's projection onto $\mathbf{t}_k$ has the following mean and variance, $\forall (k, i), 2 \leq k \leq n, 1 \leq i \leq n$:

$$\bar{t}_k = \sum_{i=1}^{n} d_i^{1/2}\tilde{u}_{ik} = 0 \tag{15.90}$$

$$\sigma_{t_k} = \frac{1}{n}\sum_{i=1}^{n} d_i \tilde{u}_{ik}^2. \tag{15.91}$$

The Random-Walk Laplacian.

This operator is not symmetric, however, its spectral properties can be easily derived from those of the normalized Laplacian using (15.7). Notice that this can be used to transform a nonsymmetric Laplacian into a symmetric one, as proposed in [53] and in [54].

Bibliography

[1] J.-S. Franco and E. Boyer, "Efficient Polyhedral Modeling from Silhouettes," *IEEE Transactions on Pattern Analysis and Machine Intelligence*, vol. 31, no. 3, pp. 414—427, 2009.

[2] J. Starck and A. Hilton, "Surface capture for performance based animation," *IEEE Computer Graphics and Applications*, vol. 27, no. 3, pp. 21–31, 2007.

[3] G. Slabaugh, B. Culbertson, T. Malzbender, and R. Schafer, "A survey of methods for volumetric scene reconstruction from photographs," in *International Workshop on Volume Graphics*, 2001, pp. 81–100.

[4] S. M. Seitz, B. Curless, J. Diebel, D. Scharstein, and R. Szeliski, "A comparison and evaluation of multi-view stereo reconstruction algorithms," in *IEEE Computer Society Conference on Computer Vision and Pattern Recognition*, 2006, pp. 519–528.

[5] D. Vlasic, I. Baran, W. Matusik, and J. Popovic, "Articulated mesh animation from multi-view silhouettes," *ACM Transactions on Graphics (Proc. SIGGRAPH)*, vol. 27, no. 3, pp. 97:1–97:9, 2008.

[6] A. Zaharescu, E. Boyer, and R. P. Horaud, "Topology-adaptive mesh deformation for surface evolution, morphing, and multi-view reconstruction," *IEEE Transactions on Pattern Analysis and Machine Intelligence*, vol. 33, no. 4, pp. 823 – 837, April 2011.

[7] Y. Chen and G. Medioni, "Object modelling by registration of multiple range images," *Image Vision Computing*, vol. 10, pp. 145–155, April 1992.

[8] P. J. Besl and N. D. McKay, "A method for registration of 3-d shapes," *IEEE Transactions on Pattern Analysis and Machine Intelligence*, vol. 14, pp. 239–256, February 1992.

[9] S. Rusinkiewicz and M. Levoy, "Efficient variants of the ICP algorithm," in *International Conference on 3D Digital Imaging and Modeling*, 2001, pp. 145–152.

[10] S. Umeyama, "An eigendecomposition approach to weighted graph matching problems," *IEEE Transactions on Pattern Analysis and Machine Intelligence*, vol. 10, no. 5, pp. 695–703, May 1988.

[11] J. H. Wilkinson, "Elementary proof of the Wielandt-Hoffman theorem and of its generalization," Stanford University, Tech. Rep. CS150, January 1970.

[12] A. Bronstein, M. Bronstein, and R. Kimmel, "Generalized multidimensional scaling: a framework for isometry-invariant partial surface matching," *Proceedings of National Academy of Sciences*, vol. 103, pp. 1168–1172, 2006.

[13] S. Wang, Y. Wang, M. Jin, X. Gu, D. Samaras, and P. Huang, "Conformal geometry and its application on 3d shape matching," *IEEE Transactions on Pattern Analysis and Machine Intelligence*, vol. 29, no. 7, pp. 1209–1220, 2007.

[14] V. Jain, H. Zhang, and O. van Kaick, "Non-rigid spectral correspondence of triangle meshes," *International Journal of Shape Modeling*, vol. 13, pp. 101–124, 2007.

[15] W. Zeng, Y. Zeng, Y. Wang, X. Yin, X. Gu, and D. Samras, "3D non-rigid surface matching and registration based on holomorphic differentials," in *European Conference on Computer Vision*, 2008, pp. 1–14.

[16] D. Mateus, R. Horaud, D. Knossow, F. Cuzzolin, and E. Boyer, "Articulated shape matching using Laplacian eigenfunctions and unsupervised point registration," in *IEEE Computer Society Conference on Computer Vision and Pattern Recognition*, 2008, pp. 1–8.

[17] M. R. Ruggeri, G. Patané, M. Spagnuolo, and D. Saupe, "Spectral-driven isometry-invariant matching of 3d shapes," *International Journal of Computer Vision*, vol. 89, pp. 248–265, 2010.

[18] Y. Lipman and T. Funkhouser, "Mobius voting for surface correspondence," *ACM Transactions on Graphics (Proc. SIGGRAPH)*, vol. 28, no. 3, pp. 72:1–72:12, 2009.

[19] A. Dubrovina and R. Kimmel, "Matching shapes by eigendecomposition of the Laplace-Beltrami operator," in *International Symposium on 3D Data Processing, Visualization and Transmission*, 2010.

[20] G. Scott and C. L. Higgins, "An Algorithm for Associating the Features of Two Images," *Biological Sciences*, vol. 244, no. 1309, pp. 21–26, 1991.

[21] L. S. Shapiro and J. M. Brady, "Feature-based correspondence: an eigenvector approach," *Image Vision Computing*, vol. 10, pp. 283–288, June 1992.

[22] B. Luo and E. R. Hancock, "Structural graph matching using the em algorithm and singular value decomposition," *IEEE Transactions on Pattern Analysis and Machine Intelligence*, vol. 23, pp. 1120–1136, October 2001.

[23] H. F. Wang and E. R. Hancock, "Correspondence matching using kernel principal components analysis and label consistency constraints," *Pattern Recognition*, vol. 39, pp. 1012–1025, June 2006.

[24] H. Qiu and E. R. Hancock, "Graph simplification and matching using commute times," *Pattern Recognition*, vol. 40, pp. 2874–2889, October 2007.

[25] M. Leordeanu and M. Hebert, "A spectral technique for correspondence problems using pairwise constraints," in *International Conference on Computer Vision*, 2005, pp. 1482–1489.

[26] O. Duchenne, F. Bach, I. Kweon, and J. Ponce, "A tensor based algorithm for high order graph matching," in *IEEE Computer Society Conference on Computer Vision and Pattern Recognition*, 2009, pp. 1980–1987.

[27] L. Torresani, V. Kolmogorov, and C. Rother, "Feature correspondence via graph matching : Models and global optimazation," in *European Conference on Computer Vision*, 2008, pp. 596–609.

[28] R. Zass and A. Shashua, "Probabilistic graph and hypergraph matching," in *IEEE Computer Society Conference on Computer Vision and Pattern Recognition*, 2008, pp. 1–8.

[29] J. Maciel and J. P. Costeira, "A global solution to sparse correspondence problems," *IEEE Transactions on Pattern Analysis and Machine Intelligence*, vol. 25, pp. 187–199, 2003.

[30] Q. Huang, B. Adams, M. Wicke, and L. J. Guibas, "Non-rigid registration under isometric deformations," *Computer Graphics Forum*, vol. 27, no. 5, pp. 1449–1457, 2008.

[31] Y. Zeng, C. Wang, Y. Wang, X. Gu, D. Samras, and N. Paragios, "Dense non-rigid surface registration using high order graph matching," in *IEEE Computer Society Conference on Computer Vision and Pattern Recognition*, 2010, pp. 382–389.

[32] Y. Sahillioglu and Y. Yemez, "3d shape correspondence by isometry-driven greedy optimization," in *IEEE Computer Society Conference on Computer Vision and Pattern Recognition*, 2010, pp. 453–458.

[33] A. M. Bronstein, M. M. Bronstein, U. Castellani, A. Dubrovina, L. J. Guibas, R. P. Horaud, R. Kimmel, D. Knossow, E. v. Lavante, M. D., M. Ovsjanikov, and A. Sharma, "Shrec 2010: robust correspondence benchmark," in *Eurographics Workshop on 3D Object Retrieval*, 2010.

[34] M. Ovsjanikov, Q. Merigot, F. Memoli, and L. Guibas, "One point isometric matching with the heat kernel," *Computer Graphics Forum (Proc. SGP)*, vol. 29, no. 5, pp. 1555–1564, 2010.

[35] A. Sharma and R. Horaud, "Shape matching based on diffusion embedding and on mutual isometric consistency," in *NORDIA workshop IEEE Computer Society Conference on Computer Vision and Pattern Recognition*, 2010.

[36] A. Sharma, R. Horaud, J. Cech, and E. Boyer, "Topologically-robust 3D shape matching based on diffusion geometry and seed growing," in *IEEE Computer Society Conference on Computer Vision and Pattern Recognition*, 2011.

[37] D. Knossow, A. Sharma, D. Mateus, and R. Horaud, "Inexact matching of large and sparse graphs using laplacian eigenvectors," in *Graph-Based Representations in Pattern Recognition*, 2009, pp. 144–153.

[38] R. P. Horaud, F. Forbes, M. Yguel, G. Dewaele, and J. Zhang, "Rigid and articulated point registration with expectation conditional maximization," *IEEE Transactions on Pattern Analysis and Machine Intelligence*, vol. 33, no. 3, pp. 587–602, 2011.

[39] M. Belkin and P. Niyogi, "Laplacian eigenmaps for dimensionality reduction and data representation," *Neural computation*, vol. 15, no. 6, pp. 1373–1396, 2003.

[40] U. von Luxburg, "A tutorial on spectral clustering," *Statistics and Computing*, vol. 17, no. 4, pp. 395–416, 2007.

[41] F. R. K. Chung, *Spectral Graph Theory*. American Mathematical Society, 1997.

[42] L. Grady and J. R. Polimeni, *Discrete Calculus: Applied Analysis on Graphs for Computational Science*. Springer, 2010.

[43] C. Godsil and G. Royle, *Algebraic Graph Theory*. Springer, 2001.

[44] A. J. Hoffman and H. W. Wielandt, "The variation of the spectrum of a normal matrix," *Duke Mathematical Journal*, vol. 20, no. 1, pp. 37–39, 1953.

[45] J. H. Wilkinson, *The Algebraic Eigenvalue Problem*. Oxford: Clarendon Press, 1965.

[46] R. A. Horn and C. A. Johnson, *Matrix Analysis*. Cambridge: Cambridge University Press, 1994.

[47] R. Burkard, *Assignment Problems*. Philadelphia: SIAM, Society for Industrial and Applied Mathematics, 2009.

[48] J. Ham, D. D. Lee, S. Mika, and B. Schölkopf, "A kernel view of the dimensionality reduction of manifolds," in *International Conference on Machine Learning*, 2004, pp. 47–54.

[49] H. Qiu and E. R. Hancock, "Clustering and embedding using commute times," *IEEE Transactions on Pattern Analysis and Machine Intelligence*, vol. 29, no. 11, pp. 1873–1890, 2007.

[50] C. M. Grinstead and L. J. Snell, *Introduction to Probability*. American Mathematical Society, 1998.

[51] D. W. Scott, "On optimal and data-based histograms," *Biometrika*, vol. 66, no. 3, pp. 605–610, 1979.

[52] M. M. Zavlanos and G. J. Pappas, "A dynamical systems approach to weighted graph matching," *Automatica*, vol. 44, pp. 2817–2824, 2008.

[53] J. Sun, M. Ovsjanikov, and L. Guibas, "A concise and provably informative multi-scale signature based on heat diffusion," in *SGP*, 2009.

[54] C. Luo, I. Safa, and Y. Wang, "Approximating gradients for meshes and point clouds via diffusion metric," *Computer Graphics Forum (Proc. SGP)*, vol. 28, pp. 1497–1508, 2009.

16

Modeling Images with Undirected Graphical Models

Marshall F. Tappen

Department of Electrical Engineering and Computer Science
University of Central Florida
Orlando, Florida, USA
Email: mtappen@eecs.ucf.edu

CONTENTS

16.1 Introduction 476
16.2 Background 476
    16.2.1 Designing the Energy Function—The Need for Graphical Models ... 477
    16.2.2 Graph Representations of Distributions 478
    16.2.3 Formally Specifying a Graphical Model 478
        16.2.3.1 General Specifications of Graphical Models 479
    16.2.4 Conditional Independence 480
    16.2.5 Factor Graphs 481
16.3 Graphical Models for Modeling Image Patches 481
16.4 Pixel-Based Graphical Models 483
    16.4.1 Stereopsis 483
        16.4.1.1 Benefit of MRF Model for Stereo 487
    16.4.2 Modeling Pixels in Natural Images 487
    16.4.3 Incorporating Observations 488
    16.4.4 Beyond Modeling Images 489
16.5 Inference in Graphical Models 489
    16.5.1 Inference in General, Tree-Structured Graphs 490
    16.5.2 Belief Propagation for Approximate Inference in Graphs with Loops 491
    16.5.3 Graph-Cut Inference 491
16.6 Learning in Undirected Graphical Models 492
    16.6.1 Implementing ML Parameter Estimation 493
        16.6.1.1 Discriminative Learning of Parameters 494
        16.6.1.2 Optimizing Parameters in Discriminative Methods—Gradient
Descent 494
        16.6.1.3 Optimizing Parameters in Discriminative Methods—Large
Margin Methods 495
16.7 Conclusion 495
    Bibliography 496

16.1 Introduction

As computer vision and image processing systems become more complex, it is increasingly important to build models in a way that makes it possible to manage this complexity. This chapter will discuss mathematical models of images that can be represented with graphs, known as *graphical models*. These models are powerful because graphs provide an powerful mechanism for describing and understanding the structure of the models. This chapter will review how these models are typically defined and used in computer vision systems.

16.2 Background

One of the first steps in creating a computer vision system is the establishment of the overall computational paradigm that will be used to compute the final solution. One of the most common and flexible ways to implement a solution is through the combination of an energy function and MAP inference.

The term MAP inference is an acronym for Maximum A Posteriori inference and has its roots in probabilistic estimation theory. For the purposes of the discussion here, the MAP inference strategy begins with the definition of a probability distribution $p(\mathbf{x}|\mathbf{y})$[1], where $\mathbf{x}$ is the vector of random variables being estimated from the observations $\mathbf{y}$. In MAP inference, the actual estimate $\mathbf{x}^*$ is found by finding the vector $\mathbf{x}^*$ that maximizes $p(\mathbf{x}|\mathbf{y})$. This is sometimes referred to as a point estimate because the estimate is limited to a single point, as opposed to actually calculating the posterior distribution $p(\mathbf{x}|\mathbf{y})$.

The connection with energy functions can be seen by expressing $p(\mathbf{x}|\mathbf{y})$ as a *Gibbs distribution*:

$$p(\mathbf{x}|\mathbf{y}) = \frac{1}{Z} \exp\left(-\sum_{\mathrm{e}} E_C(\mathbf{x}_{\mathrm{e}}; \mathbf{y})\right) \tag{16.1}$$

where $E_{\mathrm{e}}(\mathbf{x}_C; \mathbf{y})$ denotes an energy function over a set of elements of $\mathbf{x}$. This makes the sum in Equation 16.1 a sum over different sets of elements of $\mathbf{x}$. The structure of these sets is vital to the construction of the model and will be discussed in Section 16.2.3.1. The constant Z is a normalization constant that ensures that $p(\mathbf{x}|\mathbf{y})$ is a valid probability distribution.

---

[1] This distribution corresponds to the posterior in Bayesian estimation.

With this representation, it can be seen that MAP inference is equivalent to finding the vector $\mathbf{x}^*$ that maximizes the energy function

$$E(\mathbf{x}; \mathbf{y}) = \sum_{c} E_c(\mathbf{x}_c; \mathbf{y}). \tag{16.2}$$

It should be noted that this focus on MAP inference should not be construed as minimizing the usefulness of computing the full posterior distribution over possible values of $\mathbf{x}$. However, computing the MAP estimate is often much easier because it is only necessary to compute a single point.

16.2.1 Designing the Energy Function—The Need for Graphical Models

A key motivation in designing graphical models can be seen by considering an example image enhancement problem. In this motivating problem, imagine the goal is to estimate a high-quality image, denoted $\mathbf{x}$, from a corrupted observation, denoted $\mathbf{y}$. This basic formulation encompasses a number of different, challenging problems, including deblurring, denoising, and super-resolution.

Assuming that a MAP estimation strategy will be employed, the next step is to decide on the form of the distribution $p(\mathbf{x}|\mathbf{y})$. Assuming that, for simplicity, $\mathbf{x}$ represents a 2×2 image, with 256 possible states, corresponding to gray-levels, per pixel. For this type of random vector, specifying just $p(\mathbf{x})$, and ignoring the relationships with the observation $\mathbf{y}$, would require 256^4 values to be determined if $p(\mathbf{x})$ was specified in its simplest tabular form. The huge number of values is necessary because the model must account for every interaction between the four pixels.

Now consider a simplification of this model where only some interactions are considered. The model can be simplified if only the interactions between horizontal and vertical neighbors are considered. Using the labeling shown in Figure 16.1(a), only the interactions between A-B, A-C, C-D, and B-D would be modeled. Again, using a simple tabular specification of the interactions between the states at different pixels requires 256^2 values to be specified, leading to $4(256^2)$ required values to specify the distribution, which is 16384 times fewer values. Far fewer parameters are necessary because this new model ignored diagonal interactions between pairs of nodes and interactions between sets of three and four nodes. This reduction is possible because interactions between nodes like A and D are not modeled explicitly, but are handled by combination of interactions between the $A - -B$ and $B - -D$ interactions, among others. It should be remembered that this is just for a 2×2 image; the reduction in complexity grows quickly as the image becomes larger.

Of course, this simplified model is only useful if the resulting estimate $\mathbf{x}^*$ is comparable to the result from the complete model. That is the key challenge in designing a model for image processing and analysis: finding the simplest model that performs the desired task well. Graphical models have proved to be a very useful tool because

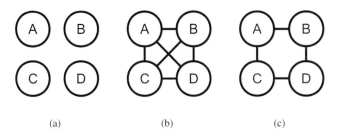

(a)   (b)   (c)

FIGURE 16.1
Each of these graphs represents the pixels in a simple 2×2 image. Each pixel is represented by a node in the graph. The edges in the graph represent how the relationships between pixels are expressed in different models. The graph in (a) describes a model that does not consider interrelationships between pixels, while the graph in (c) denotes a model that could simultaneously consider the relationship between all pixels. The graph in (b) describes a model that is more descriptive than (a), but simpler than (c). This model is built on the relationships between neighboring pixels.

they make it possible to control the complexity of the model by controlling the complexity of the interactions between the variables being estimated.

16.2.2   Graph Representations of Distributions

Graphical models are sonamed because the distribution can be represented with a graph. Each variable is represented by a node in the graph, so the 2×2 image from the previous section can be represented by the four nodes in Figure 16.1(a). For now, the edges can be informally described as representing interactions that are explicitly described in the model. (A more formal definition will follow.) The completely specified model can then be represented with the connected graph shown in Figure 16.1 (b). In this graph, there are edges between every pair of nodes because the interaction between every pair of nodes is modeled explicitly.

In contrast, the graph for the simplified model, where only the relationship between neighboring horizontal and vertical neighbors is explicitly modeled, is shown in Figure 16.1(c). Notice that the diagonal edges are missing, indicating that the relationship between these nodes is not explicitly modeled. Of course, this does not mean that the values of at nodes A and D do not affect each other when performing inference. As mentioned earlier, the model in Figure 16.1(c) indicates that they do interact, but only indirectly through the interactions between B and C. If the value at A is constrained by the model to be similar to the value at B, which is likewise constrained to be similar to the value at D, then A will be indirectly constrained to be similar to D.

16.2.3   Formally Specifying a Graphical Model

An intrinsic characteristic of a graphical model is that the graph representation is explicitly connected to the mathematical specification of a model. The graph, like Figure 16.1 describes the factorization of the distribution. The distribution over values of **x**, again temporarily ignoring **y**, for the graph in Figure 16.1(c) has the form

$$p(\mathbf{x}) = \frac{1}{Z}\psi_1(x_A, x_B)\psi_2(x_A, x_C)\psi_3(x_B, x_D)\psi_4(x_C, x_D) \tag{16.3}$$

$$= \frac{1}{Z}\exp\left[-\left(E_1(x_A, x_B) + E_2(x_A, x_C) + E_3(x_B, x_D) + E_4(x_C, x_D)\right)\right] \tag{16.4}$$

Here, we have shown the distribution as both the product of a set of potential functions, $\psi(\cdot)$ and as a Gibbs distribution with energy functions, $E_1(\cdot), \ldots, E_4(\cdot)$.

For comparison, the distribution for the model where all joint relationships are modeled would simply be[2]

$$p(\mathbf{x}) = \frac{1}{Z}\psi_1(x_A, x_B, x_C, x_D)$$

$$= \frac{1}{Z}\exp\left[-E(x_A, x_B, x_C, x_D)\right]. \tag{16.5}$$

Notice that no factorization is possible in this model because the model must have an individual energy value for every possible joint set of values.

The factorization in Equation (16.3) expresses the distribution over the whole vector **x** through a set of local relationships between neighboring pixels. This is advantageous for both specifying the model and for performing inference. This factorization is valuable because it makes it possible to construct a global model over pixels in an image by combining local relationships between pixels. It will likely be easier to construct appropriate potentials for neighboring pixels than to construct a single potential that captures all of the relationships between many different pixels. As will be discussed later, this factorization into local relationships also often makes inference more computationally efficient.

16.2.3.1   General Specifications of Graphical Models

Graphical models like the model in Equation (16.5) are referred to as undirected graphical models because they can be represented with an undirected graph[3].

The undirected graphical model is described by the potential functions $\psi(\cdot)$,

---

[2]It should be noted that the fully connected graph that matches this distribution could also correspond to a distribution that was factored similarly to the distribution in Equation (16.3), but with diagonal links. This ambiguity will be discussed in Section 16.2.5.

[3]A related family of models correspond to directed graphs. These models are also sometimes referred to as Bayes Nets and are discussed extensively in [1].

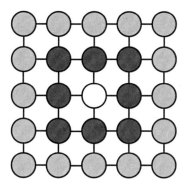

FIGURE 16.2

This graph demonstrates how the graph describing a graphical model can be analyzed to determine conditional independence relationships. The node at the center, colored in white, is conditionally independent of the light gray nodes, conditioned on the dark gray nodes. Intuitively, these relationships imply that once the values of the dark gray nodes are known, the light gray nodes provide no more information about what values the white node should take.

which are functions of subsets of nodes in the graph. The nodes included in any particular potential function correspond to a clique in the graph, which is a subset of nodes where there is an edge between each pair of nodes in the clique. Each potential in a graphical model is a function of one of these cliques.

If $\mathbf{x}_{\mathcal{C}}$ denotes a clique in the graph, then the corresponding distribution can be defined as

$$p(\mathbf{x}) = \frac{1}{Z} \prod_{\mathcal{C}} \psi_{\mathcal{C}}(\mathbf{x}_{\mathcal{C}}), \tag{16.6}$$

where the product is over all cliques $\mathcal{C}$. In this equation, the term $1/Z$ is a normalization constant and is often referred to as the partition function. This product form can also be expressed in terms of energy functions, similar to Equation (16.1), by calculating the logarithm of the potentials.

Typically, the cliques $\mathcal{C}$ are maximal cliques, which are essentially the largest possible cliques. If a clique is a maximal clique, then if any node is added the subset will no longer be a clique. The maximal clique representation can also represent any model based on nonmaximal cliques, as the maximal clique would subsume the potential functions based on nonmaximal cliques.

16.2.4   Conditional Independence

In addition to describing how the model is factorized, the graph representing a graphical model can also be analyzed to find conditional independence relationships in the model. Any node in the model is independent of all other nodes, conditioned on the nodes connected to it. Thus, in a lattice model, like the model shown in Figure 16.2, the node at the center, colored in white, is independent of the nodes colored in light gray, when conditioned on the nodes colored in dark gray. The nodes shown in dark gray can be called the Markov blanket. Intuitively, these relationships imply that once the values of the dark gray nodes are known, the light gray nodes provide no more information about what values the white node should take.

Returning to the four node graph in Figure 16.1, these conditional independence statements can be written formally in statements like

$$p(x_A | x_B, x_C, x_D) = p(x_A | x_B, x_C) \tag{16.7}$$

or

$$p(x_A, x_D | x_B, x_C, x_D) = p(x_A | x_B, x_C) p(x_D | x_B, x_C) \tag{16.8}$$

These conditional independence relationships between nodes in the graph and variables in the vector have led these models to also be commonly called Markov Random Fields.

16.2.5 Factor Graphs

A weakness of the graph representation discussed so far is that it can be ambiguous. Consider the fully connected undirected graphical model in Figure 16.3(a). This graph could represent a number of different factorizations, including

$$p(\mathbf{x}) = \frac{1}{Z} \psi_1(x_A, x_B, x_D) \psi_2(x_A, x_C, x_D) \tag{16.9}$$

or

$$p(\mathbf{x}) = \frac{1}{Z} \psi_1(x_A, x_C, x_D) \psi_2(x_A, x_B) \psi_3(x_B, x_D) \tag{16.10}$$

The factor graph representation removes this ambiguity by representing the model with a bipartite graph, with nodes explicitly representing the potentials [2]. In a factor graph, each potential is represented with a node. As variables are often represented with unfilled circles, the factors are usually represented with solid black squares. Figure 16.3(b) shows the factor graph representation of the model in Equation (16.9). In this representation, edges connect the node representing each potential to the nodes that interact in that potential.

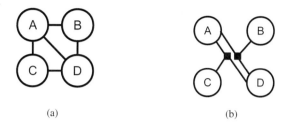

(a)   (b)

FIGURE 16.3
The graph in (a) represents the model in Equation (16.9) with the same graph that would represent a model with a different factorization. Figure (b) shows a factor graph representation that uses a bipartite graph to explicitly represent different potentials.

16.3   Graphical Models for Modeling Image Patches

Undirected graphical models are a natural tool for modeling patches of pixels in low-level image models. A good example of a patch-based models is the model used for super-resolution in [3]. In this model, a graphical model is used to estimate a high-resolution image from a low-resolution observation, $\mathbf{y}$.

In patch-based models, the image is divided into square patches, with one node in the model per patch. These patches are typically chosen by examining the observed image $\mathbf{y}$. In the model from [3], Freeman et al. use the observed low-resolution image to search a database for high-resolution patches that have a similar appearance when the resolution is reduced to match the observation. The patches can also be created dynamically, as in [4].

The system's estimate, $\mathbf{x}^*$, is created by mosaicing together the patches chosen at each node. One of the advantages of a patch-based model is that it makes it possible to model an image with a discrete-valued graphical model. A discrete-valued graphical model may be useful if it is desirable to arbitrarily specify the relationship between patches because the potentials can be expressed in a tabular form. Some of the most powerful inference algorithms discussed in Section 16.5 are also designed for graphical models.

The graph in patch-based models is typically structured as a pairwise lattice, similar to the model shown in Figure 16.2. In this model, all of the cliques consist of horizontal or vertical neighbors. In [3], the potential functions are specified by examining the overlapping regions created by making the patches slightly larger than the patches that will be mosaiced together, as shown in Figure 16.4. This overlap region can then be used to create a potential function, which can be thought of as a compatibility function that measures how compatible the possible choices of patches at two

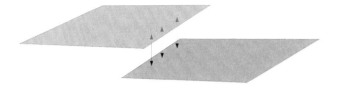

FIGURE 16.4
This figure illustrates one strategy for computing the compatibility between neighboring image patches. The resulting image will be created from mosaicing together the interior outlined regions. The compatibility between patches is measured by examining the overlapping regions outside of the outlined areas.

neighboring nodes are. In [3], the compatibility function measured by computing the squared difference between the two patches in the overlap region, as illustrated in Figure 16.4. Formally, this can be expressed as

$$\psi(x_p, x_q) = \exp\left(-\sum_{i \in \mathcal{O}} (t_i^{x_p} - t_i^{x_q})^2\right) \tag{16.11}$$

where $t_i^{x_p}$ refers to the value of the ith pixel in the patch that would be used if node p took the value x_p. This sum is computed over the overlapping region $\mathcal{O}$.

The basic function of this potential is that patches that are very similar in the overlap region will be considered compatible. Similar models have been used for other applications, such as editing photographs [5] or synthesizing texture [6]. The work in [6] further extends these models by allowing the use of nonsquare image patches.

16.4 Pixel-Based Graphical Models

The most common alternative to patches in image models is to directly model pixels. Directly modeling pixels is particularly common in stereo and image enhancement applications, where a graphical model can be used to express priors based on smoothness.

16.4.1 Stereopsis

The most difficult problem in *stereopsis* is correctly matching the pixels between two views of the scene, particularly when a pixel in one image matches with multiple pixels in the other image equally well. If multiple pixels match equally well, then other assumptions must be used to disambiguate the problem. A common assumption is that the surface is smooth. This transforms the stereopsis problem into the task of finding matches where matched pixels are both similar visually and lead to a reconstruction with smooth surfaces.

Undirected graphical models have a long history of application to this task. In many cases, the purpose of the model is also described in terms of regularizing the solution [7, 8]. One of the most common formulations of the stereo problem is through an energy function on a vector containing one variable per pixel in the graph. Each variable typically represents the disparity between the location of the pixel in the reference image and the location of the matching pixel in the other image. A common form of the energy function is

$$E(\mathbf{x}) = E_{\text{Data}}(\mathbf{x}; \mathbf{y}) + E_{\text{Smooth}}(\mathbf{x}) \qquad (16.12)$$

where $E_{\text{Data}}(\mathbf{x}; \mathbf{y})$ and $E_{\text{Smooth}}(\mathbf{x})$ are energy functions representing the visual similarity and smoothness of the surfaces, respectively [9, 10].

The term $E_{\text{Data}}(\mathbf{x}; \mathbf{y})$ measures the visual similarity between pixels matched by the disparity values in $\mathbf{x}$. This energy function is often referred to as the data term because it captures the information in the data presented to the system. It is typically based on intensity differences in the observation $\mathbf{y}$. Models for a number of different applications use a data term that captures the information in the observed data.

Because of the discretely sampled nature of images, the data term for stereo models is usually most easily expressed as a discrete-valued function, making the vector $\mathbf{x}$ a vector of discrete-valued variables. This has the added advantage of making it possible to use powerful inference techniques that will be discussed in Section 16.5.

The smoothness assumption is expressed in $E_{\text{Smooth}}(\mathbf{x})$. Although the smoothness term is also usually discrete-valued, to be compatible with the data term, it is typically based on the difference between the disparities represented. Assuming that each element of $\mathbf{x}$, x_i, is defined such that $x_i \in \{0, \dots, N-1\}$, a basic smoothness term just measures the squared difference in disparity between neighboring pixels or

$$E_{\text{Smooth}}(\mathbf{x}) = \sum_{<p,q>} (x_p - x_q)^2. \qquad (16.13)$$

In this formula, the sum is over all pairs of neighboring nodes, p and q. This is denoted in Equation (16.13) as $< p, q >$.

Equation (16.13) can be rewritten more generally with a penalty function $\rho(\cdot)$

$$E_{\text{Smooth}}(\mathbf{x}) = \sum_{<p,q>} \rho(x_p - x_q). \qquad (16.14)$$

where $\rho(z) = z^2$.

When designing a smoothness term, both the type of difference, such as first-

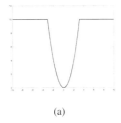 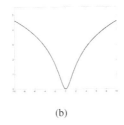 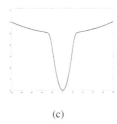

(a)   (b)   (c)

FIGURE 16.5
This figure shows three different types of potentials. The advantage of these potentials is that they combine the behavior of the quadratic and Potts models. Small differences are treated smoothly, while large differences have roughly the same penalty. This helps produce sharp edges in the results produced by the system.

or second- order difference, to use and the type of penalty function must be decided. The type of derivative influences the resulting shape. The first derivative, as in Equation (16.13), selects surfaces that are composed of fronto-parallel planes. If second-derivatives are used, the surface encouraged to be planar, but not necessarily fronto-parallel.

The other key decision is the type of penalty function. The squared-error function used in Equation (16.13) is rarely used because this type of smoothness prior is known to lead to surfaces with blurry edges. As discussed in [4], a squared error criterion prefers smaller differences across several pixels instead of a large difference at one location. The effect of this is especially noticeable at the edges of objects. Rather than having a crisp, sharp edge where an object ends, a squared error criterion will lead to blurry edges that smoothly transition to the disparity levels of different objects.

The solution to this problem is to use different potentials for implementing the smoothness prior. For discrete-valued models, a popular potential is the Potts model, which is often expressed as an energy function as

$$\rho_{\text{Potts}}(z) = \begin{cases} 0 & \text{if } z = 0 \\ T & \text{otherwise} \end{cases}, \tag{16.15}$$

where T is a constant chosen when the model is implemented.

The basic idea behind the Potts model is that neighboring pixels are encouraged to have the same disparity, but a fixed cost is assigned when they do not. The unique aspect of the Potts model is that the cost is the same no matter how great the difference in disparity. This is useful because there is typically no prior information about how separated objects should be; thus the same cost should be assigned to any change in disparity. The Potts model can also be thought of as using an L_0 norm to enforce sparsity in the horizontal and vertical gradients of the surface.

Other useful potentials can be seen as a balance of the quadratic penalty and

Potts model. Figure 16.5(a) shows the truncated quadratic potential. This model allows for smooth variations in the surface by penalizing small changes in disparity with a quadratic model, but large changes in disparity behave as a Potts model. The truncated quadratic model can be expressed similarly to the Potts model:

$$\rho_{\text{TruncQuad}}(z) = \begin{cases} z^2 & \text{if } |z < T| \\ T^2 & \text{otherwise} \end{cases}, \tag{16.16}$$

where T is a threshold determined when the model is constructed.

The truncated quadratic model works well if the model is discrete-valued. In that case, it can be used as a replacement for the Potts model. However, it may not be well suited for continuous-valued models. As will be discussed in Section 16.5, inference in continuous-valued models is often implemented with gradient-based optimization. The truncated quadratic model may be awkward if gradient-based optimization is used because it is discontinuous at the point when the difference between x_p and x_q exceeds T.

This problem can be avoided by using a potential that is continuous everywhere and is similar in shape to the truncated quadratic. Figures 16.5(b) and (c) show two examples of alternative potentials. Figure 16.5(b) shows the Lorentzian potential [11], which has the form

$$\rho_{\text{Lorentz}}(z) = \log\left(1 + \frac{1}{2}z^2\right) \tag{16.17}$$

The Lorentzian potential is useful because the rate of increase of the penalty function decreases as the difference between x_p and x_q increases. In effect, when the difference between x_p and x_q is large, increases in this difference cause relatively small increases in the penalty function. This is similar to the Potts model in that once the change in disparity is large enough, the size of the disparity change has a relatively small impact on the energy of the smoothness term.

A more flexible term is based on Gaussian Scale Mixtures [12, 13]. A Gaussian scale mixture is a mixture model where the mean of each component is zero, but the variance changes. Figure 16.5(c) shows an example of a potential based on a two-component mixture. If the energy function is not constrained to be a true Gaussian mixture, then a potential like that shown in Figure 16.5(c) can be expressed as

$$\rho_{\text{GSM}}(z) = -\log\left(\exp\left(-z^2/\sigma_1 + T_1\right) + \exp\left(-z^2/\sigma_2 + T_2\right)\right), \tag{16.18}$$

where T_1 and T_2 are constants.

The Gaussian Scale Mixture is particularly useful because multiple terms can be added to the sum in Equation (16.18). This makes it possible for the potentials to flexibly express many different shapes of potentials.

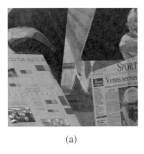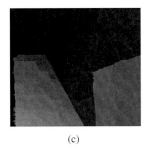

(a)   (b)   (c)

FIGURE 16.6
A stereo example that shows the benefit of the using a graphical model to estimate
stereo disparity. (a) One input image from the stereo pair. An algorithm based on
matching just using local windows will have difficulty in areas with little texture. (b)
The disparity map recovered by matching using only local windows. Notice the noise
in the disparity map in textureless areas of the scene. In these areas, the local texture
is not sufficient to match. (c) This image shows the disparity map when a smoothness term is incorporated into overall cost function. Notice how the smoothness term
eliminates noise in areas of low texture.

In addition to these potentials where the rate of change in the penalty decreases,
the $L1$ penalty is also commonly used. This penalty is defined as

$$\rho_{\text{L1}}(x_p, x_q) = |z|. \tag{16.19}$$

16.4.1.1   Benefit of MRF Model for Stereo

The usefulness of a graphical model for estimating disparity in stereo image pairs
can be seen in the example shown in Figure 16.6. Figure 16.6(b) shows the disparity
map that would be recovered from the scene shown in Figure 16.6(a) using just the
data term. Notice that in areas with little variation in texture, the disparity estimates
are quite noisy.

As described above, this noise can be ameliorated by incorporating a smoothness
term. In areas with low amounts of texture, the smoothness term effectively prop-
agates information from areas with enough texture to make matching based on the
data term possible. Without strong information from local matching, the smoothness
term will cause the estimated disparity values to be smooth with respect to values in
areas with good texture information. Figure 16.6(c) shows the disparity map recovered when a smoothness cost is incorporated. With the smoothness term added, the
noise is eliminated.

16.4.2 Modeling Pixels in Natural Images

Popular models of images have much in common with the smoothness priors for stereopsis. A popular recent model is the Field of Experts model proposed by Roth and Black [14]. The model functions as a prior on images and is combined with a data term to produce estimates. This model is based on a set of filters, $f_1, \ldots, f_{N_f}$, and weights, $w_1, \ldots, w_{N_f}$, that are combined with a Lorentzian penalty to form

$$E_{\text{FoE}}(\mathbf{x}) = \sum_p \sum_{i=1}^{N_f} -w_i \log\left(\mathbf{f}_i * \mathbf{x}_p\right), \qquad (16.20)$$

where the sum is over all pixels p. In this equation $\mathbf{x}_p$ denotes a patch of image centered at pixel p, which is the same size as the filter f_i. This makes the result of $f_i * \mathbf{x}_p$ a scalar. While the form based on energy functions is shown here, the product form of this distribution is composed of a product of factors created from the Student's t distribution.

Similar models have also been used in [15, 13]. This model can be justified in two different ways. The first justification is based on statistical properties of images. The histogram of the response of applying a filter to images has a characteristic shape. Figure 16.7(b) shows the histogram of a derivative filter applied to the image in Figure 16.7. The notable aspect of this shape is that it is sharply peaked around zero, and the tails of the distribution have more mass than would be seen in a Gaussian distribution. This type of distribution is referred to as a heavy-tailed distribution. As shown in Figure 16.7(c), the negative logarithm of this distribution has a similar shape as the Lorentzian penalty. The filters in the Field of Experts model that perform the best are larger than simple derivative filters used to generate the example in Figure 16.7, but the histogram of the response has a similar shape. Thus, the Field of Experts model can be thought of as modeling the marginal statistics of images. Improved versions of the model have also been produced by using Gaussian Scale Mixtures [13, 16].

Interestingly, the optimal filters for the Field of Experts model are similar to derivative filters. The Field of Experts model can also be viewed through the lens of smoothness, similar to the smoothness priors used in stereopsis. The Lorentzian penalty has a similar behavior as the truncated quadratic model, where small changes are penalized relative to the size of the change, but large changes have roughly the same penalty, which is not very dependent on the size of the change. This model can be seen as describing images as being made of smooth regions, with occasional large changes at edges. Because the filters used in the Field of Experts model are higher-order filters, the smoothness does not lead to flat regions in the image, but instead leads to smooth gradients.

Combined with a data term, this model has been applied to other tasks, such as in-painting [14].

(a)

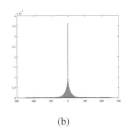
(b)

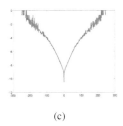
(c)

FIGURE 16.7
The structure of Field of Experts model, which is built around derivative-like filters, can be motivated by examining the statistics of natural images, such as the image shown in (a). Figure (b) shows the histogram of pixel values in the image created by filtering the image in (a) with a discrete-derivative filter. Figure (c) shows the log of the histogram in (b). The shape of this histogram is very similar to the shape of the potential functions used in the Field of Experts model. [Photo credit: Axel D from Flickr].

16.4.3 Incorporating Observations

Up this point, the issue of incorporating the observations has not been considered. At a high level, observations are typically implemented in a model in one of two ways. One approach is to model the joint distribution between the observations and the variables being estimated. In [3], the terms expressing the relationship between the observation and the hidden variables being estimated are denoted as $\phi(\cdot)$ instead of $\psi(\cdot)$. This gives a pairwise lattice model, similar to Figure 16.2, of the form

$$p(\mathbf{x}, \mathbf{y}) = \frac{1}{Z} \prod_p \phi(x_p, y_p) \prod_{<p,q>} \psi(x_p, x_q), \tag{16.21}$$

where the first product is over all pixels p and the second product is over all neighboring pairs of pixels.

Another approach is to simply make all of the factors in the distribution dependent on the observations. This is currently the most common approach because it offers the most flexibility. Models that do this are often called Conditional Random Field models or CRFs [17].

16.4.4 Beyond Modeling Images

While the focus of the previous sections have been on low-level vision models that model pixels, they have also been used to model parts in object, as in the pictorial structures model [18, 19]. In these models, the nodes in the graph represent the location of different parts. These models can be fit as generic shapes, as in [19], or can be fit to specific shape, such as the human form [20].

16.5 Inference in Graphical Models

After the model has been designed, MAP inference can be applied to recover the estimate $\mathbf{x}^*$. While graphical models offer a convenient tool for building models, there are still significant issues in finding $\mathbf{x}^*$. Inference in graphical models is an active, mathematically rigorous area of research, so this chapter will briefly survey popular approaches. More in-depth information and explanations can be found in [21, 1] and the original literature cited below. Szeliski et al. have also published a useful comparison of different inference techniques [22].

In some cases, the inference can be implemented in polynomial time using existing algorithms. If the energy function is the sum of convex quadratic terms, then the minimum energy solution $\mathbf{x}^*$, which corresponds to the MAP estimate, can be found by solving the linear system created by differentiating the energy function. Similarly, if the energy function can be expressed as the sum of absolute value penalties, then $\mathbf{x}^*$ can be found with linear programming.

For more general continuous-valued models, the inference can be implemented with general optimization techniques, so it may not be possible to guarantee that the estimate of $\mathbf{x}^*$ is the global optimum, rather than just a local optimum.

16.5.1 Inference in General, Tree-Structured Graphs

If the graph representation of the model does not have loops, then $\mathbf{x}^*$ can be recovered using a simple iterative procedure known as *Belief Propagation* [23] or the max-product algorithm.

To see how this algorithm is derived, consider a simple three-node model

$$p(\mathbf{x}) = \frac{1}{Z}\psi(x_1, x_2)\psi(x_2, x_3) \tag{16.22}$$

In MAP inference, the goal is to compute

$$\mathbf{x}^* = \arg\max_{x_1, x_2, x_3} \psi(x_1, x_2)\psi(x_2, x_3). \tag{16.23}$$

In this model, the joint max operation can be factored to make the computation more efficient:

$$\mathbf{x}^* = \arg\max_{x_2}\left[\arg\max_{x_1}\psi(x_1, x_2)\right]\left[\arg\max_{x_3}\psi(x_2, x_3)\right] \tag{16.24}$$

$$= \max_{x_2} m_{1\to2}(x_2)m_{3\to2}(x_2). \tag{16.25}$$

In Equation (16.25), the max operations have been represented as message functions. Because the max functions on the right are over x_1 and x_3, these max operations generate functions of x_2. In Equation (16.25), these functions are represented as message functions that are passed from nodes x_1 and x_3 to x_2. The leftmost max function in Equation (16.23) computes the optimal label for x_2. The optimal labels for x_1 and x_3 can be computed by messages passed from x_2.

It can be shown that a similar message-passing routine can be used to perform MAP inference in graphs where there are no loops [21, 23]. In the general operation, inference is implemented through a series of message-passing operations.

The algorithm is easiest to describe for a pairwise graph, where all cliques are pairs of nodes. For a node, x_i in the graph, inference is implemented by a series of message-passing operations between x_i and the set of its neighbors, denoted as the set $\mathcal{N}(x_i)$. At each iteration, the node x_i sends a message to each of its neighbors. The message from x_i to a neighbor x_j, denoted at $m_{i \to j}(x_j)$ is computed as

$$m_{i \to j}(x_j) = \max_{x_i} \psi(x_i, x_j) \prod_{k \in \mathcal{N}(x_i) \setminus j} m_{k \to i}(x_i), \qquad (16.26)$$

where $k \in \mathcal{N}(x_i) \setminus j$ denotes all nodes k, besides j that are neighbors of x_i.

The optimal value of x_i is computed from the messages received by its neighbors,

$$x_i^* = \arg \max_i \prod_{k \in \mathcal{N}(x_i)} m_{k \to i}(x_i) \qquad (16.27)$$

For more complicated graphs, the factor graph representation is particularly useful [2].

16.5.2 Belief Propagation for Approximate Inference in Graphs with Loops

In graphs with loops, these steps do not lead to the optimal result because the messages sent out from a node eventually loop back to that node. However, research and practice has shown that applying the belief propagation algorithm on graphs with loops will still often produce good results [24, 25].

In general, the inference in graphs with loops is an NP-Complete inference problem [10, 1]. However, recent research has also introduced other message-passing algorithms that perform better at finding good approximate solutions. In particular, the Tree Re-Weighted Belief Propagation Algorithm, or TRW algorithm, proposed by Wainwright et al. [26] and the TRW-S improvements proposed by Kolmogorov [27] performed well in the comparisons in [22].

16.5.3   Graph-Cut Inference

An alternative to the message-passing algorithms for inference is the graph-cut approach proposed by Boykov et al. in [10].

This algorithm is based on the ability to find the MAP solution for certain binary-valued graphical models by computing the min-cut of a graph [28]. While the min-cut can only be used to optimize a binary-valued graphical model, Boykov et al. introduced the α-expansion approach that uses this graph-cut procedure to compute an approximate solution. In the alpha expansion procedure, the optimization process is implemented as a series of expansion moves. Given an initial estimate of $\mathbf{x}^*$, which will be denoted $\mathbf{x}_0$, a move in [10] is a modification of this estimate that produces an estimate with a lower energy. If i is one of the possible values for variables in $\mathbf{x}$, then the expansion move modifies $\mathbf{x}_0$ by changing the value of some elements of $\mathbf{x}_0$ to have the value i.

The α-expansion move is useful because it is essentially a binary-valued graphical model. In this new model, the nodes correspond to the nodes in the original problem, while the state of each node determines whether that particular node should take the value i or not. As shown in [10], it is possible to use the graph-cut procedure to find the optimal expansion move.

Thus, the overall α-expansion approach to finding an approximate value of $\mathbf{x}^*$ is to iterate through all possible discrete values, calculating and implementing the optimal expansion move each time. Typically, moves are calculated in random order, and the moves for each state are calculated multiple times. The α-expansion algorithm is limited in that not all potentials make it possible to compute the α-expansion move using a graph cut, but recent research is proposing algorithms for overcoming this limitation [29].

16.6   Learning in Undirected Graphical Models

The factorized nature of undirected graphical models makes it convenient to specify models by hand, which has worked well for pairwise graphical models, such as in [30, 3, 31]. However, as the models become more interconnected and the number of nodes involved in the potentials increases, manually specifying the parameters becomes more difficult because the number of parameters increases. A similar problematic rise in the number of parameters also happens when the potentials involve large numbers of observed features.

A common solution to fitting models with a large number of parameters is to fit them from data examples. Learning is particularly important for low-level vision problems, due to the large number of variables in these problems [32]. Because undirected graphical models define probability distributions, the Maximum Likelihood method can be used to estimate the model parameters. Given a set of training examples, $\mathbf{t}_1, \ldots, \mathbf{t}_n$, ML parameter estimation can be implemented by minimizing the negative log-likelihood function. Assuming that the training samples are independent and identically distributed, the log-likelihood function for an undirected graphical model, evaluated on a training vector $\mathbf{t}$ can be written as

$$L(\mathbf{t}) = -\sum_c \log \psi(\mathbf{t}_c) + \log Z$$

$$= \sum_c E_c(\mathbf{t}_c) + \log Z, \qquad (16.28)$$

where Z is the normalization constant for the distribution.

The energy functions in the final line of Equation (16.28) correspond to the energy functions in the representation of this distribution as a Gibbs distribution, similar to Equation (16.1)

This can be optimized with respect to a parameter θ by differentiating Equation (16.28) and using gradient-based optimization. The derivative of Equation (16.28) with respect to a parameter θ is

$$\frac{\partial L}{\partial \theta} = \sum_c \frac{\partial E_c(\mathbf{t}_c)}{\partial \theta} + \frac{1}{Z} \int_{\mathbf{x}} -\exp\left(-\sum_c E_c(\mathbf{x}_c)\right) \sum_c \frac{\partial E_c(\mathbf{x}_c)}{\partial \theta}. \qquad (16.29)$$

The right-hand term in Equation (16.29) can be rewritten as en expectation, making the derivative

$$\frac{\partial L}{\partial \theta} = \sum_C \frac{\partial E_C(\mathbf{t}_C)}{\partial \theta} - E\left[\sum_C \frac{\partial E_C(\mathbf{x}_C)}{\partial \theta}\right]. \qquad (16.30)$$

This has an appealing, intuitive interpretation. The derivative is the difference between the derivative of energy potentials evaluated at the training example and the expected value of the derivative of the energy potential function.

16.6.1 Implementing ML Parameter Estimation

The difficulty in learning parameters for an undirected graphical model lies in computing the expectation in Equation (16.29). Like MAP inference, computing these expected values will also often be NP-complete [1].

A common solution to this problem when computing a gradient is to draw samples from the distribution defined by the current set of parameters, then use these samples to compute an approximate value of the expectation. In [33], Zhu and Mumford use Gibbs sampling to generate the samples for a model of texture.

A disadvantage of sampling is that it can be time consuming to generate the large number of samples necessary to compute an accurate approximation to the expected value. Surprisingly, it is often not necessary to compute a large number of samples. In [34], it is shown than a reasonable approximation to the gradient can be found by just computing one sample. This technique, which is described in [34] as minimizing a quantity called the Contrastive Divergence, is used to learn the Field of Experts model in [14] and fit parameter for a multiscale segmentation model in [35].

16.6.1.1 Discriminative Learning of Parameters

Discriminative methods present an alternative to the maximum likelihood approach for estimating the model parameters. To understand how discriminative methods can be applied for learning parameters, it is helpful to review this approach for the basic classification task. When implementing a classifier, the defining characteristic of a discriminative classifier is that it is not constructed by estimating the joint distributions of observations and labels. Instead, a function is directly fit for the sole goal of correctly labeling observations. This function is fit using different criterion, such as the max-margin criterion used in Support Vector Machines or the log-likelihood criterion used in logistic regression.

Likewise, discriminative methods for learning MRF parameters also avoid fitting distributions to the data. Instead, discriminative approaches define alternate criteria for estimating parameters. Since distributions will not form the basis for estimation, current discriminative methods base the learning criterion on the MAP solution of the graphical model.

The goal of a discriminative training algorithm is to make the MAP estimate produced from the graphical model as similar to the ground truth as possible. The training procedure is based on a loss function that measures how similar $\mathbf{x}^*$ is to the ground-truth value. If the graphical model is defined by an energy function $E(\mathbf{x}; \mathbf{y}, \theta)$, based on observations $\mathbf{y}$, then this training can be expressed as

$$\theta^* = \arg\min_\theta L(\mathbf{x}^*, \mathbf{t}) \tag{16.31}$$

$$\text{where } \mathbf{x}^* = \arg\min_{\mathbf{x}} E((\mathbf{x}; \mathbf{y}, \theta) \tag{16.32}$$

In this optimization, changing the parameters θ changes the model and also changes the location of the MAP estimate $\mathbf{x}^*$. The goal of the optimization is to find a value of θ that makes $\mathbf{x}^*$ as close to the ground truth $\mathbf{t}$ as possible, as measured by the loss function $L(\cdot)$.

16.6.1.2 Optimizing Parameters in Discriminative Methods—Gradient Descent

If the optimization of $\mathbf{x}^*$ can be expressed as a closed-form set of differentiable operations, then the chain rule can be used to compute the gradient of $L(\cdot)$ with respect to the parameters θ.

As pointed out in [15], if the energy function defining the model is a quadratic function, which corresponds to a graphical model that is also a Gaussian random vector, then the MAP solution, $\mathbf{x}^*$, can be found with a set of matrix multiplications and a matrix inversion operation. These are differentiable operations, so it is possible to compute the gradient vector $\partial L / \partial \theta$. While the quadratic model typically does not perform as well as models like the Field of Experts, [15] and [36] show how the training process makes it possible to learn to exploit the information in the observations to perform surprisingly well on both image enhancement and segmentation tasks.

For models where $\mathbf{x}^*$ cannot be computed analytically, [36] and [37] propose using an approximate value of the MAP solution. Both of these systems propose using an approximate value of $\mathbf{x}^*$ that is computed using some form of minimization which may not lead to a global minimum of the energy function $E(\cdot)$. If the minimization is itself a set of differentiable operations, then the chain rule can be applied to compute how this approximate $\mathbf{x}^*$ changes as the parameter vector θ changes.

These papers differ in the type of optimization used. In [36], the training is built around an upper-bound minimization strategy that fits a series of quadratic models during the optimization process. Taking a different approach, Barbu proposes using a small number of steps of gradient descent to approximate the full inference process, which leads to a very efficient learning system [37].

16.6.1.3 Optimizing Parameters in Discriminative Methods—Large Margin Methods

Just as the support vector machine uses quadratic programming to optimize the max-margin criterion, quadratic programming can be used to learn parameters in MRF models. In [38], Taskar et al. describe the M^3N method, which is one of the first margin-based methods for learning graphical model parameters. This approach poses learning as a large quadratic program and is used for a labeling application in [39].

The most influential approach recently, across all options for learning model parameters, is the cutting plane approach to training [40]. In this approach, the inference system is used as an oracle to find results that violate constraints in the training criterion. A major advantage of this approach is that it does not require that the inference process be differentiable in some way. This makes it possible to use a wide variety of structures, including models where inference is intractable [41]. Recent vision systems based on this training approach include [42, 43]. In addition, a high-quality implementation of this algorithm is available in the SVMStruct package [40].

16.7 Conclusion

Graphical models are a powerful tool for building mathematical models of images and scenes. The factorized nature of the model makes it possible to conveniently design potentials that capture properties of real-world images. Their flexibility, combined with powerful inference algorithms, has made them an invaluable tool for building complex models of images and scenes.

Bibliography

[1] D. Koller and N. Friedman, *Probabilistic Graphical Models: Principles and Techniques.* MIT Press, 2009.

[2] F. Kschischang, B. Frey, and H.-A. Loeliger, "Factor graphs and the sum-product algorithm," *IEEE Transactions on Information Theory*, vol. 47, no. 2, pp. 498 –519, Feb. 2001.

[3] W. T. Freeman, E. C. Pasztor, and O. T. Carmichael, "Learning low-level vision," *International Journal of Computer Vision*, vol. 40, no. 1, pp. 25–47, 2000.

[4] M. F. Tappen, B. C. Russell, and W. T. Freeman, "Efficient graphical models for processing images," in *Proceedings of the IEEE Conference on Computer Vision and Pattern Recognition*, vol. 2, 2004, pp. 673–680.

[5] T. S. Cho, M. Butman, S. Avidan, and W. T. Freeman, "The patch transform and its applications to image editing," in *IEEE Conference on Computer Vision and Pattern Recognition*, 2008.

[6] V. Kwatra, I. Essa, A. Bobick, and N. Kwatra, "Texture optimization for example-based synthesis," *ACM Transactions on Graphics*, vol. 24, pp. 795–802, July 2005.

[7] J. L. Marroquin, S. K. Mitter, and T. A. Poggio, "Probabilistic solution of ill-posed problems in computational vision," *American Statistical Association Journal*, vol. 82, no. 397, pp. 76–89, March 1987.

[8] R. Szeliski, "Bayesian modeling of uncertainty in low-level vision," *International Journal of Computer Vision*, vol. 5, no. 3, pp. 271–301, 1990.

[9] D. Scharstein and R. Szeliski, "A taxonomy and evaluation of dense two-frame stereo correspondence algorithms," *International Journal of Computer Vision*, vol. 47, no. 1/2/3, April-June 2002.

[10] Y. Boykov, O. Veksler, and R. Zabih, "Fast approximate energy minimization via graph cuts," *IEEE Transactions of Pattern Analysis and Machine Intelligence*, vol. 23, no. 11, pp. 1222–1239, 2001.

[11] M. J. Black and A. Rangarajan, "On the unification of line processes, outlier rejection, and robust statistics with applications in early vision," *International Journal of Computer Vision*, vol. 19, no. 1, pp. 57–92, July 1996.

[12] J. Portilla, V. Strela, M. Wainwright, and E. P. Simoncelli, "Image denoising using scale mixtures of gaussians in the wavelet domain," *IEEE Trans. Image Processing*, vol. 12, no. 11, pp. 1338–1351, November 2003.

[13] Y. Weiss and W. T. Freeman, "What makes a good model of natural images," in *Proceedings of the IEEE Conference on Computer Vision and Pattern Recognition*, 2007.

[14] S. Roth and M. Black, "Field of experts: A framework for learning image priors," in *Proceedings of the IEEE Conference on Computer Vision and Pattern Recognition*, vol. 2, 2005, pp. 860–867.

[15] M. F. Tappen, C. Liu, E. H. Adelson, and W. T. Freeman, "Learning gaussian conditional random fields for low-level vision," in *IEEE Conference on Computer Vision & Pattern Recognition*, 2007.

[16] U. Schmidt, Q. Gao, and S. Roth, "A generative perspective on mrfs in low-level vision," in *Proceedings of the IEEE Conference on Computer Vision and Pattern Recognition*, 2010, pp. 1751–1758.

[17] J. Lafferty, F. Pereira, and A. McCallum, "Conditional random fields: Probabilistic models for segmenting and labeling sequence data," in *ICML*, 2001.

[18] P. F. Felzenszwalb and D. P. Huttenlocher, "Efficient matching of pictorial structures," in *Proceedings of the IEEE Conference on Computer Vision and Pattern Recognition*, 2000.

[19] D. J. Crandall, P. F. Felzenszwalb, and D. P. Huttenlocher, "Spatial priors for part-based recognition using statistical models," in *Proceedings of the IEEE Conference on Computer Vision and Pattern Recognition*, 2005, pp. 10–17.

[20] W. Yang, Y. Wang, and G. Mori, "Recognizing human actions from still images with latent poses," in *Proceedings of the IEEE Conference on Computer Vision and Pattern Recognition*, 2010, pp. 2030–2037.

[21] C. M. Bishop, *Pattern Recognition and Machine Learning*, 1st ed. Springer, October 2007.

[22] R. Szeliski, R. Zabih, D. Scharstein, O. Veksler, V. Kolmogorov, A. Agarwala, M. F. Tappen, and C. Rother, "A comparative study of energy minimization methods for markov random fields," in *ECCV (2)*, 2006, pp. 16–29.

[23] J. Pearl, *Probabilistic Reasoning in Intelligent Systems: Networks of Plausible Inference*, 2nd ed. Morgan Kaufmann, 1988.

[24] M. F. Tappen and W. T. Freeman, "Comparison of graph cuts with belief propagation for stereo, using identical mrf parameters," in *Proceedings of the Ninth IEEE International Conference on Computer Vision (ICCV)*, 2003, pp. 900 – 907.

[25] Y. Weiss and W. T. Freeman, "On the optimality of solutions of the max-product belief-propagation algorithm in arbitrary graphs," *IEEE Transactions on Information Theory*, vol. 47, no. 2, pp. 736–744, 2001.

[26] M. J. Wainwright, T. S. Jaakkola, and A. S. Willsky, "Map estimation via agreement on (hyper)trees: Message-passing and linear-programming approaches," *IEEE Transactions on Information Theory*, vol. 51, no. 11, pp. 3697–3717, November 2005.

[27] V. Kolmogorov, "Convergent tree-reweighted message passing for energy minimization," *IEEE Transactions on Pattern Analysis and Machine Intelligence (PAMI)*, vol. 28, no. 10, pp. 1568–1583, October 2006.

[28] D. M. Greig, B. T. Porteous, and A. H. Seheult, "Exact maximum a posteriori estimation for binary images," *Journal of the Royal Statistical Society Series B*, vol. 51, pp. 271–279, 1989.

[29] V. Kolmogorov and C. Rother, "Minimizing nonsubmodular functions with graph cuts-a review," *IEEE Transactions of Pattern Analysis and Machine Intelligence*, vol. 29, no. 7, pp. 1274–1279, 2007.

[30] J. Sun, N. Zheng, and H.-Y. Shum, "Stereo matching using belief propagation," *IEEE Transactions of Pattern Analysis and Machine Intelligence*, vol. 25, no. 7, pp. 787–800, 2003.

[31] M. F. Tappen, W. T. Freeman, and E. H. Adelson, "Recovering intrinsic images from a single image," *IEEE Transactions on Pattern Analysis and Machine Intelligence*, vol. 27, no. 9, pp. 1459–1472, September 2005.

[32] M. F. Tappen, "Fundamental strategies for solving low-level vision problems," *IPSJ Transactions on Computer Vision and Applications*, 2012.

[33] S. C. Zhu, Y. Wu, and D. Mumford, "Filters, random fields and maximum entropy (frame): Towards a unified theory for texture modeling," *International Journal of Computer Vision*, vol. 27, no. 2, pp. 107–126, 1998.

[34] G. E. Hinton, "Training products of experts by minimizing contrastive divergence," *Neural Computation*, vol. 14, no. 8, pp. 1771–1800, 2002.

[35] X. He, R. Zemel, and M. Carreira-Perpinan, "Multiscale conditional random fields for image labelling," in *In 2004 IEEE Conference on Computer Vision and Pattern Recognition (CVPR)*, 2004.

[36] M. F. Tappen, "Utilizing variational optimization to learn markov random fields," in *IEEE Conference on Computer Vision and Pattern Recognition (CVPR07)*, 2007.

[37] A. Barbu, "Learning real-time MRF inference for image denoising," in *IEEE Computer Vision and Pattern Recognition*, 2009.

[38] B. Taskar, V. Chatalbashev, D. Koller, and C. Guestrin, "Learning structured prediction models: A large margin approach," in *The Twenty Second International Conference on Machine Learning (ICML-2005)*, 2005.

[39] D. Anguelov, B. Taskar, V. Chatalbashev, D. Koller, D. Gupta, G. Heitz, and A. Ng, "Discriminative learning of Markov random fields for segmentation of 3d scan data." in *Proceedings of the IEEE Conference on Computer Vision and Pattern Recognition*, 2005, pp. 169–176.

[40] T. Joachims, T. Finley, and C.-N. Yu, "Cutting-plane training of structural svms," *Machine Learning*, vol. 77, no. 1, pp. 27–59, 2009.

[41] T. Finley and T. Joachims, "Training structural SVMs when exact inference is intractable," in *International Conference on Machine Learning (ICML)*, 2008, pp. 304–311.

[42] C. Desai, D. Ramanan, and C. Fowlkes, "Discriminative models for multi-class object layout," in *Proceedings of the IEEE International Conference on Computer Vision*, 2009.

[43] M. Szummer, P. Kohli, and D. Hoiem, "Learning crfs using graph cuts," in *Proc. ECCV*, 2008, pp. 582–595.

17

Tree-Walk Kernels for Computer Vision

Zaid Harchaoui

LEAR project-team
INRIA and LJK
655, avenue de l'Europe
38330 Montbonnot Saint-Martin, FRANCE
Email: zaid.harchaoui@inria.fr

Francis Bach

SIERRA project-team
INRIA and LIENS
23, avenue d'Italie
75214 Paris Cedex 13, FRANCE
Email: francis.bach@inria.fr

CONTENTS

17.1 Introduction 500
    17.1.1 Related Work 501
17.2 Tree-Walk Kernels as Graph Kernels 502
    17.2.1 Paths, Walks, Subtrees and Tree-Walks 503
    17.2.2 Graph Kernels 504
17.3 The Region Adjacency Graph Kernel as a Tree-Walk Kernel 505
    17.3.1 Morphological Region Adjacency Graphs 506
    17.3.2 Tree-Walk Kernels on Region Adjacency Graphs 507
17.4 The Point Cloud Kernel as a Tree-Walk Kernel 510
    17.4.1 Local Kernels 511
    17.4.2 Positive Matrices and Probabilistic Graphical Models 512
    17.4.3 Dynamic Programming Recursions 516
17.5 Experimental Results 518
    17.5.1 Application to Object Categorization 518
    17.5.2 Application to Character Recognition 522
    17.5.3 Experimental Setting 523
17.6 Conlusion 524
17.7 Acknowledgments 525
    Bibliography 525

17.1 Introduction

Kernel-based methods have proved highly effective in many applications because of their wide generality. As soon as a similarity measure leading to a symmetric positive-definite kernel can be designed, a wide range of learning algorithms working on Euclidean dot-products of the examples can be used, such as support vector machines [1]. Replacing the euclidian dot products by the kernel evaluations then corresponds to applying the learning algorithm on dot products of feature maps of the examples into a higher-dimensional space. Kernel-based methods are now well-established tools for computer vision—see, e.g. [2], for a review.

A recent line of research consists of designing kernels for structured data, to address problems arising in bioinformatics [3] or text processing. Structured data are data that can be represented through a labelled discrete structure, such as strings, labelled trees, or labelled graphs [4]. They provide an elegant way of including known a priori information, by using directly the natural topological structure of objects. Using a priori knowledge through kernels on structured data can be beneficial. It usually allows one to reduce the number of training examples to reach a high-performance accuracy. It is also an elegant way to leverage existing data representations that are already well developed by experts of those domains. Last but not least, as soon as a new kernel on structured data is defined, one can bring to bear the rapidly developing kernel machinery, and in particular semi-supervised learning—see, e.g., [5]—and kernel learning —see, e.g., [6].

We propose a general framework to build positive-definite kernels between images, allowing us to leverage, e.g., shape and appearance information to build meaningful similarity measures between images. On the one hand, we introduce positive-definite kernels between appearances as coded in region adjacency graphs, with applications to classification of images of object categories, as by [7, 8]. On the other hand, we introduce positive-definite kernels between shapes as coded in point clouds, with applications to classification of line drawings, as in [9, 10, 11]. Both types of information are difficult to encode into vector features without loss of information. For instance, in the case of shapes defined through point clouds, local and global invariances with respect to rotation/translation need to be encoded in the feature space representation.

A leading principle for designing kernels between structured objects is to decompose each object into parts and to compare all parts of one object to all parts of another object [1]. Even if there is an exponential number of such decompositions, which is a common case, this is numerically possible under two conditions: (a) the object must lead itself to an efficient enumeration of subparts, and (b) the similarity function between subparts (i.e., the *local kernel*), beyond being a positive-definite kernel, must be simple enough so that the sum over a potentially exponential number of terms can be recursively performed in polynomial time through factorization.

One of the most striking instantiations of this design principle are the *string kernels* [1], which consider all substrings of a given string but still allow efficient computation in polynomial time. The same principle can also be applied to graphs: intuitively, the *graph kernels* [12, 13] consider all possible subgraphs and compare and count matching subgraphs. However, the set of subgraphs (or even the set of paths) has exponential size and cannot be efficiently described recursively. By choosing appropriate substructures, such as *walks* or *tree-walks*, and fully factorized local kernels, matrix inversion formulations [14] and efficient dynamic programming recursion allow one to sum over an exponential number of substructures in polynomial time. We first review previous work on graph kernels. Then we present our general framework for tree-walk kernels in Section 17.2. In Section 17.3, we show how one can instantiate tree-walk kernels to build a powerful positive-definite similarity measure between region adjacency graphs. Then, in Section 17.4, we show how to instantiate tree-walk kernels to build an efficient positive-definite similarity measure between point clouds. Finally, in Section 17.5, we present experimental results of our tree-walk kernels respectively on object categorization, natural image classification, and handwritten digit classification tasks. This chapter is based on two conference papers [15, 16].

17.1.1   Related Work

The underlying idea of our tree-walk kernels is to provide an expressive relaxation of graph-matching similarity scores that yet yields a positive-definite kernel. The idea of matching graphs for the purpose of image classification has a long history in computer vision, and has been investigated by many authors [17, 18, 19]. However, the general *graph matching* problem is especially hard as most of the simple operations that are simple on strings (such as matching and edit distances) are NP-hard for general undirected graphs. Namely, while exact graph matching is unrealistic in our context, inexact graph matching is NP-hard and subgraph matching is NP-complete [4]. Hence, most work so far has focused on finding ingenious approximate solutions to this challenging problem [20]. An interesting line of research consists of overcoming graph matching problems by projecting graphs on strings by the so-called seriation procedure [18], and then using string edit distance [21] as a proxy to graph edit distance.

Designing kernels for image classification is also an active research topic. Since the work of [22], who investigated image classification with kernels on color histograms, many kernels for image classification were proposed. The bag-of-pixels kernels proposed in [23, 24] compare the color distributions of two images by using kernels between probability measures, and was extended to hierarchical multi-scale settings in [24, 25]; see [8] for an in-depth review.

Graph data occur in many application domains, and kernels for attributed graphs have received increased interest in the applied machine learning literature, in particular in bioinformatics [14, 13] and computer vision. Note that in this work, we only consider kernels between graphs (each data point is a graph), as opposed to settings where kernels are used to build a neighborhood graph from all the data points of a dataset [1].

Current graph kernels can roughly be divided in two classes: the first class is composed of non-positivedefinite similarity measures based on existing techniques from the graph matching literature, that can be made positive-definite by *ad hoc* matrix transformations; this includes the edit-distance kernel [26] and the optimal assignment kernel [27, 28]. In [26], the authors propose to use the edit distance between two graphs and directly plug it into a kernel to define a similarity measure between graphs. Kernels between shock graphs for pedestrian detection were also proposed in [29], allowing them to capture the topological structure of images. Our method efficiently circumvents the graph matching problem by soft-matching tree-walks, i.e., virtual substructures of graphs, in order to obtain kernels computable in polynomial time. Our kernels keep the underlying topological structure of graphs by describing them through tree-walks, while in graph seriation the topological structure somehow fades away by only retaining one particular substring of the graph. Moreover our kernels encompasses local information, by using segment histograms as local features, as well as global information, by summing up all soft matchings of tree-walks.

Another class of graph kernels relies on a set of substructures of the graphs. The most natural ones are paths, subtrees, and more generally subgraphs; however, they do not lead to positive-definite kernels with polynomial time computation algorithms—see, in particular, NP-hardness results by [12]—and recent work has focused on larger sets of substructures. In particular, *random walk* kernels consider all possible walks and sum a local kernel over all possible walks of the graphs (with all possible lengths). With a proper length-dependent factor, the computation can be achieved by solving a large sparse linear system [14, 30], whose running time complexity has been recently reduced [31]. When considering fixed-length walks, efficient dynamic programming recursions allow to drive down the computation time, at the cost of considering a smaller feature space. This is the approach we chose to follow here. Indeed, such an approach allows extensions to other types of substructures, namely "tree-walks", whose abstract definition were first proposed by [12], that we now present. Several works later extended our approach to other settings, such as [32] for feature extraction from images and [33] for scene interpretation.

17.2 Tree-Walk Kernels as Graph Kernels

We first describe in this section the general framework in which we define tree-walk kernels. We shall then instantiate the tree-walk kernels respectively on region adjacency graphs in Section 17.3 and on point clouds in Section 17.4.

Let us consider two labelled undirected graphs $g = (\mathcal{V}_g, \epsilon_g)$ and $\mathcal{H} = (\mathcal{V}_\mathcal{H}, \epsilon_\mathcal{H})$,

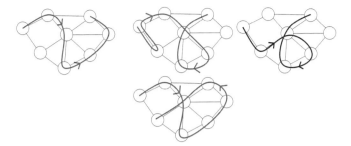

FIGURE 17.1
From left to right: path (red), 1-walk that is not a 2-walk (green), 2-walk that is not a 3-walk (blue), 4-walk (magenta).

where $\mathcal{V}_g$ and $\mathcal{V}_{\mathcal{H}}$ denote the set of vertices resp. of g and $\mathcal{H}$, ϵ_g and $\mathcal{H}$ denote the set of edges resp. of g and $\mathcal{H}$. For the sake of clarity, we shall assume in the remainder that the vertex sets have the same cardinality; yet our framework can also handle vertex sets with different cardinality. Here, both the vertices and the edges can be labelled. Two types of labels are considered: *attributes*, which are denoted $a(v) \in \mathcal{A}_g$ for vertex $v \in \mathcal{V}_g$, and $b(v') \in \mathcal{A}_{\mathcal{H}}$ for vertex $v' \in \mathcal{V}_{\mathcal{H}}$ and *positions*, which are denoted $x(v) \in \mathcal{X}_g$ and $y(v') \in \mathcal{X}_{\mathcal{H}}$. We make the distinction here between attributes and positions, since different kernels should be used to mesure similarity between attributes (respectively positions).

17.2.1 Paths, Walks, Subtrees and Tree-Walks

Given an undirected graph g with vertex set $\mathcal{V}$, a *path* is a sequence of distinct connected vertices, while a *walk* is a sequence of possibly nondistinct connected vertices. In order to prevent the walks from going back and forth too quickly, we further restrain the set of walks. Such phenomenon was termed *tottering* by [34]). For any positive integer β, we define β-walks as walks such that any $\beta + 1$ successive vertices are distinct (1-walks are regular walks); see examples in Figure 17.1. Note that when the graph g is a tree (no cycles), then the set of 2-walks is equal to the set of paths. More generally, for any graph, β-walks of length $\beta + 1$ are exactly paths of length $\beta + 1$. Note that the integer β corresponds to the "memory" of the walk, i.e., the number of past vertices it needs to "remember" before going on.

A *subtree* of g is a subgraph of g with no cycles [4]. A subtree of g can thus be seen as a connected subset of distinct nodes of G with an underlying tree structure. The notion of walk is extending the notion of path by allowing vertices to be equal. Similarly, we can extend the notion of subtrees to *tree-walks*, where equal vertices can occur. More precisely, we define an α-ary tree-walk of depth γ of g as a rooted labelled α-ary tree of depth γ whose vertices share the same labels as the labels of the corresponding vertices in g. In addition, the labels of neighbors in the tree-walk

FIGURE 17.2
(Left) binary 2-tree-walk, which is in fact a subtree, (right) binary 1-tree-walk that is not a 2-tree-walk.

must be neighbors in g (we refer to all allowed such set of labels as *consistent* labels). We assume that the tree-walks are not necessarily complete trees; i.e., each node may have less than α children. Tree-walks can be plotted on top of the original graph, as shown in Figure 17.5, and may be represented by a tree structure T over the vertex set $\{1, \ldots, |T|\}$ and a tuple of consistent but possibly nondistinct labels $I \in V^{|T|}$ (i.e., the labels of neighboring vertices in T must be neighboring vertices in g). Finally, we only consider rooted subtrees, i.e., subtrees where a specific node is identified as the root. Moreover, all the trees that we consider are unordered trees (i.e., no order is considered among siblings). Yet, it is worthwhile to note that taking into account the order becomes more important when considering subtrees of region adjacency graphs.

We can also define β-tree-walks, as tree-walks such that for each node in T, its label (which is an element of the original vertex set V) and the ones of all its descendants up to the β-th generation are all distinct. With that definition, 1-tree-walks are regular tree-walks (see Figure 17.2), and if $\alpha = 1$, we get back β-walks. From now on, we refer to the descendants up to the β-th generation as the β-descendants.

We let $\mathcal{T}_{\alpha,\gamma}$ denote the set of rooted tree structures of depth less than γ and with at most α children per node; for example, $\mathcal{T}_{1,\gamma}$ is exactly the set of chain graphs of length less than γ. For $T \in \mathcal{T}_{\alpha,\gamma}$, we denote $\mathcal{J}_\beta(T, G)$ the set of consistent labellings of T by vertices in V leading to β-tree-walks. With these definitions, a β-tree-walk of G is characterized by (a) a tree structure $T \in \mathcal{T}_{\alpha,\gamma}$ and (b) a labelling $I \in \mathcal{J}_\beta(T, G)$.

17.2.2 Graph Kernels

We assume that we are given a positive-definite kernel between tree-walks that share the same tree structure, which we refer to as the *local kernel*. This kernel depends on the tree structure T and the set of attributes and positions associated with the nodes in the tree-walks (remember that each node of g and $\mathcal{H}$ has two labels, a position and an attribute). Given a tree structure T and consistent labellings $I \in \mathcal{J}_\beta(T, G)$ and $J \in \mathcal{J}_\beta(T, H)$, we let denote $q_{T,I,J}(g, \mathcal{H})$ the value of the local kernel between two tree-walks defined by the same structure T and labellings I and J.

We can define the *tree-walk* kernel as the sum over all matching tree-walks of g and $\mathcal{H}$ of the local kernel, i.e.:

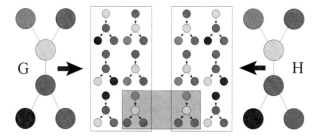

FIGURE 17.3
Graph kernels between two graphs (each color represents a different label). We display all binary 1-tree walks with a specific tree structure, extracted from two simple graphs. The graph kernels are computing and summing the local kernels between all those extracted tree-walks. In the case of the Dirac kernel (hard matching), only one pair of tree-walks is matched (for both labels and structures).

$$
k_{\alpha,\beta,\gamma}^{\mathcal{T}}(g,\mathcal{H}) = \sum_{T \in \mathcal{T}_{\alpha,\gamma}} f_{\lambda,\nu}(T) \left\{ \sum_{I \in \mathcal{J}_\beta(T,g)} \sum_{J \in \mathcal{J}_\beta(T,\mathcal{H})} q_{T,I,J}(g,\mathcal{H}) \right\}. \quad (17.1)
$$

Details on the efficient computation of the tree-walk kernel are given in the next section. When considering 1-walks (i.e., $\alpha = \beta = 1$), and letting the maximal walk length γ tend to $+\infty$, we get back the random walk kernel [12, 14]. If the kernel $q_{T,I,J}(g,\mathcal{H})$ has nonnegative values and is equal to 1 if the two tree-walks are equal, it can be seen as a soft matching indicator, and then the kernel simply counts the softly matched tree-walks in the two graphs (see Figure 17.3 for an illustration with hard matching).

We add a nonnegative penalization $f_{\lambda,\nu}(T)$ depending only on the tree structure. Besides the usual penalization of the number of nodes $|T|$, we also add a penalization of the number of leaf nodes $\ell(T)$ (i.e., nodes with no children). More precisely, we use the penalization $f_{\lambda,\nu} = \lambda^{|T|}\nu^{\ell(T)}$. This penalization, suggested by [34], is essential in our situation to prevent trees with nodes of higher degrees from dominating the sum. Yet, we shall most often drop this penalization in subsequent derivations to highlight essential calculations and to keep the notation light.

If $q_{T,I,J}(G,H)$ is obtained from a positive-definite kernel between (labelled) tree-walks, then $k_{\alpha,\beta,\gamma}^{\mathcal{T}}(g,\mathcal{H})$ also defines a positive-definite kernel. The kernel $k_{\alpha,\beta,\gamma}^{\mathcal{T}}(g,\mathcal{H})$ sums the *local kernel* $q_{T,I,J}(G,H)$ over all tree-walks of g and $\mathcal{H}$ that share the same tree structure. The number of such matching tree-walks is exponential in the depth γ, thus, in order to deal with potentially deep trees, a recursive definition is needed. The local kernel will depend on the application at hand. In the next sections, we show how one can design meaningful local kernels resp. for region adjacency graphs and point clouds.

17.3 The Region Adjacency Graph Kernel as a Tree-Walk Kernel

We propose to model appearance of images using region adjacency graphs obtained by morphological segmentation. If enough segments are used, i.e., if the image (and hence the objects of interest) is *over-segmented*, then the segmentation enables us to reduce the dimension of the image while preserving the boundaries of objects. Image dimensionality goes from millions of pixels down to hundreds of segments, with little loss of information. Those segments are naturally embedded in a planar graph structure. Our approach takes into account graph planarity. The goal of this section is to show that one can feed kernel-based learning methods with a kernel measuring appropriate region adjacency graph similarity, called the *region adjacency graph kernel*, for the purpose of image classification. Note that this kernel was previously referred to as the *segmentation graph kernel* in the original paper [15].

17.3.1 Morphological Region Adjacency Graphs

Among the many methods available for the segmentation of natural images [35, 36, 37], we chose to use the watershed transform for image segmentation [35], which allows a fast segmentation of large images into a given number of segments. First, given a color image, a grayscale gradient image is computed from oriented energy filters on the LAB representation of the image, with two different scales and eight orientations [38]. Then, the watershed transform is applied to this gradient image, and the number of resulting regions is reduced to a given number p ($p = 100$ in our experiments) using the hierarchical framework of [35]. We finally obtain p singly connected regions, together with the planar neighborhood graph. An example of segmentation is shown in Figure 17.4.

We have chosen a reasonable value of p because our similarity measure implicitly relies on the fact that images are mostly oversegmented, i.e., the objects of interest may span more than one segment, but very few segments span several objects. This has to be contrasted with the usual (and often unreached) segmentation goal of obtaining one segment per object. In this respect, we could have also used other image segmentation algorithms, such as the Superpixels algorithm [39, 40]. In this section, we always use the same number of segments; a detailed analysis of the effect of choosing a different number of segments, or a number that depends on the given image, is outside the scope of this work.

In this section, images will be represented by a *region adjacency graph*, i.e., an undirected labelled planar graph, where each vertex is one singly connected region, with edges joining neighboring regions. Since our graph is obtained from neighboring singly connected regions, the graph is planar, i.e., it can be embedded in a plane where no edge intersects [4]. The only property of planar graphs that we are going to

FIGURE 17.4
An example of a natural image (left), and its segmentation mosaic (right) obtained by using the median RGB color in each segmented region. The associated region adjacency graph is depicted in light blue

use is the fact that for each vertex in the graph, there is a natural notion of cyclic ordering of the vertices. Another typical feature of our graphs is their sparsity: indeed, the degree (number of neighbors) of each node is usually small, and the maximum degree typically does not exceed 10.

There are many ways to assign labels to regions, based on shape, color, or texture, leading to highly multivariate labels, which is to be contrasted with the usual application of structured kernels in bioinformatics or natural language processing, where labels belong to a small discrete set. Our family of kernels simply requires a simple kernel $k(\ell, \ell')$ between labels ℓ and ℓ' of different regions, which can be any positive semi-definite kernel [1], and not merely a Dirac kernel, which is commonly used for exact matching in bioinformatics applications. We could base such a kernel on any relevant visual information. In this work, we considered kernels between color histograms of each segment, as well as weights corresponding to the size (area) of the segment.

17.3.2 Tree-Walk Kernels on Region Adjacency Graphs

We now make explicit the derivation of tree-walks kernels on region adjacency graphs, in order to highlight the dynamic programming recursion that makes their computation tractable in pratice.

The tree-walk kernel between region adjacency graphs $k_{p,\alpha}(g, \mathcal{H})$ is defined as the sum over all tree-walks in G and all tree-walks in $\mathcal{H}$ (sharing the same tree structure) of the products of the local kernels between corresponding individual vertices of the subtree patterns. Note that in the context of exact matching, investigated in [41], this kernel simply counts the number of common subtree-patterns. Note also that, as long as the local kernel is a positive semi-definite kernel, the resulting tree-walk kernel is also an positive semi-definite kernel as well [1].

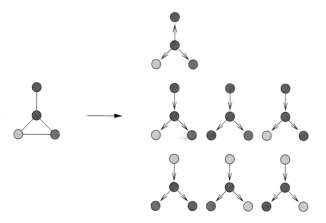

FIGURE 17.5
Examples of tree-walks from a graph. Each color represents a different label.

Local Kernels

In order to apply the tree-walk kernels to region adjacency graphs, we need to specify the kernels involved at the lower level of computation. Here, the local kernels correspond to kernels between segments.

There are plenty of choices for defining relevant features for segment comparisons, ranging from median color to sophisticated local features [42]. We chose to focus on RGB color histograms since previous work [43] thoroughly investigated their relevance for image classification, when used on the whole image without any segmentation. This allows us to fairly evaluate the efficiency of our kernels to make a smart use of segmentations for classification.

Experimental results of kernels between color histograms taken as discretized probability distributions $P = (p_i)_{i=1}^{N}$ were given in [43]. For the sake of simplicity, we shall focus here on the χ^2-kernel defined as follows. The symmetric χ^2-divergence between two distributions P and Q is defined as

$$d_\chi^2(P,Q) = \frac{1}{N} \sum_{j=1}^{N} \frac{(p_i - q_i)^2}{p_i + q_i} , \qquad (17.2)$$

and the χ^2-kernel is defined as

$$k_\chi(P,Q) = \exp(-\mu d_\chi^2(P,Q)) , \qquad (17.3)$$

with μ a free parameter to be tuned. Following results of [43] and [44], since this kernel is positive semi-definite, it can be used as a local kernel. If we denote P_ℓ the histogram of colors of a region labelled by ℓ, then it defines a kernel between labels as $k(\ell, \ell') = k_\chi(P_\ell, P_{\ell'})$.

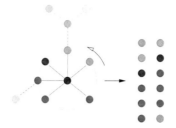

FIGURE 17.6
Enumeration of neighbor intervals of size two.

We need to deal with the too strong diagonal dominance of the obtained kernel matrices, a usual issue with kernels on structured data [3]). We propose to include a constant term λ that controls the maximal value of $k(\ell, \ell')$. We thus use the following kernel

$$k(\ell, \ell') = \lambda \exp(-\mu d_\chi^2(P_\ell, P_{\ell'})) \,,$$

with free parameters λ, μ. Note that, by doing so, we also ensure the positive semi-definiteness of the kernel.

It is natural to give more weight in the overall sum to massive segments than to tiny ones. Hence, we incorporate this into the segment kernel as

$$k(\ell, \ell') = \lambda A_\ell^\gamma A_{\ell'}^\gamma \exp(-\mu d_\chi^2(P_\ell, P_{\ell'})),$$

where A_ℓ is the area of the corresponding region, and γ is an additional free parameter in $[0, 1]$ to be tuned.

Dynamic Programming Recursion

In order to derive an efficient dynamic programming formulation, we now need to restrict the set of subtrees. Indeed, if d is an upper bound on the degrees of the vertices in g and $\mathcal{H}$, then at each depth, the q-ary tree-walk may go through any subsets of size α of the set of neighbors of a given vertex, and thus the matching complexity would be $O(d^{2\alpha})$ ($O(d^\alpha)$ for each of the two graphs). We restrict the allowed subsets of neighbors in a tree-walk, by requiring that these subsets are intervals for the natural cyclic ordering of the neighbors of a given vertex (this is possible because the graph is planar). See Figure 17.6 for an enumeration of the intervals of size $\alpha = 2$. For a given vertex r in g (respectively s in $\mathcal{H}$), we denote by $\mathcal{A}_g^\alpha(r)$ (respectively $\mathcal{A}_\mathcal{H}^\alpha(s)$) the set of nonempty intervals of length at most α around r in g (respectively around s in $\mathcal{H}$). In the rest of the section, we assume that all our subtree patterns (referred to as tree-walks from now on) are restricted to intervals of neighbors. Let $k_{p,\alpha}(g, \mathcal{H}, r, s)$ denote the sum over all tree patterns starting from vertex r in g and s in $\mathcal{H}$.

The following result shows that the recursive dynamic programming formulation that allows us to efficiently compute tree-walk kernels. In the case of the tree-walk kernel for region adjacency graphs, the final kernel writes as follows

$$k(g, \mathcal{H}) = \sum_{r \in V_g, \, s \in V_{\mathcal{H}}} k_{p,\alpha}(g, \mathcal{H}, r, s). \tag{17.4}$$

Assuming equal size intervals $\mathrm{Card}(A) = \mathrm{Card}(B)$, the kernel values $k_{p,\alpha}(g, \mathcal{H}, r, s)$ can be recursively computed through

$$k_{p,\alpha}(g, \mathcal{H}, r, s) = k(\ell_g(r), \ell_{\mathcal{H}}(s)) \sum_{A \in \mathcal{A}_g^\alpha(r), \, B \in \mathcal{A}_{\mathcal{H}}^\alpha(s)} \prod_{r' \in A, \, s' \in B} k_{p-1,\alpha}(g, \mathcal{H}, r', s').$$
$$\tag{17.5}$$

The above equation defines a dynamic programming recursion that allows us to efficiently compute the values of $k_{p,\alpha}(g, \mathcal{H}, \cdot, \cdot)$ from $p = 1$ to any desired p. We first compute $k_{1,\alpha}(g, \mathcal{H}, \cdot, \cdot)$, using that $k(g, \mathcal{H}, r, s) = k(\ell_g(r), \ell_{\mathcal{H}}(s))$ for r a segment of g and s a segment of $\mathcal{H}$. Then $k_{2,\alpha}(g, \mathcal{H}, \cdot, \cdot)$ can be computed from $k_{1,\alpha}(g, \mathcal{H}, \cdot, \cdot)$ using (17.5) with $p = 2$. And so on. Finally, note that when $\alpha = 1$ (intervals of size 1), the tree-walk kernel reduces to the walk kernel [12].

Running Time Complexity

Given labelled graphs g and $\mathcal{H}$ with n_g and $n_{\mathcal{H}}$ vertices each and maximum degrees d_g and $d_{\mathcal{H}}$, we assume that the kernel between labels $k(\ell, \ell')$ can be computed in constant time. Hence, the total cost of computing $k(\ell_g(r), \ell_{\mathcal{H}}(s))$ for all $r \in V_g$ and $s \in V_{\mathcal{H}}$ is $O(n_g n_{\mathcal{H}})$.

For walk kernels, the complexity of each recursion is $O(d_g d_{\mathcal{H}})$. Thus, computation of all q-th walk kernels for $q \leqslant p$ needs $O(p d_g d_{\mathcal{H}} n_g n_{\mathcal{H}})$ operations.

For tree-walk kernels, the complexity of each recursion is $O(\alpha^2 d_g d_{\mathcal{H}})$. Therefore, computation of all q-th α-ary tree walk kernels for $q \leqslant p$ needs $O(p \alpha^2 d_g d_{\mathcal{H}} n_g n_{\mathcal{H}})$ operations, i.e., leading to polynomial-time complexity.

We now pause in the exposition of tree-walk kernels applied to region adjacency graphs, and move on to tree-walk kernels applied to point clouds. We shall get back to tree-walk kernels on region adjacency graphs in Section 17.5.

17.4   The Point Cloud Kernel as a Tree-Walk Kernel

We propose to model shapes of images using point clouds. Indeed, we assume that each point cloud has a graph structure (most often a neighborhood graph).Then, our graph kernels consider all partial matches between two neighborhood graphs and sum over those. However, the straightforward application of graph kernels poses a major problem: in the context of computer vision, substructures correspond to matched sets of points, and dealing with local invariances by rotation and/or translation necessitates the use of a local kernel that cannot be readily expressed as a product of separate terms for each pair of points, and the usual dynamic programming and matrix inversion approaches cannot then be directly applied. We shall assume that the graphs have no self-loops. Our motivating examples are line drawings, where $\mathcal{A} = \mathbb{R}^2$ (i.e., the position is itself also an attribute). In this case, the graph is naturally obtained from the drawings by considering 4-connectivity or 8-connectivity [25]. In other cases, graphs can be easily obtained from nearest-neighbor graphs.

We show here how to design a local kernel that is not fully factorized but can be instead factorized according to the graph underlying the substructure. This is naturally done through probabilistic graphical models and the design of positive-definite kernels for covariance matrices that factorize on probabilistic graphical models. With this novel local kernel, we derive new polynomial time dynamic programming recursions.

17.4.1 Local Kernels

The local kernel is used between tree-walks that can have large depths (note that everything we propose will turn out to have linear time complexity in the depth γ). Recall here that, given a tree structure T and consistent labellings $I \in \mathcal{J}_\beta(T, G)$ and $J \in \mathcal{J}_\beta(T, H)$, the quantity $q_{T,I,J}(G, H)$ denotes the value of the local kernel between two tree-walks defined by the same structure T and labellings I and J. We use the product of a kernel for attributes and a kernel for positions. For attributes, we use the following usual factorized form $q_{\mathcal{A}}(a(I), b(J)) = \prod_{p=1}^{|I|} k_{\mathcal{A}}(a(I_p), b(J_p))$, where $k_{\mathcal{A}}$ is a positive-definite kernel on $\mathcal{A} \times \mathcal{A}$. This allows the separate comparison of each matched pair of points and efficient dynamic programming recursions. However, for our local kernel on positions, we need a kernel that *jointly* depends on the whole vectors $x(I) \in \mathsf{X}^{|I|}$ and $y(J) \in \mathsf{X}^{|J|}$, and not only on the p pairs $(x(I_p), y(J_p))$. Indeed, we do not assume that the pairs are *registered*; i.e., we do not know the matching between points indexed by I in the first graph and the ones indexed by J in the second graph.

We shall focus here on $\mathsf{X} = \mathcal{R}^d$ and *translation-invariant* local kernels, which implies that the local kernel for positions may only depend on differences $x(i) - x(i')$ and $y(j) - y(j')$ for $(i, i') \in I \times I$ and $(j, j') \in J \times J$. We further reduce these to kernel matrices corresponding to a translation-invariant positive-definite kernel $k_{\mathsf{X}}(x_1 - x_2)$. Depending on the application, k_{X} may or may not be rotation invariant. In experiments, we use the rotation-invariant Gaussian kernel of the form $k_{\mathsf{X}}(x_1, x_2) = \exp(-v\|x_1 - x_2\|^2)$.

Thus, we reduce the set of all positions in $\mathsf{X}^{|V|}$ and $\mathsf{X}^{|W|}$ to full kernel matrices $K \in \mathcal{R}^{|V| \times |V|}$ and $L \in \mathcal{R}^{|W| \times |W|}$ for each graph, defined as $K(v, v') = k_{\mathsf{X}}(x(v) - x(v'))$ (and similarly for L). These matrices are by construction symmetric positive semi-definite and, for simplicity, we assume that these matrices are positive-definite (i.e., invertible), which can be enforced by adding a multiple of the identity matrix. The local kernel will thus only depend on the submatrices $K_I = K_{I,I}$ and $L_J = L_{J,J}$, which are positive-definite matrices. Note that we use kernel matrices K and L to represent the geometry of each graph, and that we use a positive-definite kernel on such kernel matrices.

We consider the following positive-definite kernel on positive matrices K and L, the (squared) Bhattacharyya kernel k_B, defined as [45]:

$$k_B(K, L) = |K|^{1/2}|L|^{1/2} \left|\tfrac{K+L}{2}\right|^{-1}, \tag{17.6}$$

where $|K|$ denotes the determinant of K.

By taking the product of the attribute-based local kernel and the position-based local kernel, we get the following local kernel $q^0_{T,I,J}(G, H) = k_B(K_I, L_J)q_A(a(I), b(J))$. However, this local kernel $q^0_{T,I,J}(G, H)$ does not yet depend on the tree structure T and the recursion may be efficient only if $q^0_{T,I,J}(G, H)$ can be computed recursively. The factorized term $q_A(a(I), b(J))$ is not an issue. However, for the term $k_B(K_I, L_J)$, we need an approximation based on T. As we show in the next section, this can be obtained by a factorization according to the appropriate probabilistic graphical model; i.e., we will replace each kernel matrix of the form K_I by a projection onto a subset of kernel matrices, which allows efficient recursions.

17.4.2 Positive Matrices and Probabilistic Graphical Models

The main idea underlying the factorization of the kernel is to consider symmetric positive-definite matrices as covariance matrices and to look at probabilistic graphical models defined for Gaussian random vectors with those covariance matrices. The goal of this section is to show that by appropriate probabilistic graphical model techniques, we can design properly factorized approximations of Equation 17.6, namely, through Equations 17.11 and 17.12.

More precisely, we assume that we have n random variables $Z_1, \ldots, Z_n$ with probability distribution $p(z) = p(z_1, \ldots, z_n)$. Given a kernel matrix K (in our case defined as $K_{ij} = \exp(-v\|x_i - x_j\|^2)$, for positions $x_1, \ldots, x_n$), we consider jointly Gaussian distributed random variables $Z_1, \ldots, Z_n$ such that $\text{Cov}(Z_i, Z_j) = K_{ij}$. In this section, with this identification, we consider covariance matrices as kernel matrices, and vice versa.

Probabilistic Graphical Models and Junction Trees

Probabilistic graphical models provide a flexible and intuitive way of defining factorized probability distributions. Given any undirected graph Q with vertices in $\{1, \ldots, n\}$, the distribution $p(z)$ is said to factorize in Q if it can be written as a

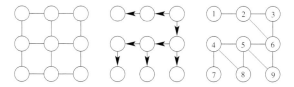

FIGURE 17.7
(Left) original graph, (middle) a single extracted tree-walk, (right) decomposable graphical model $Q_1(T)$ with added edges in red. The junction tree is a chain composed of the cliques $\{1, 2\}, \{2, 3, 6\}, \{5, 6, 9\}, \{4, 5, 8\}, \{4, 7\}$.

product of potentials over all cliques (completely connected subgraphs) of the graph Q. When the distribution is Gaussian with covariance matrix $K \in \mathcal{R}^{n \times n}$, the distribution factorizes if and only if $(K^{-1})_{ij} = 0$ for each (i, j) that is not an edge in Q [46].

In this section, we only consider *decomposable* probabilistic graphical models, for which the graph Q is *triangulated* (i.e., there exists no chordless cycle of length strictly larger than 3). In this case, the joint distribution is uniquely defined from its marginals $p_C(z_C)$ on the cliques C of the graph Q. That is, if $\mathcal{C}(Q)$ is the set of maximal cliques of Q, we can build a tree of cliques, a *junction tree*, such that

$$p(z) = \frac{\prod_{C \in \mathcal{C}(Q)} p_C(z_C)}{\prod_{C, C' \in \mathcal{C}(Q), C \sim C'} p_{C \cap C'}(z_{C \cap C'})}.$$

Figure 17.7 shows an example of a probabilistic graphical model and a junction tree. The sets $C \cap C'$ are usually referred to as *separators*, and we let $\mathcal{S}(Q)$ denote the set of such separators. Note that for a zero mean normally distributed vector, the marginals $p_C(z_C)$ are characterized by the marginal covariance matrix $K_C = K_{C,C}$. Projecting onto a probabilistic graphical model will preserve the marginal over all maximal cliques, and thus preserve the local kernel matrices, while imposing zeros in the inverse of K.

Probabilistic Graphical Models and Projections

We let $\Pi_Q(K)$ denote the covariance matrix that factorizes in Q which is closest to K for the Kullback-Leibler divergence between normal distributions. We essentially replace K by $\Pi_Q(K)$. In other words, we project all our covariance matrices onto a probabilistic graphical model, which is a classical tool in statistics [46, 47]. We leave the study of the approximation properties of such a projection (i.e., for a given K, how dense the graph should be to approximate the full local kernel correctly?) to future work—see, e.g., [48] for related results.

Practically, since our kernel on kernel matrices involves determinants, we simply need to compute $|\Pi_Q(K)|$ efficiently. For decomposable probabilistic graphical models, $\Pi_Q(K)$ can be obtained in closed form [46] and its determinant has the following simple expression:

$$\log |\Pi_Q(K)| = \sum_{C \in \mathcal{C}(Q)} \log |K_C| - \sum_{S \in \mathcal{S}(Q)} \log |K_S|. \tag{17.7}$$

The determinant $|\Pi_Q(K)|$ is thus a ratio of terms (determinants over cliques and separators), which will restrict the applicability of the projected kernels (see Proposition 1). In order to keep only products, we consider the following equivalent form: if the junction tree is rooted (by choosing any clique as the root), then for each clique but the root, a unique parent clique is defined, and we have

$$\log |\Pi_Q(K)| = \sum_{C \in \mathcal{C}(Q)} \log |K_{C|p_Q(C)}|, \tag{17.8}$$

where $p_Q(C)$ is the parent clique of Q (and $\varnothing$ for the root clique) and the conditional covariance matrix is defined, as usual, as

$$K_{C|p_Q(C)} = K_{C,C} - K_{C,p_Q(C)} K_{p_Q(C),p_Q(C)}^{-1} K_{p_Q(C),C} . \tag{17.9}$$

Probabilistic Graphical Models and Kernels

We now propose several ways of defining a kernel adapted to probabilistic graphical models. All of them are based on replacing determinants $|M|$ by $|\Pi_Q(M)|$, and their different decompositions in Equations 17.7 and 17.8. Simply using Equation 17.7, we obtain the similarity measure:

$$k_{\mathcal{B},0}^Q(K, L) = \prod_{C \in \mathcal{C}(Q)} k_{\mathcal{B}}(K_C, L_C) \prod_{S \in \mathcal{S}(Q)} k_{\mathcal{B}}(K_S, L_S)^{-1}. \tag{17.10}$$

This similarity measure turns out not to be a positive-definite kernel for general covariance matrices.

Proposition 1
For any decomposable model Q, the kernel $k_{\mathcal{B},0}^Q$ defined in Equation 17.10 is a positive-definite kernel on the set of covariance matrices K such that for all separators $S \in \mathcal{S}(Q)$, $K_{S,S} = I$. In particular, when all separators have cardinal one, this is a kernel on correlation matrices.

In order to remove the condition on separators (i.e., we want more sharing between cliques than through a single variable), we consider the rooted junction tree representation in Equation 17.8. A straightforward kernel is to compute the product of the Bhattacharyya kernels $k_{\mathcal{B}}(K_{C|p_Q(C)}, L_{C|p_Q(C)})$ for each conditional covariance matrix. However, this does not lead to a true distance on covariance matrices that factorize on Q because the set of conditional covariance matrices do not characterize entirely those distributions. Rather, we consider the following kernel:

$$k_{\mathcal{B}}^{Q}(K, L) = \prod_{C \in \mathcal{C}(Q)} k_{\mathcal{B}}^{C|p_Q(C)}(K, L); \qquad (17.11)$$

for the root clique, we define $k_{\mathcal{B}}^{R|\varnothing}(K, L) = k_{\mathcal{B}}(K_R, L_R)$ and the kernels $k_{\mathcal{B}}^{C|p_Q(C)}(K, L)$, are defined as kernels between conditional Gaussian distributions of Z_C given $Z_{p_Q(C)}$. We use

$$k_{\mathcal{B}}^{C|p_Q(C)}(K,L) = \frac{|K_{C|p_Q(C)}|^{1/2} |L_{C|p_Q(C)}|^{1/2}}{\left|\frac{1}{2} K_{C|p_Q(C)} + \frac{1}{2} L_{C|p_Q(C)} + M M^{\top}\right|}, \qquad (17.12)$$

where the additional term M is equal to $\frac{1}{2}(K_{C,p_Q(C)} K_{p_Q(C)}^{-1} - L_{C,p_Q(C)} L_{p_Q(C)}^{-1})$. This exactly corresponds to putting a prior with identity covariance matrix on variables $Z_{p_Q(C)}$ and considering the kernel between the resulting joint covariance matrices on variables indexed by $(C, p_Q(C))$. We now have a positive-definite kernel on all covariance matrices:

Proposition 2
For any decomposable model Q, the kernel $k_{\mathcal{B}}^{Q}(K, L)$ defined in Equation 17.11 and Equation 17.12 is a positive-definite kernel on the set of covariance matrices.

Note that the kernel is not invariant by the choice of the particular root of the junction tree. However, in our setting, this is not an issue because we have a natural way of rooting the junction trees (i.e, following the rooted tree-walk). Note that these kernels could be useful in domains other than computer vision.

In the next derivations, we will use the notation $k_{\mathcal{B}}^{I_1|I_2, J_1|J_2}(K, L)$ for $|I_1| = |I_2|$ and $|J_1| = |J_2|$ to denote the kernel between covariance matrices $K_{I_1 \cup I_2}$ and $L_{I_1 \cup I_2}$ adapted to the conditional distributions $I_1|I_2$ and $J_1|J_2$, defined through Equation 17.12.

Choice of Probabilistic Graphical Models

Given the rooted tree structure T of a β-tree-walk, we now need to define the probabilistic graphical model $Q_{\beta}(T)$ that we use to project our kernel matrices. A natural candidate is T itself. However, as shown later, in order to compute efficiently the kernel we simply need that the local kernel is a product of terms that only involve a node and its β-descendants. The densest graph (remember that denser graphs lead to better approximations when projecting onto the probabilistic graphical model) we may use is the following: we define $Q_{\beta}(T)$ such that for all nodes in T, the node together with all its β-descendants form a clique, i.e., a node is connected to its β-descendants and all β-descendants are also mutually connected (see Figure 17.7, for

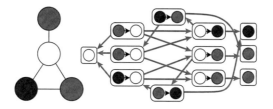

FIGURE 17.8
(Left) undirected graph G, (right) graph $G_{1,2}$.

example, for $\beta = 1$): the set of cliques is thus the set of *families* of depth $\beta + 1$ (i.e., with $\beta + 1$ generations). Thus, our final kernel is

$$k_{\alpha,\beta,\gamma}^{T}(G,H) = \sum_{T \in \mathcal{T}_{\alpha,\gamma}} f_{\lambda,\nu}(T) \left\{ \sum_{I \in \mathcal{J}_{\beta}(T,G)} \sum_{J \in \mathcal{J}_{\beta}(T,H)} k_{B}^{Q_{\beta}(T)}(K_I, L_J) q_{\mathcal{A}}(a(I), b(J)) \right\}.$$

The main intuition behind this definition is to sum local similarities over all matching subgraphs. In order to obtain a tractable formulation, we simply needed (a) to extend the set of subgraphs (to tree-walks of depth γ) and (b) to factorize the local similarities along the graphs. We now show how these elements can be combined to derive efficient recursions.

17.4.3 Dynamic Programming Recursions

In order to derive dynamic programming recursions, we follow [34] and rely on the fact that α-ary β-tree-walks of G can essentially be defined through 1-tree-walks on the augmented graph of all rooted subtrees of G of depth at most β and arity less than α. Recall that the *arity* of a tree is the maximum number of children of the root and internal vertices of the tree. We thus consider the set $V_{\alpha,\beta}$ of noncomplete rooted (unordered) subtrees of $G = (V, E)$, of depths less than β and arity less than α. Given two different rooted unordered labelled trees, they are said to be *equivalent* (or isomorphic) if they share the same tree structure, and this is denoted $\sim_t$.

On this set $V_{\alpha,\beta}$, we define a *directed* graph with edge set $E_{\alpha,\beta}$ as follows: $R_0 \in V_{\alpha,\beta}$ is connected to $R_1 \in V_{\alpha,\beta}$ if "the tree R_1 extends the tree R_0 one generation further", i.e., if and only if (a) the first $\beta - 1$ generations of R_1 are exactly equal to one of the complete subtrees of R_0 rooted at a child of the root of R_0, and (b) the nodes of depth β of R_1 are distinct from the nodes in R_0. This defines a graph $G_{\alpha,\beta} = (V_{\alpha,\beta}, E_{\alpha,\beta})$ and a neighborhood $\mathcal{N}_{G_{\alpha,\beta}}(R)$ for $R \in V_{\alpha,\beta}$ (see Figure 17.8 for an example). Similarly, we define a graph $H_{\alpha,\beta} = (W_{\alpha,\beta}, F_{\alpha,\beta})$ for the graph H. Note that when $\alpha = 1$, $V_{1,\beta}$ is the set of paths of length less than or equal to β.

For a β-tree-walk, the root with its β-descendants must have distinct vertices and thus corresponds exactly to an element of $V_{\alpha,\beta}$. We denote $k_{\alpha,\beta,\gamma}^{T}(G, H, R_0, S_0)$ the same kernel as defined in Equation 17.4.2, but restricted to tree-walks that start respectively with R_0 and S_0. Note that if R_0 and S_0 are not equivalent, then $k_{\alpha,\beta,\gamma}^{T}(G, H, R_0, S_0) = 0$.

Denote $\rho(S)$ the root of a subtree S. We obtain the following recursion between depths γ and depth $\gamma - 1$, for all $R_0 \in V_{\alpha,\beta}$ and and $S_0 \in W_{\alpha,\beta}$ such that $R_0 \sim_t S_0$:

$$k_{\alpha,\beta,\gamma}^{T}(G, H, R_0, S_0) = k_{\alpha,\beta,\gamma-1}^{T}(G, H, R_0, S_0) + \mathcal{R}_{\alpha,\beta,\gamma-1}^{T}, \qquad (17.13)$$

where $\mathcal{R}_{\alpha,\beta,\gamma-1}^{T}$ is given by

$$\mathcal{R}_{\alpha,\beta,\gamma-1}^{T} = \sum_{p=1}^{\alpha} \sum_{\substack{R_1,\ldots,R_p \in \mathcal{N}_{G_{\alpha,\beta}}(R_0) \\ R_1,\ldots,R_p \text{ disjoint}}} \sum_{\substack{S_1,\ldots,S_p \in \mathcal{N}_{H_{\alpha,\beta}}(S_0) \\ S_1,\ldots,S_p \text{ disjoint}}} \cdots$$

$$\left[\lambda \prod_{i=1}^{p} k_{\mathcal{A}}(a(\rho(R_i)), b(\rho(S_i))) \frac{k_{\mathcal{B}}^{\cup_{i=1}^{p} R_i | R_0, \cup_{i=1}^{p} S_i | S_0}(K, L)}{\prod_{i=1}^{p} k_{\mathcal{B}}^{R_i, S_i}(K, L)} \left(\prod_{i=1}^{p} k_{\alpha,\beta,\gamma-1}^{T}(G, H, R_i, S_i) \right) \right].$$

$$(17.14)$$

Note that if any of the trees R_i is not equivalent to S_i, it does not contribute to the sum. The recursion is initialized with

$$k_{\alpha,\beta,\gamma}^{T}(G, H, R_0, S_0) = \lambda^{|R_0|} \nu^{\ell(R_0)} q_{\mathcal{A}}(a(R_0), b(S_0)) k_{\mathcal{B}}(K_{R_0}, L_{S_0}) \qquad (17.15)$$

while the final kernel is obtained by summing over all R_0 and S_0, i.e., $k_{\alpha,\beta,\gamma}^{T}(G, H) = \sum_{R_0 \sim_t S_0} k_{\alpha,\beta,\gamma}^{T}(G, H, R_0, S_0)$.

Computational Complexity

The above equations define a dynamic programming recursion that allows us to efficiently compute the values of $k_{\alpha,\beta,\gamma}^{T}(G, H, R_0, S_0)$ from $\gamma = 1$ to any desired γ. We first compute $k_{\alpha,\beta,1}^{T}(G, H, R_0, S_0)$ using Equation 17.15. Then $k_{\alpha,\beta,2}^{T}(G, H, R_0, S_0)$ can be computed from $k_{\alpha,\beta,1}^{T}(G, H, R_0, S_0)$ using Equation 17.13 with $\gamma = 2$. And so on.

The complexity of computing one kernel between two graphs is linear in γ (the depth of the tree-walks), and quadratic in the size of $V_{\alpha,\beta}$ and $W_{\alpha,\beta}$. However, those sets may have exponential size in β and α in general (in particular, if graphs are densely connected). And thus, we are limited to small values (typically $\alpha \leqslant 3$ and $\beta \leqslant 6$), which are sufficient for satisfactory classification performance (in particular, higher β or α do not necessarily mean better performance). Overall, one can deal with any graph size, as long as the "sufficient statistics" (i.e., the unique local neighborhoods in $V_{\alpha,\beta}$) are not too numerous.

For example, for the handwritten digits we use in experiments, the average number of nodes in the graphs is 18 ± 4, while the average cardinality of $V_{\alpha,\beta}$ and running times[1] for one kernel evaluation are, for walk kernels of depth 24: $|V_{\alpha,\beta}| = 36$, time $= 2$ ms ($\alpha = 1$, $\beta = 2$), $|V_{\alpha,\beta}| = 37$, time $= 3$ ms ($\alpha = 1$, $\beta = 4$); and for tree kernels: $|V_{\alpha,\beta}| = 56$, time $= 25$ ms ($\alpha = 2$, $\beta = 2$), $|V_{\alpha,\beta}| = 70$, time $= 32$ ms ($\alpha = 2$, $\beta = 4$).

Finally, we may reduce the computational load by considering a set of trees of smaller arity in the previous recursions; i.e., we can consider $V_{1,\beta}$ instead of $V_{\alpha,\beta}$ with tree kernels of arity $\alpha > 1$.

17.5   Experimental Results

We present here experimental results of tree-walk kernels, respectively, on: (i) region adjacency graphs for object categorization and natural image classification, and (ii) point clouds for handwritten digits classification.

17.5.1   Application to Object Categorization

We have tested our kernels on the task of object categorization (Coil100 dataset), and natural image classification (Corel14 dataset) both in fully supervised and in semi supervised settings.

Experimental Setting

Experiments have been carried out on both `Corel14` [22] and `Coil100` datasets. Our kernels were put to the test step by step, going from the less sophisticated version to the most complex one. Indeed, we compared on a multi-class classification task performances of the usual histogram kernel (**H**), the walk-based kernel (**W**), the tree-walk kernel (**TW**), the weighted-vertex tree-walk kernel (**wTW**), and the combination of the above by multiple kernel learning (**M**). We report here their performances averaged in an outer loop of 5-fold cross-validation. The hyperparameters are tuned in an inner loop in each fold by 5-fold cross-validation (see in the sequel for further details). `Coil100` consists in a database of 7200 images of 100 *objects in a uniform background*, with 72 images per object. Data are color images of the objects taken from different angles, with steps of 5 degrees. `Corel14` is a database of 1400 *natural images* of 14 different classes, which are usually considered much

---

[1]Those do not take into account preprocessing and were evaluated on an Intel Xeon 2.33 GHz processor from MATLAB/C code, and are to be compared to the simplest recursions, which correspond to the usual random walk kernel ($\alpha = 1$, $\beta = 1$), where time $= 1$ ms.

harder to classify. Each class contains 100 images, with a non-negligible proportion of outliers.

Feature Extraction

Each image's segmentation outputs a labelled graph with 100 vertices, with each vertex labelled with the RGB-color histogram within each corresponding segment. We used 16 bins per dimension, as in [43], yielding 4096-dimensional histograms. Note that we could use LAB histograms as well. The average vertex degree was around 3 for `Coil100` and 5 for `Corel14`. In other words, region adjacency graphs are very sparsely connected.

Free Parameter Selection

For the multi-class classification task, the usual SVM classifier was used in a one-versus-all setting [1]. For each family of kernels, hyper-parameters corresponding to kernel design and the SVM regularization parameter C were learned by cross-validation, with the following usual machine learning procedure: we randomly split the full dataset in 5 parts of equal size, then we consider successively each of the 5 parts as the testing set (the outer testing fold), learning being performed on the four other parts (the outer training fold). This is in contrast to other evaluation protocols. Assume instead one is trying out different values of the free parameters on the outer training fold, computing the prediction accuracy on the corresponding testing fold, repeating five times for each of the five outer folds, and compute average performance. Then assume one selects the best hyper-parameter and report its performance. A major issue with such a protocol is that it leads to an optimistic estimation of the prediction performance [49].

It is preferable to consider each outer training fold, and split those into 5 equal parts, and learn the hyper-parameters using cross-validation on those inner folds, and use the resulting parameter, train on the full outer training fold with this set of hyperparameters (which might be different for each outer fold), and test on the outer testing fold. The prediction accuracies that we report (in particular in the boxplots of Figure 17.11) are the prediction accuracies on the outer testing folds. This two-stage approach leads to more numerous estimations of SVM parameters but provides a fair evaluation of performance.

In order to choose the values of the free parameters, we use values of free parameters on the finite grid in Figure 17.9.

The respective performances in average test error rates (on the five testing outer folds) in multi-class classification of the different kernels are listed in Figure 17.10. See Figure 17.11 for corresponding boxplots for the `Corel14` dataset.

Parameter	Values
γ	$0.0, 0.2, 0.4, 0.6, 0.8$
α	$1, 2, 3$
p	$1, 2, 3, 4, 5, 6, 7, 8, 9, 10$
C	$10^{-2}, 10^{-1}, 10^0, 10^1, 10^2, 10^3, 10^4$

FIGURE 17.9
Range of values of the free parameters.

	H	W	TW	wTW	M
Coil100	1.2%	0.8%	0.0%	0.0%	0.0%
Corel14	10.36%	8.52%	7.24%	6.12%	5.38%

FIGURE 17.10
Best test error performances of histogram, walk, tree-walk, tree-walk with weighted segments, kernels

Corel14 dataset

We have compared test error rate performances of SVM-based multi-class classification with histogram kernel (**H**), walk kernel (**W**), tree-walk kernel (**TW**), and the tree-walk kernel with weighted segments (**wTW**). Our methods, i.e., **TW** and **wTW**, clearly outperform global histogram kernels and simple walk kernels. This corroborates the efficiency of tree-walk kernels in capturing the topological structure of natural images. Our weighting scheme also seems to be reasonable.

Multiple Kernel Learning

We first tried the combination of histogram kernels with walk-based kernels, which did not yield significant performance enhancement. This suggests that histograms do not carry any supplemental information over walk-based kernels: the global histogram information is implicitly retrieved in the summation process of walk-based kernels.

We ran the multiple kernel learning (MKL) algorithm of [50] with 100 kernels (i.e., a rough subset of the set of all parameter settings with parameters taken values detailed in Figure 17.9). As seen in Figure 17.11, the performance increases as expected. It is also worth looking at the kernels that were selected. We here give results for one of the five outer cross-validation folds, where 5 kernels were selected, as displayed in Figure 17.12.

It is worth noting that various values of γ, α, and p are selected, showing that each setting may indeed capture different discriminative information from the segmented images.

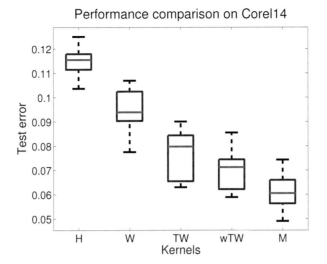

FIGURE 17.11
Test errors for supervised multi-class classification on Corel14, for **H**istogram, **W**alk, **T**ree-**W**alk, **w**eighted **T**ree-**W**alk, kernels, and optimal **M**ultiple kernel combination.

p, α, γ	$10, 3, 0.6$	$7, 1, 0.6$	$10, 3, 0.3$	$5, 3, 0.0$	$8, 1, 0.0$
η	0.12	0.17	0.10	0.07	0.04

FIGURE 17.12
Weight η of the kernels with the largest magnitude (see [50] for details on normalization of η).

Semi-Supervised Classification

Kernels allow us to tackle many tasks, from unsupervised clustering to multi-class classification and manifold learning [1]. To further explore the expressiveness of our region adjacency graph kernels, we give below the evolution of test error performances on Corel14 dataset for multi-class classification with 10% labelled examples, 10% test examples, and an increasing amount ranging from 10% to 80% of unlabelled examples. Indeed, all semi-supervised algorithms derived from statistically consistent supervised ones see their test errors falling down as the number of labelled examples is increased and the number of unlabelled examples kept constant. However, such experimental convergence of the test errors as the number of labelled examples is kept fixed and the number of unlabelled ones is increased is much less systematic [51]. We used the publicly available code for the low-density separation (LDS) algorithm of [51], since it achieved good performance on the Coil100 image dataset.

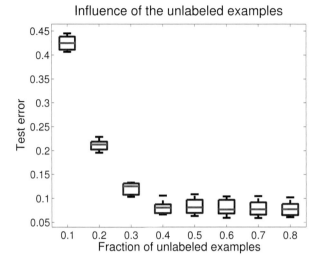

FIGURE 17.13
Test error evolution for semi-supervised multi-class classification as the number of unlabelled examples increases.

Since we are more interested in showing the flexibility of our approach than the semi-supervised learning problem in general, we simply took as a kernel the optimal kernel learned on the whole supervised multi-classification task by the multiple kernel learning method. Although this may lead to a slight overestimation of the prediction accuracies, this allowed us to bypass the kernel selection issue in semi-supervised classification, which still remains unclear and under active investigation. For each amount of unlabelled examples, as an outer loop we randomly selected 10 different splits into labelled and unlabelled examples. As an inner loop we optimized the hyperparameters, namely the regularization parameter C and ρ the cluster squeezing parameter (see [51] for details), by leave-one-out cross-validation. The boxplots in Figure 17.13 shows the variability of performances within the outer loop. Keeping in mind that the best test error performance on the Corel14 dataset of histogram kernels is around 10% for *completely supervised* multi-class classification, the results are very promising; we see that our kernel reaches this level of performance for as little as 40% unlabelled examples and 10% labelled examples.

We now present the experimental results obtained with tree-walk kernels on point clouds in a character recognition task.

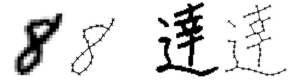

FIGURE 17.14
For digits and Chinese characters: (left) original characters, (right) thinned and sub-sampled characters.

17.5.2 Application to Character Recognition

We have tested our kernels on the task of isolated handwritten character recognition, handwritten arabic numerals (MNIST dataset), and Chinese characters (ETL9B dataset).

17.5.3 Experimental Setting

We selected the first 100 examples for the ten classes in the MNIST dataset, while for the ETL9B dataset, we selected the five hardest classes to discriminate among 3,000 classes (by computing distances between class means) and then selected the first 50 examples per class. Our learning task is to classify those characters; we use a one-versus-rest multiclass scheme with 1-norm support vector machines (see, e.g., [1]).

Feature Extraction

We consider characters as drawings in $\mathcal{R}^2$, which are sets of possibly intersecting contours. Those are naturally represented as undirected planar graphs. We have thinned and subsampled uniformly each character to reduce the sizes of the graphs (see two examples in Figure 17.14). The kernel on positions is $k_{\mathcal{X}}(x,y) = \exp(-\tau\|x - y\|^2) + \kappa\delta(x,y)$, but could take into account different weights on horizontal and vertical directions. We add the positions from the center of the bounding box as features, to take into account the global positions, i.e., we use $k_{\mathcal{A}}(x,y) = \exp(-\upsilon\|x-y\|^2)$. This is necessary because the problem of handwritten character recognition is not globally translation invariant.

Free Parameter Selection

The tree-walk kernels on point clouds have the following free parameters (shown with their possible values): arity of tree-walks ($\alpha = 1, 2$), order of tree-walks ($\beta = 1, 2, 4, 6$), depth of tree-walks ($\gamma = 1, 2, 4, 8, 16, 24$), penalization on number of nodes ($\lambda = 1$), penalization on number of leaf nodes ($\nu = .1, .01$), bandwidth for

	MNIST $\alpha = 1$	MNIST $\alpha = 2$	ETL9B $\alpha = 1$	ETL9B $\alpha = 2$
$\beta = 1$	11.6 ± 4.6	9.2 ± 3.9	36.8 ± 4.6	32 ± 8.4
$\beta = 2$	5.6 ± 3.1	5.6 ± 3.0	29.2 ± 8.8	$\mathbf{25.2 \pm 2.7}$
$\beta = 4$	$\mathbf{5.4 \pm 3.6}$	$\mathbf{5.4 \pm 3.1}$	32.4 ± 3.9	29.6 ± 4.3
$\beta = 6$	5.6 ± 3.3	6 ± 3.5	29.6 ± 4.6	28.4 ± 4.3

FIGURE 17.15
Error rates (multiplied by 100) on handwritten character classification tasks.

kernel on positions ($\tau = .05, .01, .1$), ridge parameter ($\kappa = .001$), and bandwidth for kernel on attributes ($\upsilon = .05, .01, .1$).

The first two sets of parameters ($\alpha, \beta, \gamma, \lambda, \nu$) are parameters of the graph kernel, independent of the application, while the last set (τ, κ, ν) are parameters of the kernels for attributes and positions. Note that with only a few important scale parameters (τ and ν), we are able to characterize complex interactions between the vertices and edges of the graphs. In practice, this is important to avoid considering many more distinct parameters for all sizes and topologies of subtrees. In experiments, we performed two loops of 5-fold cross-validation: in the outer loop, we consider 5 different training folds with their corresponding testing folds. On each training fold, we consider all possible values of α and β. For all of those values, we select all other parameters (including the regularization parameters of the SVM) by 5-fold cross-validation (the inner folds). Once the best parameters are found only by looking only at the training fold, we train on the whole training fold and test on the testing fold. We output the means and standard deviations of the testing errors for each testing fold.

We show in Figure 17.15 the performance for various values of α and β. We compare those favorably to three baseline kernels with hyperparameters learned by cross-validation in the same way: (a) the *Gaussian-RBF kernel* on the vectorized original images, which leads to testing errors of $11.6 \pm 5.4\%$ (MNIST) and $50.4 \pm 6.2\%$ (ETL9B); (b) the regular *random walk kernel* that sums over all walk lengths, which leads to testing errors of $8.6 \pm 1.3\%$ (MNIST) and $34.8 \pm 8.4\%$ (ETL9B); and (c) the *pyramid match kernel* [52], which is commonly used for image classification and leads here to testing errors of $10.8 \pm 3.6\%$ (MNIST) and $45.2 \pm 3.4\%$ (ETL9B). These results show that our family of kernels, which use the natural structure of line drawings, are outperforming other kernels on structured data (regular random walk kernel and pyramid match kernel) as well as the "blind" Gaussian-RBF kernel, which does not take into account explicitly the structure of images but still leads to very good performance with more training data [9]. Note that for Arabic numerals, higher arity does not help, which is not surprising since most digits have a linear structure (i.e, graphs are chains). On the contrary, for Chinese characters, which exhibit higher connectivity, best performance is achieved for binary tree-walks.

17.6 Conlusion

We have presented a family of kernels for computer vision tasks, along with two instantiations of these kernels: (i) tree-walk kernels for region adjacency graphs, and (ii) tree-walk kernels for point clouds. For (i), we have showed how one can efficiently compute kernels in polynomial time with respect to the size of the region adjacency graphs and their degrees. For (ii), we proposed an efficient dynamic programming algorithm using a specific factorized form for the local kernels between tree-walks, namely a factorization on a properly defined probabilistic graphical model. We have reported applications to object categorization and natural image classification, as well as handwritten character recognition, where we showed that the kernels were able to capture the relevant information to allow good predictions from few training examples.

17.7 Acknowledgments

This work was partially supported by grants from the Agence Nationale de la Recherche (MGA Project) and from the European Research Council (SIERRA Project), and the PASCAL 2 Network of Excellence.

Bibliography

[1] J. Shawe-Taylor and N. Cristianini, *Kernel Methods for Pattern Analysis.* Cambridge Univ. Press, 2004.

[2] C. H. Lampert, "Kernel methods in computer vision," *Found. Trends. Comput. Graph. Vis.*, vol. 4, pp. 193–285, March 2009.

[3] J.-P. Vert, H. Saigo, and T. Akutsu, *Local Alignment Kernels for Biological Sequences.* MIT Press, 2004.

[4] R. Diestel, *Graph Theory.* Springer-Verlag, 2005.

[5] O. Chapelle, B. Schölkopf, and A. Zien, Eds., *Semi-Supervised Learning (Adaptive Computation and Machine Learning).* MIT Press, 2006.

[6] F. Bach, "Consistency of the group lasso and multiple kernel learning," *Journal of Machine Learning Research*, vol. 9, pp. 1179–1225, 2008.

[7] J. Ponce, M. Hebert, C. Schmid, and A. Zisserman, *Toward Category-Level Object Recognition (Lecture Notes in Computer Science).* Springer, 2007.

[8] J. Zhang, M. Marszalek, S. Lazebnik, and C. Schmid, "Local features and kernels for classification of texture and object categories: a comprehensive study," *International Journal of Computer Vision*, vol. 73, no. 2, pp. 213–238, 2007.

[9] Y. LeCun, L. Bottou, Y. Bengio, and P. Haffner, "Gradient-based learning applied to document recognition," *Proc. IEEE*, vol. 86, no. 11, pp. 2278–2324, 1998.

[10] S. N. Srihari, X. Yang, and G. R. Ball, "Offline Chinese handwriting recognition: A survey," *Frontiers of Computer Science in China*, 2007.

[11] S. Belongie, J. Malik, and J. Puzicha, "Shape matching and object recognition using shape contexts," *IEEE Trans. PAMI*, vol. 24, no. 24, pp. 509–522, 2002.

[12] J. Ramon and T. Gärtner, "Expressivity versus efficiency of graph kernels," in *First International Workshop on Mining Graphs, Trees and Sequences*, 2003.

[13] S. V. N. Vishwanathan, N. N. Schraudolph, R. I. Kondor, and K. M. Borgwardt, "Graph kernels," *Journal of Machine Learning Research*, vol. 11, pp. 1201–1242, 2010.

[14] H. Kashima, K. Tsuda, and A. Inokuchi, "Kernels for graphs," in *Kernel Methods in Comp. Biology*. MIT Press, 2004.

[15] Z. Harchaoui and F. Bach, "Image classification with segmentation graph kernels," in *CVPR*, 2007.

[16] F. R. Bach, "Graph kernels between point clouds," in *Proceedings of the 25th international conference on Machine learning*, ser. ICML '08. New York, NY, USA: ACM, 2008, pp. 25–32.

[17] C. Wang and K. Abe, "Region correspondence by inexact attributed planar graph matching," in *Proc. ICCV*, 1995.

[18] A. Robles-Kelly and E. Hancock, "Graph edit distance from spectral seriation," *IEEE PAMI*, vol. 27, no. 3, pp. 365–378, 2005.

[19] C. Gomila and F. Meyer, "Graph based object tracking," in *Proc. ICIP*, 2003, pp. 41–44.

[20] B. Huet, A. D. Cross, and E. R. Hancock, "Graph matching for shape retrieval," in *Adv. NIPS*, 1999.

[21] D. Gusfield, *Algorithms on Strings, Trees, and Sequences*. Cambridge Univ. Press, 1997.

[22] O. Chapelle and P. Haffner, "Support vector machines for histogram-based classification," *IEEE Trans. Neural Networks*, vol. 10, no. 5, pp. 1055–1064, 1999.

[23] T. Jebara, "Images as bags of pixels," in *Proc. ICCV*, 2003.

[24] M. Cuturi, K. Fukumizu, and J.-P. Vert, "Semigroup kernels on measures," *J. Mac. Learn. Research*, vol. 6, pp. 1169–1198, 2005.

[25] S. Lazebnik, C. Schmid, and J. Ponce, "Beyond bags of features: Spatial pyramid matching for recognizing natural scene categories," in *Proc. CVPR*, 2006.

[26] M. Neuhaus and H. Bunke, "Edit distance based kernel functions for structural pattern classification," *Pattern Recognition*, vol. 39, no. 10, pp. 1852–1863, 2006.

[27] H. Fröhlich, J. K. Wegner, F. Sieker, and A. Zell, "Optimal assignment kernels for attributed molecular graphs," in *Proc. ICML*, 2005.

[28] J.-P. Vert, "The optimal assignment kernel is not positive definite, Tech. Rep. HAL-00218278, 2008.

[29] F. Suard, V. Guigue, A. Rakotomamonjy, and A. Benshrair, "Pedestrian detection using stereo-vision and graph kernels," in *IEEE Symposium on Intelligent Vehicule*, 2005.

[30] K. M. Borgwardt, C. S. Ong, S. Schönauer, S. V. N. Vishwanathan, A. J. Smola, and H.-P. Kriegel, "Protein function prediction via graph kernels." *Bioinformatics*, vol. 21, 2005.

[31] S. V. N. Vishwanathan, K. M. Borgwardt, and N. Schraudolph, "Fast computation of graph kernels," in *Adv. NIPS*, 2007.

[32] J.-P. Vert, T. Matsui, S. Satoh, and Y. Uchiyama, "High-level feature extraction using svm with walk-based graph kernel," in *Proceedings of the 2009 IEEE International Conference on Acoustics, Speech and Signal Processing*, ser. ICASSP '09, 2009.

[33] M. Fisher, M. Savva, and P. Hanrahan, "Characterizing structural relationships in scenes using graph kernels," in *ACM SIGGRAPH 2011 papers*, ser. SIGGRAPH '11, 2011.

[34] P. Mahé and J.-P. Vert, "Graph kernels based on tree patterns for molecules," *Machine Learning Journal*, vol. 75, pp. 3–35, April 2009.

[35] F. Meyer, "Hierarchies of partitions and morphological segmentation," in *Scale-Space and Morphology in Computer Vision*. Springer-Verlag, 2001.

[36] J. Shi and J. Malik, "Normalized cuts and image segmentation," *IEEE PAMI*, vol. 22, no. 8, pp. 888–905, 2000.

[37] D. Comaniciu and P. Meer, "Mean shift: a robust approach toward feature space analysis," *IEEE PAMI*, vol. 24, no. 5, pp. 603–619, 2002.

[38] J. Malik, S. Belongie, T. K. Leung, and J. Shi, "Contour and texture analysis for image segmentation," *Int. J. Comp. Vision*, vol. 43, no. 1, pp. 7–27, 2001.

[39] X. Ren and J. Malik, "Learning a classification model for segmentation," *Computer Vision, IEEE International Conference on*, vol. 1, p. 10, 2003.

[40] A. Levinshtein, A. Stere, K. N. Kutulakos, D. J. Fleet, S. J. Dickinson, and K. Siddiqi, "Turbopixels: fast superpixels using geometric flows." *IEEE Transactions on Pattern Analysis and Machine Intelligence*, vol. 31, no. 12, pp. 2290–2297, 2009.

[41] T. Gärtner, P. A. Flach, and S. Wrobel, "On graph kernels: Hardness results and efficient alternatives," in *COLT*, 2003.

[42] D. G. Lowe, "Distinctive image features from scale-invariant keypoints," *Int. J. Comp. Vision*, vol. 60, no. 2, pp. 91–110, 2004.

[43] M. Hein and O. Bousquet, "Hilbertian metrics and positive-definite kernels on probability measures," in *AISTATS*, 2004.

[44] C. Fowlkes, S. Belongie, F. Chung, and J. Malik, "Spectral grouping using the Nyström method," *IEEE PAMI*, vol. 26, no. 2, pp. 214–225, 2004.

[45] R. I. Kondor and T. Jebara, "A kernel between sets of vectors." in *Proc. ICML*, 2003.

[46] S. Lauritzen, *Graphical Models*. Oxford U. Press, 1996.

[47] D. Koller and N. Friedman, *Probabilistic Graphical Models: Principles and Techniques*. MIT Press, 2009.

[48] T. Caetano, T. Caelli, D. Schuurmans, and D. Barone, "Graphical models and point pattern matching," *IEEE Trans. PAMI*, vol. 28, no. 10, pp. 1646–1663, 2006.

[49] R. Kohavi and G. John, "Wrappers for feature subset selection," *Artificial Intelligence*, vol. 97, no. 1-2, pp. 273–324, 1997.

[50] F. R. Bach, R. Thibaux, and M. I. Jordan, "Computing regularization paths for learning multiple kernels," in *Adv. NIPS*, 2004.

[51] O. Chapelle and A. Zien, "Semi-supervised classification by low density separation," in *Proc. AISTATS*, 2004.

[52] K. Grauman and T. Darrell, "The pyramid match kernel: Efficient learning with sets of features," *J. Mach. Learn. Res.*, vol. 8, pp. 725–760, 2007.

Index

Symbols

α-β range move, 52
α-β swap, 49
α-expansion, 49

A

adjacency matrix, 444
    weighted, 444
adjunction, 147
affinities, 9
affinity, 124, 358
alpha matte, 237
alpha value, 237
anisotropic diffusion, 169, 188
applications of GED, 407
attraction, 133
attributed graph, 425
augmenting path algorithm, 99
autarky property, 41
automatic edit cost learning, 399

B

background, 112, 114
ball, 357, 359
bandwidth, 367, 368
basic energy model, 115
Bayesian, 367
belief Propagation, 490
boundaries, 3

C

catchment basin, 157
Cauchy function, 124
CCA, *see* component, 357, 359, 371, 374
centered, 355
centering
    double, 355, 356
    matrix, 356, 364

centroid, 367
children, 307
clique, 4
closing, 149
cluster, 352, 353, 374
    spectral clustering, 353
clustering, 431
CMMDS, *see* multidimensional scaling,
    356–361, 363, 365, 370, 371,
    373
co-ranking matrix, 369
color line model, 243
colorization, 191
combinatorial maps, 330
combinatorial pyramid, 337
    base level combinatorial map, 341
    implicit encoding, 341
    level, 339
    state, 339
commute-time distance, 455
commute-time embedding, 456, 458
compact partitioning, 431
component
    connected, 362, 365, 372
    curvilinear, 353, 357, 359
    principal, 353, 354
component tree, 152
composed of, 325
compositing Equation, 238
concentration, 374
conductance, 363
connecting paths, 333
constraint, 360–364
containment, 133
contains, 325
context, 267
contraction, 326

529

contraction kernel, 334
convex function, 45
coordinate, 355, 356, 361, 366–368
    centered, 355
    nonlinearly transformed, 363
coordinates, 365
cost, 36
cost function, 356, 357, 362, 364, 368, 369, 374
    edge-based, 269
    region-based, 270
covariance, 355
curse of dimensionality, 431
curvature, 127, 358
cut, 35, 99
    minimum-cut, 99
cycle, 8

D

D3P, *see* Data-Driven Decimation Process
data
    high-dimensional, 352
    representation, 352, 369
    set, 370
    visualization, 353, 366, 367, 373
data term, 33
data-driven decimation process, 316
deblurring, 106
decimation ratio, 307, 311
decoding, 354
decorrelation, 355
degree, 3
    inner, 3
    outer, 3
denoising, 104
derivative
    partial, 354, 362
diffusion, 363
Dijkstra, 359–361
dilation, 147
dimensionality
    curse of, 352
    intrinsic, 371

reduction, 352–354, 365, 369, 373, 374
    linear, 353
    nonlinear, 353
directed, 2
directed neighborhood, 321
direction, 125
discrete calculus, 6
discrete topology, 430
discretization, 427
dissimilarity, 352, 361, 373
distance, 124, 352, 353, 357, 358, 366, 368, 369, 374
    commute time, 352, 363
    commute-time, 363
    concentration, 374
    diffusion, 363
    Euclidean, 356, 358–361, 372
        squared, 356, 359
    geodesic, 352, 359, 360, 363, 372
    graph, 358–360, 373
    infinite, 360, 366
    matrix
        dense, 363
        squared, 356, 359
    preservation, 356, 358–360, 369, 374
        weighted, 360
    sorted, 369
distances, 9
distortion, 366
distribution, 366, 368, 370
    support, 367
dot product, 353, 356, 363
doubly stochastic matrices, 470
DR, *see* dimensionality, 353, 360, 361, 367
    linear, 353
dual, 114
dual graph, 328
dual graph pyramid, 336
duality, 147
dynamic, 425
dynamic programming, 509, 516

E

edge, 143
    incident, 3
edge graph, 318
edges, 2
edit costs, 396
edit operations, 395
edit path, 395
eigenproblem, 355
eigenspectrum, 360
eigenvalue, 355, 359, 361, 373
    multiplicity, 362, 363, 365
    problem, 362
eigenvector, 355, 365
    leading, 355
    trailing, 362, 365
electric lines of force, 287
EM algorithm, 462
embedding, 352, 353, 359, 361, 364, 367
    graph, 353, 358
    locally linear, 358, 364
    spiky, 373
empty self-loop, 329
encoding, 354
endpoint, 3
energy, 29, 33
equivalence class, 425
erosion, 147
error-tolerant graph matching, 393
exact graph matching, 387
exclusion, 133
expectation, 368
expectation-maximization, 367
exploratory
    analysis, 373
    data analysis, 367
    observation machine, 358, 368
extension, 155
extinction, 164

F

Fiedler value, 453
Fiedler vector, 453
filter, 149
flooding, 153

flow, 12
    maximum, 13, 14
Floyd, 359–361
force, 358
forest
    minimum spanning, 157
form, 426
function
    source, 3
    target, 3
fusion move, 51

G

Gaussian mixture model, 463
geodesic, 425
    curve, 359
    distance, *see* distance
geometric constraints, 133
Gibbs distribution, 476
global attributes, 426
gradient, 368
    descent, 356
      stochastic, 357, 368, 373, 374
Gram matrix, 353, 355, 359, 360, 363,
    365, 454
graph, 2, 20, 143, 353, 357, 361, 373, 386
    complete bipartite, 4
    directed, 2
    fully connected, 4
    higher-order, 4
    hypergraph, 4
    partial graph, 4
    adjacency, 6
    bipartite, 4
    complement, 3
    complete, 4
    connected, 8
    construction, 358, 360, 366
    digraph, 2
    distance, 358
    drawing, 367
    edge, 363
    embedding, 358, 365, 367
    induced, 155
    isomorphism, 4

layout, 367, 368
placement, 358
regular, 3
subgraph, 4
undirected, 2, 362, 366
unweighted, 2
vertex, 367
weighted, 2
graph cuts, 14, 26, 166, 266
graph edit distance, 395
graph edit distance computation, 401
graph embedding, 409
graph isomorphism, 388, 446
graph Laplacian, 208, 445
combinatorial, 445
normalized, 445
random-walk, 445
unnormalized, 445
combinatorial, 452, 453
graph search, 266
graphical model
factor graph, 481
factorization, 479
Potts model, 485
graphical models, 476
grid, 366
elasticity, 367
node, 366, 367

H
handwritten digit, 352, 370
Hausdorff metric, 427
hierarchical index, 430
hierarchy, 153
MSF, 162
hierarchy of parts, 426
higher-order energies, 43
histogram
joint, 369
histogram matching, 444, 462
Hoffman–Wielandt theorem, 447
honeycomb, 366
Hungarian algorithm, 451, 462

I

image denoising, 208
image segmentation, 266
independent set, 312
maximal, 312
maximum, 313
surviving elements, 312
indexing, 430
infimum, 143
information
context, 267
inpainting, 191
inside, 325
instabilities, 427
interpolation, 189
Isomap, 358–361, 363
isometry, 359, 360
Isotop, 358, 367, 368, 371, 374
isotropic diffusion, 186

K
K-means, 366, 367
kernel, 363

L
labeling, 27
labelings, 26
Lagrange, 354
multiplier, 354
Lagrangian, 354, 362
landmark, 372
Laplacian
eigenmaps, 358, 362, 363, 365, 371, 374
matrix, 352, 353
normalized, 362, 363, 365
unnormalized, 362, 363
lattice, 15, 143
complete, 143
learning
manifold, 353
rate, 367
likelihood, 367
LLE, 364, 365, 371, 374
local attributes, 426
local kernel, 504, 508, 511

locally linear, 363
LOGISMOS, 285–290
    cost function, 290
    electric lines of force, 287
      multiobject interaction, 288
    graph construction, 286
    interaction
      multiobject, 287
    presegmentation, 286

M

majorization, 357
manifold, 352, 353, 357, 358, 360, 366, 367
    disconnected, 360
    learning, 353
    nonconvex, 360
    smooth, 353, 359, 364
    submanifold, 352
    underlying, 371
    unfolding, 358, 360, 361
map
    self-organizing, 352, 358, 366, 367
mapping
    nonlinear, 357, 359, 374
    Sammon's nonlinear mapping, 353
    topographic, 367
    topographic mapping, 352
Markov random field, 32
mass, 358, 367
matching, 318
matrix
    p-Laplacian, 182
    constitutive, 6
    incidence, 6
    Laplacian, 7
matting, 237
matting Laplacian, 244
maximal independent edge set
    maximal matching, 318
maximal independent directed edge set, 321
maximal independent edge set, 318
maximal independent set, *see* independent set

maximal independent vertex set, 313
    candidate, 314
maximal matching, 318
maximum common subgraph, 390
maximum flow, 36
maximum-flow, 99
    breakpoint, 103
    parametric, 101
MDS, 353, *see* multidimensional scaling, 360, 370
medial axis, 426
median graph, 412
median graph via embedding, 412
meet relationship, 324
meets_each, 325
meets_exists, 325
Mercer kernel, 363
mesh, 158
metric, 424
metric-based indexing, 431
MIDES, *see* Maximal Independent Directed Edge Set
MIES, *see* Maximal Independent Edge Set
min-cut, 363
minimum common supergraph, 390
minimum cut, 13, 36
minimum spanning tree, 157
MIVS, *see* Maximal Independent Vertex Set
MNIST, 370
Moore-Penrose, 365
move-making, 49
MSF, *see* forest, minimum spanning
multidimensional scaling, 353, 369
    metric
      classical, 353, 355
multiple kernel learning, 520
MVU, *see* variance, 361, 362, 371

N

neighborhood, 358, 361, 370
    K-ary, 352, 357, 359, 360, 364, 369, 372
    ϵ-ball, 352, 357, 360, 364

graph, 360, 362–365, 367, 368
radius, 367, 368
relationship, 357, 369
network
electrical, 363
neural, 353, 366
NLDR, 353, *see* dimensionality, 357, 364, 371
NLM, *see* Sammon, 357, 359, 371
nodes, 2
adjacent, 3
isolated, 3
nonlocal, 185
nonlocal image graph, 185, 208
nonoverlapping holed pyramid, 307
nonoverlapping pyramid without hole, 308

O
object, 112, 114
octree, 19
opening, 149
operator, 177
directional derivative, 178
weighted p-Laplace anisotropic operator, 181
weighted p-Laplace isotropic operator, 180
weighted gradient, 179
difference, 178
divergence, 179
opposing metrics method, 128
optimal GED algorithms, 401
optimization, 357, 373
constrained, 361
convex, 363
heuristic, 365–368, 374
nonspectral, 374
pseudo-Newton, 374
spectral, 374
oriented, 2
origin, 3
orthogonal, 354, 355
orthogonal matrices, 460
overlapping pyramid, 308

P
P-brush, 116
parameter, 372
partial difference equations (PdE), 177
path, 8, 360, 361
augmenting, 14
shortest, 10, 359, 368, 373
PCA, 353, *see* component, 355, 356, 366, 373
kernel, 363
permutation
matrix, 369
pivot-based, 431
pixel, 370
placement
force-directed, 358
graph, 358
point, 143
point cloud, 510
point registration, 447, 462
preselected edges, 321
primal, 114
principal component analysis, *see* component
prior term, 34
probabilistic graphical model, 512
projection, 357, 366
proximal
point, 94
splitting, 94, 104
proximity, 352, 358
proximity operator, 94, 104
pseudo-Boolean functions, 43
pseudo-inverse, 365
pyramid
base, 307
level, 307
top, 307

Q
quality, 369
assessment, 369, 372, 373
criterion, 369, 370
quantitative, 424
quantum, 424

R

RAG, 326
random walk, 352, 363
random walker, 166
rank, 352, 364, 369
    ex aequo, 369
    nonreflexive, 369
    quality criteria, 352
receptive field, 307, 341
reconstruction, 354, 355, 366
    coefficient, 364
    geodesic, 154
reduction factor, *see* Decimation ratio
reduction function, 307
reduction window, 307, 311, 315
redundant double edge, 329
region adjacency graph, 19, 326, 333, 506
region quadtree, 17
regular pyramid, 307
regularization, 112, 134, 364
    p-Laplacian, 185
removal, 326
removal kernel, 336
representation
    discrete, 359, 366
    two-dimensional, 371
residual capacity, 99
residual graph, 99
resistance, 363

S

saliency map, 163
Sammon, *see* mapping, 357, 359, 369, 371, 374
scales, 426
scatterplot, 369
scribbles, 114
seeds, 114
segmentation
    cost function
        edge-based, 269
        region-based, 270
    graph-based
        3-D, 266

knee joint, 294–299
multiobject, 266
multisurface, 266
n-D, 266
optimal, 266
prostate, 291–293
retinal layers, 290–291
separation constraint, 269
smoothness constraint, 269
image, 266
LOGISMOS, 285–290
optimal multiple surface, 275
optimal single surface, 270
shape prior, 267, 281
semi-supervised image clustering, 190
semi-supervised learning, 522
semidefinite programming, 361, 373
separation constraint, 269
shape, 426
shape cell, 427
shape fitting, 168
shape interior, 426
shape matching, 442, 443, 464
    spectral, 458
shape prior, 267, 281
Shepard
    diagram, 369
shock-graph, 426
silhouette, 426
similarity, 352, 353, 358, 361, 362, 373, 374
    matrix, 373
        sparse, 358, 363, 364, 374
    preservation, 374
similarity space, 424
simple graph, 326
simple graph pyramids, 332
singular value, 355
sink, 12
smoothing term, 34
smoothness constraint, 269
SOM, 366–368, 371, 374
source, 3, 11
space
    Euclidean, 359, 360

feature, 363
high-dimensional, 352, 358, 359, 365, 367–370
latent, 357, 372
low-dimensional, 352
visualization, 370
sparse, 362–364, 373, 374
sparsity, 358
spectral graph isomorphism, 447
spectral graph matching, 442, 451
spectral graph wavelet transform, 208
spiral, 357, 370
spring, 358, 367
SSTRESS, 357
static, 425
step
    function, 367
    size, 368
stereopsis, 483
STRAIN, 356, 357
STRESS, 356, 357
stress, 357, 369
    squared, 357
structural, 424
structural change, 427
subgraph, 387
subgraph isomorphism, 388
sublinear search, 430
submodularity, 40
suboptimal GED algorithms, 403
subpath, 8
subspace, 354, 355
    linear, 353, 356, 357, 366
    nonlinear, 358, 366
superpixel, 19
supremum, 143
surviving
    directed edge, 321
    element, 312
    undirected edge, 318
    vertex, 314, 316, 319, 321
survivor, *see* surviving
Swiss roll, 357, 358, 370, 372

T

target, 3
target specification, 112, 134
targeted segmentation, 112
topology, 353
total variation, 94
    co-area formula, 96
    submodularity, 96
trail, 8
transitions, 427
tree, 10
    minimal spanning, 11
    component, 152
    spanning, 11
tree-walk, 503, 511
trimap, 113, 239

U

unfolding
    manifold, 358
    maximum variance, 358
uprooting, 163

V

variance, 373
    maximum variance unfolding, 358, 360, 363
    preservation, 355, 371
vector quantization, 366, 374
vertex, 143
vertices, 2
vicinity, 358
visualization, 353, 366, 370, 372, 373
    space, 366

W

walk, 8, 503
    closed, 8
    open, 8
watershed
    cut, 155
    power, 166
    ultrametric, 163
weight, 356, 360, 362, 363, 366
weighted graph matching, 450
weighted mean of a pair of graphs, 407
Welsch function, 124

X

XOM, 358, 368